SECOND EDITION

A
FIRST
COURSE
IN
ABSTRACT
ALGEBRA

SECOND EDITION

A FIRST COURSE IN ABSTRACT ALGEBRA

JOHN B. FRALEIGH

DEPARTMENT OF MATHEMATICS
UNIVERSITY OF RHODE ISLAND

ADDISON-WESLEY PUBLISHING COMPANY

Reading, Massachusetts • Menlo Park, California
London • Amsterdam • Don Mills, Ontario • Sydney

This book is in the

ADDISON-WESLEY SERIES IN MATHEMATICS

Consulting Editor: Lynn H. Loomis

ISBN 0-201-01984-1
GHIJKLMNOP-MA-89876543210

To my father PERCY A. FRALEIGH

Preface to the Second Edition

As in the first edition, my aim is to teach the student as much about groups, rings, and fields as is possible in a first course. During the past seven years I have received many suggestions and corrections of errors for which I am very grateful.

There seemed to be substantial agreement that equivalence relations should be presented and used: "The undergraduate major should learn about them sometime, and this is the natural course." Accordingly, I have expanded Section 0 to include a treatment of equivalence relations, and have changed the text and exercises to use them wherever they naturally occur. Consequently, instructors will probably want to spend at least half a lecture on Section 0 now, rather than assigning it as outside reading and starting right in on Section 1.

I have added to or modified some of the exercises. A few of the exercises marked with a dagger have been changed or interchanged with others, since I usually ask my students to hand them in, and I am getting tired of looking at the same ones.

I am gratified at the response to the first edition, not only from undergraduates and faculty, but from graduate students preparing for their comprehensive examinations. I hope the second edition continues to be helpful.

Kingston, Rhode Island   J.B.F.
September 1975

Preface to the First Edition

The primary objective of this work is to provide a text from which an *average* student of mathematics can acquire as much depth and comprehension in his study of abstract algebra, exclusive of linear algebra, as is possible in a first course. Since algebra frequently provides a student's first encounter with an abstract mathematical discipline, a secondary objective is to sow the seeds from which a modern mathematical attitude may grow. Mastery of this text should constitute a firm foundation for more specialized work in algebra, and also should be of great help in any further axiomatic study of mathematics.

In line with the stated secondary objective, the text begins with an introductory section dealing with the role of definitions in mathematics. This role is too seldom mentioned. To emphasize the importance of definitions, throughout the text each term appears in boldface type in its definition.

Part I is concerned with groups. The study of groups, and indeed all the work of the text, is motivated as much as possible in terms of the student's past experience with algebra. The important concept of a factor group is often a difficult one for a student to grasp. Accordingly, the treatment of factor groups and homomorphisms is delayed until the student has had time to acquire a little feeling for the group concept, and then the discussion is gradual and detailed. Part I concludes in a very unusual fashion with four sections which naively apply group theory to topology. The reasons for including this material are threefold:

1) It is desirable for the student to realize that the theory of groups is not merely a study unto itself, but that it has important uses in other areas.
2) There is nothing like sitting down and computing, to crystallize one's understanding of a topic. Computing some homology groups develops a strong, intuitive feeling for the factor group concept.
3) Algebraic topology will probably be a standard part of the undergraduate mathematics curriculum before long. These four sections should give the student a flying start in a course which many may find difficult.

For these reasons, the algebraic topology has been approached here from a geometric, rather than a purely algebraic, point of view.

ix

In Part I, several important results are presented with considerable discussion and many examples, but without proof. This may be termed a sacrilege by some mathematicians. It is my feeling that in view of the vast body of mathematics, it is important to train the student to understand and use established results without feeling that he must first carefully check every detail of the proofs. Of course, professional mathematicians have been doing this for years. Such a policy is also in line with my objective of achieving a little depth in algebra, particularly since in many schools only a single semester is allotted to the study of the subject matter of this text.

Part II deals with rings and fields. Every effort is made to point out analogies with the previous study of groups. The main emphasis of Part II is on field theory, leading up to and including Galois theory. The treatment of vector spaces is very brief, designed only to develop the concepts of linear independence and dimension which are needed for field theory. Since students often find field theory quite difficult, I have attempted to make the treatment gradual and always to make it clear what we are trying to accomplish and how we intend to do it.

Properties of the rational integers with which the student is familiar, even though he may never have seen rigorous justifications of them, are used throughout the text without comment or apology. I have found that the average student has difficulty understanding the reason for undertaking a formal study of results which he has known for years. After acquiring an overall picture of the nature of algebraic structures, he will come to view these properties in a new light. Also, this treatment is consistent with my primary objective of attempting to achieve some depth in a first course.

In view of my desire to have the student learn as much *algebra* as possible, I have chosen to treat set-theoretic material very intuitively and only as needed. ... There are two ways to acquire a working knowledge of set-theoretic material: to study it *per se*, and to plunge in and use it as it is needed. It is my experience that students find a study of "set-theoretic prerequisites" at the start of an algebra course the most discouraging part of the whole course. My approach in this matter is characteristic of my willingness throughout the text to sacrifice elegance of mathematical presentation, and sometimes even of language, for comprehension in this first course.

The text contains sufficient material for a two-semester course with an average class. *However, the sections not marked with a star were specifically designed to constitute a one-semester course. These sections are self contained; no use is made in them of starred material. They represent my attempt to present material of some depth in algebra, including Galois theory, to an average class in a single semester.* Of course, a variety of other one-semester courses can be formed from the subject matter at hand. Some starred sections are quite suitable for assignment as outside reading, especially

Sections 10, 37, 39, and 48. If there is not time to finish the field theory in the text, either Section 35, containing Kronecker's theorem, which is placed in the limelight, or Section 39 is a satisfying terminal section. In my opinion, there is little to be gained in starting Section 40 if there is not time to finish the unstarred material.

The exercises at the end of a section are usually divided into two groups by a horizontal rule. Those above the rule are recommended for the average class, and would probably be assigned by the author to his classes at U.R.I. In an attempt to make the student's transition to abstract mathematics as easy as possible, these exercises in the first group have been made largely computational. The average student is completely lost when faced with a set of exercises all starting with the words *prove* or *show*. Of course, training in proofs is important. The first group of exercises usually contains one marked with a dagger which asks for a proof. It is the author's policy to collect these marked exercises, to read them, and to make the student rewrite them, if necessary, to try to train him to write mathematics and not nonsense. The exercises in the second group usually include many asking for proofs, as well as some additional ones of a computational nature. *A star on an exercise does not denote difficulty, but rather indicates that the exercise depends on some starred material of the text.* In view of my desire to promote a good mathematical attitude, some exercises, especially near the beginning of the text, are of a somewhat metamathematical nature. Answers for, or comments concerning, almost all exercises which do not ask for proofs are given at the back of the text. Proofs requested in exercises are not given in the answers; I do not think that it is sound pedagogy to have such proofs so readily available.

A mimeographed first draft of this text was used in the spring semester, 1966, at the University of Rhode Island. I wish to express here my debt to George E. Martin, who taught one of the sections of the course. His comments and suggestions were of great value to me in the preparation of this version for publication.

Kingston, Rhode Island   J.B.F.
September 1966

Contents

PART II | RINGS AND FIELDS

0 | A Very Few Preliminaries

0.1 THE ROLE OF DEFINITIONS

Many students do not realize the great importance of definitions to mathematics. This importance stems partly from the need for mathematicians to communicate with each other regarding their work. If two people who are trying to communicate on some subject have different ideas as to what is meant by certain technical terms, there can be misunderstanding, friction, and perhaps even bloodshed. Imagine the predicament of a butcher facing an irate housewife who is trying to buy what most people call a rib roast of beef, but who insists on calling it a bottom round roast. The Utopian ideal of complete standardization of terminology unfortunately does not seem to be achievable, even among such precise beings as mathematicians. For example, when speaking of functions in mathematics, various mathematicians have given the term *range* two distinct meanings. Consequently, there is a tendency now to avoid using this ambiguous term and to use either *image* or *codomain* in its place. *In mathematics, one should strive to avoid ambiguity.*

A very important ingredient of mathematical creativity is the ability to formulate useful definitions, ones which will lead to interesting results. A mathematics student commencing graduate study may find that he spends a great deal of time discussing definitions with fellow graduate students. When the author was in graduate school, a physics graduate student once complained to him that at the evening meal the mathematics students always sat together and argued, and that the subject of their argument was always a definition. A graduate student is usually asked to give several definitions on an oral examination. If he cannot explain the meaning of a term, he probably cannot give sensible answers to questions involving that concept.

Every definition is understood to be an *if and only if* type of statement, even though it is customary to suppress the *only if*. Thus one may define: "A triangle is **isosceles** if it has two sides of equal length," really meaning that a triangle is isosceles if and only if it has two sides of equal length. Now you must not feel that you should memorize a definition word for word. The important thing is to *understand* the concept so that you can define precisely the same concept in your own words. Thus the definition "An **isosceles** triangle is one having two equal sides" is perfectly correct. The definition "An **isosceles** triangle is one having two equal angles" is also correct, for exactly the same triangles are called isosceles in all these definitions.

It is very important for you to note that if some concept has just been defined and you are asked to prove something concerning the concept, you *must* use the definition as an integral part of the proof. *Immediately after a concept is defined, the definition is the only information one has available regarding the concept.*

0.2 SETS

This basic importance of definitions to mathematics is also a structural weakness for the reason that not every concept used can be defined. Suppose, for example, one defines the term *set* by "A **set** is a well-defined collection of objects." One naturally asks what is meant by a *collection*. Perhaps then one defines: "A **collection** is an aggregate of things." What then is an *aggregate*? Now our language is finite, so after some time we will run out of new words to use and have to repeat some words already questioned. The definition is then circular and obviously worthless. Mathematicians realize that there must be some undefined or primitive concept. At the moment they have agreed that *set* shall be such a primitive concept. We shall not define *set*, but shall just hope that when such expressions as "the set of all real numbers," or "the set of all members of the United States Senate" are used, people's various ideas of what is meant are sufficiently similar to make communication feasible.

We summarize briefly some of the things we shall simply assume about sets.

1) A set S is comprised of **elements**, and if a is one of these elements, we shall denote this fact by "$a \in S$".

2) There is exactly one set with no elements. It is the **empty set** and is denoted by "$\varnothing$".

3) A set may be described either by giving a characterizing property of the elements, such as "the set of all members of the United States Senate," or by listing the elements. The standard way to describe a set by listing elements is to enclose the designations of the elements, separated by commas, in braces, e.g., $\{1, 2, 15\}$. If a set is described by a characterizing property $P(x)$ of its elements x, the brace notation "$\{x \mid P(x)\}$" is also often used, and is read, "The set of all x such that the statement $P(x)$ about x is true." Thus

$$\{2, 4, 6, 8\} = \{x \mid x \text{ is an even whole positive number} \le 8\}$$
$$= \{2x \mid x = 1, 2, 3, 4\}.$$

4) A set is **well defined**, meaning that if S is a set and a is some object, then either a is definitely in S, denoted by "$a \in S$", or a is definitely not in S, denoted by "$a \notin S$". Thus one should never say, "Consider the set S of some positive numbers," for it is not definite whether $2 \in S$ or $2 \notin S$.

On the other hand, one can consider the set T of all prime positive integers. Every positive integer is definitely either prime or not prime. Thus $5 \in T$ and $14 \notin T$. It may be hard to actually determine whether or not an object is in a set. For example, as this book goes to press it is unknown whether or not $2^{(2^{17})} + 1$ is in T. However, $2^{(2^{17})} + 1$ is certainly either prime or not prime.

It will not be feasible for the student for whom this text is intended to push every definition back to the concept of a set. The author is well aware that he is building on some very naive definitions, especially at the beginning of the text. The first definition we will meet in Section 1 says, "A **binary operation on a set** is a rule ... set." What on earth is a rule?

Throughout this text, much work will be done involving familiar sets of numbers. Let us take care of notation for these sets once and for all:

Z   is the set of all integers (i.e., whole numbers: positive, negative, and zero).

$\mathbf{Z}^+$ is the set of all positive integers. (Zero is excluded.)

Q   is the set of all rational numbers (i.e., numbers which can be expressed as quotients m/n of integers, where $n \neq 0$).

$\mathbf{Q}^+$ is the set of all positive rational numbers.

R   is the set of all real numbers.

$\mathbf{R}^+$ is the set of all positive real numbers.

C   is the set of all complex numbers.

0.3   PARTITIONS AND EQUIVALENCE RELATIONS

We just described **Q** as the set of all numbers which can be expressed as quotients m/n of integers, where $n \neq 0$. It would be incorrect to describe **Q** as the set S of all "quotient expressions" m/n for m and n in **Z** and $n \neq 0$. For surely $\frac{2}{3}$ and $\frac{4}{6}$ are distinguishable quotient expressions, but we know they represent the *same* rational number. In fact, each element of **Q** is represented by an infinite number of different elements of S. When doing arithmetic, we *identify* in our mind elements of S which represent the same rational number in **Q**.

The illustration in the preceding paragraph is typical of several situations in which we will consider different elements of a set to be arithmetically or algebraically equivalent, so that our set becomes *partitioned* into cells, each of which we may consider to be a single arithmetic or algebraic entity. If b is an element of such a set, we usually let $\bar{b}$ be the cell of all elements being identified with b. For example, in the case of our set S of formal quotients above, we have

$$\overline{2/3} = \left\{ \frac{2}{3}, \frac{-2}{-3}, \frac{4}{6}, \frac{-4}{-6}, \frac{6}{9}, \frac{-6}{-9}, \dots \right\}$$

$$= \left\{ \frac{2n}{3n} \;\middle|\; n \in \mathbf{Z} \text{ and } n \neq 0 \right\}.$$

Let us give a precise definition of such a partition.

Definition. A *partition of a set* is a decomposition of the set into cells such that every element of the set is in *exactly one* of the cells.

Two cells (or sets) having no elements in common are **disjoint**. Thus the cells in a partition of a set are disjoint.

How do we know whether two quotient expressions m/n and r/s in our set S above are in the same cell, i.e., represent the same rational number? One way to decide is to "reduce both fractions to lowest terms." This may not be easy to do; for example, 1909/4897 and 1403/3599 represent the same rational number since

$$\frac{1909}{4897} = \frac{23 \cdot 83}{59 \cdot 83} \quad \text{and} \quad \frac{1403}{3599} = \frac{23 \cdot 61}{59 \cdot 61}.$$

However, finding these factorizations, even with a hand calculator, is a somewhat tedious trial-and-error task. But as you know, in fraction arithmetic, we have $m/n = r/s$ if and only if $ms = nr$. This gives us a more efficient criterion for our problem, namely

$$(1909)(3599) = (4897)(1403) = 6870491.$$

Let "$a \sim b$" denote that a is in the same cell as b for a given partition of a set containing both a and b. Clearly the following properties are always satisfied:

$a \sim a$: The element a is in the same cell as itself.

If $a \sim b$ then $b \sim a$: If a is in the same cell as b, then b is in the same cell as a.

If $a \sim b$ and $b \sim c$, then $a \sim c$: If a is in the same cell as b and b is in the same cell as c, then a is in the same cell as c.

The theorem which follows is fundamental. It asserts that a relation $\sim$ between elements of a set which satisfies the three properties just described yields a natural partition of the set. Exhibiting a relation with these properties is frequently the most economical way of describing a partition of a set, and it is for this reason we are discussing this material now.

Theorem 0.1 *Let S be a nonempty set and let $\sim$ be a relation between elements of S which satisfies the following properties.*
1) *(Reflexive) $a \sim a$ for all $a \in S$.*
2) *(Symmetric) If $a \sim b$, then $b \sim a$.*
3) *(Transitive) If $a \sim b$ and $b \sim c$, then $a \sim c$.*
Then $\sim$ yields a natural partition of S, where

$$\bar{a} = \{x \in S \mid x \sim a\}$$

is the cell containing a for all $a \in S$. Conversely, each partition of S gives rise to a natural relation $\sim$ if $a \sim b$ is defined to mean that $a \in \bar{b}$.

Proof. We have already proved the "converse" part of the theorem.

For the direct statement, it only remains to show that the cells defined by $\bar{a} = \{x \in S \mid x \sim a\}$ do constitute a partition of S, i.e., that every element of S is in *exactly one* cell. Let $a \in S$. Then $a \in \bar{a}$ by condition (1), so a is in *at least one* cell.

Suppose now that a were in a cell $\bar{b}$ also. We need to show that $\bar{a} = \bar{b}$ as sets; this would show that $\bar{a}$ can't be in more than one cell. We do this by showing that each element of a is in $\bar{b}$ and each element of $\bar{b}$ is in $\bar{a}$. Let $x \in \bar{a}$. Then $x \sim a$. But $a \in \bar{b}$, so $a \sim b$. Then, by the transitive condition (3), $x \sim b$ so $x \in \bar{b}$. Thus $\bar{a}$ is part of $\bar{b}$. Now let $y \in \bar{b}$. Then $y \sim b$. But $a \in \bar{b}$, so $a \sim b$ and, by symmetry (2), $b \sim a$. Then by transitivity, $y \sim a$ so $y \in \bar{a}$. Hence $\bar{b}$ is part of $\bar{a}$ also, so $\bar{b} = \bar{a}$ and our proof is complete. ∎

> **Definition.** A relation $\sim$ on a set S satisfying the reflexive, symmetric, and transitive properties described in the preceding theorem is an **equivalence relation on** S. A cell $\bar{a}$ is the natural partition given by an equivalence relation is an **equivalence class**.

The symbol $\sim$ is usually reserved for an equivalence relation. We will use $\mathcal{R}$ for a relation between elements of a set S which is not necessarily an equivalence relation on S.

The term *natural*, appearing twice in the preceding theorem, has the following significance. If you start with an equivalence relation, form the partition of equivalence classes, and then consider the relation given by this partition, it is your original equivalence relation. Similarly, starting with a partition, going to the equivalence relation, and then forming the equivalence classes yields the original partition.

Example 0.1 Let us verify directly that $m/n \sim r/s$ if and only if $ms = nr$ is an equivalence relation on the set S of formal quotient expressions we considered earlier.

Reflexive: $m/n \sim m/n$ since $mn = nm$.

Symmetric: If $m/n \sim r/s$, then $ms = nr$. Consequently, $rn = sm$ so $r/s \sim m/n$.

Transitive: If $m/n \sim r/s$ and $r/s \sim u/v$, then $ms = nr$ and $rv = su$. Reordering terms and substituting, we obtain $mvs = vms = vnr = nrv = nsu = nus$. Since $s \neq 0$, we deduce that $mv = nu$, so $m/n \sim u/v$.

Each equivalence class of S is considered to *be* a rational number. ‖

Our discussion of the set S of formal quotient expressions, culminating in Example 0.1, is a special case of work we will do in Section 26.

Example 0.2 Let us define a relation $\mathcal{R}$ on the set $\mathbf{Z}$ by $n \, \mathcal{R} \, m$ if and only if $nm \geq 0$, and determine whether or not $\mathcal{R}$ is an equivalence relation.

Reflexive: $a \ \Re \ a$, since $a^2 \geq 0$ for all $a \in \mathbf{Z}$.

Symmetric: If $a \ \Re \ b$, then $ab \geq 0$, so $ba \geq 0$ and $b \ \Re \ a$.

Transitive: If $a \ \Re \ b$ and $b \ \Re \ c$, then $ab \geq 0$ and $bc \geq 0$. Thus $ab^2c = acb^2 \geq 0$. If we knew $b^2 \neq 0$, we could deduce $ac \geq 0$ whence $a \ \Re \ c$. We have to examine the case $b = 0$ separately. A moment of thought shows $-3 \ \Re \ 0$ and $0 \ \Re \ 5$ but $-3 \ \not\Re \ 5$, so the relation $\Re$ is not transitive, and hence not an equivalence relation. ‖

For each $n \in \mathbf{Z}^+$ we have a very important equivalence relation on $\mathbf{Z}$, **congruence modulo** n. For $h, \ k \in \mathbf{Z}$, we define h **congruent to** k **modulo** n, written $h \equiv k \ (\text{mod } n)$, if $h - k$ is evenly divisible by n, so that $h - k = ns$ for some $s \in \mathbf{Z}$. For example, $17 \equiv 33 \ (\text{mod } 8)$, since $17 - 33 = 8(-2)$. Equivalence classes for congruence modulo n are **residue classes modulo** n. We ask you to show that congruence modulo n is an equivalence relation and to look at some residue classes in Exercise 0.12.

EXERCISES

In Exercises 1 through 4, describe the set by listing its elements.

0.1 $\{x \in \mathbf{R} \mid x^2 = 3\}$ **0.2** $\{m \in \mathbf{Z} \mid m^2 = 3\}$

0.3 $\{m \in \mathbf{Z} \mid mn = 60$ for some $n \in \mathbf{Z}\}$ **0.4** $\{m \in \mathbf{Z} \mid m^2 - m < 115\}$

In Exercises 5 through 11, determine whether the given relation is an equivalence relation on the set. Describe the partition arising from each equivalence relation.

0.5 $n \ \Re \ m$ in $\mathbf{Z}$ if $nm > 0$ **0.6** $x \ \Re \ y$ in $\mathbf{R}$ if $x \geq y$

0.7 $x \ \Re \ y$ in $\mathbf{R}$ if $|x| = |y|$ **0.8** $x \ \Re \ y$ in $\mathbf{R}$ if $|x - y| \leq 3$

0.9 $n \ \Re \ m$ in $\mathbf{Z}^+$ if n and m have the same number of digits in the usual base ten notation.

0.10 $n \ \Re \ m$ in $\mathbf{Z}^+$ if n and m have the same final digit in the usual base ten notation.

0.11 $n \ \Re \ m$ in $\mathbf{Z}^+$ if $n - m$ is evenly divisible by 2.

0.12 Let n be a particular integer in $\mathbf{Z}^+$. Show that congruence modulo n is an equivalence relation on $\mathbf{Z}$. Describe the residue classes for $n = 1, 2, 3$.

0.13 The following is a famous incorrect argument. Find the error. "The reflexive criterion is redundant in the conditions for an equivalence relation, for from $a \sim b$ and $b \sim a$ (symmetry) we deduce $a \sim a$ by transitivity."

PART I | GROUPS

1 | Binary Operations

1.1 MOTIVATION

What constitutes the basic ingredient of algebra? The first contact of a child with algebra comes when he is taught to add and multiply numbers. Let us try to analyze what really happens here.

Suppose that you are a visitor to a strange civilization in a strange world and you are observing one of the creatures of this world drilling a class of his fellow creatures in the addition of numbers. Suppose also that you have not been told that the class is learning to add, but that you were just placed as an observer in the room where this was going on. You are asked to give a report on exactly what happens. The teacher makes noises which sound to you approximately like *gloop*, *poyt*. The class responds with *bimt*. The teacher then gives *ompt*, *gaft*, and the class responds with *poyt*. What are they doing? You cannot report that they are adding numbers, for you do not even know that the sounds are representing numbers. Of course, you do realize that there is communication going on. All you can say with any certainty is that these creatures know some rule, so that when certain pairs of things are designated in their language, one after another, like *gloop*, *poyt*, they are able to agree on a response, *bimt*. This same procedure goes on in addition drill in our first grade classes where a teacher may say, *four*, *seven*, and the class responds with *eleven*.

In our attempt to analyze addition and multiplication of numbers, we are thus led to the idea that addition is basically just a rule which people learn, enabling them to associate, with two numbers in a given order, some number as answer. Multiplication is also such a rule, but a different rule. Note finally that in playing this game with the students, the teacher has to be a little careful of what two things she gives to the class. If a first grade teacher suddenly inserts *ten*, *sky*, her poor class will be very confused. The rule is only defined for pairs of things from some specified set.

1.2 DEFINITION AND PROPERTIES

As mathematicians, let us attempt to collect the core of these basic ideas in a useful definition. As was remarked in the introductory section, we do not attempt to define a set.

Definition. A *binary operation* $*$ *on a set* is a rule which assigns to each ordered pair of elements of the set some element of the set.

The word *ordered* in this definition is very important, for it allows the possibility that the element assigned to the pair (a, b) may be different from the element assigned to the pair (b, a). Also, we were careful not to say that to each ordered pair of elements is assigned *another* or a *third* element, for we wish to permit cases such as occur in addition of numbers where $(0, 2)$ has assigned to it the number 2.

For the first few sections we shall let $a * b$ be the element assigned to the pair (a, b) by $*$. If we have several different binary operations under simultaneous discussion, we shall use subscripts or superscripts on the $*$ to distinguish them. The most important method of describing a particular binary operation $*$ on a given set is to characterize the element $a * b$ assigned to each pair (a, b) by some property defined in terms of a and b.

Example 1.1 On $\mathbf{Z}^+$, define a binary operation $*$ by $a * b$ equals the smaller of a and b or the common value if $a = b$. Thus $2 * 11 = 2, 15 * 10 = 10$, and $3 * 3 = 3$. ‖

Example 1.2 On $\mathbf{Z}^+$, define a binary operation $*'$ by $a *' b = a$. Thus $2 *' 3 = 2, 25 *' 10 = 25$, and $5 *' 5 = 5$. ‖

Example 1.3 On $\mathbf{Z}^+$, define a binary operation $*''$ by $a *'' b = (a * b) + 2$, where $*$ is defined in Example 1.1. Thus $4 *'' 7 = 6, 25 *'' 9 = 11$, and $6 *'' 6 = 8$. ‖

It probably seems to you that these examples are of no importance, but consider for a moment. Suppose you go into a store to buy a nice, large, delicious chocolate bar. Suppose you see two identical bars side by side, the wrapper of one stamped 39¢ and the wrapper of the other stamped 37¢. Of course you pick up the one stamped 37¢. Your knowledge of which one you want depends on the fact that sometime in your life you learned the binary operation $*$ of Example 1.1. *It is a very important operation.* Likewise the binary operation $*'$ of Example 1.2 is clearly dependent on the ability to distinguish order. An often cited illustration of the importance of order is the mess which would result if a person tried to put on his shoes first, and then his socks! Thus you should not be hasty about dismissing some binary operation as being of little significance. Of course, our usual operations of addition and multiplication of numbers have a practical importance well known to you.

Examples 1.1 and 1.2 were chosen to demonstrate that a binary operation may or may not depend on the order of the given pair. Thus in Example 1.1, $a * b = b * a$ for all $a, b \in \mathbf{Z}^+$, and in Example 1.2, this is not the case, for $5 *' 7 = 5$ but $7 *' 5 = 7$.

Now suppose one wishes to consider an expression of the form $a * b * c$. A binary operation $*$ enables you to combine only two elements, and here we have three. The obvious attempts to combine the three elements are to form either $(a * b) * c$ or $a * (b * c)$. With $*$ as in Example 1.1, $(2 * 5) * 9$ is computed by $2 * 5 = 2$ and then $2 * 9 = 2$. Likewise $2 * (5 * 9)$ is computed by $5 * 9 = 5$ and then $2 * 5 = 2$. Hence $(2 * 5) * 9 = 2 * (5 * 9)$, and it is easily seen that for this $*$,

$$(a * b) * c = a * (b * c),$$

so there is no ambiguity in writing $a * b * c$. But for $*''$ of Example 1.3,

$$(2 *'' 5) *'' 9 = 4 *'' 9 = 6,$$

while

$$2 *'' (5 *'' 9) = 2 *'' 7 = 4.$$

Thus $(a *'' b) *'' c$ need not equal $a *'' (b *'' c)$ and an expression $a *'' b *'' c$ may be ambiguous.

>**Definition.** A binary operation $*$ on a set S is **commutative** if (and only if) $a * b = b * a$ for all $a, b \in S$. The operation $*$ is **associative** if (and only if) $(a * b) * c = a * (b * c)$ for all $a, b, c \in S$.

As was pointed out in the introductory section, it is customary in mathematics to omit the words *and only if* from a definition. Definitions are always understood to be if and only if statements. *Theorems are not always if and only if statements, and no such convention is ever used for theorems.*

It is not difficult to show that if $*$ is associative, then longer expressions such as $a * b * c * d$ are not ambiguous. Parentheses may be inserted in any fashion for purposes of computation; the final results of two such computations will be the same.

1.3 TABLES

For a finite set, a binary operation on the set can also be defined by means of a table. The next example shows how this will be done in this text.

Example 1.4 Table 1.1 defines the binary operation $*$ on $S = \{a, b, c\}$ by the rule:

(ith *entry on the left*) $*$ (jth *entry on the top*)

                $=$ (*entry in the ith row and jth column of the answers*).

Thus $a * b = c$ and $b * a = a$, so $*$ is not commutative. $\parallel$

Table 1.1

$*$	a	b	c
a	b	c	b
b	a	c	b
c	c	b	a

The student can easily see that *a binary operation defined by a table is commutative if and only if the entries in the table are symmetric with respect to the diagonal which starts at the upper left corner of the table and terminates at the lower right corner.* We always assume that the elements of the set are listed across the top of a table in the same order as they are listed at the left.

Except for this Example 1.4, our examples of binary operations have been defined on sets of numbers. It is important to realize that binary operations may be defined on any set. Indeed, we shall be studying many important binary operations on sets whose elements are not numbers. Some of the examples to be given in a moment involve sets with *functions* as elements. It is assumed that the student has some familiarity with certain functions from calculus or other courses. We realize that the student may not understand the concept of a function at the moment and we shall say more about it later. However, we are anxious to tie in the notions being introduced with the mathematics that the student has already had.

1.4 SOME WORDS OF WARNING

The author knows from his own experience the chaos that may result if a student is given a set and asked to define some binary operation on it. Observe that in an attempt to define a binary operation * on a set *S* you must be sure that

1) *exactly one element is assigned to each possible ordered pair of elements of S,*
2) *for each ordered pair of elements of S, the element assigned to it is again in S.*

Regarding condition (1), a student will often give a rule which assigns an element of *S* to "most" ordered pairs, but for a few pairs the rule determines no element. In this event, * has **not been defined**. It may also happen that for some pairs, the rule could assign any of several elements of *S*, that is, there is ambiguity. In any case of ambiguity, * is **not well defined**. If condition (2) is violated, then *S* is **not closed under** *.

We now give several illustrations of attempts to define binary operations on sets. Some of these attempts are worthless as we point out. Since no comparison between operations will be made, we shall denote them all by *.

Example 1.5 On **Q**, "define" * by $a * b = a/b$. Here * is *not defined*, for no rational number is assigned by this rule to the pair $(2, 0)$. ||

Example 1.6 On $\mathbf{Q}^+$, define * by $a * b = a/b$. Here both conditions (1) and (2) are satisfied and * is a binary operation on $\mathbf{Q}^+$. ||

Example 1.7 On $\mathbf{Z}^+$, "define" * by $a * b = a/b$. Here condition (2) is violated, for $1 * 3$ is not in $\mathbf{Z}^+$. Thus * is not a binary operation on $\mathbf{Z}^+$, since $\mathbf{Z}^+$ is *not closed under* *. ||

Example 1.8 Let *S* be the set of all real-valued functions defined for all real numbers. Define * to give the usual sum of two functions, that is, $f * g = h$, where $h(x) = f(x) + g(x)$ for $f, g \in S$ and $x \in \mathbf{R}$. This definition of $*$ satisfies conditions (1) and (2) and gives a binary operation on S. ‖

Example 1.9 Let S be as in Example 1.8 and define $*$ to give the usual product of two functions, that is, $f * g = h$, where $h(x) = f(x)g(x)$. Again this definition is a good one and gives a binary operation on S. ‖

Example 1.10 Let S be as in Example 1.8 and "define" $*$ to give the usual quotient of f by g, that is, $f * g = h$, where $h(x) = f(x)/g(x)$. Here condition (2) is violated, for the functions in S were to be defined for *all* real numbers, and for some $g \in S$, $g(x)$ will be zero for some values of x in $\mathbf{R}$, and $h(x)$ would not be defined at those numbers in $\mathbf{R}$. For example, if $f(x) = \cos x$ and $g(x) = x^2$, then $h(0)$ is undefined, so $h \notin S$. ‖

Example 1.11 Let S be as in Example 1.8 and "define" $*$ by $f * g = h$, where h is the function greater than both f and g. This "definition" is completely worthless. In the first place, we have not defined what it means for one function to be greater than another. Even if we had, any sensible definition would result in there being many functions greater than both f and g, and $*$ would still *not be well defined*. ‖

Example 1.12 Let S be a set consisting of twenty people, no two of whom are of the same height. Define $*$ by $a * b = c$, where c is the tallest person among the twenty in S. This is a perfectly good binary operation on the set, although not a particularly interesting one. ‖

Example 1.13 Let S be as in Example 1.12 and "define" $*$ by $a * b = c$, where c is the shortest person in S who is taller than both a and b. This $*$ is *not defined*, since if either a or b is the tallest person in the set, $a * b$ is not determined. ‖

EXERCISES

1.1 Let the binary operation $*$ be defined on $S = \{a, b, c, d, e\}$ by means of Table 1.2.

a) Compute $b * d$, $c * c$ and $[(a * c) * e] * a$ from the table.

b) Compute $(a * b) * c$ and $a * (b * c)$ from the table. Can you say on the basis of this computation whether or not $*$ is associative?

c) Compute $(b * d) * c$ and $b * (d * c)$ from the table. Can you say on the basis of this computation whether or not $*$ is associative?

d) Is $*$ commutative? Why?

Table 1.2

$*$	a	b	c	d	e
a	a	b	c	b	d
b	b	c	a	e	c
c	c	a	b	b	a
d	b	e	b	e	d
e	d	b	a	d	c

1.2 Complete Table 1.3 so as to define a commutative binary operation * on $S = \{a, b, c, d\}$.

Table 1.3

*	a	b	c	d
a	a	b	c	
b	b	d		c
c	c	a	d	b
d	d			a

1.3 Table 1.4 may be completed to define an associative binary operation * on $S = \{a, b, c, d\}$. Assume this is possible and compute the missing entries.

Table 1.4

*	a	b	c	d
a	a	b	c	d
b	b	a	c	d
c	c	d	c	d
d				

1.4 Determine whether or not each of the definitions of * given below does give a binary operation on the given set. In the event that * is not a binary operation, state whether condition (1), condition (2), or both of these conditions of the text are violated.

a) On $\mathbf{Z}^+$, define * by $a * b = a - b$.
b) On $\mathbf{Z}^+$, define * by $a * b = a^b$.
c) On $\mathbf{R}$, define * by $a * b = a - b$.
d) On $\mathbf{Z}^+$, define * by $a * b = c$, where c is the smallest integer greater than both a and b.
e) On $\mathbf{Z}^+$, define * by $a * b = c$, where c is at least 5 more than $a + b$.
f) On $\mathbf{Z}^+$, define * by $a * b = c$, where c is the largest integer less than the product of a and b.

†**1.5** Prove that if * is an associative and commutative binary operation on a set S, then

$$(a * b) * (c * d) = [(d * c) * a] * b$$

for all $a, b, c, d \in S$. Assume the associative law only for triples as in the definition, i.e., assume only

$$(x * y) * z = x * (y * z)$$

for all $x, y, z \in S$.

1.6 For each binary operation $*$ defined, determine whether $*$ is commutative and whether $*$ is associative.

a) On $\mathbf{Z}$, define $*$ by $a * b = a - b$.
b) On $\mathbf{Q}$, define $*$ by $a * b = ab + 1$.
c) On $\mathbf{Q}$, define $*$ by $a * b = ab/2$.
d) On $\mathbf{Z}^+$, define $*$ by $a * b = 2^{ab}$.
e) On $\mathbf{Z}^+$, define $*$ by $a * b = a^b$.

1.7 Mark each of the following true or false.

___ a) If $*$ is any binary operation on any set S, then $a * a = a$ for all $a \in S$.
___ b) If $*$ is any commutative binary operation on any set S, then $a * (b * c) = (b * c) * a$ for all $a, b, c \in S$.
___ c) If $*$ is any associative binary operation on any set S, then $a * (b * c) = (b * c) * a$ for all $a, b, c \in S$.
___ d) The only binary operations of any importance are those defined on sets of numbers.
___ e) Mathematicians are eager to have some ambiguity in their work so that it has a better chance of being right.
___ f) Every binary operation defined on a set having exactly one element is both commutative and associative.
___ g) A binary operation on a set S assigns at least one element of S to each ordered pair of elements of S.
___ h) A binary operation on a set S assigns at most one element of S to each ordered pair of elements of S.
___ i) A binary operation on a set S assigns exactly one element of S to each ordered pair of elements of S.
___ j) A binary operation on a set S may assign more than one element of S to some ordered pair of elements of S.

1.8 Give a set different from any of those described in the examples of the text and not a set of numbers. Define two different binary operations $*$ and $*'$ on this set. Be sure that your set is *well defined*.

1.9 Let S be a set having exactly one element. How many different binary operations can be defined on S? Answer the question if S has exactly 2 elements; exactly 3 elements; exactly n elements.

1.10 How many different commutative binary operations can be defined on a set of 2 elements? on a set of 3 elements? on a set of n elements?

1.11 Observe that the binary operations $*$ and $*'$ on the set $\{a, b\}$ given by the tables

$*$	a	b
a	a	a
b	a	b

and

$*'$	a	b
a	a	b
b	b	b

provide the *same type of algebraic structure* on $\{a, b\}$, in the sense that if the table for $*'$ is rewritten

$*'$	b	a
b	b	b
a	b	a

this table for $*'$ looks just like that for $*$ with the roles of a and b interchanged.

a) Try to give a natural definition of a concept of two binary operations $*$ and $*'$ on the same set giving *algebraic structures of the same type*, which generalizes this observation.

b) How many different types of algebraic structures are given by the 16 possible different binary operations on a set of 2 elements?

2 | Groups

2.1 MOTIVATION

Let us continue the analysis of our past experience with algebra. Once the computational problems of addition and multiplication of numbers had been mastered, we were ready to apply these binary operations to the solution of problems. Often problems lead to equations involving some unknown number x which is to be determined. The simplest equations are the linear ones of the forms $a + x = b$ for the operation of addition, and $ax = b$ for multiplication. The additive linear equation always has a numerical solution, and so has the multiplicative one, provided $a \neq 0$. Indeed, the need for solutions of additive linear equations such as $5 + x = 2$ is a very good motivation for the negative numbers. Similarly, the need for rational numbers is shown by equations such as $2x = 3$ and the need for the complex number i is shown by the equation $x^2 = -1$.

It is desirable for us to be able to solve linear equations involving our binary operations. This is not possible for every binary operation, however. For example, the equation $a * x = a$ has no solution in $S = \{a, b, c\}$ for the operation $*$ of Example 1.4. Let us see just what properties of the operation of addition on the integers $\mathbf{Z}$ enable us to solve the equation $5 + x = 2$ in $\mathbf{Z}$. We must not refer to subtraction, for we are concerned with the solution phrased in terms of a single binary operation, here addition. The steps in the solution are as follows:

$$5 + x = 2, \qquad \text{given,}$$
$$-5 + (5 + x) = -5 + 2, \qquad \text{adding } -5,$$
$$(-5 + 5) + x = -5 + 2, \qquad \text{associative law,}$$
$$0 + x = -5 + 2, \qquad \text{computing } -5 + 5,$$
$$x = -5 + 2, \qquad \text{property of } 0,$$
$$x = -3, \qquad \text{computing } -5 + 2.$$

Strictly speaking, we have not shown here that -3 is a solution, but rather that it is the only possibility for a solution. To show that -3 is a solution, one merely computes $5 + (-3)$. A similar analysis could be made for the

17

equation $2x = 3$ in the rational numbers:

$$2x = 3, \qquad \text{given,}$$
$$\tfrac{1}{2}(2x) = \tfrac{1}{2}3, \qquad \text{multiplying by } \tfrac{1}{2},$$
$$(\tfrac{1}{2}2)x = \tfrac{1}{2}3, \qquad \text{associative law,}$$
$$1 \cdot x = \tfrac{1}{2}3, \qquad \text{computing } \tfrac{1}{2}2,$$
$$x = \tfrac{1}{2}3, \qquad \text{property of 1,}$$
$$x = \tfrac{3}{2}, \qquad \text{computing } \tfrac{1}{2}3.$$

Let us see what properties a set S and a binary operation $*$ on S would have to have to permit imitation of this procedure for an equation $a * x = b$ for $a, b \in S$. Basic to the procedure is the existence of an element e in S with the property that $e * x = x$ for all $x \in S$. For our additive example, 0 played the role of e, and 1 played the role for our multiplicative example. Then we need an element a' in S which has the property that $a' * a = e$. For our additive example, -5 played the role of a', and $\tfrac{1}{2}$ played the role for our multiplicative example. Finally we need the associative law. The remainder is just computation. The student will easily convince himself that in order to solve the equation $x * a = b$ (remember that $a * x$ need not equal $x * a$) one would like to have an element e in S such that $x * e = x$ for all $x \in S$ and an a' in S such that $a * a' = e$. With all of these properties of $*$ on S, we could be sure of being able to solve linear equations. These are precisely the properties of a *group*.

2.2   DEFINITION AND ELEMENTARY PROPERTIES

Definition.   A *group* $\langle G, * \rangle$ is a set G, together with a binary operation $*$ on G, such that the following axioms are satisfied:

$\mathcal{G}_1$. The binary operation $*$ is associative.

$\mathcal{G}_2$. There is an element e in G such that $e * x = x * e = x$ for all $x \in G$. This element e is an **identity element** for $*$ on G.

$\mathcal{G}_3$. For each a in G, there is an element a' in G with the property that $a' * a = a * a' = e$. The element a' is an **inverse of** a **with respect to** $*$.

Many books have another axiom for a group, namely that G is **closed under the operation** $*$, that is, $(a * b) \in G$ for all $a, b \in G$. For us, this is a consequence of our *definition* of a binary operation on G.

We should point out right now that we are going to be sloppy in notation. Observe that a group is not just a set G. A group $\langle G, * \rangle$ is comprised rather of two entities, the set G and the binary operation $*$ on G. There are *two* ingredients involved. Denoting the group by the single set symbol "G" is logically incorrect. Nevertheless, as you get further into the theory, the logical extensions of the notation "$\langle G, * \rangle$" become so unwieldy as to actually make the exposition hard to read. At some point, all authors give up and become sloppy, denoting the group by the single letter "G". We choose to recognize this and be sloppy from the start. We emphasize, however, that when you are speaking of a specific group G, you must make it clear what the group operation on G is to be, since a set could conceivably have a variety of binary operations defined on it, all giving different groups. We shall sometimes resort to the notation "$\langle G, * \rangle$" for reasons of clarity in our discussions.

Theorem 2.1 *If G is a group with binary operation $*$, then the* **left and right cancellation laws** *hold in G, that is, $a * b = a * c$ implies $b = c$, and $b * a = c * a$ implies $b = c$ for all $a, b, c \in G$.*

Proof. Suppose $a * b = a * c$. Then by $\mathcal{G}_3$, there exists a', and

$$a' * (a * b) = a' * (a * c).$$

By the associative law,

$$(a' * a) * b = (a' * a) * c.$$

By the definition of a' in $\mathcal{G}_3$, $a' * a = e$, so

$$e * b = e * c.$$

By the definition of e in $\mathcal{G}_2$,

$$b = c.$$

Similarly, from $b * a = c * a$ one can deduce that $b = c$ upon multiplication on the right by a' and use of the axioms for a group. ∎

Note how we had to use the definition of a group to prove this theorem.

Theorem 2.2 *If G is a group with binary operation $*$, and if a and b are any elements of G, then the linear equations $a * x = b$ and $y * a = b$ have unique solutions in G.*

Proof. Note that

$$
\begin{aligned}
a * (a' * b) &= (a * a') * b, &&\text{associative law,} \\
&= e * b, &&\text{definition of } a', \\
&= b, &&\text{property of } e.
\end{aligned}
$$

Thus $x = a' * b$ is a solution of $a * x = b$. In a similar fashion, $y = b * a'$ is a solution of $y * a = b$.

To show that y is unique, suppose that $y * a = b$ and $y_1 * a = b$. Then $y * a = y_1 * a$, and by Theorem 2.1, $y = y_1$. The uniqueness of x follows similarly. ∎

Of course, to prove the uniqueness in the last theorem we could have followed the procedure we used in motivating the definition of a group, showing that if $a * x = b$, then $x = a' * b$. We chose this other technique to illustrate Theorem 2.1 and also because it is a standard "trick." Note that the solutions $x = a' * b$ and $y = b * a'$ need not be the same unless $*$ is commutative.

Definition. A group G is **abelian** if its binary operation $*$ is commutative.

Let us give some examples of some sets with binary operations which give groups and also of some which do not give groups.

Example 2.1 The set $\mathbf{Z}^+$ with operation $+$ is *not* a group. There is no identity element for $+$ in $\mathbf{Z}^+$. ‖

Example 2.2 The set of all nonnegative integers (including 0) with operation $+$ is still *not* a group. There is an identity element 0, but no inverse for 2. ‖

Example 2.3 The set $\mathbf{Z}$ with operation $+$ *is* a group. All conditions of the definition are satisfied. The group is abelian. ‖

Example 2.4 The set $\mathbf{Z}^+$ with operation multiplication is *not* a group. There is an identity 1, but no inverse of 3. ‖

Example 2.5 The set $\mathbf{Q}^+$ with operation multiplication *is* a group. All conditions of the definition are satisfied. The group is abelian. ‖

Example 2.6 Define $*$ on $\mathbf{Q}^+$ by $a * b = ab/2$. Then

$$(a * b) * c = \frac{ab}{2} * c = \frac{abc}{4},$$

and likewise

$$a * (b * c) = a * \frac{bc}{2} = \frac{abc}{4}.$$

Thus $*$ is associative. Clearly,

$$2 * a = a * 2 = a$$

for all $a \in \mathbf{Q}^+$, so 2 is an identity element for $*$. Finally,

$$a * \frac{4}{a} = \frac{4}{a} * a = 2,$$

so $a' = 4/a$ is an inverse for a. Hence $\mathbf{Q}^+$ with the operation $*$ is a group. ‖

There is one other result about groups we would like to prove in this section.

Theorem 2.3 *In a group G with operation $*$, there is only one identity e such that*

$$e * x = x * e = x$$

for all $x \in G$. Likewise for each $a \in G$, there is only one element a'

such that

$$a' * a = a * a' = e.$$

In summary, the identity and inverses are unique in a group.

Proof. Suppose $e * x = x * e = x$ and also $e_1 * x = x * e_1 = x$ for all $x \in G$. We let e and e_1 compete. Now regarding e as identity, $e * e_1 = e_1$. But regarding e_1 as identity, $e * e_1 = e$. Thus

$$e_1 = e * e_1 = e,$$

and the identity of a group is unique.

Now suppose $a' * a = a * a' = e$ and $a'' * a = a * a'' = e$. Then

$$a * a'' = a * a' = e,$$

and by Theorem 2.1,

$$a'' = a',$$

so the inverse of a in a group is unique. ∎

For the student's information, we remark that algebraic structures consisting of sets with binary operations for which not all of the group axioms hold have also been studied quite extensively. Of these weaker structures, the **semigroup**, a set with an associative binary operation, has perhaps had the most attention. Recently, nonassociative structures have also been studied.

Finally, it is possible to give formally weaker axioms for a group $\langle G, * \rangle$, namely:

1) The binary operation $*$ on G is associative.
2) There exists a **left identity** e in G such that $e * x = x$ for all $x \in G$.
3) For each $a \in G$, there exists a **left inverse** a' in G such that $a' * a = e$.

From this *one-sided definition*, one can prove that the left identity is also a right identity and a left inverse is also a right inverse for the same element. Thus these axioms should not be called weaker, since they result in exactly the same structures being called groups. It is conceivable that it might be easier in some cases to check these *left axioms* than our *two-sided axioms*. Of course, by symmetry, it is clear that there are also *right axioms* for a group.

2.3 FINITE GROUPS AND GROUP TABLES

Thus far all our examples have been of infinite groups, i.e., groups where the set G has an infinite number of elements. The student may wonder whether there can be a group structure on some finite set. The answer is yes, and indeed such structures are very important.

Since a group has to have at least one element, namely the identity, a smallest set which might give rise to a group is a one-element set $\{e\}$. The only possible binary operation $*$ on $\{e\}$ is defined by $e * e = e$. The student can check at once that the three group axioms hold. The identity element is always its own inverse in every group.

Let us try to put a group structure on a set of two elements. Since one of the elements must play the role of identity element, we may as well let the set be $\{e, a\}$. Let us attempt to find a table for a binary operation $*$ on $\{e, a\}$ which gives a group structure on $\{e, a\}$. When giving a table for a group operation, we shall always list the elements in the same order across the top as down the left side, with the identity listed first, as in the following table.

$*$	e	a
e		
a		

Since e is to be the identity, so

$$e * x = x * e = x$$

for all $x \in \{e, a\}$, we are forced to fill in the table as shown, if $*$ is to give a

$*$	e	a
e	e	a
a	a	

group. Also, a must have an inverse a' such that

$$a * a' = a' * a = e.$$

In our case, a' must be either e or a. Since $a' = e$ obviously does not work, we must have $a' = a$, so we have to complete the table as below.

$*$	e	a
e	e	a
a	a	e

All the group axioms are now satisfied except possibly the associative law. We will see later in a more general situation that this operation $*$ is associative. The student is asked either to accept it here or to go through the tedious chore of checking various cases for himself.

With these examples as background, we should be able to list some necessary conditions which a table giving a binary operation on a finite set must satisfy for the operation to give a group structure on the set. There must be one element of the set, which we may as well denote by "e", that acts as identity. The condition $e * x = x$ means that the row of the table opposite e at the extreme left must contain exactly the elements appearing across the very top of the table in the same order. Similarly, the condition $x * e = x$ means that the column of the table under e at the very top must contain exactly the elements appearing at the extreme left in the same order. The fact that every element a has a right and a left inverse means that in the row opposite a at the extreme left, the element e must appear, and in the column under a at the very top, the e must appear. Thus e must appear in each row and in each column. We can do even better than this, however. By Theorem 2.2, not only the equations $a * x = e$ and $y * a = e$ have unique solutions, but also the equations $a * x = b$ and $y * a = b$. By a similar argument, this means that *each element b of the group must appear once and only once in each row and column of the table.*

Suppose conversely that a table for a binary operation on a finite set is such that there is an element acting as identity and that in each row and each column each element of the set appears exactly once. Then it can be seen that the structure is a group structure if and only if the associative law holds. If a binary operation $*$ is given by a table, the associative law is usually messy to check. If the operation $*$ is defined by some characterizing property of $a * b$, the associative law is usually easy to check. Fortunately this second case turns out to be the one most often encountered.

We saw that there was essentially only one group of two elements in the sense that if the elements are denoted by "e" and "a" with the identity e appearing first, the table must be as follows.

$*$	e	a
e	e	a
a	a	e

Suppose that a set has three elements. As before, we might as well let the set be $\{e, a, b\}$. For e to be an identity, a binary operation $*$ on this set has to have a table of the form shown in Table 2.1. This leaves four places to be filled in. The student can quickly see that Table 2.1 must be completed as shown in Table 2.2 if each row and each column are to contain each

Table 2.1

$*$	e	a	b
e	e	a	b
a	a		
b	b		

Table 2.2

$*$	e	a	b
e	e	a	b
a	a	b	e
b	b	e	a

element exactly once. Again the student is asked to accept without proof
the fact that this operation is associative, so that * does give a group structure
on $G = \{e, a, b\}$.

Now suppose that G' is any other group of three elements and imagine
a table for G' with identity element appearing first. Since our filling out of
the table for $G = \{e, a, b\}$ could be done in only one way, we see that if we
rename the identity of G' by "e", the next element of G' listed by "a", and
the last element by "b", the resulting table for G' must be the same as the
one we had for G. In other words, the *structural* features are the same for
the two groups, and one group can be made to look exactly like the other
by a renaming of the elements. *Thus any two groups of three elements are
structurally the same.* This is our introduction to the concept of *isomorphism*.
The groups G and G' are *isomorphic*. This concept is sometimes a bit sticky
for the student. We say no more about it now, but we shall make it more
precise later.

EXERCISES

2.1 For each binary operation * defined on a set below, say whether or not * gives
a group structure on the set. If no group results, give the first axiom in the order
$\mathcal{G}_1, \mathcal{G}_2, \mathcal{G}_3$ that does not hold.

a) Define * on **Z** by $a * b = ab$.

b) Define * on **Z** by $a * b = a - b$.

c) Define * on $\mathbf{R}^+$ by $a * b = ab$.

d) Define * on **Q** by $a * b = ab$.

e) Define * on the set of all nonzero real numbers by $a * b = ab$.

f) Define * on **C** by $a * b = a + b$.

2.2 Consider our axioms $\mathcal{G}_1$, $\mathcal{G}_2$, and $\mathcal{G}_3$ for a group. We gave them in the order
$\mathcal{G}_1\mathcal{G}_2\mathcal{G}_3$. Conceivable other orders to state the axioms are $\mathcal{G}_1\mathcal{G}_3\mathcal{G}_2$, $\mathcal{G}_2\mathcal{G}_1\mathcal{G}_3$,
$\mathcal{G}_2\mathcal{G}_3\mathcal{G}_1$, $\mathcal{G}_3\mathcal{G}_1\mathcal{G}_2$, and $\mathcal{G}_3\mathcal{G}_2\mathcal{G}_1$. Of these six possible orders, exactly three are
acceptable for a definition. Which orders aren't acceptable, and why? (Remember
this. Most instructors ask the student to define a group on at least one test.)

2.3 Show by computation and by Theorem 2.3 that if G is a group with binary
operation *, then for all $a, b \in G$, we have $(a * b)' = b' * a'$. What is a similar
expression for $(a * b' * c)'$?

2.4 Proceed as follows to show that there are two possible different types of group structures on a set of four
elements. Let the set be $\{e, a, b, c\}$, with e the identity
element for the group operation. A group table would
then have to start in the manner shown in Table 2.3.
The square indicated by the question mark can't be filled
in with a. It must be filled in either with the identity e
or with an element different from both e and a. In this
latter case, it is no loss of generality to assume that this
element is b. If this square is filled in with e, the table

Table 2.3

*	e	a	b	c
e	e	a	b	c
a	a	?		
b	b			
c	c			

can then be completed in two ways to give a group. Find these two tables. (You need not check the associative laws.) If this square is filled in with b, then the table can only be completed in one way to give a group. Find this table. (Again you need not check the associative law.) Of the three tables you now have, two give the same type of group structure. Determine which two tables these are, and show how the elements in one table would have to be renamed for these two tables to be the same. Are all groups of 4 elements commutative?

†**2.5** Show that if G is a finite group with identity e and with an even number of elements, then there is $a \neq e$ in G such that $a * a = e$.

2.6 Mark each of the following true or false.

    a) A group may have more than one identity element.
    b) Any two groups of three elements are isomorphic.
    c) In a group, each linear equation has a solution.
    d) The proper attitude toward a definition is to memorize it so that you can reproduce it word for word as in the text.
    e) Any definition a person gives for a group is correct provided that everything which is a group by his definition is also a group by the definition in the text.
    f) Any definition a person gives for a group is correct provided that everything that satisfies his definition satisfies the one in the text and conversely.
    g) Every finite group of at most three elements is abelian.
    h) An equation of the form $a * x * b = c$ always has a unique solution in a group.
    i) The empty set can be considered to be a group.
    j) The text has as yet given no examples of groups which are not abelian.

2.7 Give a table for a binary operation on the set $\{e, a, b\}$ of three elements satisfying axioms $\mathcal{G}_2$ and $\mathcal{G}_3$ for a group but not axiom $\mathcal{G}_1$.

2.8 According to Exercise 1.9, there are 16 possible binary operations on a set of 2 elements. How many of these give a structure of a group? How many of the 19,683 possible binary operations on a set of 3 elements give a group structure?

2.9 Let S be the set of all real numbers except -1. Define $*$ on S by

$$a * b = a + b + ab.$$

a) Show that $*$ gives a binary operation on S.
b) Show that $\langle S, * \rangle$ is a group.
c) Find the solution of the equation $2 * x * 3 = 7$ in S.

2.10 Let $\mathbf{R}^*$ be the set of all real numbers except 0. Define $*$ on $\mathbf{R}^*$ by $a * b = |a|b$.

a) Show that $*$ gives an associative binary operation on $\mathbf{R}^*$.
b) Show that there is a left identity for $*$ and a right inverse for each element in $\mathbf{R}^*$.
c) Is $\mathbf{R}^*$ with this binary operation a group?
d) Explain the significance of this exercise.

2.11 If $*$ is a binary operation on a set S, an element x of S is an **idempotent for** $*$ if $x * x = x$. Prove that a group has exactly one idempotent element. (You may use any theorems proved so far in the text.)

2.12 Show that every group G with identity e and such that $x * x = e$ for all $x \in G$ is abelian. (If you get stuck, you will find this shown in the course of Example 18.5.)

2.13 Prove that a set G, together with a binary operation $*$ on G satisfying the left axioms (1), (2), and (3) given just prior to subheading 2.3, is a group.

2.14 Prove that a nonempty set G, together with an associative binary operation $*$ on G satisfying

$$a * x = b \text{ and } y * a = b \text{ have solutions in } G \text{ for all } a, b \in G,$$

is a group. [*Hint:* Use Exercise 2.13.]

2.15 The following "definitions" of a group are taken verbatim, including spelling and punctuation, from students' examination papers. Criticize them.

a) A group G is a set of elements together with a binary operation $*$ such that the following conditions are satisfied

1) $*$ is associative
2) There exists an $e \in G$ such that

$$e * x = x * e = x = \text{identity.}$$

3) For every $a \in G$ there exists an a' (inverse) such that

$$a \cdot a' = a' \cdot a = e$$

b) A group is a set G such that

1) The operation on G is associative.
2) there is an identity element (e) in G.
3) for every $a \in G$, there is an a' (inverse for each element)

c) A group is a set with a binary operation such

1) the binary operation is defined
2) an inverse exists
3) an identity element exists

d) A set G is called a group over the binery operation $*$ such that for all $a, b \in G$

1) Binary operation $*$ is associative under addition
2) there exist an element $\{e\}$ such that

$$a * e = e * a = e$$

3) Fore every element a there exists an element a' such that

$$a * a' = a' * a = e$$

3|Subgroups

3.1 NOTATION AND TERMINOLOGY

It is time to explain some conventional notation and terminology used in group theory. The algebraist as a rule does not use a special symbol "$*$" to denote a binary operation different from the usual addition and multiplication. He sticks with the conventional additive or multiplicative notation and even calls the operation *addition* or *multiplication*, depending on the symbol used. The symbol for addition is of course "$+$", and usually multiplication is denoted by juxtaposition without a dot, if no confusion results. Thus in place of the notation "$a * b$", we shall be using either "$a + b$" to be read "the *sum* of a and b", or "ab" to be read "the *product* of a and b". There is a sort of gentlemen's agreement that the symbol "$+$" should be used only to designate commutative operations. The algebraist feels very uncomfortable when he sees "$a + b \neq b + a$". For this reason when developing our group theory in a general situation where the operation may or may not be commutative, we shall always use multiplicative notation.

Table 3.1			
	1	a	b
1	1	a	b
a	a	b	1
b	b	1	a

Table 3.2			
$+$	0	a	b
0	0	a	b
a	a	b	0
b	b	0	a

Mathematicians frequently use the symbol "0" to denote an additive identity and the symbol "1" to denote a multiplicative identity, even though they may not be actually denoting our numbers 0 and 1. Of course, if a person is also talking about numbers at the same time, so that confusion would result, symbols such as "e" or "u" are used. Thus a table for a group of three elements might be one such as Table 3.1, or, since such a group is commutative, the table might look like Table 3.2. In general situations we shall continue to use "e" to denote the identity element of a group.

It is customary to denote the inverse of an element a in a group by "a^{-1}" in multiplicative notation and by "$-a$" in additive notation. From now on we shall be using these notations in place of the symbol "a'".

27

Let us explain one more term which is used so often that it merits a special definition.

Definition. If G is a finite group, then the *order* $|G|$ *of* G is the number of elements in G. In general, for any finite set S, $|S|$ is the number of elements in S.

Finally, in place of the phrase *with the binary operation of* we shall use the word *under*, so that "the group **R** with the binary operation of addition" becomes "the group **R** under addition."

3.2 SUBSETS AND SUBGROUPS

The student may have noticed that we sometimes have had groups contained within larger groups. For example, the group **Z** under addition is contained within the group **Q** under addition, which in turn is contained within the group **R** under addition. When we view the group $\langle \mathbf{Z}, + \rangle$ as contained in the group $\langle \mathbf{R}, + \rangle$, it is very important to notice that the operation $+$ on integers n and m as elements of $\langle \mathbf{Z}, + \rangle$ produces the same element $n + m$ as would result if you were to think of n and m as elements in $\langle \mathbf{R}, + \rangle$. Thus we should *not* regard the group $\langle \mathbf{Q}^+, \cdot \rangle$ as contained in $\langle \mathbf{R}, + \rangle$, even though $\mathbf{Q}^+$ is contained in **R** as a set. In this instance, $2 \cdot 3 = 6$ in $\langle \mathbf{Q}^+, \cdot \rangle$, while $2 + 3 = 5$ in $\langle \mathbf{R}, + \rangle$. We are requiring not only that the set of one group be contained in the set of the other, but also that the group operation on the smaller set assign the same element to each ordered pair from this smaller set as is assigned by the group operation of the larger set. Let us give a sequence of definitions to make these ideas really precise.

Definition. A set B is a *subset of a set* A, denoted by "$B \subseteq A$" or "$A \supseteq B$", if every element of B is in A. The notations "$B \subset A$" or "$A \supset B$" will be used for $B \subseteq A$ but $B \neq A$.

Note that according to this definition, for any set A, A itself and $\emptyset$ are both subsets of A. These are trivial cases.

Definition. If A is any set, then $\emptyset$ and A are *improper subsets of* A. Any other subset of A is a *proper subset of* A.

Definition. Let G be a group and let S be a subset of G. If for every $a, b \in S$ it is true that the product ab computed in G is also in S, then the group operation of G is *closed on* S. The binary operation on S thus defined is the *induced operation on* S *from* G.

We can now make the concept of one group being contained in another precise.

Definition. If H is a subset of a group G such that the group operation of G is closed on H, and if H is itself a group under this induced operation, then H is a *subgroup of* G. We shall let "$H \leq G$" or "$G \geq H$" denote that H is a subgroup of G, and "$H < G$" or "$G > H$" shall mean $H \leq G$ but $H \neq G$.

Thus $\langle \mathbf{Z}, + \rangle < \langle \mathbf{R}, + \rangle$ but $\langle \mathbf{Q}^+, \cdot \rangle$ is *not* a subgroup of $\langle \mathbf{R}, + \rangle$, even though as sets, $\mathbf{Q}^+ \subset \mathbf{R}$. Every group G has as *trivial* subgroups G itself and $\{e\}$, where e is the identity element of G.

Definition. If G is a group, then G and $\{e\}$ are ***improper subgroups of*** G. All other subgroups are ***proper subgroups***.

We turn to some illustrations.

Example 3.1 $\mathbf{Q}^+$ under multiplication is a proper subgroup of $\mathbf{R}^+$ under multiplication. ∥

Example 3.2 There are two different types of group structures of order 4 (see Exercise 2.4). We describe them by their group tables (Tables 3.3 and 3.4). The group V is the **Klein 4-group,** and the notation "V" comes from the German word, *Viergruppe*.

The only proper subgroup of $\mathbf{Z}_4$ is $\{0, 2\}$. Note that $\{0, 3\}$ is *not* a subgroup of $\mathbf{Z}_4$, since the operation $+$ is *not closed* on $\{0, 3\}$. For example, $3 + 3 = 2$, and $2 \notin \{0, 3\}$. On the other hand, the group V has three proper subgroups, $\{e, a\}$, $\{e, b\}$, and $\{e, c\}$. Here $\{e, a, b\}$ is *not* a subgroup, since the operation of V is not closed on $\{e, a, b\}$. For example, $ab = c$, and $c \notin \{e, a, b\}$. ∥

Table 3.3

$\mathbf{Z}_4$:

+	0	1	2	3
0	0	1	2	3
1	1	2	3	0
2	2	3	0	1
3	3	0	1	2

Table 3.4

V:

	e	a	b	c
e	e	a	b	c
a	a	e	c	b
b	b	c	e	a
c	c	b	a	e

It is often useful to draw a *lattice diagram* of the subgroups of a group. In such a diagram, a line running downward from a group G to a group H means that H is a subgroup of G. Thus the larger group is placed nearer the top of the diagram. Figure 3.1 contains the lattice diagrams for the groups $\mathbf{Z}_4$ and V of Example 3.2.

Note that if $H \leq G$ and $a \in H$, then by Theorem 2.2 the equation $ax = a$ must have a unique solution, namely the identity element of H. But this equation can also be viewed as one in G, and we see that this unique solution must also be the identity e of G. A similar argument then applied to the equation $ax = e$, viewed in both H and G, shows that the inverse a^{-1} of a in G is also the inverse of a in the subgroup H.

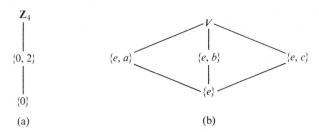

Fig. 3.1. (a) Lattice diagram for $\mathbf{Z}_4$. (b) Lattice diagram for V.

It is convenient to have a routine criterion for determining whether a subset of a group G is a subgroup of G. The next theorem gives such a criterion. While more compact criteria are available, involving only one condition, we prefer this more transparent one for a first course.

Theorem 3.1   *A subset H of a group G is a subgroup of G if and only if*

1) *the binary operation of G is closed on H,*
2) *the identity e of G is in H,*
3) *for all $a \in H$ it is true that $a^{-1} \in H$ also.*

Proof. The fact that if $H \leq G$ then (1), (2), and (3) must hold follows at once from the definition of a subgroup and from the remarks preceding the statement of the theorem.

Conversely, suppose H is a subset of a group G such that (1), (2), and (3) hold. By (2), we have at once that $\mathcal{G}_2$ is satisfied. Also $\mathcal{G}_3$ is satisfied by (3). It remains to check the associative axiom, $\mathcal{G}_1$. But surely for all $a, b, c \in H$ it is true that $(ab)c = a(bc)$ in H, for we may actually view this as an equation in G, where the associative law holds. Hence $H \leq G$. ∎

3.3   CYCLIC SUBGROUPS

We remarked in Example 3.2 that $\{0, 3\}$ is not a subgroup of $\mathbf{Z}_4$. Let us see how big a subgroup H of $\mathbf{Z}_4$ would have to be if it contained 3. It would have to contain the identity 0 and the inverse of 3, which is 1. Also H would have to contain $3 + 3$, which is 2. Hence the only subgroup of $\mathbf{Z}_4$ containing 3 is $\mathbf{Z}_4$ itself.

Let us imitate this reasoning in a general situation. As we remarked before, for a general argument we always use multiplicative notation. Let G be a group and let $a \in G$. A subgroup of G containing a must, by Theorem 3.1, contain aa, which we denote by "a^2". Then it must contain a^2a, which we denote by "a^3". In general, it must contain a^n, the result of computing products of a and itself for n factors for every positive integer n. (In additive notation we would denote this by "na".) These positive integral powers of a do give a set closed under multiplication. It is possible, however, that the inverse of a is not in this set. Of course, a subgroup containing a

must also contain a^{-1}, and then $a^{-1}a^{-1}$, which we denote by "a^{-2}", and, in general, it must contain a^{-m} for all $m \in \mathbf{Z}^+$. It must contain the identity $e = aa^{-1}$. For obvious symbolic reasons, we agree to let a^0 be e. Summarizing, we have shown that *a subgroup of G containing a must contain all elements a^n (or na for additive groups) for all $n \in \mathbf{Z}$.* That is, a subgroup containing a must contain $\{a^n \mid n \in \mathbf{Z}\}$. Observe that these powers a^n of a need not be distinct. For example, in the group V of Example 3.2,

$$a^2 = e, \qquad a^3 = a, \qquad a^4 = e, \qquad a^{-1} = a, \quad \text{etc.}$$

It is easy to see that our usual law of exponents, $a^m a^n = a^{m+n}$ for $m, n \in \mathbf{Z}$, holds. For $m, n \in \mathbf{Z}^+$, it is clear. We illustrate another type of case by an example:

$$a^{-2}a^5 = a^{-1}a^{-1}aaaaa = a^{-1}(a^{-1}a)aaaa = a^{-1}eaaaa = a^{-1}(ea)aaa$$
$$= a^{-1}aaaa = (a^{-1}a)aaa = eaaa = (ea)aa = aaa = a^3.$$

The boring details of a careful proof for the general case are left to the student who does not mind the tedium. We have almost proved the next theorem.

Theorem 3.2 *Let G be a group and let $a \in G$. Then*

$$H = \{a^n \mid n \in \mathbf{Z}\}$$

is a subgroup of G and is the smallest subgroup of G which contains a.

Proof. We check the three conditions for a subset of a group to give a subgroup which were given by Theorem 3.1. Since $a^r a^s = a^{r+s}$ for $r, s \in \mathbf{Z}$, we see that the product in G of two elements of H is again in H. Thus H is closed under the group operation of G. Also $a^0 = e$, so $e \in H$, and for $a^r \in H$, $a^{-r} \in H$ and $a^{-r}a^r = e$. Hence all the conditions are satisfied, and $H \leq G$.

Our arguments prior to the statement of the theorem showed that any subgroup of G containing a must contain H, so H is surely the smallest subgroup of G containing a. ∎

Definition. The group H of Theorem 3.2 is the **cyclic subgroup of G generated by** a, and will be denoted by "$\langle a \rangle$".

Definition. An element a of a group G **generates** G and is a **generator for** G if $\langle a \rangle = G$. A group G is **cyclic** if there is some element a in G which generates G.

Example 3.3 Let $\mathbf{Z}_4$ and V be the groups of Example 3.2. Then $\mathbf{Z}_4$ is cyclic and both 1 and 3 are generators, that is,

$$\langle 1 \rangle = \langle 3 \rangle = \mathbf{Z}_4.$$

However, V is *not* cyclic, for $\langle a \rangle$, $\langle b \rangle$, and $\langle c \rangle$ are proper subgroups of two elements. Of course $\langle e \rangle$ is the improper subgroup of one element. ‖

Example 3.4 The group $\mathbf{Z}$ under addition is a cyclic group. Both 1 and -1 are generators for the group. $\|$

Example 3.5 Consider the group $\mathbf{Z}$ under addition. Let us find $\langle 3 \rangle$. Here the notation is additive, and $\langle 3 \rangle$ must contain

$$3, \quad 3 + 3 = 6, \quad 3 + 3 + 3 = 9, \quad \text{etc.},$$
$$0, \quad -3, \quad -3 + -3 = -6, \quad -3 + -3 + -3 = -9, \quad \text{etc.}$$

In other words, the cyclic subgroup generated by 3 consists of all multiples of 3, positive, negative, and zero. We denote this subgroup by "$3\mathbf{Z}$" as well as by "$\langle 3 \rangle$". In a similar way, we shall let $n\mathbf{Z}$ be the cyclic subgroup $\langle n \rangle$ of $\mathbf{Z}$. Note that $6\mathbf{Z} < 3\mathbf{Z}$. $\|$

EXERCISES

3.1 Determine which of the following subsets of the complex numbers are subgroups under addition of the group $\mathbf{C}$ of complex numbers under addition.

a) $\mathbf{R}$ b) $\mathbf{Q}^+$ c) $7\mathbf{Z}$

d) The set $i\mathbf{R}$ of pure imaginary numbers including 0

e) The set $\pi\mathbf{Q}$ of rational multiples of π f) The set $\{\pi^n \mid n \in \mathbf{Z}\}$

3.2 A number of groups are given below. Give a *complete* list of all relations of one group being a subgroup of one of the groups listed.

$G_1 = \mathbf{Z}$ under addition
$G_2 = 12\mathbf{Z}$ under addition
$G_3 = \mathbf{Q}^+$ under multiplication
$G_4 = \mathbf{R}$ under addition
$G_5 = \mathbf{R}^+$ under multiplication
$G_6 = \{\pi^n \mid n \in \mathbf{Z}\}$ under multiplication
$G_7 = 3\mathbf{Z}$ under addition
$G_8 = $ the set of all integral multiples of 6 under addition
$G_9 = \{6^n \mid n \in \mathbf{Z}\}$ under multiplication

3.3 Write at least 5 elements of each of the following cyclic groups.

a) $25\mathbf{Z}$ under addition

b) $\{(\frac{1}{2})^n \mid n \in \mathbf{Z}\}$ under multiplication

c) $\{\pi^n \mid n \in \mathbf{Z}\}$ under multiplication

3.4 Which of the following groups are cyclic? For each cyclic group, give all the generators of the group.

$G_1 = \langle \mathbf{Z}, + \rangle \qquad G_2 = \langle \mathbf{Q}, + \rangle \qquad G_3 = \langle \mathbf{Q}^+, \cdot \rangle \qquad G_4 = \langle 6\mathbf{Z}, + \rangle$

$G_5 = \{6^n \mid n \in \mathbf{Z}\}$ under multiplication
$G_6 = \{a + b\sqrt{2} \mid a, b \in \mathbf{Z}\}$ under addition

3.5 Study the structure of the table for the group $\mathbf{Z}_4$ of Example 3.2.

a) By analogy, complete Table 3.5 to give a cyclic group $\mathbf{Z}_6$ of six elements. (You need not prove the associative law.)

Table 3.5

$\mathbf{Z}_6$:	+	0	1	2	3	4	5
	0	0	1	2	3	4	5
	1	1	2	3	4	5	0
	2	2					
	3	3					
	4	4					
	5	5					

b) Compute the subgroups $\langle 1 \rangle$, $\langle 2 \rangle$, $\langle 3 \rangle$, $\langle 4 \rangle$, and $\langle 5 \rangle$ of the group $\mathbf{Z}_6$ given in part (a).

c) Which elements are generators for the group $\mathbf{Z}_6$ of part (a)?

†**3.6** Prove that if G is an abelian group with identity e, then all elements x of G satisfying the equation $x^2 = e$ form a subgroup H of G.

3.7 Mark each of the following true or false.

— a) The associative law holds in every group.
— b) There may be a group in which the cancellation law fails.
— c) Every group is a subgroup of itself.
— d) Every group has exactly two improper subgroups.
— e) In every cyclic group, every element is a generator.
— f) This text has still given no example of a group which is not abelian.
— g) Every set of numbers which is a group under addition is also a group under multiplication.
— h) A subgroup may be defined as a subset of a group.
— i) $\mathbf{Z}_4$ is a cyclic group.
— j) Every subset of every group is a subgroup under the induced operation.

3.8 Find the flaw in the following argument: "Condition (2) of Theorem 3.1 is redundant, since it can be derived from (1) and (3), for let $a \in H$. Then $a^{-1} \in H$ by (3), and by (1), $aa^{-1} = e$ is an element of H which gives (2)."

3.9 Show that a nonempty subset H of a group G is a subgroup of G if and only if $ab^{-1} \in H$ for all $a, b \in H$. (This is one of the *more compact criteria* referred to prior to Theorem 3.1.)

3.10 Prove that a cyclic group with only one generator can have at most two elements.

3.11 Show that if $a \in G$, where G is a finite group with identity e, then there exists $n \in \mathbf{Z}^+$ such that $a^n = e$.

3.12 Repeat Exercise 3.6 for the general situation of the set H of all solutions x of the equation $x^n = e$ for a fixed integer $n \geq 1$ in an abelian group G with identity e.

3.13 Let G be a group and let a be a fixed element of G. Show that

$$H_a = \{x \in G \mid xa = ax\}$$

is a subgroup of G.

3.14 Generalizing Exercise 3.13, let S be any subset of a group G.

a) Show that $H_S = \{x \in G \mid xs = sx \text{ for all } s \in S\}$ is a subgroup of G.

b) In reference to part (a), the subgroup H_G is the **center of** G. Show that H_G is an abelian group.

3.15 Let the binary operation of a group G be closed on a nonempty finite subset H of G. Show that H is a subgroup of G.

3.16 For sets H and K, we define the **intersection** $H \cap K$ by

$$H \cap K = \{x \mid x \in H \text{ and } x \in K\}.$$

Show that if $H \leq G$ and $K \leq G$, then $H \cap K \leq G$.

3.17 Show by means of an example that it is possible for the quadratic equation $x^2 = e$ to have more than two solutions in some group G with identity e.

4 | Permutations I

4.1 FUNCTIONS AND PERMUTATIONS

In this section and the one that follows, we consider groups whose elements are entities called *permutations*. These groups will provide us with our first examples of groups that are not abelian. Indeed, we shall show in a later section that any group is structurally the same as some group of permutations. Unfortunately, this result, which sounds very powerful, does not turn out to be particularly useful to us.

The student is probably familiar with the idea of a permutation of a set as a rearrangement of the elements of the set. Thus for the set $\{1, 2, 3, 4, 5\}$, a rearrangement of the elements could be given schematically as in Fig. 4.1, resulting in the new arrangement $\{4, 2, 5, 3, 1\}$. Let us think of this schematic diagram in Fig. 4.1 as a carrying or a *mapping* of each element listed in the left column into a single (not necessarily different) element from the same set listed at the right. Thus 1 is carried into 4, 2 is mapped into 2, etc. Furthermore, to be a permutation of the set, this mapping must be such that each element appears in the right column once and only once. For example, the diagram in Fig. 4.2 does *not* give a permutation, for 3 appears twice while 1 does not appear at all in the right column. We shall be defining a permutation as such a mapping. However, the general idea of assigning to each element of some set an element from the same or possibly from a different set will arise so often in our work that we first give a separate definition for this concept. The concept is that of a *function*, a term which the student has already encountered.

$1 \rightarrow 4$	$1 \rightarrow 3$
$2 \rightarrow 2$	$2 \rightarrow 2$
$3 \rightarrow 5$	$3 \rightarrow 4$
$4 \rightarrow 3$	$4 \rightarrow 5$
$5 \rightarrow 1$	$5 \rightarrow 3$
Fig. 4.1	**Fig. 4.2**

Definition. A *function* or *mapping* ϕ *from a set A into a set B* is a rule which assigns to each element a of A exactly one element b of B. We say that ϕ *maps a into b*, and that ϕ *maps A into B*.

35

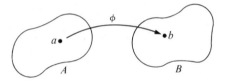

Fig. 4.3

The classical notation to denote that ϕ maps a into b is

$$\text{``}\phi(a) = b\text{''}.$$

However, to conform with the trend among some enlightened algebraists, we shall usually use the notation

$$\text{``}a\phi = b\text{''}.$$

The notation "$a^\phi = b$" is also found in the literature. The element b is the **image of** a **under** ϕ. The fact that ϕ maps A into B will be symbolically expressed by

$$\text{``}\phi\colon A \to B\text{''}.$$

It may help the student to visualize a function in terms of Fig. 4.3. As to the three possible notations for a being mapped into b by ϕ given after the definition, the student is acquainted with the notation "$\phi(a) = b$" from previous courses. More and more algebraists are using the notations "$a\phi = b$" and "$a^\phi = b$" for the following reason. If ϕ and ψ are functions with $\phi\colon A \to B$ and $\psi\colon B \to C$, then there is a natural function mapping A into C, as illustrated in Fig. 4.4. That is, you can get from A to C via B, using the functions ϕ and ψ. This function mapping A into C is the **composite function** consisting of ϕ followed by ψ. In classical notation, $\phi(a) = b$ and $\psi(b) = c$, so

$$\psi(\phi(a)) = c,$$

and one denotes the composite function by "$\psi\phi$". The symbol "$\psi\phi$" for ϕ followed by ψ then has to be read from right to left. In the more recent notations, we have $a\phi = b$ and $b\psi = c$ with

$$a(\phi\psi) = (a\phi)\psi = c,$$

or $a^\phi = b$ and $b^\psi = c$ with

$$a^{(\phi\psi)} = (a^\phi)^\psi = c.$$

Fig. 4.4

Thus the composite function in these notations is $\phi\psi$ and can be read from left to right. We suggest that the student read the notations "$a\phi = b$" and "$a^\phi = b$" as "the image of a under ϕ is b". You should realize that all this talk is not about the concept involved, but is about notation. However, a bad choice of notation may greatly hinder the development of a mathematical theory.

Returning to permutations, according to our definition, we see that the assignment of Fig. 4.2 is a function from $\{1, 2, 3, 4, 5\}$ into itself. We do not wish to call this a permutation, however. We need to pick out those functions such that each element of the set has exactly one element mapped into it. There is again a terminology for a more general situation.

Definition. A function from a set A into a set B is **one to one** if each element of B has at most one element of A mapped into it, and is **onto** B if each element of B has at least one element of A mapped into it.

In terms of Fig. 4.3, a function $\phi: A \to B$ is one to one if each $b \in B$ has *at most one* arrow coming into it. To say that ϕ is onto B is to say that every $b \in B$ has *at least one* arrow coming into it. Since we will be often proving that certain functions are one to one or onto or both, it is worth outlining the technique always used.

1) To show that ϕ is one to one, you show that $a_1\phi = a_2\phi$ implies $a_1 = a_2$.
2) To show that ϕ is onto B, you show that for each $b \in B$, there exists $a \in A$ such that $a\phi = b$.

Finally, we remark that for $\phi: A \to B$, the set A is the **domain of** ϕ, the set B is the **codomain of** ϕ, and the set $\{a\phi \mid a \in A\}$ is the **image of A under** ϕ.

Since for a permutation of a set A we want each element of A to have both at most one element of A and at least one element of A mapped into it, we arrive at the following definition.

Definition. A *permutation of a set* A is a function from A into A which is both one to one and onto. In other words, a permutation of A is a one-to-one function from A onto A.

One sometimes writes

$$\text{``}\phi: A \xrightarrow[\text{onto}]{\text{1-1}} B\text{''}$$

for a one-to-one function ϕ mapping A onto B.

The student is urged to take a little time to study and to try to understand these ideas. It will make things easier for him throughout the course. The terminology is still standard, although another terminology is coming more and more into vogue, having been spread by the disciples of N. Bourbaki. While we shall not use this terminology, we give it here so that you may realize its meaning if you run across it elsewhere. In the new terminology, a one-to-one map is an **injection,** an onto map is a **surjection,** and a map which is both one to one and onto is a **bijection.**

4.2 GROUPS OF PERMUTATIONS

A natural binary operation, *permutation multiplication*, is defined on the permutations of a set. Let A be a set, and let σ and τ be permutations of A so that σ and τ are both one-to-one functions mapping A onto A. The composite function $\sigma\tau$ as illustrated in Fig. 4.4, with $B = C = A$, $\phi = \sigma$, and $\psi = \tau$, gives a mapping of A into A. Now $\sigma\tau$ will be a permutation if it is one to one and onto A. We are using the notation of writing functions on the right so that $\sigma\tau$ can be read from left to right. Let us show that $\sigma\tau$ is one to one. If

$$a_1(\sigma\tau) = a_2(\sigma\tau),$$

then

$$(a_1\sigma)\tau = (a_2\sigma)\tau,$$

and since τ is given to be one to one, we know that $a_1\sigma = a_2\sigma$. But, then, since σ is one to one, this gives $a_1 = a_2$. Hence $\sigma\tau$ is one to one. To show that $\sigma\tau$ is onto A, let $a \in A$. Since τ is onto A, there exists $a' \in A$ such that $a'\tau = a$. Since σ is onto A, there exists $a'' \in A$ such that $a' = a''\sigma$. Then

$$a = a'\tau = (a''\sigma)\tau = a''(\sigma\tau),$$

so $\sigma\tau$ is onto A.

To illustrate, suppose that

$$A = \{1, 2, 3, 4, 5\}$$

and that σ is the permutation given by Fig. 4.1. We write σ in a more standard notation as

$$\sigma = \begin{pmatrix} 1 & 2 & 3 & 4 & 5 \\ 4 & 2 & 5 & 3 & 1 \end{pmatrix},$$

so that $1\sigma = 4$, $2\sigma = 2$, etc. Let

$$\tau = \begin{pmatrix} 1 & 2 & 3 & 4 & 5 \\ 3 & 5 & 4 & 2 & 1 \end{pmatrix}.$$

Then

$$\sigma\tau = \begin{pmatrix} 1 & 2 & 3 & 4 & 5 \\ 4 & 2 & 5 & 3 & 1 \end{pmatrix}\begin{pmatrix} 1 & 2 & 3 & 4 & 5 \\ 3 & 5 & 4 & 2 & 1 \end{pmatrix} = \begin{pmatrix} 1 & 2 & 3 & 4 & 5 \\ 2 & 5 & 1 & 4 & 3 \end{pmatrix}.$$

For example,

$$1(\sigma\tau) = (1\sigma)\tau = 4\tau = 2.$$

We now show that the collection of all permutations of a nonempty set A forms a group under this permutation multiplication.

Theorem 4.1 *Let A be a nonempty set, and let S_A be the collection of all permutations of A. Then S_A is a group under permutation multiplication.*

Proof. We have three axioms to check. Since permutations are functions, in order to show for permutations σ, τ, and μ that

$$(\sigma\tau)\mu = \sigma(\tau\mu),$$

we have to show that each composite function maps each $a \in A$ into the same image in A. That is, we must show that

$$a[(\sigma\tau)\mu] = a[\sigma(\tau\mu)]$$

for all $a \in A$. We have

$$a[(\sigma\tau)\mu] = [a(\sigma\tau)]\mu = [(a\sigma)\tau]\mu = (a\sigma)(\tau\mu) = a[\sigma(\tau\mu)].$$

Thus $(\sigma\tau)\mu$ and $\sigma(\tau\mu)$ map each $a \in A$ into the same element $[(a\sigma)\tau]\mu$ and hence are the same permutation. Since we made no use of the fact that σ, τ, and μ are one to one and onto, we have actually proved that *function composition is associative.* Hence $\mathcal{G}_1$ is satisfied.

The permutation ι such that $a\iota = a$ for all $a \in A$ obviously acts as identity. Therefore $\mathcal{G}_2$ is satisfied.

For a permutation σ, define σ^{-1} to be the permutation which reverses the direction of the mapping σ, that is, $a\sigma^{-1}$ is to be the element a' of A such that $a = a'\sigma$. The existence of exactly one such element a' is a consequence of the fact that, as a function, σ is both one to one and onto. It is clear that

$$a\iota = a = a'\sigma = (a\sigma^{-1})\sigma = a(\sigma^{-1}\sigma)$$

and also that

$$a'\iota = a' = a\sigma^{-1} = (a'\sigma)\sigma^{-1} = a'(\sigma\sigma^{-1}),$$

so that $\sigma^{-1}\sigma$ and $\sigma\sigma^{-1}$ are both the permutation ι. Thus $\mathcal{G}_3$ is satisfied. ∎

There was nothing in our definition of a permutation to require that the set A be finite. However, most of our examples of permutation groups will be concerned with permutations of finite sets. Clearly, if A and B both have the same number of elements, then the group of all permutations of A has the same structure as the group of all permutations of B. One group can be obtained from the other by just renaming elements. This is again the concept of *isomorphic groups* mentioned in Section 2 and about which more will be said later.

Definition. If A is the finite set $\{1, 2, \ldots, n\}$, then the group of all permutations of A is the **symmetric group on n letters**, and is denoted by "S_n".

Note that S_n has $n!$ elements, where

$$n! = n(n - 1)(n - 2) \cdots (3)(2)(1).$$

4.3 TWO IMPORTANT EXAMPLES

Example 4.1 An interesting example for us is the group S_3 of $3! = 6$ elements. Let the set A be $\{1, 2, 3\}$. We list the permutations of A and assign to each a subscripted Greek letter for a name. The reasons for the choice of names and for the shading in the table will be clear later. Let

$$\rho_0 = \begin{pmatrix} 1 & 2 & 3 \\ 1 & 2 & 3 \end{pmatrix}, \qquad \mu_1 = \begin{pmatrix} 1 & 2 & 3 \\ 1 & 3 & 2 \end{pmatrix},$$

$$\rho_1 = \begin{pmatrix} 1 & 2 & 3 \\ 2 & 3 & 1 \end{pmatrix}, \qquad \mu_2 = \begin{pmatrix} 1 & 2 & 3 \\ 3 & 2 & 1 \end{pmatrix};$$

$$\rho_2 = \begin{pmatrix} 1 & 2 & 3 \\ 3 & 1 & 2 \end{pmatrix}, \qquad \mu_3 = \begin{pmatrix} 1 & 2 & 3 \\ 2 & 1 & 3 \end{pmatrix}.$$

The student can check that the multiplication table given in Fig. 4.5 is correct. Note that this group is not abelian! It is our first such example. We have seen that any group of at most 4 elements is abelian. Later we will see that a group of 5 elements is also abelian. Thus this is a "smallest" possible example of a nonabelian group. ∥

	ρ_0	ρ_1	ρ_2	μ_1	μ_2	μ_3
ρ_0	ρ_0	ρ_1	ρ_2	μ_1	μ_2	μ_3
ρ_1	ρ_1	ρ_2	ρ_0	μ_2	μ_3	μ_1
ρ_2	ρ_2	ρ_0	ρ_1	μ_3	μ_1	μ_2
μ_1	μ_1	μ_3	μ_2	ρ_0	ρ_2	ρ_1
μ_2	μ_2	μ_1	μ_3	ρ_1	ρ_0	ρ_2
μ_3	μ_3	μ_2	μ_1	ρ_2	ρ_1	ρ_0

Fig. 4.5

There is a natural correspondence between the elements of S_3 in Example 4.1 and the ways in which two copies of an equilateral triangle with vertices 1, 2, and 3 (see Fig. 4.6) can be placed, one covering the other. For this reason, S_3 is also the **group D_3 of symmetries of an equilateral triangle**. Naively, we used ρ_i for *rotations* and μ_i for *mirror images* in bisectors of angles. The notation "D_3" stands for the third dihedral group. The **nth dihedral group D_n** is the group of symmetries of the regular n-gon.

Example 4.2 Let us form the dihedral group D_4 of permutations corresponding to the ways that two copies of a square with vertices 1, 2, 3, and 4 can be placed, one covering the other (see Fig. 4.7). D_4 will then

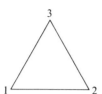

Fig. 4.6

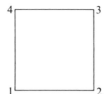

Fig. 4.7

be the **group of symmetries of the square.** It is also called the "**octic group.**" Again we choose seemingly arbitrary notation and use shading which we shall explain later. Naively, we are using ρ_i for *rotations*, μ_i for *mirror images* in perpendicular bisectors of sides, and δ_i for *diagonal flips*. There are eight permutations involved here. Let

$$\rho_0 = \begin{pmatrix} 1 & 2 & 3 & 4 \\ 1 & 2 & 3 & 4 \end{pmatrix}, \qquad \mu_1 = \begin{pmatrix} 1 & 2 & 3 & 4 \\ 2 & 1 & 4 & 3 \end{pmatrix},$$

$$\rho_1 = \begin{pmatrix} 1 & 2 & 3 & 4 \\ 2 & 3 & 4 & 1 \end{pmatrix}, \qquad \mu_2 = \begin{pmatrix} 1 & 2 & 3 & 4 \\ 4 & 3 & 2 & 1 \end{pmatrix},$$

$$\rho_2 = \begin{pmatrix} 1 & 2 & 3 & 4 \\ 3 & 4 & 1 & 2 \end{pmatrix}, \qquad \delta_1 = \begin{pmatrix} 1 & 2 & 3 & 4 \\ 3 & 2 & 1 & 4 \end{pmatrix},$$

$$\rho_3 = \begin{pmatrix} 1 & 2 & 3 & 4 \\ 4 & 1 & 2 & 3 \end{pmatrix}, \qquad \delta_2 = \begin{pmatrix} 1 & 2 & 3 & 4 \\ 1 & 4 & 3 & 2 \end{pmatrix}.$$

The student can check that the table for D_4 given in Fig. 4.8 is correct. Note that D_4 is again nonabelian. This group is simply beautiful. It will provide us with nice examples for almost all concepts we will introduce in group theory. Look at the lovely symmetries in that table!

	ρ_0	ρ_1	ρ_2	ρ_3	μ_1	μ_2	δ_1	δ_2
ρ_0	ρ_0	ρ_1	ρ_2	ρ_3	μ_1	μ_2	δ_1	δ_2
ρ_1	ρ_1	ρ_2	ρ_3	ρ_0	δ_2	δ_1	μ_1	μ_2
ρ_2	ρ_2	ρ_3	ρ_0	ρ_1	μ_2	μ_1	δ_2	δ_1
ρ_3	ρ_3	ρ_0	ρ_1	ρ_2	δ_1	δ_2	μ_2	μ_1
μ_1	μ_1	δ_1	μ_2	δ_2	ρ_0	ρ_2	ρ_1	ρ_3
μ_2	μ_2	δ_2	μ_1	δ_1	ρ_2	ρ_0	ρ_3	ρ_1
δ_1	δ_1	μ_2	δ_2	μ_1	ρ_3	ρ_1	ρ_0	ρ_2
δ_2	δ_2	μ_1	δ_1	μ_2	ρ_1	ρ_3	ρ_2	ρ_0

Fig. 4.8

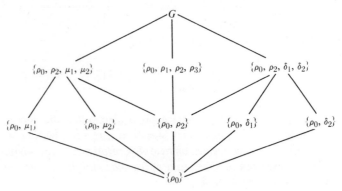

Fig. 4.9. Lattice diagram for D_4.

Finally, we give in Fig. 4.9 the lattice diagram for the subgroups of D_4. The student can verify that it is correct. ‖

EXERCISES

4.1 Consider the three permutations in S_6:

$$\sigma = \begin{pmatrix} 1 & 2 & 3 & 4 & 5 & 6 \\ 3 & 1 & 4 & 5 & 6 & 2 \end{pmatrix},$$

$$\tau = \begin{pmatrix} 1 & 2 & 3 & 4 & 5 & 6 \\ 2 & 4 & 1 & 3 & 6 & 5 \end{pmatrix},$$

$$\mu = \begin{pmatrix} 1 & 2 & 3 & 4 & 5 & 6 \\ 5 & 2 & 4 & 3 & 1 & 6 \end{pmatrix}.$$

Compute

a) $\sigma\tau$,   b) $\sigma\tau^2$,   c) $\sigma^2\mu$,   d) $\tau\sigma^{-2}$,   e) $\sigma\tau\sigma^{-1}$.

4.2 Which of the functions given below from **R** into **R** are permutations of **R**?

a) $f_1: \mathbf{R} \rightarrow \mathbf{R}$ defined by $f_1(x) = x + 1$
b) $f_2: \mathbf{R} \rightarrow \mathbf{R}$ defined by $f_2(x) = x^2$
c) $f_3: \mathbf{R} \rightarrow \mathbf{R}$ defined by $f_3(x) = -x^3$
d) $f_4: \mathbf{R} \rightarrow \mathbf{R}$ defined by $f_4(x) = e^x$
e) $f_5: \mathbf{R} \rightarrow \mathbf{R}$ defined by $f_5(x) = x^3 - x^2 - 2x$

4.3 Consider the group S_3 of Example 4.1.

a) Find the cyclic subgroups $\langle \rho_1 \rangle$, $\langle \rho_2 \rangle$, and $\langle \mu_1 \rangle$ of S_3.
b) Find *all* subgroups, proper and improper, of S_3 and give the lattice diagram for them.

4.4 Give the multiplication table for the cyclic subgroup of S_5 generated by

$$\rho = \begin{pmatrix} 1 & 2 & 3 & 4 & 5 \\ 2 & 4 & 5 & 1 & 3 \end{pmatrix}.$$

There will be six elements. Let them be ρ, ρ^2, ρ^3, ρ^4, ρ^5, and $\rho^0 = \rho^6$. Is this group isomorphic to S_3?

†**4.5** Let A be a set and let a be a fixed element of A. Let T_a be the set of all permutations of A having the property that $a\sigma = a$. Show that T_a is a subgroup of the group S_A of all permutations of A which is given by Theorem 4.1.

4.6 Mark each of the following true or false.

— a) A permutation is a one-to-one function.
— b) A function is a permutation if and only if it is one to one.
— c) A function from a finite set onto itself must be one to one.
— d) The text has still given no example of a group which is nonabelian.
— e) Every subgroup of an abelian group is abelian.
— f) Every element of a group generates a cyclic subgroup of the group.
— g) The symmetric group S_{10} has ten elements.
— h) The symmetric group S_3 is cyclic.
— i) It is a good idea to refuse to accept mathematical results even though they have been checked by several competent mathematicians until you have gone through the details of the proofs for yourself.
— j) Every group is isomorphic to some group of permutations.

4.7 Show by an example that every proper subgroup of a nonabelian group may be abelian.

4.8 In analogy with Examples 4.1 and 4.2, consider a regular plane n-gon for $n \geq 3$. Each way that two copies of such an n-gon can be placed, with one covering the other, corresponds to a certain permutation of the vertices. The set of these permutations is a group, the **nth dihedral group D_n**, under permutation multiplication. Find the order of this group D_n. Argue *geometrically* that this group has a subgroup having just half as many elements as the whole group has.

4.9 Consider a cube which exactly fills a certain cubical box. As in Examples 4.1 and 4.2, the ways in which the cube can be placed into the box correspond to a certain group of permutations of the vertices of the cube. This group is the **group of rigid motions of the cube**. (It should not be confused with the *group of symmetries of the cube*, which will be discussed in the exercises of Section 10.) How many elements does this group have? Argue *geometrically* that this group has at least three different subgroups of order 4 and at least four different subgroups of order 3.

4.10 Let A be a set and let $\sigma \in S_A$. For a fixed $a \in A$, the set

$$\mathcal{O}_{a,\sigma} = \{a\sigma^n \mid n \in \mathbf{Z}\}$$

is the **orbit of a under σ**. Find the orbits of 1 under each of the permutations in Exercise 4.1.

4.11 Referring to the concept defined in Exercise 4.10, show that if for $a, b \in A$, $\mathcal{O}_{a,\sigma}$ and $\mathcal{O}_{b,\sigma}$ have an element in common, then $\mathcal{O}_{a,\sigma} = \mathcal{O}_{b,\sigma}$.

4.12 If A is a set, then a subgroup H of S_A is **transitive on A** if for each $a, b \in A$ there exists $\sigma \in H$ such that $a\sigma = b$. Show that if A is a nonempty finite set, then there exists a finite cyclic subgroup H of S_A with $|H| = |A|$ which is transitive on A.

4.13 Referring to Exercises 4.10 and 4.12, show that for $\sigma \in S_A$, $\langle \sigma \rangle$ is transitive on A if and only if $\mathcal{O}_{a,\sigma} = A$ for some $a \in A$.

4.14 Show that S_n is a nonabelian group for $n \geq 3$.

4.15 Strengthening Exercise 4.14, show that if $n \geq 3$, then the only element σ of S_n satisfying $\sigma \gamma = \sigma \gamma$ for all $\gamma \in S_n$ is $\sigma = \iota$, the identity permutation.

5 | Permutations II

5.1 CYCLES AND CYCLIC NOTATION

There is another notation for a permutation which is often used. Suppose that n elements $a_1, a_2, \ldots, a_n$ are distributed evenly on the circumference of a circle, as shown in Fig. 5.1. Suppose the circle is then rotated $2\pi/n$ radians counterclockwise so that a_1 is carried into the position formerly occupied by a_2, a_2 into that occupied by a_3, etc. This gives a one-to-one mapping of the set $\{a_1, a_2, \ldots, a_n\}$ onto itself, namely the permutation

$$\begin{pmatrix} a_1 & a_2 & \cdots & a_n \\ a_2 & a_3 & \cdots & a_1 \end{pmatrix}.$$

This permutation is a **cycle of length** n and for it we introduce a new, more compact notation

$$\text{``}(a_1, a_2, \ldots, a_n)\text{''}.$$

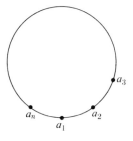

The new notation is **cyclic notation.** Under $(a_1, a_2, \ldots, a_n)$, each element is mapped onto the one following except the last one which is mapped onto the first. We will only use this notation for permutations of a finite set.

Fig. 5.1

Consider a permutation σ of a set S and let A be the subset of S consisting of all those elements which are moved by the permutation, that is,

$$A = \{x \in S \mid x\sigma \neq x\}.$$

Clearly, σ defines a natural permutation σ_A on the set A, where for $a \in A$, $a\sigma_A = a\sigma$. Suppose that σ_A is a cycle; let us suppose that

$$\sigma_A = (a_1, a_2, \ldots, a_n).$$

By abuse of notation, we shall write

$$\text{``}\sigma = (a_1, a_2, \ldots, a_n)\text{''},$$

where it is understood that σ maps onto itself any element of S which is not one of the a_i for $i = 1, 2, \ldots, n$. Of course the set S has to be known.

45

Example 5.1 If $S = \{1, 2, 3, 4, 5\}$, then

$$(1, 3, 5, 4) = \begin{pmatrix} 1 & 2 & 3 & 4 & 5 \\ 3 & 2 & 5 & 1 & 4 \end{pmatrix}.$$

Observe that

$$(1, 3, 5, 4) = (3, 5, 4, 1) = (5, 4, 1, 3) = (4, 1, 3, 5). \; \|$$

Of course since cycles are special types of permutations, they can be multiplied just as any two permutations. The product of two cycles need not again be a cycle, however.

Example 5.2 Let $(1, 4, 5, 6)$ and $(2, 1, 5)$ be cycles in the group S_6 of all permutations of $\{1, 2, 3, 4, 5, 6\}$. Then

$$(1, 4, 5, 6)(2, 1, 5) = \begin{pmatrix} 1 & 2 & 3 & 4 & 5 & 6 \\ 4 & 1 & 3 & 2 & 6 & 5 \end{pmatrix}$$

and

$$(2, 1, 5)(1, 4, 5, 6) = \begin{pmatrix} 1 & 2 & 3 & 4 & 5 & 6 \\ 6 & 4 & 3 & 5 & 2 & 1 \end{pmatrix}.$$

Neither of these permutations is a cycle. $\|$

The cycles in a collection of cycles are **disjoint** if there is no element of S which appears in the notations for two different cycles of the collection, i.e., if no element of S is moved by two different cycles of the collection. In terms of mappings, every element of S must be mapped onto itself by all but at most one of the cycles in the collection, if they are to be disjoint.

Let us agree that any cycle of length 1 represents the identity permutation.

We are going to show that any permutation of a finite set is a product of disjoint cycles. The proof we shall give of this is constructive, i.e., the steps of the proof can be used to actually find a representation of a given permutation as a product of disjoint cycles. We believe the student will get more understanding from this proof than from a more formal and elegant induction argument. We first illustrate the technique with an example.

Example 5.3 Consider the permutation

$$\begin{pmatrix} 1 & 2 & 3 & 4 & 5 & 6 \\ 6 & 5 & 2 & 4 & 3 & 1 \end{pmatrix}.$$

Let us write it as a product of disjoint cycles. First, 1 is moved to 6 and then 6 to 1, giving the cycle $(1, 6)$. Then starting with 2, 2 is moved to 5, which is moved to 3, which is moved to 2, or $(2, 5, 3)$. This takes care of all elements but 4, which is left fixed. Thus

$$\begin{pmatrix} 1 & 2 & 3 & 4 & 5 & 6 \\ 6 & 5 & 2 & 4 & 3 & 1 \end{pmatrix} = (1, 6)(2, 5, 3).$$

Multiplication of *disjoint* cycles is clearly commutative, so the order of the factors (1, 6) and (2, 5, 3) is not important. ‖

Theorem 5.1 *Every permutation σ of a finite set S is a product of disjoint cycles.*

Proof. It is no loss of generality to assume that $S = \{1, 2, 3, \ldots, n\}$. Consider the elements
$$1, \, 1\sigma, \, 1\sigma^2, \, 1\sigma^3, \ldots$$
Since S is finite, these elements cannot all be distinct. Let $1\sigma^r$ be the first term in the sequence which has appeared previously. Then $1\sigma^r = 1$, for if $1\sigma^r = 1\sigma^s$, with $0 < s < r$, we would have $1\sigma^{r-s} = 1$, with $r - s < r$, contradicting our choice of r. Let
$$\tau_1 = (1, \, 1\sigma, \, 1\sigma^2, \ldots, \, 1\sigma^{r-1}).$$
We see that τ_1 has the same effect as σ on all elements of S appearing in this cyclic notation for τ_1.

Let i be the first element of S not appearing in this cyclic notation for τ_1. Repeating the above argument with the sequence
$$i, \, i\sigma, \, i\sigma^2, \ldots,$$
we arrive at a cycle τ_2. Now τ_2 and τ_1 are disjoint, for if they had any element j of S in common, they would be identical, since each cycle could be constructed by repeated application of the permutation σ starting at j.

Continuing, we pick the first element in S not appearing in the cyclic notations of either τ_1 or τ_2 and construct τ_3, etc. Since S is finite, this process must terminate with some τ_m. The product
$$\tau_1 \tau_2 \cdots \tau_m$$
then clearly has the same effect on each element of S as σ does, so
$$\sigma = \tau_1 \tau_2 \cdots \tau_m. \quad \blacksquare$$

The student can easily convince himself that the representation of a permutation as a product of disjoint cycles, none of which is the identity permutation, is unique up to the order of the factors.

5.2 EVEN AND ODD PERMUTATIONS

Definition. A cycle of length 2 is a ***transposition***.

Thus a transposition leaves all elements but two fixed, and maps each of these onto the other. A computation shows that
$$(a_1, a_2, \ldots, a_n) = (a_1, a_2)(a_1, a_3) \cdots (a_1, a_n).$$
Therefore any cycle is a product of transpositions. We then have the following as a corollary to Theorem 5.1.

Corollary. *Any permutation of a finite set of at least two elements is a product of transpositions.*

Naively, this corollary just states that any rearrangement of n objects can be achieved by successively interchanging pairs of them.

Example 5.4 Following the remarks prior to the corollary, we see that $(1, 6)(2, 5, 3)$ is the product $(1, 6)(2, 5)(2, 3)$ of transpositions. ‖

We have seen that every permutation of a finite set with at least 2 elements is a product of transpositions. The transpositions may not be disjoint, and a representation of the permutation in this way is not unique. For example, one can always insert at the beginning the transposition (a, b) twice, since $(a, b)(a, b)$ is the identity permutation. What is true is that the number of transpositions used to represent a given permutation must either always be even or always be odd. This is an important fact, and the usual proof of it, which may be found in the first edition of this text, involves a rather artificial construction. In 1971 William I. Miller published a proof which we like much better, and which we give here.†

Theorem 5.2 *The number of transportations whose product is a given permutation of a finite set is either always even or always odd.*

Proof. It is no loss of generality to let the set be $S = \{1, 2, \ldots, n\}$, and we are assuming that $n \geq 2$ so that transpositions exist.

We first study the special case of the identity permutation ι. Of course ι can be expressed as a product of an even number of transpositions, for $\iota = (1, 2)(1, 2)$. We must show that if

$$\iota = \tau_1 \tau_2 \cdots \tau_k, \tag{1}$$

where each τ_i is a transposition, then k must be even. Choose any integer m which appears in one of the transpositions in (1) and let τ_j be the first transposition, counting from left to right, in which m occurs. We can't have $j = k$, or ι would not leave m fixed. Now $\tau_j \tau_{j+1}$ must have the form on the left-hand side of one of the following easily verified identities.

$$
\begin{aligned}
(m, x)(m, x) &= \iota \\
(m, x)(m, y) &= (x, y)(m, x) \\
(m, x)(y, z) &= (y, z)(m, x) \\
(m, x)(x, y) &= (x, y)(m, y)
\end{aligned}
\tag{2}
$$

If we substitute the correct identity in (2) for $\tau_j \tau_{j+1}$ in (1), we either reduce the number k of transpositions in (1) by two, or shift the first occurrence of m one step to the right. We repeat this process until m is eliminated from

---

† William I. Miller, "Even and Odd Permutations," *Mathematics Associations of Two-Year Colleges,* Journal 5 (Spring, 1971), p. 32.

the expression (1); recall that m can't appear for the first time in the final transposition, so eventually the situation in the first identity of (2) must occur to eliminate m completely. We then choose another integer in S appearing in our reduced (1), and eliminate it from (1) by a similar process, and continue until the right-hand side of (1) is reduced to a sequence $\iota \cdots \iota$. Since the number k was either left unchanged or reduced by two at each substitution of an identity from (2), we see that k must have been even.

It is now easy to prove the theorem from our special case for ι. Suppose that

$$\sigma = \tau_1 \tau_2 \cdots \tau_r = \tau_1' \tau_2' \cdots \tau_s'.$$

Since every transposition is its own inverse, we obtain

$$\iota = \sigma\sigma^{-1} = \tau_1 \tau_2 \cdots \tau_r (\tau_1' \tau_2' \cdots \tau_s')^{-1} = \tau_1 \tau_2 \cdots \tau_r \tau_s' \cdots \tau_2' \tau_1'.$$

Our special case now shows that $r + s$ is an even number, so r and s are either both odd numbers or both even numbers. ∎

Definition. A permutation of a finite set is **even** or **odd** according to whether it can be expressed as the product of an even number of transpositions or the product of an odd number of transpositions, respectively.

5.3 THE ALTERNATING GROUPS

We claim that for $n \geq 2$, the number of even permutations in S_n is the same as the number of odd permutations, that is, S_n is split equally and both numbers are $(n!)/2$. To show this, let A_n be the set of even permutations in S_n and B_n, the set of odd permutations for $n \geq 2$. We proceed to define a one-to-one function from A_n onto B_n. This is exactly what is needed to show that A_n and B_n have the same number of elements.

Let τ be any fixed transposition in S_n; it exists since $n \geq 2$. Suppose that $\tau = (1, 2)$. We define a function

$$\lambda_\tau : A_n \to B_n$$

by

$$\sigma \lambda_\tau = \tau\sigma,$$

that is, $\sigma \in A_n$ is mapped into $(1, 2)\sigma$ by λ_τ. Observe that since σ is even, the permutation $(1, 2)\sigma$ appears as a product of a $(1 +$ even number), or odd number, of transpositions, so $(1, 2)\sigma$ is indeed in B_n. If for σ and $\mu \in A_n$ it is true that $\sigma\lambda_\tau = \mu\lambda_\tau$, then

$$(1, 2)\sigma = (1, 2)\mu,$$

and since S_n is a group, we have $\sigma = \mu$. Thus λ_τ is a one-to-one function. Finally,

$$\tau = (1, 2) = \tau^{-1},$$

so if $\rho \in B_n$, then

$$\tau^{-1}\rho \in A_n,$$

and

$$(\tau^{-1}\rho)\lambda_\tau = \tau(\tau^{-1}\rho) = \rho.$$

Thus λ_τ is onto B_n. Hence the number of elements in A_n is the same as the number in B_n since there is a one-to-one correspondence between the elements of the sets.

Note that the product of two even permutations is again even. Also since $n \geq 2$, S has 2 elements a and b and $\iota = (a, b)(a, b)$ is an even permutation. Finally, note that if σ is expressed as a product of transpositions, the product of the same transpositions taken in just the opposite order is σ^{-1}. Thus if σ is an even permutation, σ^{-1} must also be even. Referring to Theorem 3.1, we see that we have proved:

Theorem 5.3 *If $n \geq 2$, the collection of all even permutations of a finite set of n elements forms a subgroup of order $n!/2$ of the symmetric group S_n.*

Definition. The subgroup of S_n consisting of the even permutations of n letters is the **alternating group A_n on n letters.**

Both S_n and A_n are very important groups. We have already mentioned without proof that every finite group is structurally identical to some subgroup of S_n for some n. The importance of A_n will appear later.

EXERCISES

5.1 Compute the indicated product of cycles which are permutations of $\{1, 2, 3, 4, 5, 6, 7, 8\}$.

a) $(1, 4, 5)(7, 8)(2, 5, 7)$ b) $(1, 3, 2, 7)(4, 8, 6)$

c) $(1, 2)(4, 7, 8)(2, 1)(7, 2, 8, 1, 5)$

5.2 Express each of the following permutations of $\{1, 2, 3, 4, 5, 6, 7, 8\}$ as a product of disjoint cycles, and then as a product of transpositions.

a) $\begin{pmatrix} 1 & 2 & 3 & 4 & 5 & 6 & 7 & 8 \\ 8 & 2 & 6 & 3 & 7 & 4 & 5 & 1 \end{pmatrix}$ b) $\begin{pmatrix} 1 & 2 & 3 & 4 & 5 & 6 & 7 & 8 \\ 3 & 6 & 4 & 1 & 8 & 2 & 5 & 7 \end{pmatrix}$

c) $\begin{pmatrix} 1 & 2 & 3 & 4 & 5 & 6 & 7 & 8 \\ 3 & 1 & 4 & 7 & 2 & 5 & 8 & 6 \end{pmatrix}$

†**5.3** Show that for every subgroup H of S_n for $n \geq 2$, either all the permutations in H are even or exactly half of them are even.

5.4 Which of the permutations in S_3 of Example 4.1 are even permutations? Give the table for the alternating group A_3.

5.5 An element a of a group G with identity e has **order** $r > 0$ if $a^r = e$ and no smaller positive power of a is the identity. Consider the group S_8 as in Exercise 5.1.

a) What is the order of the cycle $(1, 4, 5, 7)$?

b) State a theorem suggested by part (a).

c) What is the order of $\sigma = (4, 5)(2, 3, 7)$? of $\tau = (1, 4)(3, 5, 7, 8)$?

d) Find the order of each of the permutations given in Exercise 5.2 by looking at its decomposition into a product of disjoint cycles.

e) State a theorem suggested by parts (c) and (d). [*Hint:* The important words you are looking for are *least common multiple.*]

5.6 Mark each of the following true or false.

___ a) Every permutation is a cycle.

___ b) Every cycle is a permutation.

___ c) The definition of even and odd permutations could have been given equally well before Theorem 5.2.

___ d) It would be terrible for a student to ever refer to an even or odd permutation without a thorough understanding of every bit of the proof of Theorem 5.2.

___ e) A_5 has 120 elements.

___ f) S_n is not cyclic for any $n \geq 1$.

___ g) A_3 is a commutative group.

___ h) S_7 is isomorphic to the subgroup of all those elements of S_8 which leave the number 8 fixed.

___ i) Any undergraduate mathematics major should be able to learn to multiply permutations.

___ j) The odd permutations in S_8 form a subgroup of S_8.

5.7 Let σ be a permutation of a set A. We shall say "σ **moves** $a \in A$" if $a\sigma \neq a$. If A is a finite set, how many elements are moved by a cycle $\sigma \in S_A$ of length n?

5.8 Let A be an infinite set. Let H be the set of all $\sigma \in S_A$ which move (see Exercise 5.7) only a finite number of elements of A. Show that H is a subgroup of S_A.

5.9 Let A be an infinite set. Let K be the set of all $\sigma \in S_A$ which move (see Exercise 5.7) at most 50 elements of A. Is K a subgroup of S_A? Why?

5.10 Give a more elegant proof of Theorem 5.1 than the one given in the text, by making an induction argument on the number of elements moved (see Exercise 5.7) by σ.

5.11 Consider S_n for a fixed $n \geq 2$ and let σ be a fixed odd permutation. Show that every odd permutation in S_n is a product of σ and some permutation in A_n.

5.12 Show that if σ is a cycle, then σ^2 is a cycle, provided that the length of σ is an odd integer.

5.13 Following the line of thought opened by Exercise 5.12, complete the following with a condition involving n and r so that the resulting statement is a theorem:

"If σ is a cycle of length n, then σ^r is also a cycle if and only if . . ."

5.14 Let G be a group and let a be a fixed element of G. Show that the map $\lambda_a: G \rightarrow G$, given by $g\lambda_a = ag$ for $g \in G$, is a permutation of the set G.

5.15 Referring to Exercise 5.14, show that $H = \{\lambda_a \mid a \in G\}$ is a subgroup of S_G, the group of all permutations of G.

5.16 Referring to Exercise 4.12, show that H of Exercise 5.15 is transitive on the set G. [*Hint:* This is an immediate corollary of one of the theorems in Section 2.]

6 | Cyclic Groups

6.1 ELEMENTARY PROPERTIES

Recall the following facts from Section 3:

If G is a group and $a \in G$, then

$$H = \{a^n \mid n \in \mathbf{Z}\}$$

is a subgroup of G (Theorem 3.2). This group is the **cyclic subgroup of** G **generated by** a. Also, given a group G and an element $a \in G$, if

$$G = \{a^n \mid n \in \mathbf{Z}\},$$

then a is a **generator of** G and the group $G = \langle a \rangle$ is **cyclic**.

It is the purpose of this section to classify all cyclic groups and all subgroups of cyclic groups.

Theorem 6.1 *Every cyclic group is abelian.*

Proof. Let G be a cyclic group and let a be a generator of G so that

$$G = \langle a \rangle = \{a^n \mid n \in \mathbf{Z}\}.$$

If g_1 and g_2 are any two elements of G, there exist integers r and s such that $g_1 = a^r$ and $g_2 = a^s$. Then

$$g_1 g_2 = a^r a^s = a^{r+s} = a^{s+r} = a^s a^r = g_2 g_1,$$

so G is abelian. ∎

We shall continue to use multiplicative notation for our general work on cyclic groups, even though they are abelian.

There is a weak but very important converse of Theorem 6.1 which will be discussed in greater detail later. Namely, it can be shown that every "sufficiently small" abelian group can be built up from cyclic groups in a certain fashion. Thus cyclic groups are very fundamental in the study of abelian groups. Cyclic groups are sort of elementary types of abelian groups. One would hope that a portion of an elementary type is again an elementary type. The next theorem shows that this is indeed so. We first give a seemingly trivial but very fundamental number-theoretic lemma.

52

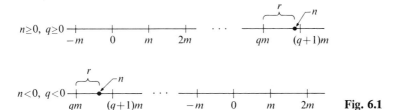

$n \geq 0,\ q \geq 0$

$n < 0,\ q < 0$

Fig. 6.1

Lemma 6.1 (Division Algorithm for **Z**). *If m is a positive integer and n is any integer, then there exist unique integers q and r such that*

$$n = mq + r \quad \text{and} \quad 0 \leq r < m.$$

Proof. We give an intuitive diagrammatic explanation. On the "*real x-axis*" of analytic geometry, mark off the multiples of m and the position of n. Now n falls either on a multiple qm of m and r can be taken as 0, or n falls between two multiples of m. If the latter is the case, let qm be the first multiple of m to the left of n. Then r is as shown on the diagrams in Fig. 6.1. Note that $0 \leq r < m$. After a little thought, the uniqueness of q and r is clear from the diagrams. ∎

Theorem 6.2 *A subgroup of a cyclic group is cyclic.*

Proof. Let G be cyclic generated by a and let H be a subgroup of G. If $H = \{e\}$, then $H = \langle e \rangle$ is cyclic. If $H \neq \{e\}$, then $a^n \in H$ for some $n \in \mathbf{Z}^+$. Let $m \in \mathbf{Z}^+$ be minimal such that $a^m \in H$.

We claim that $c = a^m$ generates H, that is,

$$H = \langle a^m \rangle = \langle c \rangle.$$

We must show that every $b \in H$ is a power of c. Since $b \in H$ and $H \leq G$, we have $b = a^n$ for some n. Find q and r such that

$$n = mq + r \quad \text{for} \quad 0 \leq r < m$$

by Lemma 6.1. Then

$$a^n = a^{mq+r} = (a^m)^q a^r,$$

so

$$a^r = (a^m)^{-q} a^n.$$

Now since $a^n \in H$, $a^m \in H$, and H is a group, both $(a^m)^{-q}$ and a^n are in H, and thus

$$(a^m)^{-q} a^n \in H, \quad \text{that is,} \quad a^r \in H.$$

Since m was the smallest positive integer such that $a^m \in H$ and $0 \leq r < m$, we must have $r = 0$. Thus $n = qm$ and

$$b = a^n = (a^m)^q = c^q,$$

so b is a power of c. ∎

Theorem 6.2 is for some reason one which the student is often asked to prove on a test in the course, in an oral examination for a master's degree, etc. It is easy, but will test whether the student has any ability in understanding and devising proofs. Note first that since the theorem concerns cyclic groups and we have proved practically nothing about them yet, the *definition* of a cyclic group *must* be used. That is, you want to show that a subgroup H of a cyclic group G is cyclic. You *must* realize that G cyclic means that there exists $a \in G$ such that every element of G is of the form a^n for $n \in \mathbf{Z}$, and that you have to pull out of thin air one element c of H which will do the same for H. *This all comes from just the definition of a cyclic group.* The ingenuity comes in defining c. But it is a natural choice, being about the only element of H you can get hold of in terms of a. Then you have to show that c works, i.e., that c generates H.

As noted in Examples 3.4 and 3.5, $\mathbf{Z}$ under addition is cyclic and for a positive integer n, the set $n\mathbf{Z}$ of all multiples of n is a subgroup of $\mathbf{Z}$ under addition, the cyclic subgroup generated by n. Theorem 6.2 shows that these cyclic subgroups are the only subgroups of $\mathbf{Z}$ under addition. We state this as a corollary.

Corollary. *The subgroups of $\mathbf{Z}$ under addition are precisely the groups $n\mathbf{Z}$ under addition for $n \in \mathbf{Z}$.*

6.2 THE CLASSIFICATION OF CYCLIC GROUPS

Let G be a cyclic group with generator a. We consider two cases.

CASE I. *G has an infinite number of elements, i.e., the order of G is infinite.*

In this case we claim that no two distinct exponents h and k can give equal elements a^h and a^k of G. For suppose $a^h = a^k$ and say $h > k$. Then

$$a^h a^{-k} = a^{h-k} = e,$$

the identity, and $h - k > 0$. Let m be the smallest positive integer such that $a^m = e$ (note the similarity with the construction in the proof of Theorem 6.2). We claim that G would then only have the distinct elements $e, a, a^2, \ldots, a^{m-1}$. For let $a^n \in G$, and find q and r such that

$$n = mq + r \qquad \text{for} \quad 0 \le r < m$$

by Lemma 6.1. Then

$$a^n = a^{mq+r} = (a^m)^q a^r = e^q a^r = a^r$$

for $0 \le r < m$. This would mean that G would be finite, contradicting our assumption for Case I. *Thus all powers of a are distinct.*

Suppose that G' is another infinite cyclic group with generator b. Clearly, by renaming b^n by "a^n", G' can be made to look exactly like G, that is, the groups are isomorphic. This will be done again carefully in the next section.

Thus all infinite cyclic groups are just alike except for the names of the elements and operations. We will take **Z** with the operation of addition as a prototype of any infinite cyclic group. From now on in Part I, "the group **Z**" will always mean "the group **Z** under addition".

Example 6.1 It may seem quite odd to you that **Z** and 3**Z**, both being infinite cyclic groups under addition, are structurally identical although 3**Z** < **Z**. You may say that $1 \in \mathbf{Z}$ but $1 \notin 3\mathbf{Z}$, so how can they be structurally the same? But names don't matter, and if you rename 1 by "3", 2 by "6" and in general n by "$3n$", you have converted **Z** into 3**Z** as an additive group. ‖

CASE II. *G has finite order.*

In this case, not all positive powers of a generator a of G can be distinct, so for some h and k we must have $a^h = a^k$. Following the argument of Case I, there exists an integer m such that $a^m = e$ and no smaller positive power of a is e. The group G then consists of the distinct elements $e, a, a^2, \ldots, a^{m-1}$.

Since it is usual to use "n" for the order of a general finite cyclic group, we shall change notation, setting $m = n$ for what follows.

Example 6.2 It is nice to visualize the elements $e = a^0, a^1, a^2, \ldots, a^{n-1}$ of a cyclic group of order n as being distributed evenly on the circumference of a circle (see Fig. 6.2). The element a^h is located h of these equal units counterclockwise along the circumference, measured from the bottom where $e = a^0$ is located. To multiply a^h and a^k diagrammatically, you start from a^h and go k additional units around counterclockwise. To see arithmetically where you end up, find q and r such that

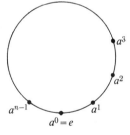

$$h + k = nq + r \quad \text{for} \quad 0 \le r < n.$$

The nq takes you all the way around the circle q times, and you then wind up at a^r. ‖

Fig. 6.2

Definition. Let n be a fixed positive integer and let h and k be any integers. The number r such that

$$h + k = nq + r \quad \text{for} \quad 0 \le r < n$$

is the ***sum of h and k modulo n.***

Theorem 6.3 *The set* $\{0, 1, 2, \ldots, n - 1\}$ *is a cyclic group* $\mathbf{Z}_n$ *of* n *elements under addition modulo* n.

Congruence modulo n was discussed in Section 0; we see that if $h + k = r$ in $\mathbf{Z}_n$, then for addition in **Z** we have $h + k \equiv r \pmod{n}$.

The proof of this theorem is easy and would be a tedious bore for us to write out and for you to read. You have to check $\mathcal{G}_1$, $\mathcal{G}_2$ and $\mathcal{G}_3$. Check them mentally. Think of the diagram in Fig. 6.3 as explained in Example 6.2. This really just amounts to renaming the element a^h of Example 6.2 by "h".

Thus there is a cyclic group of order n for every positive integer n. Throughout Part I, we will let $\mathbf{Z}_n$ be the group given by Theorem 6.3. Just as in the infinite case, it is clear that if G and G' are two cyclic groups of n elements each and generators a and b respectively, then by renaming b^r by "a^r", G' can be made to look just like G, that is, *any two cyclic groups of the same finite order are isomorphic.*

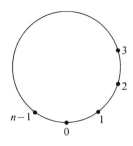

Fig. 6.3

6.3 SUBGROUPS OF FINITE CYCLIC GROUPS

We have completed our classification of cyclic groups and turn now to their subgroups. The corollary of Theorem 6.2 gives us complete information about subgroups of infinite cyclic groups. Let us give the basic theorem regarding generators of subgroups for the finite cyclic groups.

> **Theorem 6.4** *Let G be a cyclic group with n elements and generated by a. Let $b \in G$ and let $b = a^s$. Then b generates a cyclic subgroup H of G containing n/d elements, where d is the greatest common divisor, abbreviated gcd, of n and s.*

Proof. That b generates a cyclic subgroup H of G is known from Theorem 3.2. We need show only that H has n/d elements. Following the argument of our Case I above, H has as many elements as the smallest power of b which gives the identity. Now $b = a^s$, and $b^m = e$ if and only if $(a^s)^m = a^{ms} = e$, or if and only if n divides ms. What is the smallest value of m such that n divides ms? If d is the largest number dividing both n and s, then writing $n = d(n/d)$, the d factor of n will divide the s factor of ms, and no prime factors of n/d can be absorbed by s also. Thus all of n/d must be absorbed in m, so the smallest such m is $m = (n/d)$. ∎

Example 6.3 Consider $\mathbf{Z}_{12}$ which has a generator $a = 1$. Since the greatest common divisor (gcd) of 3 and 12 is 3, $3 = 3 \cdot 1$ generates a subgroup of $\frac{12}{3} = 4$ elements, namely

$$\langle 3 \rangle = \{0, 3, 6, 9\}.$$

Since the gcd of 8 and 12 is 4, 8 generates a subgroup of $\frac{12}{4} = 3$ elements, namely

$$\langle 8 \rangle = \{0, 4, 8\}.$$

Since the gcd of 12 and 5 is 1, 5 generates a subgroup of $\frac{12}{1}$ = 12 elements, that is, 5 is a generator of the whole group Z_{12}. ‖

The following corollary is immediate from the theorem.

Corollary. *If a is a generator of a finite cyclic group G of order n, then the other generators of G are the elements of the form a^r, where r is **relatively prime** to n, that is, where the greatest common divisor of r and n is 1.*

Example 6.4 Let us find all subgroups of Z_{18} and give their lattice diagram. All subgroups are cyclic. By the corollary of Theorem 6.4, 1, 5, 7, 11, 13 and 17 are all generators of Z_{18}. Starting with 2,

$$\langle 2 \rangle = \{0, 2, 4, 6, 8, 10, 12, 14, 16\}$$

is of order 9 and has as generators elements of the form $h2$, where h is relatively prime to 9, namely $h = 1, 2, 4, 5, 7$, and 8, so $h2 = 2, 4, 8, 10, 14$, and 16. The element 6 of $\langle 2 \rangle$ generates $\{0, 6, 12\}$ and 12 also is a generator of this subgroup.

We have thus far found all sub-groups generated by 0, 1, 2, 4, 5, 6, 7, 8, 10, 11, 12, 13, 14, and 16. This leaves just 3, 9, and 15 to consider.

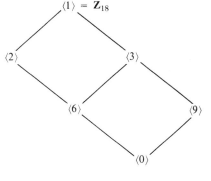

$$\langle 3 \rangle = \{0, 3, 6, 9, 12, 15\},$$

and 15 also generates this group of order 6, since $15 = 5 \cdot 3$ and the gcd of 5 and 6 is 1. Finally,

$$\langle 9 \rangle = \{0, 9\}.$$

The lattice diagram for these sub-groups of Z_{18} is given in Fig. 6.4.

Fig. 6.4. Lattice diagram for Z_{18}.

This example is very easy; we are afraid we wrote it out in such hideous detail that it may look hard to you. The exercises will give you some practice along these lines. ‖

EXERCISES

6.1 Find the number of generators of cyclic groups of orders 6, 8, 12, and 60.

†6.2 Show that a group which has only a finite number of subgroups must be a finite group.

6.3 Find the number of elements in each of the indicated cyclic groups.

a) The cyclic subgroup of Z_{30} generated by 25
b) The cyclic subgroup of Z_{42} generated by 30
c) The cyclic subgroup $\langle i \rangle$ of the group C^* of nonzero complex numbers under multiplication
d) The cyclic subgroup of the group C^* of part (c) generated by $(1 + i)/\sqrt{2}$
e) The cyclic subgroup of the group C^* of part (c) generated by $1 + i$

6.4 For each of the following groups, find all the subgroups and give the lattice diagram for these subgroups.

a) Z_{12} b) Z_{36} c) Z_8

6.5 Find all orders of subgroups of Z_6, Z_8, Z_{12}, Z_{60}, and Z_{17}.

6.6 Mark each of the following true or false.

— a) Every cyclic group is abelian.
— b) Every abelian group is cyclic.
— c) Q under addition is a cyclic group.
— d) Every element of every cyclic group generates the group.
— e) There is at least one abelian group of every finite order >0.
— f) Every group of order ≤ 4 is cyclic.
— g) This business of defining and proving things in mathematics is a sort of a game with rules which are not very explicitly formulated.
— h) S_3 is a cyclic group.
— i) A_3 is a cyclic group.
— j) Any undergraduate mathematics major should be able to learn to add integers modulo a fixed positive integer n.

6.7 Show by a counterexample that the following "converse" of Theorem 6.2 is not a theorem: "If a group G is such that every proper subgroup is cyclic, then G is cyclic."

6.8 Let p and q be prime numbers. Find the number of generators of the cyclic group Z_{pq}.

6.9 Let p be a prime number. Find the number of generators of the cyclic group Z_{p^r}, where r is an integer ≥ 1.

6.10 Show that in a finite cyclic group G of order n, the equation $x^m = e$ has exactly m solutions x in G for each positive integer m which divides n.

6.11 With reference to Exercise 6.10, what is the situation if $1 < m < n$ and m does not divide n?

6.12 Show that Z_p has no proper subgroups if p is a prime number.

6.13 Let G be an abelian group and let H and K be finite cyclic subgroups with $|H| = r$ and $|K| = s$.

a) Show that if r and s are relatively prime, then G contains a cyclic subgroup of order rs.

b) Generalizing part (a), show that G contains a cyclic subgroup of order the least common multiple of r and s.

7 | Isomorphism

7.1 DEFINITION AND ELEMENTARY PROPERTIES

We now come to the business of making more mathematically precise the idea that two groups, G and G', are structurally the same or *isomorphic*. Recall that the idea that G and G' are isomorphic means that the groups are identical except for the names of the elements and operations. Thus we should be able to obtain G' from G by renaming an element x in G with the name of a certain element x' in G'. That is, to each $x \in G$ is assigned its counterpart $x' \in G'$. This is really nothing but a *function* ϕ with domain G. Clearly two different elements x and y in G should have two different counterparts $x' = x\phi$ and $y' = y\phi$ in G', that is, the function ϕ must be one to one. Also every element of G' must be the counterpart of some element of G, that is, the function ϕ must be onto G'. This renames the elements. Finally, if the groups are to be structurally the same and if for the moment we denote the group operation of G by "$*$" and that of G' by "$*'$", then the counterpart of $x * y$ should be $x' *' y'$, or $(x * y)\phi$ should be $(x\phi) *' (y\phi)$. Usually we drop the notations "$*$" and "$*'$" for the operations and use multiplicative notations, that is,

$$\text{``}(xy)\phi = (x\phi)(y\phi)\text{''}.$$

Note that the multiplication xy on the left side in $(xy)\phi = (x\phi)(y\phi)$ is the multiplication in G, whereas the multiplication $(x\phi)(y\phi)$ on the right is in G'. We collect these ideas in a definition.

Definition. An ***isomorphism of a group G with a group G'*** is a one-to-one function ϕ mapping G onto G' such that for all x and y in G,

$$(xy)\phi = (x\phi)(y\phi).$$

The groups G and G' are then ***isomorphic***. The usual notation is "$G \simeq G'$".

Let us prove a theorem which is very obvious if we consider an isomorphism to be a renaming of one group so that it is just like another. We shall prove it from our definition of an isomorphism, of course.

Theorem 7.1 *If $\phi: G \rightarrow G'$ is an isomorphism of G with G', and e is the identity of G, then $e\phi$ is the identity in G'. Also,*

$$a^{-1}\phi = (a\phi)^{-1} \qquad \text{for all} \quad a \in G.$$

Loosely, an isomorphism maps the identity onto the identity and inverses onto inverses.

Proof. Let $x' \in G'$. Since ϕ is onto, there is $x \in G$ such that $x\phi = x'$. Then

$$x' = x\phi = (ex)\phi = (e\phi)(x\phi) = (e\phi)x'.$$

Likewise

$$x' = x\phi = (xe)\phi = (x\phi)(e\phi) = x'(e\phi).$$

Thus for every $x' \in G'$, we have

$$(e\phi)x' = x' = x'(e\phi),$$

so $e\phi$ is the identity of G'.

Also for $a \in G$, we have

$$e\phi = (a^{-1}a)\phi = (a^{-1}\phi)(a\phi).$$

Likewise

$$e\phi = (aa^{-1})\phi = (a\phi)(a^{-1}\phi).$$

Thus $a^{-1}\phi = (a\phi)^{-1}$. ∎

7.2 HOW TO SHOW THAT GROUPS ARE ISOMORPHIC

In the past, some of the author's students have had a hard time understanding and using the concept of isomorphism. We introduced it several sections before we made it precise in the hope that you would really comprehend the importance and meaning of the concept. Regarding its use, we now give an outline showing how the mathematician would proceed from the definition to show that two groups, G and G', are isomorphic.

STEP 1. *Define the function ϕ which gives the isomorphism of G with G'.* Now this means that you have to describe, in some fashion, what $x\phi$ is to be in G' for every $x \in G$.

STEP 2. *Show that ϕ is a one-to-one function.*

STEP 3. *Show that ϕ is onto G'.*

STEP 4. *Show that $(xy)\phi = (x\phi)(y\phi)$.* This is just a question of computation. One computes both sides of the equation and sees whether or not they are the same.

We illustrate this technique with an example.

Example 7.1 Let us show that $\mathbf{R}$ under addition is isomorphic to $\mathbf{R}^+$ under multiplication.

STEP 1. For $x \in \mathbf{R}$, define $x\phi = e^x$. This gives a mapping $\phi: \mathbf{R} \to \mathbf{R}^+$.

STEP 2. If $x\phi = y\phi$, then $e^x = e^y$, so $x = y$. Thus ϕ is one to one.

STEP 3. If $r \in \mathbf{R}^+$, then

$$(\ln r)\phi = e^{\ln r} = r,$$

where $(\ln r) \in \mathbf{R}$. Thus ϕ is onto $\mathbf{R}^+$.

STEP 4. For $x, y \in \mathbf{R}$, we have

$$(x + y)\phi = e^{x+y} = e^x e^y = (x\phi)(y\phi). \;\|$$

We illustrate this technique again in a theorem.

Theorem 7.2 *Any infinite cyclic group G is isomorphic to the group $\mathbf{Z}$ of integers under addition.*

Proof. We suppose that G has a generator a and use multiplicative notation for the operation in G. Thus

$$G = \{a^n \mid n \in \mathbf{Z}\}.$$

Our argument in Case I, Section 6, for infinite cyclic groups showed that the elements a^n of G are all distinct, that is, $a^n \neq a^m$ if $n \neq m$.

STEP 1. Define $\phi: G \rightarrow \mathbf{Z}$ by $a^n \phi = n$ for all $a^n \in G$.

STEP 2. If $a^n \phi = a^m \phi$, then $n = m$ and $a^n = a^m$. Thus ϕ is one to one.

STEP 3. For any $n \in \mathbf{Z}$, the element $a^n \in G$ is mapped onto n by ϕ. Thus ϕ is onto $\mathbf{Z}$.

STEP 4. Now $(a^n a^m)\phi = a^{n+m}\phi = n + m$. (Note that the binary operation here was in the group G.) It remains to compute $(a^n\phi) + (a^m\phi)$, the "$+$" appearing since the operation in $\mathbf{Z}$ is addition. But $(a^n\phi) + (a^m\phi)$ is again $n + m$. Thus $(a^n a^m)\phi = (a^n\phi) + (a^m\phi)$. $\blacksquare$

The preceding proof was certainly very easy. The student should be sure he understands the steps involved.

It is immediate that every group G is isomorphic to itself; the identity function ι defined by $g\iota = g$ for all $g \in G$ shows this. If G is isomorphic to G', then G' is isomorphic to G; the function $\phi^{-1}: G' \rightarrow G$ for an isomorphism $\phi: G \rightarrow G'$ shows this (see Exercise 7.6). Finally, if G is isomorphic to G' and G' is isomorphic to G'', then G is isomorphic to G''; if $\phi: G \rightarrow G'$ and $\psi: G' \rightarrow G''$ are isomorphisms, then the composite function $\phi\psi$ shows this (see Exercise 7.7). The reader should recognize that we have shown that the property of being isomorphic is an equivalence relation on a collection of groups. In view of Theorem 0.1, this means that *given a nonempty collection of groups, you can always partition the collection into cells (equivalence classes) such that any two groups in the same cell are isomorphic, and no two groups in different cells are isomorphic.*

We have seen that any two groups of order 3 are isomorphic. We express this by saying that *there is only one group of order 3 up to isomorphism.*

Example 7.2 There is only one group of order 1, one of order 2, and one of order 3 up to isomorphism. We saw in Example 3.2 that there are exactly two different groups of order 4 up to isomorphism, the group Z_4 and the Klein 4-group V. There are at least two different groups of order 6 up to isomorphism, namely Z_6 and S_3. ‖

7.3 HOW TO SHOW THAT GROUPS ARE NOT ISOMORPHIC

We turn now to a topic discussed in few algebra texts, namely:

> *How do you demonstrate that two groups G and G' are not isomorphic, if this is the case?*

This would mean that there is no one-to-one function ϕ from G onto G' with the property $(xy)\phi = (x\phi)(y\phi)$. In general, it is clearly not feasible to try every possible one-to-one function to find out whether it has the above property, except in the case where there are *no* one-to-one functions. This is the case, for example, if G and G' are of finite order and have different numbers of elements.

Example 7.3 Z_4 and S_6 are *not* isomorphic. There is no one-to-one function from Z_4 onto S_6. ‖

In the infinite case, it is not always clear whether or not there are any one-to-one onto functions. For example, you may think that Q has "more" elements than Z, but your instructor can show you in five minutes (ask him to!) that *there are lots of one-to-one functions from Z onto Q*. However, it is true that R *has too many elements to be put into a one-to-one correspondence with Z*. This will take your instructor only another five minutes to show you.

Example 7.4 Z under addition is *not* isomorphic with R under addition, for there is no one-to-one function from Z onto R. ‖

In the event that there are one-to-one mappings of G onto G', one usually shows that the groups are not isomorphic (if this is the case) by showing that one group has some algebraic property that the other does not possess, i.e., they are not structurally the same. An **algebraic property of a group** is one whose definition is just in terms of the binary operation of the group, and does not depend on the names or some other nonstructural characteristics of the elements. Clearly, an algebraic property of a group G must be mirrored by a corresponding "renamed" algebraic property in any isomorphic group.

Example 7.5 You can't say that Z and $3Z$ under addition are not isomorphic because $17 \in Z$ and $17 \notin 3Z$. This is *not* an algebraic property but rather has to do with just the names of the elements. Actually Z and $3Z$ are isomorphic under the map $\phi: Z \rightarrow 3Z$, where $n\phi = 3n$. ‖

Example 7.6 You can't say that Z and Q, both under addition, are not isomorphic because $\frac{1}{2} \in Q$ and $\frac{1}{2} \notin Z$. But you can say that they are not isomorphic because Z is cyclic and Q is not. ‖

Example 7.7 The group $\mathbf{Q}^*$ of nonzero elements of $\mathbf{Q}$ under multiplication is not isomorphic to the group $\mathbf{R}^*$ of nonzero elements of $\mathbf{R}$ under multiplication. One argument is that there is no one-to-one correspondence between them. Another is that every element in $\mathbf{R}^*$ is the cube of some element of $\mathbf{R}^*$, that is, for $a \in \mathbf{R}^*$, the equation $x^3 = a$ has a solution in $\mathbf{R}^*$. This is not true for $\mathbf{Q}^*$; for example, the equation $x^3 = 2$ has no solution in $\mathbf{Q}^*$. ‖

Example 7.8 The group $\mathbf{R}^*$ of nonzero real numbers under multiplication is not isomorphic to the group $\mathbf{C}^*$ of nonzero complex numbers under multiplication. Every element of $\mathbf{R}^*$ generates an infinite cyclic subgroup except for 1 and -1, which generate subgroups of orders 1 and 2, respectively. But in $\mathbf{C}^*$, i generates the cyclic subgroup $\{i, -1, -i, 1\}$ of order 4. Or, for another argument, the equation $x^2 = a$ has a solution x in $\mathbf{C}^*$ for every $a \in \mathbf{C}^*$, but $x^2 = -1$ has no solution in $\mathbf{R}^*$. ‖

Example 7.9 The group $\mathbf{R}^*$ of nonzero real numbers under multiplication is not isomorphic to the group $\mathbf{R}$ of real numbers under addition. An equation $x + x = a$ always has a solution in $\langle \mathbf{R}, + \rangle$ for every $a \in \mathbf{R}$, but the corresponding equation $x \cdot x = a$ does not always have a solution in $\langle \mathbf{R}^*, \cdot \rangle$, for example if $a = -1$. ‖

7.4 CAYLEY'S THEOREM

Look at any group table in the text. Note how each row of the table gives a permutation of the set of elements of the group, as listed at the top of the table. Similarly, each column of the table gives a permutation of the group set, as listed at the left of the table. In view of these observations, it is not surprising that at least every finite group G is isomorphic to a subgroup of the group S_G of all permutations of G. The same is true for infinite groups also; Cayley's theorem states that *every* group is isomorphic to some group consisting of permutations under permutation multiplication. This is a nice and intriguing result, although we shall have no important use for it. However, the theorem is a classic of group theory and appears in almost all algebra texts. Moreover, it is the first theorem we come to of any complexity; it brings together various ideas and techniques to which the student has been exposed separately before. Every student should know the statement of Cayley's theorem. We mark the proof with a star, indicating that we do not consider this result to be of basic importance for this text.

To make it easier for the student to follow the proof, we outline the steps to be taken. Starting with any given group G, we shall proceed as follows:

STEP 1. Find a set G' of permutations which is a candidate for forming a group under permutation multiplication isomorphic to G.

STEP 2. Prove that G' is a group under permutation multiplication.

STEP 3. Define a mapping $\phi : G \to G'$, and show that ϕ is an isomorphism of G with G'.

Theorem 7.3 (Cayley). *Every group is isomorphic to a group of permutations.*

*Proof. Let G be a given group.

STEP 1. Our first job is to find a set G' of permutations which is a candidate to form a group isomorphic to G. Think of G as just a set, and let S_G be the group of all permutations of G given by Theorem 4.1. (Note that in the finite case if G has n elements, S_G has $n!$ elements. Thus, in general, S_G is clearly too big to be isomorphic to G.) We define a certain subset of S_G. For $a \in G$, let ρ_a be the mapping of G into G given by

$$x\rho_a = xa$$

for $x \in G$. (We can think of ρ_a as meaning *right multiplication by a*.) If $x\rho_a = y\rho_a$, then $xa = ya$, so $x = y$ by Theorem 2.1. Thus ρ_a is a one-to-one function. Also, if $y \in G$, then

$$(ya^{-1})\rho_a = (ya^{-1})a = y,$$

so ρ_a maps G onto G. Since $\rho_a\colon G \to G$ is both one to one and onto G, ρ_a is a permutation of G, that is, $\rho_a \in S_G$. Let

$$G' = \{\rho_a \mid a \in G\}.$$

STEP 2. We claim that G' is a subgroup of S_G. We must show that G' is closed under permutation multiplication, contains the identity permutation, and contains an inverse for each of its elements. First, we claim that

$$\rho_a\rho_b = \rho_{ab}.$$

To show that these functions are the same, we must show that they have the same action on each $x \in G$. Now

$$x(\rho_a\rho_b) = (x\rho_a)\rho_b = (xa)\rho_b = (xa)b = x(ab) = x\rho_{ab}.$$

Thus $\rho_a\rho_b = \rho_{ab}$, so G' is closed under multiplication. Clearly for all $x \in G$,

$$x\rho_e = xe = x,$$

where e is the identity element of G, so ρ_e is the identity permutation ι of S_G and is in G'. Since $\rho_a\rho_b = \rho_{ab}$, we have

$$\rho_a\rho_{a^{-1}} = \rho_{aa^{-1}} = \rho_e,$$

and also

$$\rho_{a^{-1}}\rho_a = \rho_e.$$

Hence

$$(\rho_a)^{-1} = \rho_{a^{-1}},$$

so $(\rho_a)^{-1} \in G'$. Thus G' is a subgroup of S_G.

STEP 3. It remains for us now to prove that G is isomorphic to this group

G' which we have described. Define $\phi \colon G \to G'$ by

$$a\phi = \rho_a$$

for $a \in G$. If $a\phi = b\phi$, then ρ_a and ρ_b must be the same permutation of G. In particular,

$$e\rho_a = e\rho_b,$$

so $ea = eb$, and $a = b$. Thus ϕ is one to one. It is immediate that ϕ is onto G' by the definition of G'. Finally, $(ab)\phi = \rho_{ab}$, while

$$(a\phi)(b\phi) = \rho_a\rho_b.$$

But we saw above that ρ_{ab} and $\rho_a\rho_b$ are the same permutation of G. Thus

$$(ab)\phi = (a\phi)(b\phi). \quad \blacksquare$$

For the proof of the theorem, we could have considered equally well the permutations λ_a of G defined by

$$x\lambda_a = ax$$

for $x \in G$. (We can think of λ_a as meaning *left multiplication by a*.) These permutations would have formed a subgroup G'' of S_G, again isomorphic to G but under the map $\psi \colon G \to G''$ defined by

$$a\psi = \lambda_{a^{-1}}.$$

Definition. The group G' in the proof of Theorem 7.3 is the *right regular representation of G*, and the group G'' in the preceding comment is the *left regular representation of G*.

Table 7.1			
	e	a	b
e	e	a	b
a	a	b	e
b	b	e	a

Table 7.2			
	ρ_e	ρ_a	ρ_b
ρ_e	ρ_e	ρ_a	ρ_b
ρ_a	ρ_a	ρ_b	ρ_e
ρ_b	ρ_b	ρ_e	ρ_a

Example 7.10 Let us compute the right regular representation of the group given by the group table, Table 7.1. By "compute," we mean give the elements of the right regular representation and the group table. Here the elements are

$$\rho_e = \begin{pmatrix} e & a & b \\ e & a & b \end{pmatrix}, \qquad \rho_a = \begin{pmatrix} e & a & b \\ a & b & e \end{pmatrix}, \qquad \text{and} \qquad \rho_b = \begin{pmatrix} e & a & b \\ b & e & a \end{pmatrix}.$$

The table is just like the original table with x renamed "ρ_x", as seen in

Table 7.2. *This "renaming" is the basic idea of an isomorphism.* For example,

$$\rho_a \rho_b = \begin{pmatrix} e & a & b \\ a & b & e \end{pmatrix} \begin{pmatrix} e & a & b \\ b & e & a \end{pmatrix} = \begin{pmatrix} e & a & b \\ e & a & b \end{pmatrix} = \rho_e. \ \|$$

For a finite group given by a group table, ρ_a is the permutation of the elements corresponding to their order in the column under a at the very top, and λ_a is the permutation corresponding to the order of the elements in the row opposite a at the extreme left. The notations "ρ_a" and "λ_a" were chosen to suggest right and left multiplication by a, respectively.

EXERCISES

7.1 Give two arguments showing that Z_4 is not isomorphic to the Klein 4-group V of Example 3.2.

7.2 Partition the following collection of groups into subcollections of isomorphic groups, as discussed following Theorem 7.2. Here a $*$ superscript means all non-zero elements of the set.

Z under addition	S_2
Z_6	**R*** under multiplication
Z_2	$\mathbf{R}^+$ under multiplication
S_6	**Q*** under multiplication
17**Z** under addition	**C*** under multiplication
Q under addition	The subgroup $\langle \pi \rangle$ of **R*** under multiplication
3**Z** under addition	The subgroup G of S_8 generated by $(1, 3, 4)(2, 6)$
R under addition	

†**7.3** Let $\langle G, \cdot \rangle$ be a group. Consider the binary operation $*$ on the set G defined by

$$a * b = b \cdot a$$

for $a, b \in G$. Show that $\langle G, * \rangle$ is a group and that $\langle G, * \rangle$ is actually isomorphic to $\langle G, \cdot \rangle$. [*Hint:* Consider the map ϕ with $a\phi = a^{-1}$ for $a \in G$.]

Comments: This is one instance where the notations "$\langle G, \cdot \rangle$" and "$\langle G, * \rangle$" are *very* handy. See the discussion following the definition of a group. Note that if G is finite, then we obtain the group table for $\langle G, * \rangle$ from the group table for $\langle G, \cdot \rangle$ by reading this table from the top to the left instead of from the left to the top.

7.4 An isomorphism of a group with itself is an **automorphism of the group**. How many automorphisms are there of Z_2? of Z_6? of Z_8? of **Z**? of Z_{17}? [*Hint:* Use Exercise 7.3.]

7.5 Mark each of the following true or false.

___ a) Any two groups of order 3 are isomorphic.
___ b) There is, up to isomorphism, only one cyclic group of a given finite order.
___ c) Any two finite groups with the same number of elements are isomorphic.
___ d) Every isomorphism is a one-to-one function.
___ e) Every one-to-one function between groups is an isomorphism.
___ f) The property of being cyclic (or not being cyclic, as the case may be) is an algebraic property of a group.

  — g) The concept of an algebraic property was not very precisely described in this section.

  — h) An abelian group can't be isomorphic to a nonabelian group.

  — i) An additive group can't be isomorphic to a multiplicative group.

  — j) $\mathbf{R}$ under addition is isomorphic to a group of permutations.

7.6 Let $\phi: G \rightarrow G'$ be an isomorphism of a group G with a group G'. Show that the map $\phi^{-1}: G' \rightarrow G$, defined by $x'\phi^{-1} = x$ such that $x\phi = x'$ for $x' \in G'$, is a well-defined function, and is an isomorphism of G' with G.

7.7 Let $\phi: G \rightarrow G'$ be an isomorphism of a group G with a group G', and $\psi: G' \rightarrow G''$ an isomorphism of G' with a group G''. Show that $\phi\psi: G \rightarrow G''$ is an isomorphism of G with G''.

7.8 Let G be a group and let g be a fixed element of G. Show that the map i_g, such that $xi_g = gxg^{-1}$ for $x \in G$, is an isomorphism of G with itself, i.e., an automorphism of G (see Exercise 7.4).

7.9 Give a formal proof (e.g., as was done in Theorem 7.1) of the statement that if ϕ is an isomorphism of a group G with a group G', and H is a subgroup of G, then

$$H\phi = \{h\phi \mid h \in H\}$$

is a subgroup of G'. (This is obvious from the motivation of our definition of an isomorphism, but it will do you good to try to write down a formal proof based on just the *definition* of an isomorphism.)

7.10 Prove in a manner similar to that used in Theorem 7.2 that every finite cyclic group of order n is isomorphic to $\mathbf{Z}_n$.

7.11 Let G be a cyclic group with generator a, and let G' be a group isomorphic to G. If $\phi: G \rightarrow G'$ is an isomorphism, show that, for every $x \in G$, $x\phi$ is completely determined by the value $a\phi$.

7.12 Let $\langle S, * \rangle$ be the group of all real numbers except -1 under the operation $*$ defined by $a * b = a + b + ab$ (see Exercise 2.9). Show that $\langle S, * \rangle$ is isomorphic to the group $\mathbf{R}^*$ of nonzero real numbers under multiplication. Actually define an isomorphism $\psi: \mathbf{R}^* \rightarrow S$.

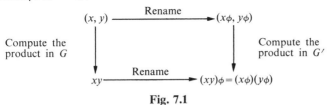

Fig. 7.1

7.13 Let ϕ be an isomorphism of a group G with a group G'. If for $x \in G$ we view "$x\phi$" as a new name for x or consider $x\phi$ as x renamed, then the condition $(xy)\phi = (x\phi)(y\phi)$ corresponds to the statement that the diagram in Fig. 7.1 is commutative. By "**the diagram is commutative**," we mean that starting in the upper left corner and following the path to the lower right corner, given by (arrow down) (arrow across), gives the same result as following the path (arrow across) (arrow down). Illustrating with the isomorphism ψ of our answer to Exer-

cise 7.12, if we consider $x\psi$ as x renamed for $x \in \mathbf{R}^*$, we get the diagram in Fig. 7.2
for $x = 2$ and $y = 5$.

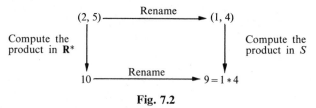

Fig. 7.2

a) Starting with the group $\mathbf{R}^*$ under multiplication, suppose we rename x by
   "$x - 4$" for $x \in \mathbf{R}^*$. Let S_1 be the set of all real numbers except -4. Define
   $*_1$ on S_1 such that $\langle S_1, *_1 \rangle$ is isomorphic to $\mathbf{R}^*$ under multiplication by means
   of this renaming.

b) Repeat part (a) with a renaming of $x \in \mathbf{R}^*$ by "$x - t$" for fixed $t \in \mathbf{R}$. First
   determine the required set S_2.

c) Repeat part (a) with a renaming of $x \in \mathbf{R}^*$ by "$x^3 + 1$". First determine
   the required set S_3.

*7.14 Compute the left regular representation of $\mathbf{Z}_4$. Compute the right regular
representation of S_3 using the notation of Example 4.1.

8 | Direct Products

8.1 EXTERNAL DIRECT PRODUCTS

Let us take a moment to review our present stock pile of groups. Starting with finite groups, we have the cyclic group $\mathbf{Z}_n$, the symmetric group S_n, and the alternating group A_n for each positive integer n. We also have the octic group D_4 of Example 4.2 and the Klein 4-group V. Of course we know that subgroups of these groups exist, and indeed Cayley's theorem applied to finite groups shows that every finite group is isomorphic to a subgroup of some S_n. But we have no easy way of computing all subgroups of a given group. Turning to infinite groups, we have groups consisting of sets of numbers under the usual addition or multiplication, as for example, $\mathbf{Z}$ and $\mathbf{R}$ under addition.

One purpose of this section is to show a constructive method for using known groups as building blocks to form more groups. The Klein 4-group will be recovered from the cyclic groups. As we shall describe in the next section, employing this procedure with the cyclic groups will give us a large class of abelian groups which includes all abelian groups of finite order. Let us start with a set-theoretic definition.

Definition. The *Cartesian product of sets* $S_1, S_2, \ldots, S_n$ is the set of all ordered n-tuples $(a_1, a_2, \ldots, a_n)$, where $a_i \in S_i$. The Cartesian product is denoted by either

$$\text{``}S_1 \times S_2 \times \cdots \times S_n\text{''}$$

or by

$$\text{``}\bigtimes_{i=1}^{n} S_i\text{''}.$$

One can also define the Cartesian product of an infinite number of sets, but the definition is considerably more sophisticated and we shall not need it.

Now let $G_1, G_2, \ldots, G_n$ be groups, and let us use multiplicative notation for all the group operations. Regarding the G_i as sets, we can form $\bigtimes_{i=1}^{n} G_i$. Let us show that one can make $\bigtimes_{i=1}^{n} G_i$ into a group by means of a binary operation of *multiplication by components*. We point out again that we are being sloppy in our use of the same notation for a group as for the set of elements of the group.

69

Theorem 8.1 *Let* $G_1, G_2, \ldots, G_n$ *be groups. For* $(a_1, a_2, \ldots, a_n)$ *and* $(b_1, b_2, \ldots, b_n)$ *in* $\bigtimes_{i=1}^{n} G_i$, *define* $(a_1, a_2, \ldots, a_n)(b_1, b_2, \ldots, b_n)$ *to be* $(a_1 b_1, a_2 b_2, \ldots, a_n b_n)$. *Then* $\bigtimes_{i=1}^{n} G_i$ *is a group, the* **external direct product of the groups** G_i, *under this binary operation.*

Proof. Note that since $a_i \in G_i$, $b_i \in G_i$, and G_i is a group, we have $a_i b_i \in G_i$. Thus the definition of the binary operation on $\bigtimes_{i=1}^{n} G_i$ given in the statement of the theorem makes sense, i.e., the binary operation is closed on $\bigtimes_{i=1}^{n} G_i$.

The associative law in $\bigtimes_{i=1}^{n} G_i$ is thrown back onto the associative law in each component as follows:

$$(a_1, a_2, \ldots, a_n)[(b_1, b_2, \ldots, b_n)(c_1, c_2, \ldots, c_n)]$$
$$= (a_1, a_2, \ldots, a_n)(b_1 c_1, b_2 c_2, \ldots, b_n c_n)$$
$$= (a_1(b_1 c_1), a_2(b_2 c_2), \ldots, a_n(b_n c_n))$$
$$= ((a_1 b_1)c_1, (a_2 b_2)c_2, \ldots, (a_n b_n)c_n)$$
$$= (a_1 b_1, a_2 b_2, \ldots, a_n b_n)(c_1, c_2, \ldots, c_n)$$
$$= [(a_1, a_2, \ldots, a_n)(b_1, b_2, \ldots, b_n)](c_1, c_2, \ldots, c_n).$$

If e_i is the identity element in G_i, then clearly, with multiplication by components, $(e_1, e_2, \ldots, e_n)$ is an identity in $\bigtimes_{i=1}^{n} G_i$. An inverse of $(a_1, a_2, \ldots, a_n)$ is $(a_1^{-1}, a_2^{-1}, \ldots, a_n^{-1})$; just compute the product by components. Hence $\bigtimes_{i=1}^{n} G_i$ is a group. ∎

In the event that the operation in each G_i is commutative, we sometimes use additive notation in $\bigtimes_{i=1}^{n} G_i$ and refer to $\bigtimes_{i=1}^{n} G_i$ as the "**external direct sum of the groups** G_i". The notation "$\sum_{i=1}^{n} G_i$" is sometimes used in this case in place of "$\bigtimes_{i=1}^{n} G_i$". We leave as an exercise the trivial proof that an external direct product of abelian groups is again abelian.

It is easily seen that if the set S_i has r_i elements for $i = 1, \ldots, n$, then $\bigtimes_{i=1}^{n} S_i$ has $r_1 r_2 \cdots r_n$ elements, for in an n-tuple, there are r_1 choices for the first component from S_1, and for each of these there are r_2 choices for the next component from S_2, etc.

Example 8.1 Consider the group $\mathbf{Z}_2 \times \mathbf{Z}_3$ which has $2 \cdot 3 = 6$ elements, namely $(0, 0)$, $(0, 1)$, $(0, 2)$, $(1, 0)$, $(1, 1)$, and $(1, 2)$. We claim that $\mathbf{Z}_2 \times \mathbf{Z}_3$ is cyclic. It is only necessary to find a generator. Let us try $(1, 1)$. Here the operations in $\mathbf{Z}_2$ and $\mathbf{Z}_3$ are written additively, so we do the same in the external direct product $\mathbf{Z}_2 \times \mathbf{Z}_3$.

$$(1, 1) = (1, 1)$$
$$2(1, 1) = (1, 1) + (1, 1) = (0, 2)$$
$$3(1, 1) = (1, 1) + (1, 1) + (1, 1) = (1, 0)$$
$$4(1, 1) = 3(1, 1) + (1, 1) = (1, 0) + (1, 1) = (0, 1)$$
$$5(1, 1) = 4(1, 1) + (1, 1) = (0, 1) + (1, 1) = (1, 2)$$
$$6(1, 1) = 5(1, 1) + (1, 1) = (1, 2) + (1, 1) = (0, 0)$$

Thus $(1, 1)$ generates all of $\mathbf{Z}_2 \times \mathbf{Z}_3$. Since there is only one cyclic group of a given order up to isomorphism, we have $(\mathbf{Z}_2 \times \mathbf{Z}_3) \simeq \mathbf{Z}_6$. ‖

Example 8.2 Consider $Z_3 \times Z_3$. This is a group of 9 elements. We claim that $Z_3 \times Z_3$ is *not* isomorphic to Z_9. We need only show that $Z_3 \times Z_3$ is not cyclic. Since the addition is by components, and since in Z_3 every element added to itself three times gives the identity, the same is true in $Z_3 \times Z_3$. Thus, no element can generate the group, for a generator added to itself successively could only give the identity after 9 summands. We have found another group of order 9. A similar argument shows that $Z_2 \times Z_2$ is not cyclic. Thus $Z_2 \times Z_2$ must be isomorphic to the Klein 4-group. ‖

The preceding examples illustrate the following theorem:

Theorem 8.2 *The group $Z_m \times Z_n$ is isomorphic to Z_{mn} if and only if m and n are relatively prime, i.e., the gcd of m and n is 1.*

Proof. Consider the cyclic subgroup of $Z_m \times Z_n$ generated by $(1, 1)$ as described by Theorem 3.2. As our previous work has shown, the order of this cyclic subgroup is the smallest "power" of $(1, 1)$ which gives the identity $(0, 0)$. Here taking a "power" of $(1, 1)$ in our additive notation will involve adding $(1, 1)$ to itself repeatedly. Under addition by components, the first component $1 \in Z_m$ yields 0 only after m summands, $2m$ summands, etc., and the second component $1 \in Z_n$ yields 0 only after n summands, $2n$ summands, etc. For them to yield 0 simultaneously, the number of summands must be a multiple of both m and n. The smallest number which is a multiple of both m and n will be mn if and only if the gcd of m and n is 1; in this case $(1, 1)$ generates a cyclic subgroup of order mn which is the order of the whole group. ∎

It is clear that this theorem can be extended to a product of more than two factors by an induction argument. We state this as a corollary without going through the details of the proof.

Corollary. *The group $\bigtimes_{i=1}^{n} Z_{m_i}$ is cyclic and isomorphic to $Z_{m_1 m_2 \cdots m_n}$ if and only if the numbers m_i for $i = 1, \cdots, n$ are such that the gcd of any two of them is 1.*

Example 8.3 The preceding corollary shows that if n is written as a product of powers of distinct prime numbers, as in

$$n = (p_1)^{n_1}(p_2)^{n_2} \cdots (p_r)^{n_r},$$

then Z_n is isomorphic to

$$Z_{(p_1)^{n_1}} \times Z_{(p_2)^{n_2}} \times \cdots \times Z_{(p_r)^{n_r}}.$$

In particular, Z_{72} is isomorphic to $Z_8 \times Z_9$. ‖

We have had several occasions to use the concept of the smallest positive power of an element of a group which gives the identity. It is time we introduced the usual terminology for this.

Definition. Let G be a group and $a \in G$. If there is a positive integer n such that $a^n = e$, the least such positive integer n is the **order of** a. If no such n exists, then a is of **infinite order**.

It is immediate that *if a is an element of a group G, the order of a is the same as the order of the cyclic subgroup generated by a.* This is a very useful fact to remember.

Theorem 8.3 *Let $(a_1, a_2, \ldots, a_n) \in X_{i=1}^{n} G_i$. If a_i is of finite order r_i in G_i, then the order of $(a_1, a_2, \ldots, a_n)$ in $X_{i=1}^{n} G_i$ is equal to the least common multiple of all the r_i.*

Proof. This follows by a repetition of the argument used in the proof of Theorem 8.2. For a power of $(a_1, a_2, \ldots, a_n)$ to give $(e_1, e_2, \ldots, e_n)$, the power must simultaneously be a multiple of r_1, so that this power of the first component a_1 will yield e_1, a multiple of r_2, so that this power of the second component a_2 will yield e_2, etc. ∎

It is obvious that if $X_{i=1}^{n} G_i$ is an external direct product of groups G_i, the subset

$$\overline{G}_i = \{(e_1, e_2, \ldots, e_{i-1}, a_i, e_{i+1}, \ldots, e_n) \mid a_i \in G_i\},$$

that is, the set of all n-tuples with the identity elements in all places but the ith, is a subgroup of $X_{i=1}^{n} G_i$. It is also clear that this subgroup $\overline{G}_i$ is naturally isomorphic to G_i under the correspondence given by the projection mapping π_i, where

$$(e_1, e_2, \ldots, e_{i-1}, a_i, e_{i+1}, \ldots, e_n)\pi_i = a_i.$$

The group G_i is mirrored in the ith component of the elements of $\overline{G}_i$, and the e_j in the other components just ride along. We consider $X_{i=1}^{n} G_i$ to be the *internal direct product* of these subgroups $\overline{G}_i$. The terms *internal* and *external*, as applied to a direct product of groups, just reflect whether or not (respectively) you are regarding the component groups as subgroups of the product group. After this section, we shall usually omit the words *external* and *internal* and just say *direct product*. Which we mean will be clear from the context. The next starred heading gives a careful treatment of the internal direct product for those who wish to see it. A basic set-theoretic definition and an equally basic group-theoretic theorem will be needed. We give these here, since we shall have other use for them later.

Definition. Let $\{S_i \mid i \in I\}$ be a collection of sets. Here I may be any set of indices. The **intersection** $\bigcap_{i \in I} S_i$ **of the sets** S_i is the set of all elements which are in all the sets S_i, that is,

$$\bigcap_{i \in I} S_i = \{x \mid x \in S_i \text{ for all } i \in I\}.$$

If I is finite, $I = \{1, 2, \ldots, n\}$, we may denote $\bigcap_{i \in I} S_i$ by

$$\text{"}S_1 \cap S_2 \cap \cdots \cap S_n\text{"}.$$

Theorem 8.4 *The intersection of subgroups H_i of a group G for $i \in I$ is again a subgroup of G.*

Proof. Let us show closure. Let $a \in \bigcap_{i \in I} H_i$ and $b \in \bigcap_{i \in I} H_i$, so that $a \in H_i$ for all $i \in I$ and $b \in H_i$ for all $i \in I$. Then $ab \in H_i$ for all $i \in I$, since H_i is a group. Thus $ab \in \bigcap_{i \in I} H_i$.

Since H_i is a subgroup for all $i \in I$, we have $e \in H_i$ for all $i \in I$, and hence $e \in \bigcap_{i \in I} H_i$.

Finally, for $a \in \bigcap_{i \in I} H_i$, we have $a \in H_i$ for all $i \in I$, so $a^{-1} \in H_i$ for all $i \in I$, which implies that $a^{-1} \in \bigcap_{i \in I} H_i$. ∎

*8.2   INTERNAL DIRECT PRODUCTS

Definition. Let a group G have subgroups H_i for $i = 1, \ldots, n$. G is the ***internal direct product of the subgroups*** H_i if the map $\phi\colon \bigtimes_{i=1}^{n} H_i \to G$ given by

$$(h_1, h_2, \ldots, h_n)\phi = h_1 h_2 \cdots h_n$$

is an isomorphism.

Note that under this isomorphism ϕ, the subgroup $\overline{H}_i$ of $\bigtimes_{i=1}^{n} H_i$ is mapped in a natural way onto H_i. In view of the isomorphism appearing in this definition, everything we observe for either an external direct product or an internal direct product has an immediate interpretation for the other.

Theorem 8.5 *If G is the internal direct product of subgroups $H_1, H_2, \ldots, H_n$, then each $g \in G$ can be uniquely written as $g = h_1 h_2 \cdots h_n$, where $h_i \in H_i$.*

Proof. Using the isomorphism of our definition, we need only show that the corresponding statement is true for the external direct product $\bigtimes_{i=1}^{n} H_i$ with respect to its subgroups $\overline{H}_i$, which are naturally isomorphic to H_i. We have to show that every element $(h_1, h_2, \ldots, h_n)$ of $\bigtimes_{i=1}^{n} H_i$ can be uniquely written as a product

$$(a_1, e_2, \ldots, e_n)(e_1, a_2, \ldots, e_n) \cdots (e_1, e_2, \ldots, a_n),$$

where $a_i \in H_i$. This is obvious. We must have $a_i = h_i$. ∎

The preceding definition and theorem suggest that it would be interesting to examine products of elements from various subgroups of a group. For the rest of this section we work with only two subgroups of a group, although the definitions and theorems can be generalized to more than two subgroups.

Let H and K be subgroups of a group G. We want to look at $\{hk \mid h \in H, k \in K\}$, which we denote by "$HK$". Unfortunately, *this set HK need not be a subgroup of G*, since $h_1 k_1 h_2 k_2$ need not be of the form hk. Of course, if G is abelian, or even if just each element h of H **commutes** with each element k of K, that is, $hk = kh$, then

$$h_1 k_1 h_2 k_2 = h_1 h_2 k_1 k_2 = h_3 k_3,$$

where $h_3 = h_1 h_2$ and $k_3 = k_1 k_2$ are elements of H and K, respectively. It is easily checked that in this case we have a subgroup, for $ee = e$ and $(hk)^{-1} = k^{-1}h^{-1} = h^{-1}k^{-1}$.

Let us try to recover a subgroup in the noncommutative case. Note that there is always at least one subgroup of G containing HK, namely G itself.

Definition. Let H and K be subgroups of a group G. The *join* $H \vee K$ *of* H *and* K is the intersection of all subgroups of G containing $HK = \{hk \mid h \in H, k \in K\}$.

Clearly, this intersection will be the smallest possible subgroup of G containing HK, and if elements in H and K commute, in particular, if G is abelian, we have $H \vee K = HK$. Note that since $h = he$ and $k = ek$, $H \subseteq HK$ and $K \subseteq HK$, so $H \leq H \vee K$ and $K \leq H \vee K$. But clearly, $H \vee K$ would be contained in any subgroup containing both H and K. Thus we see that $H \vee K$ *is the smallest subgroup of* G *containing both* H *and* K.

We conclude with a standard theorem for which we will have use in later starred sections.

Theorem 8.6 *A group* G *is the internal direct product of subgroups* H *and* K *if and only if*

1) $G = H \vee K$,
2) $hk = kh$ for all $h \in H$ and $k \in K$,
3) $H \cap K = \{e\}$.

Proof. Let G be the internal direct product of H and K. We claim that (1), (2), and (3) are obvious if one will regard G as isomorphic to the external direct product of H and K under the map ϕ, with $(h, k)\phi = hk$. Under this map,

$$\overline{H} = \{(h, e) \mid h \in H\}$$

corresponds to H, and

$$\overline{K} = \{(e, k) \mid k \in K\}$$

corresponds to K. Then (1), (2), and (3) follow immediately from the corresponding assertions regarding $\overline{H}$ and $\overline{K}$ in $H \times K$, which are obvious.

Conversely, let (1), (2), and (3) hold. We must show that the map ϕ of the external direct product $H \times K$ into G, given by $(h, k)\phi = hk$, is an isomorphism. The map ϕ has already been defined.

Suppose

$$(h_1, k_1)\phi = (h_2, k_2)\phi.$$

Then $h_1 k_1 = h_2 k_2$; consequently $h_2^{-1} h_1 = k_2 k_1^{-1}$. But $h_2^{-1} h_1 \in H$ and $k_2 k_1^{-1} \in K$, and they are the same element and thus in $H \cap K = \{e\}$ by (3). Therefore, $h_2^{-1} h_1 = e$ and $h_1 = h_2$. Likewise, $k_1 = k_2$, so $(h_1, k_1) = (h_2, k_2)$. This shows that ϕ is one to one.

The fact that $hk = kh$ by (2) for all $h \in H$ and $k \in K$ means that

$$HK = \{hk \mid h \in H, k \in K\}$$

is a group, for we have seen that this is the case if elements of H commute with those of K. Thus by (1), $HK = H \vee K = G$, so ϕ is onto G.

Finally,

$$[(h_1, k_1)(h_2, k_2)]\phi = (h_1 h_2, k_1 k_2)\phi = h_1 h_2 k_1 k_2,$$

while

$$[(h_1, k_1)\phi][(h_2, k_2)\phi] = h_1 k_1 h_2 k_2.$$

But by (2) we have $k_1 h_2 = h_2 k_1$. Thus

$$[(h_1, k_1)(h_2, k_2)]\phi = [(h_1, k_1)\phi][(h_2, k_2)\phi]. \quad \blacksquare$$

EXERCISES

8.1 List the 8 elements of $\mathbf{Z}_2 \times \mathbf{Z}_4$. Find the order of each of the elements. Is this group cyclic?

8.2 What is the largest order among the orders of all the cyclic subgroups of $\mathbf{Z}_6 \times \mathbf{Z}_8$? of $\mathbf{Z}_{12} \times \mathbf{Z}_{15}$?

†8.3 Prove that an external direct product of abelian groups is abelian.

8.4 Find all proper subgroups of $\mathbf{Z}_2 \times \mathbf{Z}_2$.

8.5 Neglecting the order of the factors, write direct products of two or more groups of the form $\mathbf{Z}_n$ so that the resulting product is isomorphic to $\mathbf{Z}_{60}$ in as many ways as possible.

8.6 Fill in the blanks.

a) The cyclic subgroup of $\mathbf{Z}_{24}$ generated by 18 has order _____.
b) $\mathbf{Z}_3 \times \mathbf{Z}_4$ is of order _____.
c) The element (4, 2) of $\mathbf{Z}_{12} \times \mathbf{Z}_8$ has order _____.
d) The Klein 4-group is isomorphic to $\mathbf{Z}$ __ $\times \mathbf{Z}$ __.
e) $\mathbf{Z}_2 \times \mathbf{Z} \times \mathbf{Z}_4$ has _____ elements of finite order.

8.7 Mark each of the following true or false.

— a) If G_1 and G_2 are any groups, then $G_1 \times G_2$ is always isomorphic to $G_2 \times G_1$.
— b) Computation in an external direct product of groups is easy if you know how to compute in each component group.
— c) One can only form an external direct product of groups of finite order.
— d) A group of prime order could not be the internal direct product of two proper subgroups.
— e) $\mathbf{Z}_2 \times \mathbf{Z}_4$ is isomorphic to $\mathbf{Z}_8$.
— f) $\mathbf{Z}_2 \times \mathbf{Z}_4$ is isomorphic to S_8.
— g) $\mathbf{Z}_3 \times \mathbf{Z}_8$ is isomorphic to S_4.
— h) Every element in $\mathbf{Z}_4 \times \mathbf{Z}_8$ has order 8.
— i) The order of $\mathbf{Z}_{12} \times \mathbf{Z}_{15}$ is 60.
— j) Any undergraduate mathematics major should be able to understand the concept of the direct product of groups.

8.8 Find all proper subgroups of $Z_2 \times Z_2 \times Z_2$.

8.9 Find all subgroups of $Z_2 \times Z_2 \times Z_4$ which are isomorphic to the Klein 4-group.

8.10 Give an example illustrating that not every abelian group is the internal direct product of two proper subgroups. (See Exercise 8.16 for a corresponding example for nonabelian groups.)

8.11 Let G be an abelian group. Let H be the subset of G consisting of the identity e together with all elements of G of order 2. Show that H is a subgroup of G.

8.12 Following up the idea of Exercise 8.11, will H always be a subgroup for every abelian group G if H consists of the identity e together with all elements of G of order 3? of order 4? For what positive integers n will H always be a subgroup for every abelian group G, if H consists of the identity e together with all elements of G of order n? Compare with Exercise 3.12.

8.13 Find a counterexample for Exercise 8.11 with the hypothesis that G is abelian deleted.

8.14 You sometimes hear people carelessly say, "An intersection of groups is a group." Why is it incorrect to say this?

*8.15 State a theorem similar to the theorem of Exercise 8.3, but for the case of a group which is the internal direct product of subgroups.

*8.16 Show that S_3 of Example 4.1 is not the internal direct product of the subgroups

$$H = \{\rho_0, \rho_1, \rho_2\} \qquad \text{and} \qquad K = \{\rho_0, \mu_1\}.$$

*8.17 Let $n = rs$, where r and s are relatively prime integers, i.e., the gcd of r and s is 1. Show that Z_n is the internal direct product of its cyclic subgroups $\langle r \rangle$ and $\langle s \rangle$.

*8.18 Consider the subgroups $H = \langle 2 \rangle$ and $K = \langle 6 \rangle$ of Z_{12}. Find HK and $H \vee K$.

*8.19 Consider the subgroups

$$H = \{\rho_0, \mu_1\} \qquad \text{and} \qquad K = \{\rho_0, \mu_2\}$$

of S_3 of Example 4.1. Find HK and $H \vee K$.

*8.20 Consider the subgroups

$$H = \{\rho_0, \delta_1\} \qquad \text{and} \qquad K = \{\rho_0, \delta_2\}$$

of the group D_4 of symmetries of the square in Example 4.2. Find HK and $H \vee K$.

*8.21 Let G be a group. Let h and k be elements of G which commute and are of relatively prime orders r and s. Apply Theorem 8.6 to the subgroups $\langle h \rangle$ and $\langle k \rangle$ of $\langle h \rangle \vee \langle k \rangle$ to show that hk is of order rs.

9 | Finitely Generated Abelian Groups

This is the first of several sections in this text in which important results are presented without proof. Few texts do this, for such a practice seems to be regarded by many mathematicians as a pedagogical sacrilege. They say, perhaps correctly, that the purpose of mathematics is to prove theorems. But this author feels that, more and more, there has to be a limit to the number of times one expects to see theorems "re-proved." The body of mathematical literature is so great now that if every mathematician insisted on checking every last detail in the proof of every theorem he used, he would never get to important frontiers of the subject. Professional mathematicians just don't operate this way, but for some peculiar reason many seem to feel that all students should. The results we shall present without proof are well within the limits of the student's understanding, but it would be hard to get to these fascinating facts in an average level one-semester undergraduate course if we insisted on giving complete proofs of them. *They have all been checked by many mathematicians much more competent than either the author or the student.* Some results in this section will be proved from preceding material. Others will be "proved" from newly stated results. With these comments out of the way, let us get on with our heresy.

9.1 GENERATORS AND TORSION

The first concept we define is of great importance to us. Contrary to our previous practice, we shall first give an elegant definition and then explain it naively in a theorem. Recall that by Theorem 8.4, an intersection of subgroups of a group is again a group. Let G be a group, and let a_i be elements of G for $i \in I$, where I is some indexing set. There is at least one subgroup of G containing all the a_i, namely G itself. Obviously, *the intersection of all subgroups of G containing all the a_i is the smallest subgroup of G containing all the a_i for $i \in I$.*

> **Definition.** Let G be a group and let $a_i \in G$ for $i \in I$. The smallest subgroup of G containing $\{a_i \mid i \in I\}$ is the **subgroup generated by** $\{a_i \mid i \in I\}$. If this subgroup is all of G, then $\{a_i \mid i \in I\}$ **generates** G and the a_i are **generators of G**. If there is a finite set $\{a_i \mid i \in I\}$ which generates G, then G is **finitely generated**.

77

Note that this definition is consistent with our previous definition of a generator for a cyclic group. Be sure to note that the next theorem is stated and proved for *any* group G, rather than only for abelian groups.

Theorem 9.1 *If G is a group and $a_i \in G$ for $i \in I$, then the subgroup H of G generated by $\{a_i \mid i \in I\}$ has as elements precisely those elements of G which are finite products of integral powers of the a_i, where powers of a fixed a_i may occur several times in the product.*

Proof. The fact that we may have to have powers of a fixed a_i occur several times in a product is due to the fact that G is not assumed to be abelian. If G is abelian, then $(a_1)^{-3}(a_2)^5(a_1)^7$ could be simplified to $(a_1)^4(a_2)^5$, but this may not be true in the nonabelian case.

Let K denote the set of all finite products of integral powers of the a_i. Clearly, $K \subseteq H$. We need only observe that K is a subgroup, and then, since H is the smallest subgroup containing a_i for $i \in I$, we will be done. It is obvious that a product of elements in K is again in K. Since $(a_1)^0 = e$, we have $e \in K$. For every element k in K, if you form from the product giving k, a new product with the order of the a_i reversed and the opposite sign on all exponents, you have k^{-1}, which is thus in K. For example,

$$[(a_1)^3(a_2)^2(a_1)^{-7}]^{-1} = (a_1)^7(a_2)^{-2}(a_1)^{-3},$$

which is again in K. ∎

Example 9.1 $\mathbf{Z} \times \mathbf{Z}_2$ is generated by $\{(1, 0), (0, 1)\}$. ‖

Although we are concerned chiefly with abelian groups in this section, we shall consistently use multiplicative notation in our general discussions. It is our experience that the student understands the notation "a^n" more readily than "na". In the latter notation, he tends to make the error of thinking of n as an element of the group.

Definition. A group G is a *torsion group* if every element in G is of finite order. G is *torsion free* if no element other than the identity is of finite order.

Theorem 9.2 *In an abelian group G, the set T of all elements of G of finite order is a subgroup of G, the **torsion subgroup** of G.*

Proof. We use multiplicative notation. Let a and b be elements of T. Then there are positive integers m and n such that $a^m = b^n = e$. Now

$$(ab)^{mn} = a^{mn}b^{mn},$$

since G is abelian, so

$$(ab)^{mn} = a^{mn}b^{mn} = (a^m)^n(b^n)^m = e^n e^m = e.$$

Thus ab is of finite order and hence in T. This shows that T is closed under group multiplication.

Of course, e is of finite order and consequently in T.

Finally, if $a \in T$ and $a^m = e$, then

$$e = e^m = (aa^{-1})^m = a^m(a^{-1})^m = e(a^{-1})^m = (a^{-1})^m,$$

so a^{-1} is of finite order and thus in T. ∎

Example 9.2 Every finite group is a torsion group, while $\mathbf{Z}$ under addition is torsion free. If we consider $\mathbf{Z} \times \mathbf{Z}_2$, the element $(1, 0)$ is not of finite order, but the element $(0, 1)$ is of order 2. It is clear that $T = \{(0, 0), (0, 1)\}$ is the torsion subgroup of $\mathbf{Z} \times \mathbf{Z}_2$. ∥

9.2   THE FUNDAMENTAL THEOREM

We now state some results without giving their proofs.

Lemma 9.1 *If G is a finitely generated abelian group with a torsion subgroup T, then G is an (internal) direct product $T \times F$ for some subgroup F of G which is torsion free.*

Proof. See the literature, for example, Chapter 2 of Hungerford [9]. ∎

Lemma 9.2 *A finitely generated torsion free abelian group F is isomorphic to $\mathbf{Z} \times \mathbf{Z} \times \cdots \times \mathbf{Z}$ for some number m of factors. The number m, the* **betti number** *of F, is unique.*

Proof. See the literature, for example, Chapter 2 of Hungerford [9]. ∎

Lemma 9.3 *A finite abelian group T is isomorphic to two different types of direct products of cyclic groups as follows:*

1) *T is isomorphic to a product*

$$\mathbf{Z}_{(p_1)^{r_1}} \times \mathbf{Z}_{(p_2)^{r_2}} \times \cdots \times \mathbf{Z}_{(p_n)^{r_n}},$$

where the p_i are primes and are not necessarily distinct. This direct product of cyclic groups of prime power order isomorphic to T is unique except for a rearrangement of the factors.

2) *T is isomorphic to a direct product*

$$\mathbf{Z}_{m_1} \times \mathbf{Z}_{m_2} \times \cdots \times \mathbf{Z}_{m_r},$$

where m_i divides m_{i+1}. The numbers m_i, the **torsion coefficients** *of T, are unique.*

Proof. See the literature, for example, Chapter 2 of Hungerford [9]. ∎

The terms *betti number* and *torsion coefficients* come from algebraic topology, where they play an important role.

The fact that the primes p_i appearing in (1) of Lemma 9.3 may not be distinct is unavoidable. We have seen, for example, that $\mathbf{Z}_5 \times \mathbf{Z}_5 \times \mathbf{Z}_9$ is not isomorphic to $\mathbf{Z}_{25} \times \mathbf{Z}_9$, for the former group has elements at most of order 45, while the latter is cyclic of order 225.

Just think for a moment of the significance and tremendous power of Lemma 9.3. *It gives us a description, up to isomorphism, of all finite abelian groups.*

Let us describe a method of finding a group, expressed as in (2) of Lemma 9.3, which is isomorphic to a given direct product of cyclic groups of prime power order. For each prime appearing in the order of the group, write the subscripts in the direct product involving that prime in a row in order of increasing magnitude. Keep the right-hand ends of the rows aligned. Thus starting with $Z_2 \times Z_4 \times Z_3 \times Z_3 \times Z_5$, we form the array

$$
\begin{array}{cc}
2 & 4 \\
3 & 3 \\
 & 5
\end{array}
$$

Then take the product of the numbers in each column, getting in this case the numbers 6 and 60. Then $Z_2 \times Z_4 \times Z_3 \times Z_3 \times Z_5$ is isomorphic to $Z_6 \times Z_{60}$. Likewise, $Z_2 \times Z_2 \times Z_2 \times Z_3 \times Z_3 \times Z_5$ gives rise to the array

$$
\begin{array}{ccc}
2 & 2 & 2 \\
 & 3 & 3 \\
 & & 5 \\
\hline
2 & 6 & 30
\end{array}
$$

so it is isomorphic to $Z_2 \times Z_6 \times Z_{30}$. We give no formal proof of the validity of this algorithm. From the theory we developed in Section 8, especially Theorem 8.2, the student can easily see why it works.

Example 9.3 Let us find all abelian groups (up to isomorphism) of order 360. First we express 360 as a product of prime powers $2^3 3^2 5$. Then using (1) of Lemma 9.3, we get as possibilities

1. $Z_2 \times Z_2 \times Z_2 \times Z_3 \times Z_3 \times Z_5$,
2. $Z_2 \times Z_4 \times Z_3 \times Z_3 \times Z_5$,
3. $Z_2 \times Z_2 \times Z_2 \times Z_9 \times Z_5$,
4. $Z_2 \times Z_4 \times Z_9 \times Z_5$,
5. $Z_8 \times Z_3 \times Z_3 \times Z_5$,
6. $Z_8 \times Z_9 \times Z_5$.

Thus there are six (up to isomorphism) different abelian groups of order 360. We write these six cases below as in (2) of Lemma 9.3, keeping the groups (up to isomorphism) in the same order. That is, the first group listed above is isomorphic to the first listed below, etc. This is easy to verify from the remarks preceding this example.

1. $Z_2 \times Z_6 \times Z_{30}$ 2. $Z_6 \times Z_{60}$
3. $Z_2 \times Z_2 \times Z_{90}$ 4. $Z_2 \times Z_{180}$
5. $Z_3 \times Z_{120}$ 6. Z_{360} ‖

Except for the knowledge that a subgroup of a finitely generated abelian group is again finitely generated, for a proof of which we again refer the student to the literature, our main theorem follows immediately from the preceding lemmas.

Theorem 9.3 (Fundamental Theorem of Finitely Generated Abelian Groups). *Every finitely generated abelian group G is isomorphic to a direct product of cyclic groups in the form*

1) $$\mathbf{Z}_{(p_1)^{r_1}} \times \mathbf{Z}_{(p_2)^{r_2}} \times \cdots \times \mathbf{Z}_{(p_n)^{r_n}} \times \mathbf{Z} \times \mathbf{Z} \times \cdots \times \mathbf{Z},$$

where the p_i are primes, not necessarily distinct, and also in the form

2) $$\mathbf{Z}_{m_1} \times \mathbf{Z}_{m_2} \times \cdots \times \mathbf{Z}_{m_r} \times \mathbf{Z} \times \mathbf{Z} \times \cdots \times \mathbf{Z},$$

where m_i divides m_{i+1}.

*In both cases the direct product is unique except for possible rearrangement of the factors, i.e., the number (**betti number** of G) of factors of **Z** is unique, the **torsion coefficients** m_i of G are unique, and prime powers $(p_i)^{r_i}$ are unique.*

Proof. By the preceding lemmas and remarks. ∎

Do you really comprehend the significance of these results? Among other things, they give us complete information about all finite abelian groups.

*9.3 APPLICATIONS

We conclude this section with a sampling of the many theorems which we could now prove regarding abelian groups. Some follow from work in Section 8; for others we will need our powerful Theorem 9.3.

Definition. A group G is ***decomposable*** if it is isomorphic to a direct product of two proper subgroups. Otherwise G is ***indecomposable***.

Theorem 9.4 *The finite indecomposable abelian groups are exactly the cyclic groups with order a power of a prime.*

Proof. Let G be a finite indecomposable abelian group. Then by Theorem 9.3 (or Lemma 9.3), G is isomorphic to a direct product of cyclic groups of prime power order. Since G is indecomposable, this direct product must consist of just one cyclic group of prime power order.

Conversely, let p be a prime. Our work in Section 8 shows that $\mathbf{Z}_{p^r}$ is indecomposable, for if $\mathbf{Z}_{p^r}$ were isomorphic to $\mathbf{Z}_{p^i} \times \mathbf{Z}_{p^j}$, where $i + j = r$, then every element would have an order at most $p^{\max(i,j)} < p^r$. ∎

Theorem 9.5 *If m divides the order of a finite abelian group G, then G has a subgroup of order m.*

Proof. By Theorem 9.3 (or Lemma 9.3), we can think of G as being

$$\mathbf{Z}_{(p_1)^{r_1}} \times \mathbf{Z}_{(p_2)^{r_2}} \times \cdots \times \mathbf{Z}_{(p_n)^{r_n}},$$

where not all primes p_i need be distinct. Since $(p_1)^{r_1}(p_2)^{r_2} \cdots (p_n)^{r_n}$ is the order of G, then m must be of the form $(p_1)^{s_1}(p_2)^{s_2} \cdots (p_n)^{s_n}$, where $0 \leq s_i \leq r_i$. By Theorem 6.4, $(p_i)^{r_i-s_i}$ generates a cyclic subgroup of $\mathbf{Z}_{(p_i)^{r_i}}$ of order equal to the quotient of $(p_i)^{r_i}$ by the gcd of $(p_i)^{r_i}$ and $(p_i)^{r_i-s_i}$. But the gcd of $(p_i)^{r_i}$ and $(p_i)^{r_i-s_i}$ is $(p_i)^{r_i-s_i}$. Thus $(p_i)^{r_i-s_i}$ generates a cyclic subgroup of $\mathbf{Z}_{(p_i)^{r_i}}$ of order

$$[(p_i)^{r_i}]/[(p_i)^{r_i-s_i}] = (p_i)^{s_i}.$$

Recalling that "$\langle a \rangle$" denotes the cyclic subgroup generated by a, we see that

$$\langle (p_1)^{r_1-s_1} \rangle \times \langle (p_2)^{r_2-s_2} \rangle \times \cdots \times \langle (p_n)^{r_n-s_n} \rangle$$

is the required subgroup of order m. ∎

Theorem 9.6 *If m is a square free integer, that is, m is not divisible by the square of any prime, then every abelian group of order m is cyclic.*

Proof. Let G be an abelian group of square free order m. Then by Theorem 9.3 (or Lemma 9.3), G is isomorphic to

$$\mathbf{Z}_{(p_1)^{r_1}} \times \mathbf{Z}_{(p_2)^{r_2}} \times \cdots \times \mathbf{Z}_{(p_n)^{r_n}},$$

where $m = (p_1)^{r_1}(p_2)^{r_2} \cdots (p_n)^{r_n}$. Since m is square free, we must have all $r_i = 1$ and all p_i distinct primes. The corollary of Theorem 8.2 then shows that G is isomorphic to $\mathbf{Z}_{p_1 p_2 \cdots p_n}$, so G is cyclic. ∎

EXERCISES

9.1 Find all abelian groups (up to isomorphism) of order 720; of order 1089. Express them in both forms (1) and (2) of Lemma 9.3, and pair up isomorphic groups of forms (1) and (2).

9.2 Find the torsion coefficients and the betti number of the group

$$\mathbf{Z} \times \mathbf{Z}_6 \times \mathbf{Z} \times \mathbf{Z} \times \mathbf{Z}_{12} \times \mathbf{Z}_{10}.$$

9.3 How many abelian groups (up to isomorphism) are there of order 24? of order 25? of order (24)(25)?

†**9.4** Following the idea suggested in Exercise 9.3, let m and n be relatively prime positive integers. Show that if there are (up to isomorphism) r abelian groups of order m and s of order n, then there are (up to isomorphism) rs abelian groups of order mn.

9.5 Use Exercise 9.4 to determine the number of abelian groups (up to isomorphism) of order $(10)^5$.

9.6 Mark each of the following true or false.

___ a) Every abelian group of prime order is cyclic.

___ b) Every abelian group of prime power order is cyclic.

___ c) $\mathbf{Z}_8$ is generated by $\{4, 6\}$.

___ d) $\mathbf{Z}_8$ is generated by $\{4, 5, 6\}$.

___ e) All finite abelian groups are classified up to isomorphism by Lemma 9.3.

- f) A student can never understand the meaning and significance of Theorem 9.3 without seeing the whole proof in detail.
- g) Every abelian group of order divisible by 5 contains a cyclic subgroup of order 5.
- h) Every abelian group of order divisible by 4 contains a cyclic subgroup of order 4.
- i) Every abelian group of order divisible by 6 contains a cyclic subgroup of order 6.
- j) Every finite abelian group has a betti number of 0.

9.7 Find the subgroup of Z_{12} generated by $\{2, 3\}$. Generated by $\{4, 6\}$. Generated by $\{8, 6, 10\}$.

9.8 Find the order of the torsion subgroup of $Z_4 \times Z \times Z_3$; of $Z_{12} \times Z \times Z_{12}$.

9.9 Let G be an abelian group of order 72.

a) Can you say how many subgroups of order 8 G has? Why?
b) Can you say how many subgroups of order 4 G has? Why?

9.10 Prove that if a finite abelian group has order a power of a prime p, then the order of every element in the group is a power of p.

9.11 For what positive integers n is it true that the only abelian groups of order n are cyclic?

9.12 Let p and q be distinct prime numbers. How does the number (up to isomorphism) of abelian groups of order p^r compare to the number (up to isomorphism) of abelian groups of order q^r?

9.13 (For students knowing a bit about complex numbers, especially DeMoivre's theorem.) Find the torsion subgroup T of the multiplicative group C^* of nonzero complex numbers.

9.14 Show that S_n is generated by $\{(1, 2), (1, 2, 3, \ldots, n)\}$. [*Hint:* Show that as r varies, $(1, 2, 3, \ldots, n)^{n-r}(1, 2)(1, 2, 3, \ldots, n)^r$ gives all the transpositions $(1, 2)$, $(2, 3)$, $(3, 4)$, $\ldots$, $(n - 1, n)$, $(n, 1)$. Then show that any transposition is a product of some of these transpositions and use the corollary of Theorem 5.1.]

9.15 What is the least number of elements which can be used to generate S_3 of Example 4.1? the group D_4 of symmetries of the square in Example 4.2? the group $Z_2 \times Z_2 \times Z_2$?

9.16 What is wrong with the following argument.

"By Exercise 9.14, S_n can be generated by 2 elements. By Cayley's theorem, every finite group is isomorphic to some subgroup of some S_n. Therefore, every finite group can be generated by 2 elements."

Note that the third part of Exercise 9.15 shows that this conclusion is false.

9.17 Let G, H, and K be finitely generated abelian groups. Show that if $G \times K \simeq H \times K$, then $G \simeq H$.

10 | Groups in Geometry and Analysis

We interrupt our purely algebraic study to give a rather vague indication of the significance of the group concept in geometry and analysis. Since we are not trying to teach either geometry or analysis, our discussion as it pertains to these fields will not be very precise.

*10.1 GROUPS IN GEOMETRY

Definition. By a *"transformation of a set A"*, a geometer means a permutation of the set, i.e., a one-to-one function of A onto itself.

According to Theorem 4.1, the transformations of a set form a group under transformation (permutation) multiplication. Remember that this multiplication is nothing but function composition. That is, if ϕ and ψ are transformations of A, then the product $\mu = \phi\psi$ is defined by $a\mu = (a\phi)\psi$ for $a \in A$.

Felix Klein (1849–1925) gave a famous definition of a *geometry* in an address (the *"Erlanger Programm"*) in 1872 when he accepted a chair at the University of Erlangen. From the present-day geometer's point of view, Klein's definition is not inclusive enough, but it will serve for our purpose.

Definition (*Klein*). A *geometry* is the study of those properties of a space (set) which remain invariant under some fixed subgroup of the full transformation group.

Let us illustrate this definition as it applies to the classical Euclidean geometry of the Euclidean line, plane, 3-space, etc. Here we have sets for which there is an idea of *distance between elements* defined. If we let $d(x, y)$ be the distance between two elements x and y, then we can talk about transformations which preserve distance.

Definition. If A is a set on which an idea of distance is defined, a transformation ϕ of A is an *isometry* if $d(x, y) = d(x\phi, y\phi)$, that is, if ϕ preserves distance.

It is clear that the subset of the full transformation group consisting of all isometries of the set is a subgroup. The *Euclidean geometry* of the line,

84

plane, 3-space, etc. is exactly the study of those properties left invariant under the group of isometries. Thus in Euclidean geometry, one can talk about the concepts of the length of a line segment, the size of an angle, and the number of sides of a polygon, for these are all invariant under an isometry.

Let us describe some of the isometries of the Euclidean plane. A **translation** of the plane is a transformation which moves each point a fixed distance in a fixed direction. In terms of coordinates, a translation $\tau_{(a,b)}$ moves a point (x, y) to $(x + a, y + b)$. Clearly,

$$\tau_{(a,b)}\tau_{(c,d)} = \tau_{(a+c,b+d)}.$$

It is immediate that the translations form a subgroup of the group of isometries isomorphic to $\mathbf{R} \times \mathbf{R}$ under addition. A **rotation** $\rho_{(P,\theta)}$ is a transformation which rotates the plane about the point P counterclockwise through the angle θ, where $0 \le \theta < 2\pi$. The rotations do not form a subgroup of the isometries, for $\rho_{(P,\theta_1)}\rho_{(Q,\theta_2)}$ is not a rotation if $P \ne Q$ and $\theta_1 + \theta_2 = 2\pi$. But clearly,

$$\rho_{(P,\theta_1)}\rho_{(P,\theta_2)} = \rho_{(P,\theta_1+\theta_2 \bmod 2\pi)},$$

where

$$(\theta_1 + \theta_2 \bmod 2\pi) = \begin{cases} \theta_1 + \theta_2 & \text{if } (\theta_1 + \theta_2) < 2\pi, \\ \theta_1 + \theta_2 - 2\pi & \text{if } (\theta_1 + \theta_2) \ge 2\pi. \end{cases}$$

The rotations about a fixed point P do form a subgroup of the isometries. This group is not isomorphic to $\mathbf{R}$ under addition, for it has cyclic subgroups of finite order. For example, $\rho_{(P,\pi/2)}$ generates a cyclic subgroup of order 4. All rotations about a fixed point actually form a group isomorphic to the multiplicative group of complex numbers of absolute value 1. Finally, a **reflection** in the plane is a function μ which carries each point of a fixed line l into itself and every point not on the line into the mirror image point straight across l and the same distance from l, as indicated in Fig. 10.1. It can be shown that the translations, rotations, and reflections generate (in the sense of Section 9) the whole isometry group of the plane. Actually this set of generators is much larger than is necessary. *The reflections alone can be shown to generate the isometry group, every plane isometry being expressible as a product of at most three reflections.* The student will have no difficulty in convincing himself, for example, that a translation can be written as the product of two reflections in lines which are perpendicular to the direction of the translation and of distance apart equal to half the length of the translation. We refer the interested student to Coxeter [44].

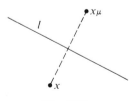

Fig. 10.1

The group S_3 given in Example 4.1 has a nice geometric interpretation. Consider an equilateral triangle as shown in Fig. 10.2. We let

$$\rho_1 = \begin{pmatrix} 1 & 2 & 3 \\ 2 & 3 & 1 \end{pmatrix},$$

where "ρ" denotes a rotation, and we can think of this as a rotation counter-clockwise through $2\pi/3$ radians. Similarly,

$$\rho_2 = \begin{pmatrix} 1 & 2 & 3 \\ 3 & 1 & 2 \end{pmatrix}$$

is a rotation counterclockwise through $4\pi/3$ radians, and

$$\rho_0 = \begin{pmatrix} 1 & 2 & 3 \\ 1 & 2 & 3 \end{pmatrix}$$

is a rotation through 0 radians. Also,

$$\mu_1 = \begin{pmatrix} 1 & 2 & 3 \\ 1 & 3 & 2 \end{pmatrix},$$

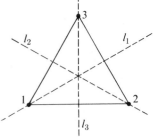

Fig. 10.2

where "μ" denotes a mirror image, corresponds to a reflection of the triangle in the line l_1, and in general our μ_i of Example 4.1 was a reflection in the line l_i. A reflection μ_i corresponds to a turning over of the triangle about an axis l_i. The reason for our choice of the notation in Example 4.1 is clear. Look again at the table in Fig. 4.5 of Example 4.1. Note that it divides itself into four blocks which we brought out by shading. This block arrangement is shown again in Fig. 10.3. Algebraically, the table in Fig. 10.3 gives a group of order 2. Geometrically, this table means that a product of two rotations is a rotation, a product of a rotation and a reflection is a reflection, and a product of two reflections is a rotation. This splitting up of a group into blocks which form a group by themselves will be the next topic in our algebraic study.

	ρ terms	μ terms
ρ terms	ρ terms	μ terms
μ terms	μ terms	ρ terms

Fig. 10.3

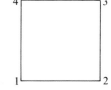

Fig. 10.4

Your first guess might be that the symmetries of a square should give S_4, but be careful. The permutation

$$\begin{pmatrix} 1 & 2 & 3 & 4 \\ 1 & 3 & 4 & 2 \end{pmatrix}$$

is not an isometry of the square in Fig. 10.4, for the distance from vertex 1 to vertex 2 would become the longer distance from vertex 1 to vertex 3. The group of symmetries of the square, or the octic group, was computed in Example 4.2. The computation of the dihedral group of symmetries of the regular n-gon in the plane for $n \geq 3$ is left for the exercises at the end of this section. Note that the group of symmetries of the regular n-gon for $n \geq 3$ is S_n only for $n = 3$.

One can recover S_4 geometrically as the group of symmetries of the regular tetrahedron, each face of which is an equilateral triangle, as shown in Fig. 10.5. You may say that you can't achieve the permutation

$$\begin{pmatrix} 1 & 2 & 3 & 4 \\ 1 & 3 & 2 & 4 \end{pmatrix}$$

by a rigid motion, but just as you have to go out of the plane to turn the equilateral triangle over and to obtain all of S_3, you have to go out of 3-space to 4-space to "turn over" the tetrahedron and to get all of S_4. This permutation amounts to a reflection in the plane containing the line through vertices 1 and 4 and perpendicular to the line through vertices 2 and 3.

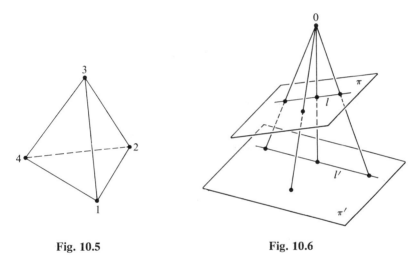

Fig. 10.5   Fig. 10.6

We give two more illustrations of Klein's definition. If we augment the usual plane $\mathbf{R} \times \mathbf{R}$ by adding a *line at infinity* containing one point for each direction in $\mathbf{R} \times \mathbf{R}$, we obtain the **projective plane**. In the projective plane, there are no such things as parallel lines, for two lines which are parallel in $\mathbf{R} \times \mathbf{R}$ are defined as intersecting at the point on the line at infinity corresponding to their common direction in the projective plane. Now let π and π' be two such projective planes with their usual $\mathbf{R} \times \mathbf{R}$ portions viewed as part of 3-space and not necessarily parallel as 3-space planes. A **projection of π onto π'** is a mapping of π onto π' by means of projection from a point not in either plane, or projection by parallel rays. Projection from a point is illustrated in Fig. 10.6. By means of two projections, we can project π onto π' and then π' back onto π, the composition giving a transformation of π onto itself. The **group of projective transformations of π** is the subgroup of all transformations of π generated by the type of mappings of π onto itself just described. Clearly, lines are carried into lines and quadrilaterals into quadrilaterals, that is, these are concepts of *projective geometry*. However, distance is not preserved, so distance is not a projective geometry concept.

One can show, however, that the so called *cross ratio*

$$\frac{(CA/CB)}{(DA/DB)}$$

of four points on a line, as shown in Fig. 10.7, is an invariant of the projective group. This cross ratio is about the only numerical quantity one can get hold of, and thus plays a very important role in projective geometry.

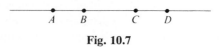

Fig. 10.7

Finally, if for a given space we have enough of an idea of when points are near each other so that we can talk about *continuous transformations* (such a set is a *topological space*), then we can define **topology** as the study of those properties of such spaces which are invariant under the group of all transformations which are continuous and whose inverses also are continuous. The Euclidean line, plane, 3-space, etc. are all topological spaces. Vaguely, a continuous transformation with a continuous inverse is one that we can achieve by bending, stretching and twisting the space without tearing or cutting it. For Euclidean spaces these transformations include all isometries, and for projective planes they include all the projective transformations. Topologically, one can't tell the difference between a football and a basketball, for one can be deformed, without being torn, to look just like the other. Similarly, a square and a circle are topologically the same. However, you can't make a solid chocolate mint disk look like a lifesaver candy without tearing a hole in it. Thus these are topologically different. Topology is by far the most active field of all the geometries at the present time.

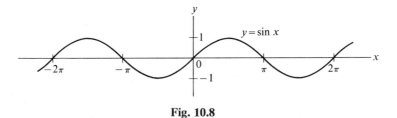

Fig. 10.8

*10.2   GROUPS IN ANALYSIS

Turning to analysis, we describe very briefly a few situations in which groups arise naturally. Of course analysis works with subgroups of the complex numbers primarily, so they appear in that obvious way. In addition, consider the function f of one real variable given by $f(x) = \sin x$. It has a well-known graph which is shown in Fig. 10.8. Also

$$\sin x = \sin (x + 2\pi n)$$

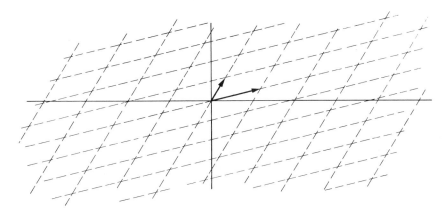

Fig. 10.9

for every integer n, that is, the function *sine* of one real variable is invariant under a transformation of its domain by an element of the infinite cyclic subgroup $\langle 2\pi \rangle$ of the translation group of **R**. Such a function of one real variable which is invariant under a transformation of its domain by an element of an infinite cyclic group is a **periodic function**.

Recall that the complex numbers can be viewed as filling up the Euclidean plane. A function on the plane which is invariant under an element of a group of transformations generated by two translations in different (not opposite) directions, as in Fig. 10.9, is a **doubly periodic function**. The group here is isomorphic to **Z** $\times$ **Z**. An **elliptic function** is defined as a doubly periodic meromorphic function of a complex variable. The term *meromorphic function* is explained in a first undergraduate course in functions of a complex variable. An elliptic function is thus known everywhere if it is known in one fundamental region, i.e., one of the diamond-shaped regions in Fig. 10.9. Both the trigonometric and elliptic functions are special cases of **automorphic functions**, which are functions invariant under a discrete group of transformations. The term *discrete* means roughly that no two group elements are close together.

Finally, in *measure theory* one likes to assign a numerical size to certain "nice" subsets of a set. If the set has a group structure $\langle G, \cdot \rangle$ and a is an element of the group, it is often convenient to have a measure such that the size of a subset S of G is the same as the size of the subset

$$aS = \{a \cdot s \mid s \in S\}.$$

By analogy with **R** and the operation of addition, one often calls aS a "**left translation of S by a**"; similarly, Sa is a **right translation of S by a**. A **left invariant measure** is a measure such that the size of aS is the same as the size of S for every $a \in G$ and every nice set S. A **right invariant measure** is similarly defined. Thus for **R** under addition, our ordinary idea of size (length) for nice sets (intervals) is a left and also a right invariant measure. Similarly, our usual idea of area in the plane $\mathbf{C}$ (or $\mathbf{R} \times \mathbf{R}$) under addition is a left and also a right invariant measure. However, our idea of area in $\mathbf{C}^*$, the group of nonzero complex numbers under multiplication, i.e., the plane minus the origin, is neither a left nor a right invariant measure. For example, if S is the inside of the circle of radius 1 about the origin minus the origin, it has area π, while $2S$, which has radius 2, has area 4π. However, Haar has shown that there is a left invariant measure (and also a right invariant measure), **Haar measure**, on every locally compact topological group.[†] The classification *locally compact topological group* covers many natural groups composed of complex numbers including $\mathbf{C}^*$ above.

EXERCISES

*10.1 Review Exercise 4.8, or do it now if you did not work it out before.

*10.2 Show that the nth dihedral group D_n of Exercise 4.8 can be generated by two elements. Argue geometrically.

*10.3 Review Exercise 4.9, or do it now if you did not work it out before.

*10.4 Show that the group of rigid motions of the cube given in Exercise 4.9 can be generated by two elements. Argue geometrically.

*10.5 Consider the group of *all* symmetries (isometries) of the cube. This group includes all rigid motions and also all reflections of the cube. What is the order of the group? Show that this group can be generated by three elements.

*10.6 Consider the *finite affine geometry* of four points A, B, C, D and six lines AB, AC, AD, BC, BD, CD, as indicated schematically in Fig. 10.10. Here each line contains just two points. A **collineation of an affine geometry** is a one-to-one map of the set of points onto itself which carries lines onto lines. (It is not necessary to know what a point or a line really is to do this exercise. It is based on your naive idea that a line is composed of points, etc.)

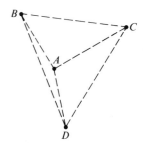

Fig. 10.10

a) Show that *every* one-to-one map of the set of points of this four-point affine geometry onto itself is a collineation.

b) Show that for any affine geometry, the collineations form a group, the **affine group**, under function composition.

c) To what group we have seen before is the affine group for the geometry of Fig. 10.10 isomorphic?

*10.7 Following the ideas of Exercise 10.6, consider the affine geometry of 9 points and 12 lines, each containing 3 points, given schematically in Fig. 10.11.

[†] A. Haar, "Der Massbegriff in der Theorie der kontinuierlichen Gruppen," *Ann. of Math.* (2), **34**, 147–169 (1933).

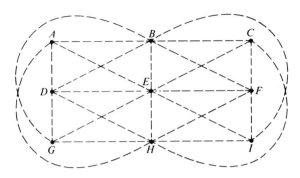

Fig. 10.11

a) Show that not every one-to-one map of the set of points onto itself is a collineation for this nine-point geometry.

b) Show that a collineation in this nine-point geometry is completely determined by its values on any three points not all on the same line.

c) What is the order of the affine group, i.e., the group of all collineations, for this nine-point geometry?

d) Consider the subgroup H of the affine group of part (c) which consists of those collineations leaving each point on the line ABC fixed. What is the order of this subgroup? To what group we have defined before is this subgroup H isomorphic?

11 | Groups of Cosets

11.1 INTRODUCTION

You probably noticed that the 36 entries in the table for S_3 in Example 4.1 divide themselves naturally into four blocks, each block consisting just of terms ρ_i or of terms μ_i. These blocks were brought out in Fig. 4.5 by shading. Thus the group S_3 itself was partitioned into two cells B_ρ and B_μ of equal size and the set $\{B_\rho, B_\mu\}$ forms a group with a table derived from that in Fig. 4.5 and shown here in Fig. 11.1. This partitioning of a group into cells such that the set of cells itself forms a group is a notion of basic importance in algebra. Let us call an element of a cell a "**representative of the cell.**" The equation

$$B_\rho B_\mu = B_\mu$$

means that *any* representative of B_ρ multiplied by *any* representative of B_μ gives some representative of B_μ.

	B_ρ	B_μ
B_ρ	B_ρ	B_μ
B_μ	B_μ	B_ρ

Fig. 11.1

Passing to the general case, we would like to determine precise conditions under which a group G can be partitioned into cells B_i such that *any* representative of a fixed cell B_r multiplied by *any* representative of another fixed cell B_s always produces a representative of one and the same cell B_t, which we would then consider as the product $B_r B_s$. The cell product $B_r B_s$ is *defined* to be the cell B_t obtained by multiplying representatives of B_r and B_s, and in order to have the binary operation of cell multiplication on $\{B_i\}$ *well defined*, as explained in Section 1, the final cell B_t containing the product of the representatives must be the same no matter which representatives one chooses from B_r and from B_s. This binary operation of cell multiplication on the set $\{B_i\}$ is the **operation induced on $\{B_i\}$ by the operation of G**. *Only if this operation is well defined does it make sense to ask whether the set $\{B_i\}$ is a group under the operation.*

Theorem 11.1 *If a group G can be partitioned into cells with the induced operation described above well defined, and if the cells form a group under this induced operation, then the cell containing the identity e of G must be a subgroup of G.*

Proof. Let G be partitioned into cells with the induced operation well defined and giving a group, and let B_e be the cell containing the identity. In computing B_eB_r we can take *any* representatives from B_e and from B_r and compute their product in G. Let us choose $e \in B_e$ and say $r \in B_r$. Then $er = r$ and $r \in B_r$, so $B_eB_r = B_r$. Similarly $B_rB_e = B_r$. Thus B_e must act as the identity cell in our group of cells. Therefore,

$$B_eB_e = B_e,$$

which shows, if we choose all possible representatives, that B_e is closed under the multiplication of the group G.

By its definition, B_e contains e.

Let $a \in B_e$. Now a^{-1} is in some cell B_k. Since B_e is the identity cell, we know $B_eB_k = B_k$. Choosing representatives $a \in B_e$ and $a^{-1} \in B_k$ and using them to compute B_eB_k, we see that we must have $B_eB_k = B_e$. Thus $B_k = B_e$, and $a^{-1} \in B_e$.

Hence B_e is a subgroup of G. ∎

11.2   COSETS

Suppose a group G can be partitioned into cells with the induced operation well defined and giving a group. Let B_e be the cell containing the identity. The preceding theorem shows that B_e is a subgroup of G. Let B_a be the cell containing $a \in G$. The equation $B_aB_e = B_a$ shows, if we choose the representative $a \in B_a$ and all representatives of B_e, that the set

$$aB_e = \{ax \mid x \in B_e\}$$

must be contained in B_a. This suggests that these *translations* or *cosets* aB_e of a subgroup B_e might be important.

Definition. Let H be a subgroup of a group G, and let $a \in G$. The **left coset aH of H** is the set $\{ah \mid h \in H\}$. The **right coset Ha** is similarly defined.

We have seen that if G can be partitioned into cells with the induced operation well defined and giving a group, then

$$aB_e \subseteq B_a.$$

Let $a^{-1} \in B_k$. Then $B_kB_a = B_e$, so choosing representatives $a^{-1} \in B_k$ and any $x \in B_a$, we have $a^{-1}x \in B_e$. Thus $a^{-1}x = b$ and $x = ab$, where $b \in B_e$. This shows that

$$B_a \subseteq aB_e,$$

so

$$B_a = aB_e.$$

Of course, we also have $B_a = B_ea$ by a similar argument. We summarize these results in a theorem.

Theorem 11.2 *If a group G can be partitioned into cells with the induced operation well defined and giving a group, then the cells must be precisely the left (and also the right) cosets of a subgroup of G. In particular, every left coset must be a right coset.*

Example 11.1 Let us determine how the left cosets of $3\mathbf{Z}$ as a subgroup of $\mathbf{Z}$ under addition look. The notation here is additive. Of course $3\mathbf{Z} = 0 + 3\mathbf{Z}$ is itself a left coset. Another left coset is $1 + 3\mathbf{Z}$. Upon a moment's reflection, we can see that $1 + 3\mathbf{Z}$ consists of all integers which leave a remainder of 1 when divided by 3 in the sense of Lemma 6.1. Likewise the left coset $2 + 3\mathbf{Z}$ consists of all integers leaving a remainder of 2 when divided by 3. Since Lemma 6.1 shows that the remainder of any integer divided by 3 is an integer r, where $0 \le r < 3$, the only possibilities are 0, 1, and 2. Thus these are all the left cosets. ‖

We might ask whether, given a subgroup H of a group G, the left (or right) cosets of H always do give a partition of G into distinct cells. Of course, any such partition corresponds to an equivalence relation on G by Theorem 0.1. Note that $b \in aH$ if and only if $b = ah$ for some $h \in H$, or if and only if $a^{-1}b \in H$. This suggests that we examine the relation $a \sim b$ if and only if $a^{-1}b \in H$ to see if it is an equivalence relation. As indicated in the statement of the theorem which follows, this relation is given a special notation.

Theorem 11.3 *Let H be a subgroup of a group G. The relations*

$$a \equiv_\ell b \ (mod\ H) \qquad if\ and\ only\ if\ a^{-1}b \in H$$

and

$$a \equiv_r b \ (mod\ H) \qquad if\ and\ only\ if\ ab^{-1} \in H$$

are equivalence relations on G, **left congruence modulo** *H and* **right congruence modulo** *H respectively. The equivalence classes of left (right) congruence modulo H are the left (right) cosets of H. All cosets of H have the same number of elements.*

Proof. We prove the statement for left congruence and left cosets. Similar proofs hold for right congruence and right cosets. Let us first show that left congruence modulo H is an equivalence relation.

Reflexive: $a \equiv_\ell a \ (\text{mod } H)$ since $a^{-1}a = e$ is in H.

Symmetric: If $a \equiv_\ell b \ (\text{mod } H)$, then $a^{-1}b \in H$, and H is a subgroup so $(a^{-1}b)^{-1} = b^{-1}a$ is in H also; consequently $b \equiv_\ell a \ (\text{mod } H)$.

Transitive: If $a \equiv_\ell b \ (\text{mod } H)$ and $b \equiv_\ell c \ (\text{mod } H)$, then $a^{-1}b \in H$ and $b^{-1}c \in H$. Since H is a subgroup, $(a^{-1}b)(b^{-1}c) = a^{-1}c$ is in H, so $a \equiv_\ell c$ (mod H).

Thus left congruence modulo H is an equivalence relation.

The equivalence class $\bar{a}$ containing a is easily computed as

$$\begin{aligned}
\bar{a} &= \{x \in G \mid x \equiv_\ell a \pmod{H}\} = \{x \in G \mid a^{-1}x \in H\} \\
&= \{x \in G \mid a^{-1}x = h \in H\} = \{x \in G \mid x = ah \text{ for some } h \in H\} \\
&= aH.
\end{aligned}$$

Thus the equivalence classes of left congruence modulo H are precisely the left cosets of H.

To show that any two left cosets have the same number of elements, consider the map $\lambda_a \colon H \to aH$ given by $h\lambda_a = ah$. This map λ_a is clearly onto aH. If h_1 and h_2 are in H, and $ah_1 = ah_2$, then $h_1 = h_2$ by the group cancellation law. Thus λ_a maps H one to one and onto aH. Hence every left coset has the same number of elements as H, and therefore all have the same number of elements. ∎

We now have a pretty good idea of how a partitioning of a group into disjoint cells looks if the induced operation is well defined and gives a group. The cells are always cosets (left and right cosets are the same in this case by Theorem 11.2) of some subgroup. We shall drop the term *cell* from now on and always use the term *coset* instead.

Unfortunately, the left cosets of a subgroup of a group do not always form a group under the induced operation. *The difficulty lies in the fact that the induced operation may not be well defined.* We give an example of this.

Example 11.2 Consider the group S_3 of Example 4.1 and the subgroup $H = \{\rho_0, \mu_1\}$. To find the left cosets of a subgroup H of a finite group G, one hunts for an element a in G not in H and finds the left coset aH. Then hunting for $b \in G$ with b not in either H or aH, one finds a new left coset bH. Continuing this process, one finds all left cosets of H in G. For our example, it is easy to see that the left cosets of $H = \{\rho_0, \mu_1\}$ in S_3 are

$$\begin{aligned}
H &= \{\rho_0, \mu_1\}, \\
\rho_1 H &= \{\rho_1, \mu_2\}, \\
\rho_2 H &= \{\rho_2, \mu_3\}.
\end{aligned}$$

Let us write the group table for S_3 again but with the elements in the order

$$\rho_0, \mu_1 \mid \rho_1, \mu_2 \mid \rho_2, \mu_3.$$

In Fig. 11.2, we shade lightly squares containing elements of the coset $\{\rho_1, \mu_2\}$ and heavily those containing elements of $\{\rho_2, \mu_3\}$. One glance at this table will show whether the induced operation is well defined and these left cosets form a group. We see that it doesn't work. The product of two blocks does not always produce elements of just one shade, i.e., elements in just one coset. By multiplying elements in the left coset $\{\rho_0, \mu_1\}$ by elements in the left coset $\{\rho_1, \mu_2\}$, one can get all the elements in $\{\rho_1, \rho_2, \mu_2, \mu_3\}$, which is not a left coset. *The induced operation is not well defined on these left cosets.* ‖

	ρ_0	μ_1	ρ_1	μ_2	ρ_2	μ_3
ρ_0	ρ_0	μ_1	ρ_1	μ_2	ρ_2	μ_3
μ_1	μ_1	ρ_0	μ_3	ρ_2	μ_2	ρ_1
ρ_1	ρ_1	μ_2	ρ_2	μ_3	ρ_0	μ_1
μ_2	μ_2	ρ_1	μ_1	ρ_0	μ_3	ρ_2
ρ_2	ρ_2	μ_3	ρ_0	μ_1	ρ_1	μ_2
μ_3	μ_3	ρ_2	μ_2	ρ_1	μ_1	ρ_0

Fig. 11.2

	0	3	1	4	2	5
0	0	3	1	4	2	5
3	3	0	4	1	5	2
1	1	4	2	5	3	0
4	4	1	5	2	0	3
2	2	5	3	0	4	1
5	5	2	0	3	1	4

Fig. 11.3

Example 11.3 Let us see whether the group $\mathbf{Z}_6$ can be partitioned into a group of left cosets of the subgroup $H = \{0, 3\}$. Here, using additive notation, our left cosets are

$$H = \{0, 3\},$$
$$1 + H = \{1, 4\},$$
$$2 + H = \{2, 5\}.$$

The table for $\mathbf{Z}_6$ with elements in the order

$$0, 3 \mid 1, 4 \mid 2, 5,$$

again with shaded squares according to the coset containing the element, is given in Fig. 11.3. It works! Isn't that lovely? Such a delightful display of symmetry should make shivers of joy run up and down the spine of anyone with any mathematical sensitivity. ‖

	Even	Odd
Even	Even	Odd
Odd	Odd	Even

Fig. 11.4

A group of left cosets formed from a group G gives some information about G. You don't know precisely what the product of any two elements of G is if you only know the left coset group, but you do know the *type* of element resulting from the product of two *types* of elements. This is the power of the concept. We have illustrated it with S_3, where elements are of two types, type ρ (rotations) and type μ (reflections). Another analysis is that those of type ρ are even permutations and those of type μ are odd permutations. The group given by the two left cosets of A_3 in S_3 then has a simple interpretation in terms of classifying products of permutations as *even* or *odd* as shown in Fig. 11.4. It is not surprising that subgroups whose left cosets form a group play a fundamental role in the theory of groups.

To complete this discussion, it would only remain to characterize exactly the types of subgroups of a group G whose cosets do form a group under the induced operation. We delay this until the next section, mainly because this is very important material which we wish the student to have lots of time to assimilate.

11.3 APPLICATIONS

Instead of putting the finishing touches on this theory now, we are going to prove several lovely results about finite groups which follow very easily from the work we have done thus far.

Theorem 11.4 (Lagrange). *Let G be a group of finite order n and H a subgroup of G. The order of H divides the order of G.*

Proof. Let H have m elements. Consider the collection of left cosets of H. By Theorem 11.3, these left cosets are disjoint, have the same number m of elements as H has, and every element of G is in some left coset. Thus, if there are r left cosets, we must have $n = rm$, so m divides n. ∎

Note that this elegant and important theorem comes from the simple counting done in Theorem 11.3, which showed that all cosets have the same number of elements. *Never underestimate a theorem that counts something.*

Our Theorem 9.3 shows at once that any *abelian* group of prime order is cyclic. But as a corollary to Theorem 11.4, we have

Corollary. *Every group of prime order is cyclic.*

Proof. Let G be of prime order p, and let a be an element of G different from the identity. Then the cyclic subgroup $\langle a \rangle$ of G generated by a has at least two elements, a and e. But by Theorem 11.4, the order $m \geq 2$ of $\langle a \rangle$ must divide the prime p. Thus we must have $m = p$ and $\langle a \rangle = G$, so G is cyclic. ∎

Thus there is only one group (up to isomorphism) of a given prime order. Now doesn't this elegant result follow easily from the theorem of Lagrange, a *counting* theorem? *Never underestimate a theorem that counts something.* The preceding corollary is a favorite examination question.

Theorem 11.5 *The order of an element of a finite group divides the order of the group.*

Proof. Remembering that the order of an element is the same as the order of the cyclic subgroup generated by the element, we see that this theorem follows directly from Theorem 11.4. ∎

Definition. Let H be a subgroup of a group G of finite order. The **index** $(G : H)$ **of H in G** is equal to $|G|/|H|$.

Thus the index of H in G is the number of left (or right) cosets of H.

We state a basic theorem concerning indices of subgroups, and leave the proof to the exercises (see Exercise 11.8).

Theorem 11.6 *If H and K are subgroups of a finite group G such that $K \leq H \leq G$, then $(G : K) = (G : H)(H : K)$.*

Theorem 11.4 shows that if there is a subgroup H of a finite group G, then the order of H divides the order of G. The student may wonder whether the converse is true. That is, if G is a group of order n, and m divides n, is there always a subgroup of order m? The answer is no, although from Theorem 9.3 it follows easily that the converse is true for abelian groups (see Theorem 9.5). However, A_4 can be shown to have no subgroup of order 6, which gives a counterexample for nonabelian groups.

EXERCISES

11.1 Find the number of left cosets of each of the following subgroups:

a) The subgroup $\langle 18 \rangle$ of $\mathbf{Z}_{36}$.
b) The subgroup $\langle 1 \rangle \times \langle 0 \rangle \times \langle 0 \rangle$ of $\mathbf{Z}_3 \times \mathbf{Z}_2 \times \mathbf{Z}_4$.
c) The subgroup $\langle 0 \rangle \times \langle 1 \rangle \times \langle 2 \rangle$ of $\mathbf{Z}_3 \times \mathbf{Z}_2 \times \mathbf{Z}_4$.

11.2 Three subgroups of the group of symmetries of the square of Example 4.2 are $H_1 = \{\rho_0, \rho_2\}$, $H_2 = \{\rho_0, \mu_1\}$, and $H_3 = \{\rho_0, \rho_1, \rho_2, \rho_3\}$. For each of these subgroups, write the table for the group of symmetries of the square in an order exhibiting the left cosets as we did in Examples 11.2 and 11.3, and determine whether or not the induced operation is well defined and whether or not the left cosets form a group. The use of a different color for each coset will make your tables very striking.

11.3 Write the left cosets of the cyclic subgroup $\langle (1, 2) \rangle$ of $\mathbf{Z}_2 \times \mathbf{Z}_4$, and determine whether or not the induced operation is well defined and whether or not the left cosets form a group. See Exercise 11.2 regarding the use of colors.

11.4 How many groups are there of order 17, up to isomorphism?

†11.5 Show that there are the same number of left as right cosets of a subgroup H of a group G, that is, exhibit a one-to-one map of the collection of left cosets onto the collection of right cosets. (Note that this result is obvious by counting for finite groups. Your proof must hold for any group.)

11.6 Mark each of the following true or false.

— a) Every subgroup of every group has left cosets.
— b) The number of left cosets of a subgroup of a finite group divides the order of the group.
— c) Every group of prime order is abelian.
— d) One cannot have left cosets of a finite subgroup of an infinite group.
— e) A subgroup of a group is a left coset of itself.
— f) Only subgroups of finite groups can have left cosets.
— g) A_n is of index 2 in S_n.
— h) The theorem of Lagrange is a nice result.

— i) Every finite group contains an element of every order which divides the order of the group.

— j) Every finite cyclic group contains an element of every order which divides the order of the group.

11.7 Show that a group with at least two elements but with no proper subgroups must be finite and of prime order.

11.8 Prove Theorem 11.6. [*Hint:* Let $\{a_iH \mid i = 1, \ldots, r\}$ be the collection of distinct left cosets of H in G and $\{b_jK \mid j = 1, \ldots, s\}$ be the collection of distinct left cosets of K in H. Show that

$$\{(a_ib_j)K \mid i = 1, \ldots, r; j = 1, \ldots, s\}$$

is the collection of distinct left cosets of K in G.]

11.9 Show that if H is a subgroup of an abelian group G, then every left coset of H is also a right coset of H.

11.10 Show that if H is a subgroup of index 2 in a finite group G, then every left coset of H is also a right coset of H.

11.11 Taking our notation for S_3 in Example 4.1, try to decide, without writing out the table, whether or not the induced operation is well defined on the left cosets of the subgroup

$$\{(\rho_0, 0), (\rho_0, 3), (\rho_1, 0), (\rho_1, 3), (\rho_2, 0), (\rho_2, 3)\}$$

of $S_3 \times Z_6$ and whether or not they form a group. How about left cosets of

$$\{(\rho_0, 0), (\rho_0, 3), (\mu_1, 0), (\mu_1, 3)\} \text{ ?}$$

11.12 Show that the left cosets of the subgroup $\{0\} \times Z_2$ of $Z \times Z_2$ form a group isomorphic to Z under the induced operation on left cosets.

11.13 Show that if a group G with identity e has finite order n, then $a^n = e$ for all $a \in G$.

11.14 Show that every left coset of the subgroup Z of the additive group of real numbers contains exactly one representative x in R such that $0 \leq x < 1$.

11.15 Show that the function *sine* assigns the same value to each representative of any fixed left coset of the subgroup $\langle 2\pi \rangle$ of the additive group R of real numbers. (Thus *sine* induces a well-defined function on the set of cosets; the value of the function on a coset is obtained when we choose a representative x of the coset and compute $\sin x$.)

11.16 Show that a finite cyclic group of order n has exactly one subgroup of each order d dividing n, and that these are all the subgroups.

11.17 The **Euler phi-function** is defined for positive integers n by $\varphi(n) = s$, where s is the number of positive integers less than or equal to n which are relatively prime to n. Use Exercise 11.16 to show that

$$n = \sum_{d \mid n} \varphi(d),$$

the sum being taken over all positive integers d dividing n. [*Hint:* Note that the number of generators of Z_d is $\varphi(d)$ by the corollary of Theorem 6.4.]

11.18 Let G be a finite group. Show that if for each positive integer m the number of solutions x of the equation $x^m = e$ in G is at most m, then G is cyclic. [*Hint:* Use Theorem 11.5 and Exercise 11.17 to show that G must contain an element of order $n = |G|$.]

12 | Normal Subgroups and Factor Groups

12.1   CRITERIA FOR THE EXISTENCE OF A COSET GROUP

We now return to the problem of deciding for what subgroups H of a group G the left (right) cosets will form a group under the induced operation. The crucial thing is whether or not the induced operation is well defined.

> **Lemma 12.1**   *If H is a subgroup of G, and if the induced operation of coset multiplication on left (right) cosets of H is well defined, then the collection of left (right) cosets of H is a group under this induced coset multiplication.*

Proof. Recall that a coset product $(aH)(bH)$ of left cosets was defined as the coset containing the product of *any* two representatives, one from aH and one from bH. We are assuming that this coset product is well defined, i.e., independent of the choice of representatives of aH and bH. We can thus take x as a representative of the coset xH for purposes of computation.

The group axioms for the collection of cosets are thrown back on the group axioms in G, since multiplication of cosets is defined in terms of multiplication of elements of G. To prove the associative law we must show that

$$aH[(bH)(cH)] = [(aH)(bH)]cH.$$

Computing, we have

$$aH[(bH)(cH)] = (aH)(bcH) = a(bc)H.$$

Also

$$[(aH)(bH)](cH) = (abH)(cH) = (ab)cH.$$

But by the associative law in G, we know that

$$a(bc) = (ab)c,$$

so coset multiplication is also associative. Likewise eH acts as identity, since the representative e acts as identity in G, and the inverse of aH is $a^{-1}H$. ∎

> **Theorem 12.1**   *If H is a subgroup of a group G, then the operation of induced multiplication is well defined on the left (right) cosets of H if and only if every left coset is a right coset.*

Proof. Suppose the induced operation is well defined. The left (right) cosets then form a group by Lemma 12.1, and by Theorem 11.2 the left cosets are the same as the right cosets.

Conversely, let us assume every left coset aH is also a right coset. Since $g \in gH$, and the right coset containing g is Hg, where $g \in G$, we are assuming that $gH = Hg$ as sets for all $g \in G$. We wish to show that if one attempts to define $(aH)(bH)$ by multiplying representatives, the left coset in which the product of the representatives is found is the same for all choices of these representatives. To this end, suppose a_1 and a_2 are representatives of aH, and b_1 and b_2 are representatives of bH. We have to show that $a_1 b_1$ and $a_2 b_2$ are in the same left coset of H. Since $aH = a_1 H = a_2 H$ and $bH = b_1 H = b_2 H$, we can write $a_1 = a_2 h_1$ and $b_1 = b_2 h_2$ for some h_1 and h_2 in H. Then

$$a_1 b_1 = a_2 h_1 b_2 h_2.$$

Now since $b_2 H = H b_2$ by assumption, we have $h_1 b_2 = b_2 h_3$ for some $h_3 \in H$. Thus

$$a_1 b_1 = a_2 b_2 h_3 h_2, \qquad \text{so} \qquad a_1 b_1 \in a_2 b_2 H,$$

that is, $a_1 b_1$ and $a_2 b_2$ are in the same left coset. This shows that the induced operation is well defined. ∎

12.2   INNER AUTOMORPHISMS AND NORMAL SUBGROUPS

We now know that cosets of a subgroup H of a group G form a group under the induced operation of coset multiplication if and only if $gH = Hg$ for every $g \in G$. We can rewrite $gH = Hg$ as $gHg^{-1} = H$ for all $g \in G$, where, of course,

$$gHg^{-1} = \{ghg^{-1} \mid h \in H\}.$$

To get more insight into the meaning of gHg^{-1}, we are led to the idea of studying for each $g \in G$ the mapping $i_g \colon G \to G$ given by

$$xi_g = gxg^{-1}.$$

Definition. An isomorphism of a group G with itself is an ***automorphism of the group*** G.

Theorem 12.2 *For each* $g \in G$, *the mapping* $i_g \colon G \to G$ *given by* $xi_g = gxg^{-1}$ *is an automorphism of* G, *the* **inner automorphism of** G **under conjugation by** g.

Proof. We must show that i_g is an isomorphism of G with itself. The map is defined, so we show that it is one to one, onto, and that

$$(xy)i_g = (xi_g)(yi_g).$$

First, for one to one, if $xi_g = yi_g$, then $gxg^{-1} = gyg^{-1}$, so $x = y$ by the group cancellation law. For onto, if $x \in G$, then

$$(g^{-1}xg)i_g = g(g^{-1}xg)g^{-1} = x.$$

Finally,

$$(xy)i_g = gxyg^{-1},$$

and also,

$$(xi_g)(yi_g) = (gxg^{-1})(gyg^{-1}) = gxyg^{-1}. \blacksquare$$

Since an automorphism of G is an isomorphism of G with itself, it just interchanges names of elements of G, preserving all algebraic structural features. That is, i_g will map a subgroup of G one to one onto a (possibly different) subgroup of G, an element of order n of G onto a (possibly different) element of order n in G, etc. Now to say that $gH = Hg$ is the same as saying that

$$H = gHg^{-1} = \{ghg^{-1} \mid h \in H\}.$$

This means that i_g carries H onto itself. It does not mean that necessarily $hi_g = h$ for all $h \in H$, but rather that $hi_g \in H$ for all $h \in H$. That is, $gHg^{-1} = H$ means that i_g gives a *permutation* of H.

Definition. A subgroup H of a group G is a **normal** (or **invariant**) **subgroup of G** if $gHg^{-1} = H$ for all $g \in G$, that is, if H is left invariant by every inner automorphism of G.

Thus normal subgroups are precisely those important subgroups of a group having the property that, for the cosets (left and right are the same), the induced operation is well defined, and the cosets form a group.

It is worth noting that if H is a subgroup of G such that $gHg^{-1} \subseteq H$ for all $g \in G$, then $gHg^{-1} = H$ for all $g \in G$, that is, H is then a normal subgroup of G. For if $gHg^{-1} \subseteq H$ for all $g \in G$, then

$$H \subseteq g^{-1}Hg = g^{-1}H(g^{-1})^{-1}$$

for all $g^{-1} \in G$, so $gHg^{-1} = H$ for all $g \in G$. *Thus to show that a subgroup H of a group G is a normal subgroup, one usually shows that $ghg^{-1} \in H$ for all $h \in H$ and all $g \in G$.*

Theorem 12.3 *Every subgroup of an abelian group is a normal subgroup.*

Proof. This is easy since if H is a subgroup of an abelian group G, then for all $g \in G$ and $h \in H$, we have

$$ghg^{-1} = gg^{-1}h = eh = h. \blacksquare$$

Example 12.1 Example 11.2 shows that $\{\rho_0, \mu_1\}$ is not a normal subgroup of S_3 of Example 4.1. Indeed,

$$\rho_1\mu_1(\rho_1)^{-1} = \rho_1\mu_1\rho_2 = \mu_3 \notin \{\rho_0, \mu_1\}.$$

Here i_{ρ_1} maps $\{\rho_0, \mu_1\}$ onto the subgroup $\{\rho_0, \mu_3\}$. $\|$

Definition. Two subgroups H and K of a group G are **conjugate** if $H = aKa^{-1}$ for some $a \in G$, that is, if one is mapped onto the other by some inner automorphism of G.

Thus Example 12.1 shows that $\{\rho_0, \mu_1\}$ and $\{\rho_0, \mu_3\}$ are conjugate subgroups of S_3 of Example 4.1.

12.3 FACTOR GROUPS

Definition. If N is a normal subgroup of a group G, the group of cosets of N under the induced operation is the **factor group of G modulo N**, and is denoted by "G/N". The cosets are **residue classes of G modulo N**.

Example 12.2 Referring to Example 11.1, we see that since $\mathbf{Z}$ is abelian, $3\mathbf{Z}$ is a normal subgroup, and thus $\mathbf{Z}/3\mathbf{Z}$ is the factor group of the three residue classes

$$0 + 3\mathbf{Z}, \qquad 1 + 3\mathbf{Z}, \qquad 2 + 3\mathbf{Z}.$$

This is a group of order 3, and hence is cyclic and isomorphic to $\mathbf{Z}_3$. ‖

Example 12.3 The subgroup $n\mathbf{Z}$ is normal in $\mathbf{Z}$ for all $n \in \mathbf{Z}^+$. There are n residue classes:

$$0 + n\mathbf{Z}, \qquad 1 + n\mathbf{Z}, \qquad \dots, \qquad (n-1) + n\mathbf{Z}.$$

Since we perform the addition of residue classes by choosing representatives, it is immediate that the map $\phi_n : \mathbf{Z}/n\mathbf{Z} \to \mathbf{Z}_n$ given by

$$(m + n\mathbf{Z})\phi_n = m \qquad \text{for} \quad 0 \le m < n$$

is an isomorphism. ‖

By abuse of notation, we may sometimes write "$\mathbf{Z}/n\mathbf{Z} = \mathbf{Z}_n$", and think of $\mathbf{Z}_n$ as the additive group of residue classes of $\mathbf{Z}$ modulo $\langle n \rangle$, or again by abuse of notation, the group of residue classes of $\mathbf{Z}$ modulo n. Recall from Section 0 that two integers a and b are congruent modulo n, denoted by "$a \equiv b \pmod{n}$," if and only if n divides $a - b$. This is precisely the criterion that a and b be in the same coset of $\mathbf{Z}/n\mathbf{Z}$.

We point out here that the formation of $\mathbf{Z}/n\mathbf{Z}$ is the *elegant* approach to the demonstration of the existence of a cyclic group of order n, as opposed to our naive approach in Theorem 6.3.

Example 12.4 Let us compute the factor group $(\mathbf{Z}_4 \times \mathbf{Z}_6)/\langle (0, 1) \rangle$. Here $\langle (0, 1) \rangle$ is the cyclic subgroup H of $\mathbf{Z}_4 \times \mathbf{Z}_6$ generated by $(0, 1)$. Thus

$$H = \{(0, 0), (0, 1), (0, 2), (0, 3), (0, 4), (0, 5)\}.$$

Since $\mathbf{Z}_4 \times \mathbf{Z}_6$ has 24 elements and H has 6 elements, all cosets of H must have 6 elements and $(\mathbf{Z}_4 \times \mathbf{Z}_6)/H$ must have order 4. Since $\mathbf{Z}_4 \times \mathbf{Z}_6$

is abelian, so is $(\mathbf{Z}_4 \times \mathbf{Z}_6)/H$ (remember, you compute in a factor group by means of representatives from the original group).   By *computing* $(\mathbf{Z}_4 \times \mathbf{Z}_6)/H$, we mean: determine this finite abelian group according to our classification in the Fundamental Theorem of Abelian Groups.   In additive notation, the cosets are

$$ H = (0, 0) + H, \qquad (1, 0) + H, \qquad (2, 0) + H, \qquad (3, 0) + H. $$

Since we can compute by choosing the representatives $(0, 0)$, $(1, 0)$, $(2, 0)$, and $(3, 0)$, it is clear that $(\mathbf{Z}_4 \times \mathbf{Z}_6)/H$ is isomorphic to $\mathbf{Z}_4$. Note that this is what you would expect, since in a factor group modulo H, everything in H becomes the identity, i.e., you are essentially "setting everything in H equal to zero". Thus the whole second factor $\mathbf{Z}_6$ of $\mathbf{Z}_4 \times \mathbf{Z}_6$ is collapsed, leaving just the first factor $\mathbf{Z}_4$. This is an example of a general situation given in Theorem 12.7. ‖

Example 12.5 Let us compute the factor group $(\mathbf{Z}_4 \times \mathbf{Z}_6)/\langle(0, 2)\rangle$. Now $(0, 2)$ generates the subgroup

$$ H = \{(0, 0), (0, 2), (0, 4)\} $$

of $\mathbf{Z}_4 \times \mathbf{Z}_6$ of order 3. Here the first factor $\mathbf{Z}_4$ of $\mathbf{Z}_4 \times \mathbf{Z}_6$ is left alone. The $\mathbf{Z}_6$ factor, on the other hand, is essentially collapsed by a subgroup of order 3, giving a factor group in the second factor of order 2 which must be isomorphic to $\mathbf{Z}_2$. Thus $(\mathbf{Z}_4 \times \mathbf{Z}_6)/\langle(0, 2)\rangle$ is isomorphic to $\mathbf{Z}_4 \times \mathbf{Z}_2$. ‖

Example 12.6 Let us compute the factor group $(\mathbf{Z}_4 \times \mathbf{Z}_6)/\langle(2, 3)\rangle$. *Be careful!* There is a great temptation to say that you are setting the 2 of $\mathbf{Z}_4$ and the 3 of $\mathbf{Z}_6$ both equal to zero, so that $\mathbf{Z}_4$ is collapsed to a factor group isomorphic to $\mathbf{Z}_2$ and $\mathbf{Z}_6$ to one isomorphic to $\mathbf{Z}_3$, giving a total factor group isomorphic to $\mathbf{Z}_2 \times \mathbf{Z}_3$. *This is wrong!* Note that

$$ H = \langle(2, 3)\rangle = \{(0, 0), (2, 3)\} $$

is of order 2, so $(\mathbf{Z}_4 \times \mathbf{Z}_6)/\langle(2, 3)\rangle$ has order 12, not 6. You see, "setting $(2, 3)$ equal to zero" does not make $(2, 0)$ and $(0, 3)$ equal to zero individually, so the factors don't collapse separately.

The possible abelian groups of order 12 are $\mathbf{Z}_4 \times \mathbf{Z}_3$ and $\mathbf{Z}_2 \times \mathbf{Z}_2 \times \mathbf{Z}_3$, and we must decide to which one our factor group is isomorphic. These two groups are most easily distinguished in that $\mathbf{Z}_4 \times \mathbf{Z}_3$ has an element of order 4, and $\mathbf{Z}_2 \times \mathbf{Z}_2 \times \mathbf{Z}_3$ does not. We claim that the coset $(1, 0) + H$ is of order 4 in the factor group $(\mathbf{Z}_4 \times \mathbf{Z}_6)/H$. To find the smallest power of a coset giving the identity in a factor group modulo H, we must, by choosing representatives, find the smallest power of a representative which is in the subgroup H. Clearly,

$$ 4(1, 0) = (1, 0) + (1, 0) + (1, 0) + (1, 0) = (0, 0) $$

is the first time that $(1, 0)$ added to itself gives an element of H. Thus $(\mathbf{Z}_4 \times \mathbf{Z}_6)/\langle(2, 3)\rangle$ has an element of order 4 and is isomorphic to $\mathbf{Z}_4 \times \mathbf{Z}_3$ or $\mathbf{Z}_{12}$. ‖

Example 12.7 Let us compute (i.e., classify as in Theorem 9.3) the group $(\mathbf{Z} \times \mathbf{Z})/\langle(1, 1)\rangle$. We may visualize $\mathbf{Z} \times \mathbf{Z}$ as the points in the plane with both coordinates integers, as indicated by the dots in Fig. 12.1. The subgroup $\langle(1, 1)\rangle$ consists of those points which lie on the "45° line through the origin," shown dashed in the figure. The coset $(1, 0) + \langle(1, 1)\rangle$ consists of those dots on the 45° line through the point $(1, 0)$, also shown dashed in the figure. Continuing, we easily see that each coset consists of those dots lying on one of the 45° lines dashed in the figure. We may choose the representatives

$$\ldots, (-3, 0), (-2, 0), (-1, 0), (0, 0), (1, 0), (2, 0), (3, 0), \ldots$$

of these cosets to compute in the factor group. Since these representatives correspond precisely to the points of $\mathbf{Z}$ on the x-axis, we see at once that the factor group $(\mathbf{Z} \times \mathbf{Z})/\langle(1, 1)\rangle$ is isomorphic to $\mathbf{Z}$. ‖

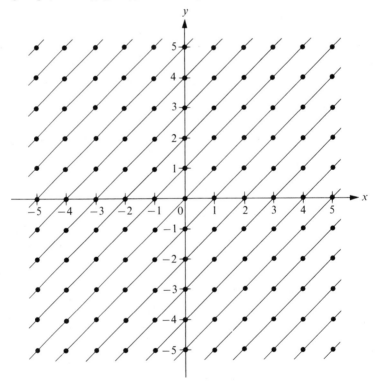

Fig. 12.1

12.4 SIMPLE GROUPS

As we mentioned in the preceding section, one feature of a factor group is that it gives crude information about the structure of the whole group. Of course, sometimes there may be no proper normal subgroups. For example, Theorem 11.4 shows that a group of prime order can have no proper subgroups of any sort.

Example 12.8 It is clear that the improper subgroups G and $\{e\}$ of a group G are both normal subgroups. Obviously G/G is the trivial group of one element while $G/\{e\}$ is isomorphic to G under the natural map carrying $g\{e\}$ into g for each $g \in G$. These factor groups are worthless so far as telling anything further about the structure of G is concerned. ‖

Definition. A group is *simple* if it has no proper normal subgroups.

The determination of all simple groups of certain types, such as all simple groups of finite order, is a very important problem on which mathematicians are actively working. We know that a group of prime order is simple. The next theorem is also classical.

Theorem 12.4 *The alternating group A_n is simple for $n \geq 5$.*

Proof. See the literature. ∎

We will have an important use for Theorem 12.4 in the very last section of this text. There are many simple groups other than those given above. For example, A_5 is of order 60 and A_6 is of order 360, and there is a simple group of nonprime order, namely 168, between these orders.

*12.5 APPLICATIONS

For the remainder of this section we try to further illustrate various aspects of factor groups (proving subgroups normal, computation with factor groups, their significance, uses, etc.). To illustrate how easy it is to compute in a factor group if you can compute in the whole group, we prove the following theorem.

Theorem 12.5 *A factor group of a cyclic group is cyclic.*

Proof. Let G be cyclic with generator a, and let N be a normal subgroup of G. We claim the coset aN generates G/N. We must compute all powers of aN. But this amounts to computing, in G, all powers of the representative a and all these powers give all elements in G. Hence the powers of aN certainly give all cosets of N and G/N is cyclic. ∎

Note that in forming a factor group of G modulo a subgroup N, you are essentially putting every element in G which is in N equal to e, for N forms your new identity in the factor group. This indicates another use for factor groups. Suppose, for example, that you are studying the structure of a non-abelian group G. Since Theorem 9.3 gives complete information about the structure of all sufficiently small abelian groups, it might be of interest to try to form an abelian group as much like G as possible, *an abelianized version of G*, by starting with G and then requiring that $ab = ba$ for all a and b in your new group structure. To require that $ab = ba$ is to say that $aba^{-1}b^{-1} = e$ in your new group. An element $aba^{-1}b^{-1}$ in a group is a **commutator of the group**. Thus we wish to attempt to form an abelianized version of G by replacing every commutator of G by e. By the first observation of this paragraph, we should then attempt to form the factor group of G modulo the smallest normal subgroup we can find which contains all commutators of G.

Theorem 12.6 *The set of all commutators $aba^{-1}b^{-1}$ of a group G generates a normal subgroup G' (the* **commutator subgroup**) *of G, and G/G' is abelian. Furthermore, G/N is abelian if and only if $G' \leq N$.*

Proof. The commutators certainly generate a subgroup G'; we must show that it is normal in G. Note that the inverse $(aba^{-1}b^{-1})^{-1}$ of a commutator is again a commutator, namely $bab^{-1}a^{-1}$. Also $e = eee^{-1}e^{-1}$ is a commutator. Theorem 9.1 then shows that G' consists precisely of all finite products of commutators. For $x \in G'$, we must show that $gxg^{-1} \in G'$ for all $g \in G$, or that if x is a product of commutators, so is gxg^{-1} for all $g \in G$. By inserting $e = g^{-1}g$ between each product of commutators occurring in x, we see that it is sufficient to show for each commutator $cdc^{-1}d^{-1}$ that $g(cdc^{-1}d^{-1})g^{-1}$ is in G'. But

$$
\begin{aligned}
g(cdc^{-1}d^{-1})g^{-1} &= (gcdc^{-1})(e)(d^{-1}g^{-1}) \\
&= (gcdc^{-1})(g^{-1}d^{-1}dg)(d^{-1}g^{-1}) \\
&= [(gc)d(gc)^{-1}d^{-1}][dgd^{-1}g^{-1}],
\end{aligned}
$$

which is in G'. Thus G' is normal in G.

The rest of the theorem is obvious if you have acquired the proper feeling for factor groups. One doesn't visualize it this way, but writing it out, that G/G' is abelian follows from

$$
\begin{aligned}
(aG')(bG') = abG' &= ab(b^{-1}a^{-1}ba)G' \\
&= (abb^{-1}a^{-1})baG' = baG' = (bG')(aG').
\end{aligned}
$$

Furthermore, if G/N is abelian, then $(a^{-1}N)(b^{-1}N) = (b^{-1}N)(a^{-1}N)$, that is, $aba^{-1}b^{-1}N = N$, so $aba^{-1}b^{-1} \in N$, and $G' \leq N$. Finally, if $G' \leq N$, then

$$
\begin{aligned}
(aN)(bN) = abN &= ab(b^{-1}a^{-1}ba)N \\
&= (abb^{-1}a^{-1})baN = baN = (bN)(aN). \quad \blacksquare
\end{aligned}
$$

Theorem 12.7 *If G is the internal direct product of subgroups H and K, then H and K are normal subgroups of G. Also G/H is isomorphic to K in a natural way.*

Proof. We can regard G as isomorphic to the external direct product $H \times K$, and need to show that $\bar{H} = \{(h, e) \mid h \in H\}$ is normal in $H \times K$, and that $(H \times K)/\bar{H}$ is isomorphic to $\bar{K} = \{(e, k) \mid k \in K\}$.

For normality, we need to show that

$$(h, k)\bar{H}(h, k)^{-1} = \bar{H}$$

for all $(h, k) \in H \times K$. But

$$(h, k)(h_1, e)(h, k)^{-1} = (h, k)(h_1, e)(h^{-1}, k^{-1})$$
$$= (hh_1h^{-1}, kek^{-1}) = (hh_1h^{-1}, e),$$

and $(hh_1h^{-1}, e) \in \bar{H}$. Thus $\bar{H}$ is normal in $H \times K$. The cosets of $\bar{H}$ are clearly all of the form $(e, k)\bar{H}$ for $k \in K$. It is obvious that the map

$$\phi: \bar{K} \to (H \times K)/\bar{H} \qquad \text{given by} \qquad (e, k)\phi = (e, k)\bar{H}$$

is an isomorphism. ∎

EXERCISES

12.1 Fill in the blanks.

a) The factor group $\mathbf{Z}_6/\langle 3 \rangle$ is of order _____.

b) The factor group $(\mathbf{Z}_4 \times \mathbf{Z}_{12})/(\langle 2 \rangle \times \langle 2 \rangle)$ is of order _____.

c) The factor group $(\mathbf{Z}_4 \times \mathbf{Z}_{12})/\langle (2, 2) \rangle$ is of order _____.

d) The coset $5 + \langle 4 \rangle$ is of order _____ in the factor group $\mathbf{Z}_{12}/\langle 4 \rangle$.

e) The coset $26 + \langle 12 \rangle$ is of order _____ in the factor group $\mathbf{Z}_{60}/\langle 12 \rangle$.

12.2 Show that A_n is a normal subgroup of S_n and compute S_n/A_n, that is, find a known group to which S_n/A_n is isomorphic.

12.3 Compute (i.e., classify as in Theorem 9.3) the following:

a) $(\mathbf{Z} \times \mathbf{Z})/\langle (0, 1) \rangle$.

b) $(\mathbf{Z} \times \mathbf{Z})/\langle (1, 2) \rangle$.

c) $(\mathbf{Z} \times \mathbf{Z} \times \mathbf{Z})/\langle (1, 1, 1) \rangle$.

12.4 This exercise illustrates the fact that if G contains two isomorphic normal subgroups H and K, then G/H need not be isomorphic to G/K. It depends on how H and K are *embedded* in G.

a) Compute $(\mathbf{Z}_2 \times \mathbf{Z}_4)/\langle (1, 0) \rangle$. Note that $\langle (1, 0) \rangle$ is cyclic of order 2. *Compute* means discover to which of the two (up to isomorphism) groups of order 4 this factor group is isomorphic.

b) Repeat part (a) with $(\mathbf{Z}_2 \times \mathbf{Z}_4)/\langle (0, 2) \rangle$.

c) Repeat part (a) with $(\mathbf{Z}_2 \times \mathbf{Z}_4)/\langle (1, 2) \rangle$.

12.5 Find all subgroups of S_3 of Example 4.1 which are conjugate to $\{\rho_0, \mu_1\}$.

†**12.6** Prove that the torsion subgroup T of an abelian group G is a normal subgroup of G, and that G/T is torsion free. Do not use Lemma 9.1, but proceed directly from the definitions of a torsion group and a normal subgroup.

12.7 Mark each of the following true or false.

— a) It makes sense to speak of the factor group G/N if and only if N is a normal subgroup of the group G.

— b) Every subgroup of an abelian group G is a normal subgroup of G.

— c) An inner automorphism of an abelian group must be just the identity map.

— d) Every factor group of a finite group is again of finite order.

— e) Every factor group of a torsion group is a torsion group.

— f) Every factor group of a torsion free group is torsion free.

— g) Every factor group of an abelian group is abelian.

— h) Every factor group of a nonabelian group is nonabelian.

— i) $\mathbf{Z}/n\mathbf{Z}$ is cyclic of order n.

— j) $\mathbf{R}/n\mathbf{R}$ is cyclic of order n, where $n\mathbf{R} = \{nr \mid r \in \mathbf{R}\}$ and $\mathbf{R}$ is under addition.

12.8 Describe all subgroups of order ≤ 4 of $\mathbf{Z}_4 \times \mathbf{Z}_4$, and in each case classify the factor group of $\mathbf{Z}_4 \times \mathbf{Z}_4$ modulo the subgroup as in Theorem 9.3, part (1). That is, describe the subgroup and say that the factor group of $\mathbf{Z}_4 \times \mathbf{Z}_4$ modulo the subgroup is isomorphic to $\mathbf{Z}_2 \times \mathbf{Z}_4$, or whatever the case may be. [*Hint:* $\mathbf{Z}_4 \times \mathbf{Z}_4$ has six different cyclic subgroups of order 4. Describe them by giving a generator, such as the subgroup $\langle(1, 0)\rangle$. There is one subgroup of order 4 that is isomorphic to the Klein 4-group. There are three subgroups of order 2.]

12.9 Let H be a normal subgroup of a finite group G, and let $m = (G : H)$. Show that $a^m \in H$ for every $a \in G$.

12.10 Show that an intersection of normal subgroups of a group G is again a normal subgroup of G.

12.11 Show that it makes sense to speak of the smallest normal subgroup of a group G which contains a fixed subset S of G. [*Hint:* Use Exercise 12.10.]

12.12 Show that if a finite group G has exactly one subgroup H of a given order, then H is a normal subgroup of G.

12.13 Show that if a finite group G contains a proper subgroup of index 2 in G, then G is not simple.

12.14 Show that if H and N are subgroups of a group G, and N is normal in G, then $H \cap N$ is normal in H. Show by an example that $H \cap N$ need not be normal in G.

12.15 Let G be a group containing at least one subgroup of a fixed finite order s. Show that the intersection of all subgroups of G of order s is a normal subgroup of G. [*Hint:* Use the fact that if H has order s, then so does xHx^{-1} for all $x \in G$.]

12.16 a) Show that all automorphisms of a group G form a group under function composition.

b) Show that the inner automorphisms of a group G form a normal subgroup of the group of all automorphisms of G under function composition. [*Warning:* Be sure to show that the inner automorphisms do form a subgroup.]

12.17 Show that the set of all $g \in G$ such that $i_g: G \to G$ is the identity inner automorphism i_e is a normal subgroup of a group G.

12.18 Let G be a group. Show that the relation $a \sim b$ if and only if $a = gbg^{-1}$ for some $g \in G$ is an equivalence relation on G. Some equivalence classes contain only one element c. Characterize those elements c.

12.19 Let G be a group. Show that the relation of $A \sim B$ if and only if A and B are conjugate subgroups of G, so that $A = gBg^{-1}$ for some $g \in G$, is an equivalence relation on the collection of all subgroups of G. Some equivalence classes may contain a single subgroup K. Characterize those subgroups K.

*12.20 Find the commutator subgroup G' of the group D_4 of symmetries of the square in Example 4.2.

*12.21 a) Show that if N is a normal subgroup of G, and H is any subgroup of G, then $HN = NH = N \vee H$.

b) Show that if N and M are normal subgroups of G, then NM is also a normal subgroup of G.

*12.22 Show that if H and K are normal subgroups of a group G such that $H \cap K = \{e\}$, then $hk = kh$ for all $h \in H$ and $k \in K$. [*Hint:* Consider the commutator $hkh^{-1}k^{-1} = (hkh^{-1})k^{-1} = h(kh^{-1}k^{-1}).$]

13|Homomorphisms

13.1 DEFINITION AND ELEMENTARY PROPERTIES

An isomorphism of a group G with a group G' was defined as a one-to-one map ϕ of G onto G' such that for all a and b in G, $(ab)\phi = (a\phi)(b\phi)$. If we drop the condition that ϕ be one to one and onto, just retaining $(ab)\phi = (a\phi)(b\phi)$, the map ϕ is then a *homomorphism*. Homomorphisms are closely connected with factor groups, as we shall see.

Definition. A map ϕ of a group G into a group G' is a **homomorphism** if

$$(ab)\phi = (a\phi)(b\phi)$$

for all elements a and b in G.

Let us examine the idea behind the requirement $(ab)\phi = (a\phi)(b\phi)$ for a homomorphism $\phi: G \rightarrow G'$. This requirement is the only one which distinguishes a homomorphism from just any map of G into G'. It asserts that ϕ is a *structure relating map*. The algebraic structure of G is completely determined by the binary operation on G, and that of G' is completely determined by the binary operation on G'. In the condition $(ab)\phi = (a\phi)(b\phi)$, the operation ab on the left-hand side takes place in G, while the operation $(a\phi)(b\phi)$ on the right-hand side takes place in G'. The homomorphism condition thus relates the structure of G to that of G'.

Example 13.1 The natural map γ of $\mathbf{Z}$ into $\mathbf{Z}_n$ given by $m\gamma = r$, where r is the remainder (in the sense of Lemma 6.1) of m when divided by n, is a homomorphism. We need to observe that

$$(s + t)\gamma = s\gamma + t\gamma.$$

If

$$(1) \quad s = q_1 n + r_1 \quad\quad \text{and} \quad\quad (2) \quad t = q_2 n + r_2$$

for $0 \leq r_i < n$, then $s\gamma = r_1$ and $t\gamma = r_2$. Thus

$$s\gamma + t\gamma = (r_1 + r_2) \text{ modulo } n.$$

That is, if $r_1 + r_2 = q_3 n + r_3$ for $0 \leq r_3 < n$, then

$$s\gamma + t\gamma = r_3.$$

Adding equations (1) and (2) above, we get

$$s + t = (q_1 + q_2)n + r_1 + r_2 = (q_1 + q_2 + q_3)n + r_3$$

and $0 \leq r_3 < n$. Thus

$$(s + t)\gamma = r_3$$

also.

Viewing $\mathbf{Z}_n$ as the group $\mathbf{Z}/n\mathbf{Z}$ of residue classes modulo n, we see that γ assigns to each element of $\mathbf{Z}$ the coset or residue class in which it appears modulo n. This is an example of a general situation described in the next theorem. ∥

Theorem 13.1 *If N is a normal subgroup of a group G, then the canonical (or natural) map* $\gamma: G \rightarrow G/N$ *given by* $a\gamma = aN$ *for* $a \in G$ *is a homomorphism.*

Proof. This is an immediate consequence of the definition of coset multiplication in terms of multiplication of representatives, for

$$(ab)\gamma = abN = (aN)(bN) = (a\gamma)(b\gamma). \; \blacksquare$$

Definition. The **kernel of a homomorphism** ϕ of a group G into a group G' is the set of all elements of G mapped onto the identity element of G' by ϕ.

Example 13.2 For the canonical map $\mathbf{Z} \rightarrow \mathbf{Z}_n$ given in Example 13.1, the kernel is $n\mathbf{Z}$. Note that $n\mathbf{Z}$ is a normal subgroup of $\mathbf{Z}$, and that $\mathbf{Z}/n\mathbf{Z}$ is isomorphic to $\mathbf{Z}_n$. ∥

The preceding example illustrates the general connection between homomorphisms and factor groups, which we shall state and prove in Theorem 13.3.

Definition. Let ϕ be a mapping of a set X into a set Y, and let $A \subseteq X$ and $B \subseteq Y$. The **image** $A\phi$ **of** A **in** Y **under** ϕ is $\{a\phi \mid a \in A\}$. The **inverse image** $B\phi^{-1}$ **of** B **in** X is $\{x \in X \mid x\phi \in B\}$.

The next theorem gives some structural features preserved under a homomorphism.

Theorem 13.2 *Let* ϕ *be a homomorphism of a group G into a group G'. If e is the identity in G, then* $e\phi$ *is the identity in G', and if* $a \in G$, *then* $a^{-1}\phi = (a\phi)^{-1}$. *If H is a subgroup of G, then* $H\phi$ *is a subgroup of G', and H normal in G implies that* $H\phi$ *is normal in* $G\phi$. *Going the other way, if K' is a subgroup of G', then* $K'\phi^{-1}$ *is a subgroup of G, and K' normal in* $G\phi$ *implies that* $K'\phi^{-1}$ *is normal in G. Loosely, subgroups correspond to subgroups and normal subgroups to normal subgroups under a homomorphism.*

Proof. Let ϕ be a homomorphism of G into G'. Then

$$a\phi = (ae)\phi = (a\phi)(e\phi).$$

Thus $e\phi$ must be the identity e' in G'. The equation

$$e\phi = (aa^{-1})\phi = (a\phi)(a^{-1}\phi)$$

then shows that $a^{-1}\phi = (a\phi)^{-1}$.

Let H be a subgroup of G, and let $a\phi$ and $b\phi$ be any two elements in $H\phi$. Then $(ab)\phi = (a\phi)(b\phi)$, so $(a\phi)(b\phi) \in H\phi$, that is, $H\phi$ is closed under the operation of G'. The fact that $e' = e\phi$ and $a^{-1}\phi = (a\phi)^{-1}$ completes the proof that $H\phi$ is a subgroup of $G\phi$. Suppose that H is normal in G, and let $g\phi \in G\phi$. Now

$$(g\phi)(h\phi)(g\phi)^{-1} = (g\phi)(h\phi)(g^{-1}\phi) = (ghg^{-1})\phi.$$

Since $ghg^{-1} \in H$, we have $(ghg^{-1})\phi \in H\phi$. Thus $H\phi$ is normal in $G\phi$.

Going the other way, let K' be a subgroup of G'. Suppose a and b are in $K'\phi^{-1}$. Then $(ab)\phi = (a\phi)(b\phi)$ and $(a\phi)(b\phi) \in K'$, so $ab \in K'\phi^{-1}$. Also K' must contain the identity $e\phi$, so $e \in K'\phi^{-1}$. If $a \in K'\phi^{-1}$, then $a\phi \in K'$, so $(a\phi)^{-1} \in K'$. But $(a\phi)^{-1} = a^{-1}\phi$, so we must have $a^{-1} \in K'\phi^{-1}$. Hence $K'\phi^{-1}$ is a subgroup of G. If K' is a normal subgroup of $G\phi$, then for $b \in K'\phi^{-1}$ and $g \in G$, we have

$$(gbg^{-1})\phi = (g\phi)(b\phi)(g\phi)^{-1},$$

and $(g\phi)(b\phi)(g\phi)^{-1}$ is in K', so $gbg^{-1} \in K'\phi^{-1}$. Thus $K'\phi^{-1}$ is normal in G. ∎

Theorem 13.2 may seem complex to the student, but it is very easy, and the proof is really mechanical. It would be an excellent exercise for the student to write out the whole proof without referring to the text. This will test whether or not he really understands the definitions involved.

13.2   THE FUNDAMENTAL HOMOMORPHISM THEOREM

Theorem 13.2 shows, in particular, that *for a homomorphism $\phi: G \rightarrow G'$, the kernel $K = \{e'\}\phi^{-1}$ is a normal subgroup of G.* We are now in a position to prove our main theorem.

Theorem 13.3 (Fundamental Homomorphism Theorem). *Let ϕ be a homomorphism of a group G into a group G' with kernel K. Then $G\phi$ is a group, and there is a canonical (natural) isomorphism of $G\phi$ with G/K.*

Proof. We saw in Theorem 13.2 that $G\phi$ is a group, for G is a special case of a subgroup of G. Let $aK \in G/K$, and let us attempt to define a map $\psi: G/K \rightarrow G\phi$ by

$$(aK)\psi = a\phi.$$

We thus define the map ψ on a coset by choosing a representative a of the coset; our first job is to show that ψ is well defined, i.e., independent of the choice of representative. To this end, let $b \in aK$. We must show that $a\phi = b\phi$. But $b \in aK$ means that $b = ak_1$ for $k_1 \in K$ so $a^{-1}b = k_1$. Then

$$e' = k_1\phi = (a^{-1}b)\phi = (a^{-1}\phi)(b\phi) = (a\phi)^{-1}(b\phi).$$

Hence

$$b\phi = (a\phi)e' = a\phi.$$

Thus ψ is well defined.

To show that ψ is one to one, suppose that $(aK)\psi = (bK)\psi$. Then $a\phi = b\phi$, so

$$e' = (a\phi)^{-1}(b\phi) = (a^{-1}\phi)(b\phi) = (a^{-1}b)\phi.$$

Thus $a^{-1}b \in K$, by definition of K. But $a^{-1}b \in K$ implies that $b \in aK$, so $bK = aK$. Thus ψ is one to one.

It is obvious that ψ is onto $G\phi$.

The equation

$$[(aK)(bK)]\psi = (abK)\psi = (ab)\phi = (a\phi)(b\phi) = [(aK)\psi][(bK)\psi]$$

completes the proof that ψ is an isomorphism.

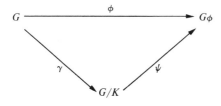

Fig. 13.1

The map ψ is a canonical (or natural) map in the sense that if γ is the canonical homomorphism $\gamma: G \to G/K$ of Theorem 13.1, then

$$\phi = \gamma\psi.$$

One says that the diagram in Fig. 13.1 is *commutative*. ∎

Theorem 13.3 is one which students often have trouble understanding. It really says that for a homomorphism ϕ of a group G, the image, except for the names of the elements, is just G/K, where K is the kernel, and that the homomorphism ϕ is essentially the canonical map $\gamma: G \to G/K$. In other words, Theorem 13.1 describes, in a sense, all homomorphisms.

We tell the student here once and for all that any time you have a homomorphism, two things are of paramount importance: the image and the kernel.

Theorem 13.1 and Theorem 13.3 show that homomorphisms correspond to factor groups in a natural way. Namely, for each factor group G/N, there is a homomorphism $\gamma: G \to G/N$ with kernel N. Conversely, for each homomorphism $\phi: G \to G'$, the image $G\phi$ is essentially G/K, where K is the kernel of ϕ. "Essentially" means up to a canonical isomorphism.

Example 13.3 The student having some knowledge of complex number theory will see that the map $\phi: \mathbf{R} \to \mathbf{C}^*$ given by

$$x\phi = \cos x + i \sin x$$

is a homomorphism, where $\mathbf{R}$ is under addition and $\mathbf{C}^*$ is the multiplicative group of nonzero complex numbers. Note that $\cos x + i \sin x = 1$ if and only if $x = 2\pi n$ for some integer n. Thus the kernel of the homomorphism is the cyclic subgroup $\langle 2\pi \rangle$ of $\mathbf{R}$.

Theorem 13.3 shows that $\mathbf{R}/\langle 2\pi \rangle$ is isomorphic to $\mathbf{R}\phi$, which is the multiplicative group of complex numbers of absolute value 1, that is, the complex numbers on the unit circle. This isomorphism can be visualized geometrically. Every coset of $\mathbf{R}/\langle 2\pi \rangle$ has exactly one representative ≥ 0 and $< 2\pi$. Thus the interval $0 \leq x < 2\pi$ can be visualized as $\mathbf{R}/\langle 2\pi \rangle$, and if bent around so that the *open end* of the interval at 2π joins onto the *closed end* at 0, it forms a circle. The addition in $\mathbf{R}/\langle 2\pi \rangle$, viewed in this way as a circle, is just arc length (or central angle) addition, which is exactly what happens if two complex numbers on the unit circle are multiplied. ‖

13.3 APPLICATIONS

Definition. A **maximal normal subgroup of a group** G is a normal subgroup M not equal to G such that there is no proper normal subgroup N of G properly containing M.

Theorem 13.4 *M is a maximal normal subgroup of G if and only if G/M is simple.*

Proof. Let M be a maximal normal subgroup of G. Consider the canonical homomorphism $\gamma: G \to G/M$ given by Theorem 13.1. Now γ^{-1} of any proper normal subgroup of G/M would be a proper normal subgroup of G properly containing M. But M is maximal so this can't happen. Thus G/M must be simple.

Conversely, Theorem 13.2 shows that if N is a normal subgroup of G properly containing M, then $N\gamma$ is normal in G/M. If also $N \neq G$, then

$$N\gamma \neq G/M \qquad \text{and} \qquad N\gamma \neq \{M\}.$$

Thus, if G/M is simple so that no such $N\gamma$ can exist, no such N can exist, and M is maximal. ∎

Note that a homomorphism ϕ gives an isomorphism of the domain of ϕ with the image of ϕ if and only if ϕ is a one-to-one map.

Theorem 13.5 *A homomorphism ϕ of a group G is a one-to-one function if and only if the kernel of ϕ is $\{e\}$.*

Proof. Of course, if the map ϕ is one to one, the kernel is just $\{e\}$, since we know that $e\phi$ is the identity e' of the image.

Conversely, suppose that the kernel is $\{e\}$. If $a\phi = b\phi$, then

$$e' = e\phi = (a\phi)^{-1}(b\phi) = (a^{-1}\phi)(b\phi) = (a^{-1}b)\phi,$$

so $a^{-1}b$ is in the kernel. Since the kernel is $\{e\}$, we must have $a^{-1}b = e$, so $a = b$. Thus ϕ is one to one. ∎

In view of Theorem 13.5, we revise the list of steps a mathematician usually goes through to exhibit an isomorphism.

STEP 1. *Define the map.*

STEP 2. *Prove that the map is a homomorphism.*

STEP 3. *Prove that the kernel of the map is $\{e\}$. It is then known that the map gives an isomorphism of the domain with the image.*

While we shall not use this terminology, we state for the student's information that a homomorphism $\phi\colon G \to G'$ which is a one-to-one map is a **monomorphism**, and ϕ is an **epimorphism** if it is onto G'.

*13.4   THE FIRST ISOMORPHISM THEOREM

There are several standard *isomorphism theorems*. We prove one of them.

> ***Theorem 13.6*** (First Isomorphism Theorem). *Let H and K be normal subgroups of a group G, with $K \leq H$. Then there is a natural isomorphism of G/H with $(G/K)/(H/K)$.*

Proof. We have seen in Theorem 13.2 that if H is a normal subgroup of G and γ is a homomorphism of G, then $H\gamma$ is a normal subgroup of the image $G\gamma$. Taking $\gamma\colon G \to G/K$ to be the canonical homomorphism of G onto G/K, we see that we can regard $H\gamma$ as a normal subgroup of G/K. Since $K \leq H$, $H\gamma$ is precisely H/K. This is the sense in which $(G/K)/(H/K)$ is to be understood. Now consider the map $\phi\colon G \to (G/K)/(H/K)$ given by

$$a\phi = (aK)(H/K)$$

for $a \in G$. Clearly,

$$(ab)\phi = (abK)(H/K) = [(aK)(bK)](H/K)$$
$$= [(aK)(H/K)][(bK)(H/K)] = (a\phi)(b\phi),$$

so ϕ is a homomorphism. The kernel of ϕ consists of those $x \in G$ such that $x\phi = H/K$. These are just the elements of H. Then Theorem 13.3 shows that G/H is naturally isomorphic to $(G/K)/(H/K)$. ∎

A nice way of viewing Theorem 13.6 is to regard the canonical map $\gamma_H\colon G \to G/H$ as being *factored* via a normal subgroup K of G, $K \leq H \leq G$, to give

$$\gamma_H = \gamma_K \gamma_{H/K}$$

up to a natural isomorphism, as illustrated in Fig. 13.2. Another way of visualizing this theorem is to use the lattice diagram in Fig. 13.3, where each group is a normal subgroup of G and is contained in the one above it. *The larger the normal subgroup, the smaller the factor group.* Thus you can think of G collapsed by H, that is, G/H, as being smaller than G collapsed by K. Theorem 13.6 states that you can collapse G all the way down to G/H in two steps. First collapse to G/K, and then using H/K, collapse this to

$(G/K)/(H/K)$. The total result is the same (up to isomorphism) as collapsing G by H.

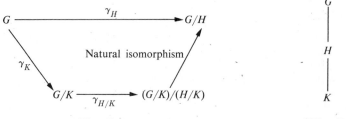

Fig. 13.2   Fig. 13.3

EXERCISES

13.1 Determine which of the following maps are homomorphisms. A superscript * denotes nonzero elements. If the map is a homomorphism, describe the image and the kernel.

a) $\phi: \mathbf{Z} \to \mathbf{R}$ under addition given by $n\phi = n$

b) $\phi: \mathbf{R} \to \mathbf{Z}$ under addition given by $x\phi =$ the greatest integer $\leq x$

c) $\phi: \mathbf{R}^* \to \mathbf{R}^*$ under multiplication given by $x\phi = |x|$

d) $\phi: \mathbf{Z}_6 \to \mathbf{Z}_2$ given by $x\phi =$ the remainder of x when divided by 2, as in Lemma 6.1

e) $\phi: \mathbf{Z}_9 \to \mathbf{Z}_2$ given by $x\phi =$ the remainder of x when divided by 2, in the sense of Lemma 6.1

$\dagger$**13.2** Let a group G be generated by $\{a_i \mid i \in I\}$, where I is some indexing set and $a_i \in G$. Let $\phi: G \to G'$ be a homomorphism of G into a group G'. Show that the value of ϕ on every element of G is completely determined by the values $a_i\phi$. Thus, for example, a homomorphism of a cyclic group is completely determined by the value of the homomorphism on a generator of the group. [*Hint:* Use Theorem 9.1 and, of course, the definition of a homomorphism.]

13.3 How many homomorphisms are there of $\mathbf{Z}$ onto $\mathbf{Z}$? of $\mathbf{Z}$ into $\mathbf{Z}_2$? of $\mathbf{Z}$ onto $\mathbf{Z}_2$? [*Hint:* Use Exercise 13.2. See also Exercise 13.11.]

13.4 How many homomorphisms are there of $\mathbf{Z}$ into $\mathbf{Z}_8$? of $\mathbf{Z}$ onto $\mathbf{Z}_8$? [*Hint:* Use Exercise 13.2. See also Exercise 13.11.]

13.5 How many homomorphisms are there of $\mathbf{Z}_{12}$ onto $\mathbf{Z}_5$? of $\mathbf{Z}_{12}$ into $\mathbf{Z}_6$? of $\mathbf{Z}_{12}$ onto $\mathbf{Z}_6$? of $\mathbf{Z}_{12}$ into $\mathbf{Z}_{14}$? of $\mathbf{Z}_{12}$ into $\mathbf{Z}_{16}$? [*Hint:* Use Exercise 13.2.]

13.6 What can one say about homomorphisms of a simple group?

13.7 Mark each of the following true or false.

___ a) A_n is a normal subgroup of S_n.

___ b) Every isomorphism is also a homomorphism.

___ c) Every homomorphism is an isomorphism.

___ d) A homomorphism is an isomorphism of the domain with the image if and only if the kernel consists of the group of the identity element alone.

___ e) The image of a group of 6 elements under some homomorphism may have 4 elements.

— f) The image of a group of 6 elements under a homomorphism may have 12 elements.

— g) There is a homomorphism of some group of 6 elements into some group of 12 elements.

— h) There is a homomorphism of some group of 6 elements into some group of 10 elements.

— i) All homomorphisms of a group of prime order are in some sense trivial.

— j) It is not possible to have a homomorphism of some infinite group into some finite group.

13.8 How many homomorphisms are there of $Z_2 \times Z_2$ into Z_2? of $Z_2 \times Z_2$ onto Z_2? of $Z_2 \times Z_2$ into Z_6? of $Z_2 \times Z_2$ into $Z_2 \times Z_2 \times Z_2$? of $Z_2 \times Z_2$ into $Z_2 \times Z_2 \times Z_4$? [*Hint:* Use Exercise 13.2.]

13.9 The **sign of an even permutation** is $+1$ and the **sign of an odd permutation** is -1. Observe that the map $\mathrm{sgn}_n: S_n \to \{1, -1\}$ defined by

$$\mathrm{sgn}_n(\sigma) = \text{sign of } \sigma$$

is a homomorphism of S_n onto the group $\{1, -1\}$ under multiplication. What is the kernel?

13.10 For groups G_1 and G_2, consider the map $\pi_1: (G_1 \times G_2) \to G_1$ given by $(x, y)\pi_1 = x$. Show that π_1 is a homomorphism. What is the kernel? To what group is the kernel isomorphic?

13.11 Let G be any group, and let a be any element of G. Let $\phi: Z \to G$ be defined by $n\phi = a^n$. Show that ϕ is a homomorphism. Describe the image and the possibilities for the kernel of ϕ. [*Comment:* Using this exercise and Theorem 13.3, we get the *elegant proof* that every cyclic group is isomorphic to Z/nZ for some nonnegative integer n, and also the *elegant proof* that every element of a group generates a cyclic subgroup of the group.]

13.12 Let G be a finite abelian group of order n, and let r be a positive integer relatively prime to n.

a) Show that the map $\phi_r: G \to G$ given by $a\phi_r = a^r$ is an isomorphism of G onto itself. (Follow the outline given after Theorem 13.5.)

b) Deduce that the equation $x^r = a$ always has a unique solution in a finite abelian group G if r is relatively prime to the order of G. What can happen if r is not relatively prime to the order of G?

13.13 Show that if G, G', and G'' are groups and if $\phi: G \to G'$ and $\psi: G' \to G''$ are homomorphisms, then the composite function $\phi\psi: G \to G''$ is a homomorphism.

13.14 Let G be a group, and let $\mathcal{I}_G$ be the group of inner automorphisms of G given by Exercise 12.16. Show that the map $\phi: G \to \mathcal{I}_G$ given by $g\phi = i_{g^{-1}}$ is a homomorphism of G onto $\mathcal{I}_G$. Show that the kernel (the **center of** G) is

$$\{a \in G \mid ax = xa \text{ for all } x \in G\}.$$

Describe when ϕ is an isomorphism.

13.15 Let G_1 and G_2 be groups, and let $\phi_1: G_1 \to G_2$ and $\phi_2: G_2 \to G_1$ be homomorphisms such that $\phi_1\phi_2 = \phi_2\phi_1 = \iota$, where ι is the identity map, that is,

$\phi_1\phi_2\colon G_1 \to G_1$ and $\phi_2\phi_1\colon G_2 \to G_2$ are both the identity map. Show that both ϕ_1 and ϕ_2 are isomorphisms of G_1 with G_2, and that $\phi_1 = (\phi_2)^{-1}$.

13.16 Let G and G' be groups, and let H and H' be normal subgroups of G and G', respectively. Let ϕ be a homomorphism of G into G'. Show that ϕ induces a natural homomorphism $\phi_*\colon (G/H) \to (G'/H')$ if $H\phi \subseteq H'$. (This fact is used constantly in algebraic topology.)

*13.17 (*Second Isomorphism Theorem*). Let G be a group and N and H two sub-groups of G with N a normal subgroup of G. Show that $(H \vee N)/N$ is isomorphic to $H/(H \cap N)$. [*Hint:* Recall from Exercise 12.19 that $H \vee N = HN$. Define $\phi\colon HN \to H/(H \cap N)$ by $(hn)\phi = h(H \cap N)$. Show that ϕ is well defined and is a homomorphism onto $H/(H \cap N)$; find the kernel and apply Theorem 13.3.]

14 | Series of Groups

14.1 SUBNORMAL AND NORMAL SERIES

This section and the next are concerned with more results which give insight into the structure of groups. These will be of interest chiefly in the nonabelian case. They also hold for abelian groups, but are not too important for us there because of our strong Theorem 9.3. Many of our illustrations will be taken from abelian groups, however, for ease of computation.

> **Definition.** A *subnormal* (or *subinvariant*) *series of a group* G is a finite sequence $H_0, H_1, \ldots, H_n$ of subgroups of G such that $H_i < H_{i+1}$ and H_i is a normal subgroup of H_{i+1} with $H_0 = \{e\}$ and $H_n = G$. A *normal* (or *invariant*) *series of* G is a finite sequence $H_0, H_1, \ldots, H_n$ of normal subgroups of G such that $H_i < H_{i+1}$, $H_0 = \{e\}$, and $H_n = G$.

Note that for abelian groups the notions of subnormal and normal series coincide, since every subgroup is normal. A normal series is always subnormal, but the converse need not be true. We defined a subnormal series before a normal series, since the concept of a subnormal series is more important for our work.

Example 14.1 Two examples of normal series of $\mathbf{Z}$ under addition are

$$\{0\} < 8\mathbf{Z} < 4\mathbf{Z} < \mathbf{Z}$$

and

$$\{0\} < 9\mathbf{Z} < \mathbf{Z}. \;\|$$

Example 14.2 Consider the group D_4 of symmetries of the square in Example 4.2. The series

$$\{\rho_0\} < \{\rho_0, \mu_1\} < \{\rho_0, \rho_2, \mu_1, \mu_2\} < D_4$$

is a subnormal series, as the student can easily check. It is not a normal series, since $\{\rho_0, \mu_1\}$ is not normal in D_4. $\|$

> **Definition.** A subnormal (normal) series $\{K_j\}$ is a *refinement of a subnormal* (*normal*) *series* $\{H_i\}$ of a group G if $\{H_i\} \subseteq \{K_j\}$, that is, if each H_i is one of the K_j.

Example 14.3 The series

$$\{0\} < 72\mathbf{Z} < 24\mathbf{Z} < 8\mathbf{Z} < 4\mathbf{Z} < \mathbf{Z}$$

is a refinement of the series

$$\{0\} < 72\mathbf{Z} < 8\mathbf{Z} < \mathbf{Z}.$$

Two new terms, $4\mathbf{Z}$ and $24\mathbf{Z}$, have been inserted. ∥

Of interest in studying the structure of G are the factor groups H_{i+1}/H_i. These are defined for both normal and subnormal series, since H_i is normal in H_{i+1} in either case.

Definition. Two subnormal (normal) series $\{H_i\}$ and $\{K_j\}$ of the same group G are *isomorphic* if there is a one-to-one correspondence between the collections of factor groups $\{H_{i+1}/H_i\}$ and $\{K_{j+1}/K_j\}$ such that corresponding factor groups are isomorphic.

Clearly, two isomorphic subnormal (normal) series must have the same number of groups.

Example 14.4 The two series of $\mathbf{Z}_{15}$,

$$\{0\} < \langle 5 \rangle < \mathbf{Z}_{15}$$

and

$$\{0\} < \langle 3 \rangle < \mathbf{Z}_{15},$$

are isomorphic. Both $\mathbf{Z}_{15}/\langle 5 \rangle$ and $\langle 3 \rangle/\{0\}$ are isomorphic to $\mathbf{Z}_5$, and $\mathbf{Z}_{15}/\langle 3 \rangle$ is isomorphic to $\langle 5 \rangle/\{0\}$, or to $\mathbf{Z}_3$. ∥

14.2   THE JORDAN-HÖLDER THEOREM

The following theorem is fundamental to the theory.

Theorem 14.1 *Two subnormal (normal) series of a group G have isomorphic refinements.*

Proof. See the literature. ∎

The proof of Theorem 14.1 is really not too difficult. However, the writer knows from experience that some students get lost in the proof and then tend to feel that they can't understand the theorem. We don't prove it, even though many students could follow it. Let us, however, illustrate this theorem.

Example 14.5 Let us try to find isomorphic refinements of the series

$$\{0\} < 8\mathbf{Z} < 4\mathbf{Z} < \mathbf{Z}$$

and

$$\{0\} < 9\mathbf{Z} < \mathbf{Z}$$

given in Example 14.1. Consider the refinement

$$\{0\} < 72\mathbf{Z} < 8\mathbf{Z} < 4\mathbf{Z} < \mathbf{Z}$$

of $\{0\} < 8\mathbf{Z} < 4\mathbf{Z} < \mathbf{Z}$ and the refinement

$$\{0\} < 72\mathbf{Z} < 18\mathbf{Z} < 9\mathbf{Z} < \mathbf{Z}$$

of $\{0\} < 9\mathbf{Z} < \mathbf{Z}$. In both cases the refinements have four factor groups isomorphic to $\mathbf{Z}_4$, $\mathbf{Z}_2$, $\mathbf{Z}_9$, and $72\mathbf{Z}$ or $\mathbf{Z}$. The *order* in which the factor groups occur is different to be sure. ‖

We now come to the real meat of the theory.

Definition. A subnormal series $\{H_i\}$ of a group G is a **composition series** if all the factor groups H_{i+1}/H_i are simple. A normal series $\{H_i\}$ of G is a **principal** or **chief series** if all the factor groups H_{i+1}/H_i are simple.

Note that for abelian groups, the concepts of composition and principal series coincide. Also, since every normal series is subnormal, every principal series is a composition series for any group, abelian or not.

Example 14.6 We claim that $\mathbf{Z}$ has no composition (and also no principal) series. For if

$$\{0\} = H_0 < H_1 < \cdots < H_{n-1} < H_n = \mathbf{Z}$$

is a subnormal series, H_1 must be of the form $r\mathbf{Z}$ for some $r \in \mathbf{Z}^+$. But then H_1/H_0 is isomorphic to $r\mathbf{Z}$, which is infinite cyclic with many proper normal subgroups, for example $2r\mathbf{Z}$. Thus $\mathbf{Z}$ has no composition (and also no principal) series. ‖

Example 14.7 The series

$$\{e\} < A_n < S_n$$

for $n \geq 5$ is a composition series (and also a principal series) of S_n, since $A_n/\{e\}$ is isomorphic to A_n, which is simple for $n \geq 5$, and S_n/A_n is isomorphic to $\mathbf{Z}_2$, which is simple. Likewise the two series given in Example 14.4 are composition series (and also principal series) of $\mathbf{Z}_{15}$. They are isomorphic, as shown in that example. This illustrates our main theorem, which will be stated shortly. ‖

Observe that by Theorem 13.4 H_{i+1}/H_i is simple if and only if H_i is a maximal normal subgroup of H_{i+1}. Thus for a composition series, each H_i must be a maximal normal subgroup of H_{i+1}. *To form a composition series of a group G, we just hunt for a maximal normal subgroup H_{n-1} of G, then for a maximal normal subgroup of H_{n-1}, etc. If this process terminates in a finite number of steps, we have a composition series.* Note that by Theorem 13.4, a composition series cannot have any further refinement. *To form a principal series, we have to hunt for a maximal normal subgroup H_{n-1} of G, then for a maximal normal subgroup of H_{n-1} which is also normal in G, etc.* The main theorem is as follows.

Theorem 14.2 (Jordan-Hölder). *Any two composition (principal) series of a group G are isomorphic.*

Proof. Let $\{H_i\}$ and $\{K_j\}$ be two composition (principal) series of G. By Theorem 14.1, they have isomorphic refinements. But since all factor groups are already simple, Theorem 13.4 shows that neither series has any further refinement. Thus $\{H_i\}$ and $\{K_j\}$ must already be isomorphic. ∎

Theorem 14.3 *If G has a composition (principal) series, and if N is a proper normal subgroup of G, then there exists a composition (principal) series containing N.*

Proof. The series

$$\{e\} < N < G$$

is both a subnormal and a normal series. Since G has a composition series $\{H_i\}$, then by Theorem 14.1, there is a refinement of $\{e\} < N < G$ to a subnormal series isomorphic to a refinement of $\{H_i\}$. But as a composition series, $\{H_i\}$ can have no further refinement. Thus $\{e\} < N < G$ can be refined to a subnormal series all of whose factor groups are simple, i.e., to a composition series. A similar argument holds if we start with a principal series $\{K_j\}$ of G. ∎

Example 14.8 A composition (and also a principal) series of $\mathbf{Z}_4 \times \mathbf{Z}_9$ containing $\langle (0, 1) \rangle$ is

$$\{(0, 0)\} < \langle (0, 3) \rangle < \langle (0, 1) \rangle < \langle 2 \rangle \times \langle 1 \rangle < \langle 1 \rangle \times \langle 1 \rangle = \mathbf{Z}_4 \times \mathbf{Z}_9. \;\|$$

The next definition is basic to the last section of this text, which deals with solutions of polynomial equations in terms of radicals.

Definition. A group G is ***solvable*** if it has a composition series $\{H_i\}$ such that all factor groups H_{i+1}/H_i are abelian.

By the Jordan-Hölder theorem, we see that for a solvable group, *every* composition series $\{H_i\}$ must have abelian factor groups H_{i+1}/H_i.

Example 14.9 The group S_3 is solvable, for the composition series

$$\{e\} < A_3 < S_3$$

has factor groups isomorphic to $\mathbf{Z}_3$ and $\mathbf{Z}_2$, which are abelian. The group S_5 is not solvable, for since A_5 is simple, the series

$$\{e\} < A_5 < S_5$$

is a composition series, and $A_5/\{e\}$, which is isomorphic to A_5, is not abelian. *This group A_5 of order 60 can be shown to be the smallest group which is not solvable.* This fact is closely connected with the fact that a polynomial equation of degree 5 is not in general solvable by radicals, but a polynomial equation of degree ≤ 4 is. $\|$

*14.3 THE CENTER AND THE ASCENDING CENTRAL SERIES

For no good reason, we give one other type of series for a group. However, the concept of the *center* of a group will play a role in the next section, so you should not skip the following definition and Theorem 14.4, if you wish to understand the Sylow theory.

Definition. The *center of a group* G is the set of all $a \in G$ such that $ax = xa$ for all $x \in G$, that is, the set of all elements of G which commute with every element of G.

Theorem 14.4 *The center of a group is a normal subgroup of the group.*

Proof. The proof of this is so easy and instructive that we save it for an exercise in this section. ∎

If you have the table for a finite group G, it is easy to find the center. An element a will clearly be in the center of G if and only if the elements in the row opposite a at the extreme left are given in the same order as the elements in the column under a at the very top of the table.

Now let G be a group, and let $Z(G)$ be the center of G. Since by Theorem 14.4, $Z(G)$ is normal in G, we can form the factor group $G/Z(G)$ and find the center $Z(G/Z(G))$ of this factor group. Since $Z(G/Z(G))$ is normal in $G/Z(G)$, if $\gamma: G \to G/Z(G)$ is the canonical map, then by Theorem 13.2, $[Z(G/Z(G))]\gamma^{-1}$ is a normal subgroup $Z_1(G)$ of G. We can then form the factor group $G/Z_1(G)$ and find its center, take $(\gamma_1)^{-1}$ of it to get $Z_2(G)$, etc.

Definition. The series

$$\{e\} \leq Z(G) \leq Z_1(G) \leq Z_2(G) \leq \cdots$$

described in the preceding discussion is the *ascending central series of the group* G.

Example 14.10 The center of S_3 is just the identity $\{\rho_0\}$. Thus the ascending central series of S_3 is

$$\{\rho_0\} \leq \{\rho_0\} \leq \{\rho_0\} \leq \cdots$$

The center of the group D_4 of symmetries of the square in Example 4.2 is $\{\rho_0, \rho_2\}$. (Do you remember that we said that this group would give us nice examples of almost anything we discussed?) Since $G/\{\rho_0, \rho_2\}$ is of order 4 and hence abelian, its center is all of $G/\{\rho_0, \rho_2\}$. Thus the ascending central series of D_4 is

$$\{\rho_0\} \leq \{\rho_0, \rho_2\} \leq D_4 \leq D_4 \leq D_4 \leq \cdots \; \|$$

EXERCISES

14.1 Give isomorphic refinements of the two normal series $\{0\} < 60\mathbf{Z} < 20\mathbf{Z} < \mathbf{Z}$ and $\{0\} < 245\mathbf{Z} < 49\mathbf{Z} < \mathbf{Z}$ of $\mathbf{Z}$ under addition.

14.2 Find all composition series of Z_{60} and show that they are indeed all isomorphic.

14.3 Find all composition series of $Z_5 \times Z_5$.

14.4 Find all composition series of $S_3 \times Z_2$.

†14.5 Show that if

$$H_0 = \{e\} < H_1 < H_2 < \cdots < H_n = G$$

is a subnormal (normal) series for a group G, and if H_{i+1}/H_i is of finite order s_{i+1}, then G is of finite order $s_1 s_2 \cdots s_n$.

14.6 Mark each of the following true or false.

___ a) Every normal series is also subnormal.

___ b) Every subnormal series is also normal.

___ c) Every principal series is a composition series.

___ d) Every composition series is a principal series.

___ e) Every abelian group has exactly one composition series.

___ f) Every finite group has a composition series.

___ g) A group is solvable if and only if it has a composition series with simple factor groups.

___ h) S_7 is a solvable group.

___ i) The Jordan-Hölder theorem has some similarity with the *Fundamental Theorem of Arithmetic*, which states that every positive integer greater than 1 can be factored into a product of primes uniquely up to order.

___ j) Every finite group of prime order is solvable.

14.7 Show that an infinite abelian group can have no composition series. [*Hint:* Use Exercise 14.5, together with the fact that an infinite abelian group always has a proper normal subgroup.]

14.8 Find a composition series of $S_3 \times S_3$. Is $S_3 \times S_3$ solvable?

14.9 Show that a finite direct product of solvable groups is solvable.

14.10 Is the group D_4 of symmetries of the square in Example 4.2 solvable?

*14.11 Find the center of $S_3 \times Z_4$.

*14.12 Prove that the center of a group is a normal subgroup of the group. (*Warning:* Don't forget that you have to prove that it is a subgroup before you can prove that it is normal.)

*14.13 Find the ascending central series of $S_3 \times Z_4$.

*14.14 Show that a subgroup K of a solvable group G is solvable. [*Hint:* Let $H_0 = \{e\} < H_1 < \cdots < H_n = G$ be a composition series for G. Show that the distinct groups among $K \cap H_i$ for $i = 0, \ldots, n$ form a composition series for K. Observe that

$$(K \cap H_i)/(K \cap H_{i-1}) \simeq [H_{i-1}(K \cap H_i)]/[H_{i-1}],$$

by Exercise 13.17, with $H = K \cap H_i$ and $N = H_{i-1}$, and that $H_{i-1}(K \cap H_i) \leq H_i$.]

*14.15 Let $H_0 = \{e\} < H_1 < \cdots < H_n = G$ be a composition series for a group G. Let N be a normal subgroup of G, and suppose that N is a simple group.

Show that the distinct groups among H_0, H_iN for $i = 0, \ldots, n$ also form a composition series for G. [*Hint:* H_iN is a group by Exercise 12.19. Show that $H_{i-1}N$ is normal in H_iN. By Exercise 13.17,

$$(H_iN)/(H_{i-1}N) \simeq H_i/[H_i \cap (H_{i-1}N)],$$

and the latter group is isomorphic to

$$[H_i/H_{i-1}]/[(H_i \cap (H_{i-1}N))/H_{i-1}],$$

by Theorem 13.6. But H_i/H_{i-1} is simple.]

*14.16 Let G be a group, and let $H_0 = \{e\} < H_1 < \cdots < H_n = G$ be a composition series for G. Let N be a normal subgroup of G, and let $\gamma \colon G \to G/N$ be the canonical map. Show that the distinct groups among $H_i\gamma$ for $i = 0, \ldots, n$ form a composition series for G/N. [*Hint:* Observe that the map

$$\psi \colon H_iN \to (H_i\gamma)/(H_{i-1}\gamma)$$

defined by

$$(h_in)\psi = ((h_in)\gamma)(H_{i-1}\gamma)$$

is a homomorphism with kernel $H_{i-1}N$. By Theorem 13.3,

$$(H_i\gamma)/(H_{i-1}\gamma) \simeq (H_iN)/(H_{i-1}N).$$

Proceed via Exercise 13.17, as shown in the hint for Exercise 14.15.]

*14.17 Prove that a homomorphic image of a solvable group is solvable. [*Hint:* Apply Exercise 14.16 to get a composition series for the homomorphic image. The hints of Exercises 14.15 and 14.16 then show how the factor groups of this composition series in the image look.]

15 | The Sylow Theorems

In this section we present three theorems, Theorems 15.2, 15.3, and 15.4, which are basic to the study of the structure of finite groups. While it is our intent to prove Theorem 15.2 and part of Theorem 15.3, the proofs giving a further indication of the power and uses of the factor group concept, the average student should not feel upset if he gets lost in the details of the proofs. We believe that it is even advisable for the student to read the statements of these theorems and then the applications and examples before he attempts the proofs, if he attempts them at all. The author feels that an understanding of these results and their applications is more important than a formal following through of the proofs. For these reasons, the proofs are placed at the end of the section.

*15.1 CONJUGATE CLASSES AND THE CLASS EQUATION

Definition. Let a be any element of a group G. An element b of G is *conjugate to* a if $b = xax^{-1}$ for some $x \in G$.

We shall now show that the relation of conjugacy is an equivalence relation on G. In view of Theorem 0.1, conjugacy thus yields a partition of G into cells; two elements are in the same cell if and only if they are conjugate.

Theorem 15.1 *Let G be a group. The relation $a \sim b$ if and only if b is conjugate to a is an equivalence relation.*

Proof. We check the three conditions for an equivalence relation.

Reflexive: $a = eae^{-1}$ so $a \sim a$.

Symmetric: If $a \sim b$, then $b = xax^{-1}$ for some $x \in G$. Then $x^{-1}bx = x^{-1}b(x^{-1})^{-1} = a$ and $x^{-1} \in G$, so $b \sim a$.

Transitive: If $a \sim b$ and $b \sim c$, then $b = xax^{-1}$ and $c = yby^{-1}$ for some $x, y \in G$. Then $c = y(xax^{-1})y^{-1} = (yx)a(yx)^{-1}$ and $yx \in G$, so $a \sim c$. ∎

Definition. The equivalence class containing an element a of a group G under the relation of conjugacy is the *conjugate class $C[a]$ of a*.

Of course, $C[a] = \{xax^{-1} \mid x \in G\}$.

128

Example 15.1 In an abelian group,

$$xax^{-1} = xx^{-1}a = ea = a,$$

so $C[a] = \{a\}$. The student can check that in S_3 of Example 4.1,

$$C[\mu_1] = \{\mu_1, \mu_2, \mu_3\}. \parallel$$

Note that $ax = xa$ for all $x \in G$ if and only if $xax^{-1} = a$ for all $x \in G$. Thus a is in the center of G if and only if $C[a] = \{a\}$.

Example 15.2 It is easy to check that for S_3 of Example 4.1, the conjugate classes are

$$\{\rho_0\}, \quad \{\rho_1, \rho_2\}, \quad \{\mu_1, \mu_2, \mu_3\}. \parallel$$

We come now to our first main theorem. Remember that $C[a] = \{a\}$ if and only if a is in the center of the group G. Example 15.2 illustrates that the conjugate classes of a group need not contain the same number of elements. It also illustrates the important fact that the number of elements in each conjugate class divides the order of the group if the group is finite. Of course if G is finite, then the order of G is the sum of the numbers of elements in the conjugate classes. We summarize all these facts in a theorem.

Theorem 15.2 *Let G be a group of finite order n, and let c be the number of elements in the center of G. Suppose furthermore that in addition to the trivial conjugate classes of one element each, forming the center, there are r other conjugate classes each of which contains more than one element. Let the number of elements in the ith such class be n_i for $i = 1, \ldots, r$. Then*

$$n = c + n_1 + n_2 + \cdots + n_r,$$

and c and each n_i for $i = 1, \ldots, r$ are divisors of n.

The equation $n = c + n_1 + n_2 + \cdots + n_r$ of Theorem 15.2 is the **class equation of G**. Since the order c of the center divides the order of a finite group by the theorem of Lagrange, all we must do to complete the proof of Theorem 15.2 is to show that n_i divides n for $i = 1, \ldots, r$. This is deferred until the end of the section.

Example 15.3 Example 15.2 showed that the class equation of S_3 is

$$6 = 1 + 2 + 3. \parallel$$

*15.2 THE SYLOW THEOREMS

The next theorem is obvious for abelian groups, by Theorem 9.3. The theorem holds for any finite group, and it becomes our strongest theorem on the existence of certain types of subgroups of a general finite group. A partial proof of this theorem is given at the end of this section.

Theorem 15.3 (First Sylow Theorem). *Let G be a group of finite order $n = p^r m$, where p is a prime number and p does not divide m. Then G contains a subgroup H_i of every order p^i for $1 \leq i \leq r$. Also, every subgroup K_i of G of order p^i is a normal subgroup of some subgroup K_{i+1} of order p^{i+1} for $1 \leq i < r$.*

Theorem 15.4 (Second and Third Sylow Theorems). *If the order of a finite group G is $p^r m$, where p is a prime not dividing m, then all subgroups of G of order p^r are conjugate, and the number of them is congruent to 1 modulo p and divides the order of G.*

Proof. See the literature. ∎

Note that Theorem 15.4 is basically a counting theorem as is Theorem 15.2. *Never underestimate a theorem which counts something.*

Example 15.4 The subgroups of order 2 of S_3 in Example 4.1 are

$$\{\rho_0, \mu_1\}, \quad \{\rho_0, \mu_2\}, \quad \{\rho_0, \mu_3\}.$$

Note that there are 3 of them and that $3 \equiv 1 \pmod 2$, that is, 3 and 1 have the same remainder when divided by 2. Also 3 divides 6, the order of S_3. It is easily seen that

$$\{\rho_0, \mu_1\} i_{\rho_1} = \{\rho_0, \mu_3\} \qquad \text{and} \qquad \{\rho_0, \mu_1\} i_{\rho_2} = \{\rho_0, \mu_2\},$$

where $x i_{\rho_j} = \rho_j x (\rho_j)^{-1}$, thus illustrating that they are all conjugate. ‖

Example 15.5 Let us use these Sylow theorems to show that no group of order 15 is simple. Let G have order 15. We claim that G has a normal subgroup of order 5. By Theorem 15.3, G has at least one subgroup of order 5, and by Theorem 15.4, the number of such subgroups is congruent to 1 modulo 5 and divides 15. Since 1, 6, and 11 are the only positive numbers less than 15 which are congruent to 1 modulo 5, and since among these only the number 1 divides 15, we see that G has exactly one subgroup H of order 5. But for each $g \in G$, the inner automorphism i_g of G with $x i_g = g x g^{-1}$ maps H onto a subgroup $g H g^{-1}$ again of order 5. Hence we must have $g H g^{-1} = H$ for all $g \in G$, so H is a normal subgroup of G. Therefore, G is not simple. ‖

We trust that Example 15.5 gives the student some inkling of the power of Theorems 15.3 and 15.4. *Never underestimate a theorem which counts something.*

*15.3 APPLICATIONS TO p-GROUPS

We give some standard definitions and some applications of these important results. More applications will appear in the next section.

Definition. A group G is a ***p-group*** if every element in G has order a power of the prime p. A subgroup of a group G is a ***p-subgroup of G*** if the subgroup is itself a p-group.

Definition. A *Sylow p-subgroup of a group* G is a maximal p-subgroup of G, that is, a p-subgroup contained in no larger p-subgroup of G.

Theorem 15.5 *A finite group* G *is a p-group if and only if its order is a power of* p.

Proof. If G is a p-group, let the order of G be $p^r m$ for p not dividing m. If q were a prime dividing m, then by Theorem 15.3, G would have a subgroup of prime order q which would be a cyclic subgroup with an element of order $q \neq p$ as generator. This would contradict our assumption that G is a p-group. Thus we must have $m = 1$, and the order of G is p^r.

Conversely, if G has order a power of p, then the fact that the order of an element divides the order of the group (Theorem 11.5) shows that G is a p-group. ∎

Corollary. *For a group* G *of finite order* n, *the Sylow p-subgroups are exactly the subgroups of order* p^r, *where* p^r *is the highest power of* p *dividing* n.

Proof. By Theorem 15.5, the Sylow p-subgroups are among the subgroups of G of order a power of p. Theorem 15.3 shows that a maximal such subgroup has order p^r. ∎

*15.4   SOME PROOFS

We shall prove Theorem 15.2 and part of Theorem 15.3. Our task in proving Theorem 15.2 is to show that the number of elements in each conjugate class of a group divides the order of the group if the group has finite order (see Example 15.1 and Example 15.2). Our technique will be to show that the number of elements in a conjugate class is the same as the index of a certain subgroup in the whole group. The next definition and theorem get hold of the subgroup for us.

Definition. If a is an element of a group G, the *normalizer* $N[a]$ *of* a *in* G is $\{x \in G \mid xax^{-1} = a\}$, that is, the set of all $x \in G$ such that a is left fixed under the inner automorphism $i_x : G \to G$ of Theorem 12.2.

Theorem 15.6 *If* a *is an element of a group* G, *then* $N[a]$ *is a subgroup of* G. *Also* $xax^{-1} = yay^{-1}$ *for* x *and* y *in* G *if and only if* x *and* y *are in the same left coset of* $N[a]$.

Proof. If b and c are in $N[a]$, then $bab^{-1} = a$ and $cac^{-1} = a$. Thus, by substituting, we get $b(cac^{-1})b^{-1} = a$, so

$$(bc)a(bc)^{-1} = a,$$

that is, $bc \in N[a]$. Also $eae^{-1} = a$, showing that $e \in N[a]$. Note that if $b \in N[a]$, then $bab^{-1} = a$, so $ba = ab$ and

$$a = b^{-1}ab = b^{-1}a(b^{-1})^{-1}.$$

Thus $b \in N[a]$ implies that $b^{-1} \in N[a]$. Therefore $N[a]$ is a subgroup of G.

Suppose that $xax^{-1} = yay^{-1}$ for certain elements x and y in G. Then $y^{-1}xax^{-1}y = a$, so

$$(y^{-1}x)a(y^{-1}x)^{-1} = a,$$

that is, $y^{-1}x \in N[a]$. But then $x \in yN[a]$, so x and y are in the same left coset of $N[a]$.

Conversely, suppose that x and y are in the same left coset $sN[a]$ of $N[a]$. Then $x = sb$ and $y = sc$ for b and c in $N[a]$. Therefore,

$$xax^{-1} = (sb)a(sb)^{-1} = sbab^{-1}s^{-1} = sas^{-1}$$

for $bab^{-1} = a$, since $b \in N[a]$. Likewise $yay^{-1} = sas^{-1}$, so $xax^{-1} = yay^{-1}$ if x and y are in the same left coset. $\blacksquare$

Corollary (Proof of Theorem 15.2). *If a is an element of a finite group G, then the number of elements in $C[a]$ is a divisor of the order of G.*

Proof. Theorem 15.6 shows that the number of elements in $C[a]$ is equal to the number of left cosets of the subgroup $N[a]$, that is, the index $(G : N[a])$. That this is a divisor of the order of G follows at once from the theorem of Lagrange applied to $N[a]$ as a subgroup of G. $\blacksquare$

If one defined

$$N[S] = \{x \in G \mid xSx^{-1} = S\}$$

for a subset S of a group G, then Theorem 15.6 would be valid with a replaced by S. In particular, if H is a subgroup of G, then $N[H]$ is clearly the largest possible subgroup of G having H as normal subgroup. This is the reason for the name "*normalizer*."

Partial Proof of Theorem 15.3. We shall prove that a finite group G of order $p^r m$, where p does not divide m, contains a chain $H_1 < H_2 < \cdots < H_r$ of subgroups, where H_i is of order p^i and is normal in H_{i+1} for $1 \leq i < r$. We will *not* show that *every* subgroup K_i of order p^i for $1 \leq i < r$ is a normal subgroup of some group K_{i+1} of order p^{i+1}. See Hungerford [9], pp. 88–96, for elegant proofs of the Sylow theorems. We use induction on the order n of G. For $n < 6$, G is abelian and the theorem follows from Theorem 9.3. Let us suppose that the theorem is true for all groups of order less than n. Let c be the order of the center $Z(G)$ of G. We consider two cases.

CASE I. Suppose that p divides c. Since $Z(G)$ is abelian, Theorem 9.3 shows that $Z(G)$ has a cyclic subgroup P of order p. Also $P \leq Z(G)$ shows that P is normal in G. Then G/P is of order n/p, and by the induction hypothesis, our theorem is true in G/P, that is, there exist subgroups H'_i of order p^i for $1 \leq i < r$ in G/P, with H'_i normal in H'_{i+1} for $1 \leq i < r - 1$. If $\gamma: G \to G/P$ is the canonical homomorphism, then by Theorem 13.2, the groups $(H'_{i-1})\gamma^{-1} = H_i$ give our desired sequence of subgroups H_i of G, for $|H_i| = p^i$, since P is of order p. We take $H_1 = P$ in this case.

CASE II. Suppose p does not divide c. Then the class equation

$$n = c + n_1 + n_2 + \cdots + n_r$$

of Theorem 15.2 shows that p can't divide n_j for some j, for if p were to divide all n_j and n, then p would divide c. Let $h \in G$ be such that p does not divide the order $n_j > 1$ of $C[h]$, the conjugate class of h. But $n_j = (G : N[h])$ and $|G| = (G : N[h])|N[h]|$, so p^r divides the order of $N[h]$. Since $n_j > 1$, $N[h]$ is of order $< n$, and by our induction hypothesis, it has the required subgroups H_i, which are of course also subgroups of G. ∎

EXERCISES

*15.1 Fill in the blanks.

a) A Sylow 3-subgroup of a group of order 12 has order _____.
b) A Sylow 3-subgroup of a group of order 54 has order _____.
c) A group of order 24 must have either _____ or _____ Sylow 2-subgroups. (Use only the information given in Theorem 15.4.)
d) A group of order $255 = (3)(5)(17)$ must have either _____ or _____ Sylow 3-subgroups and _____ or _____ Sylow 5-subgroups. (Use only the information given in Theorem 15.4.)
e) A group of order 215 can have at most _____ conjugate classes.

*15.2 Let D_4 be the group of symmetries of the square in Example 4.2.

a) Find the decomposition of D_4 into conjugate classes.
b) Write the class equation for D_4.

*15.3 Find all Sylow 3-subgroups of S_4 and demonstrate that they are all conjugate.

†*15.4 Prove Cauchy's theorem, which states that if a prime p divides the order of a finite group G, then G has an element of order p. You may use any theorems we have given.

*15.5 Show that every group of order 45 has a normal subgroup of order 9.

*15.6 Mark each of the following true or false.

__ a) Any two Sylow p-subgroups of a finite group are conjugate.
__ b) Theorem 15.4 shows that a group of order 15 has only one Sylow 5-subgroup.
__ c) Every Sylow p-subgroup of a finite group has order a power of p.
__ d) Every p-subgroup of every finite group is a Sylow p-subgroup.
__ e) Every finite abelian group has exactly one Sylow p-subgroup for each prime p dividing the order of G.
__ f) The conjugate class of every element of every group is always a subgroup of the group.
__ g) A finite group of order n is abelian if and only if it has n conjugate classes.
__ h) A Sylow p-subgroup of a finite group G is normal in G if and only if it is the only Sylow p-subgroup of G.
__ i) The elements in the center of a finite group G are all conjugate to each other.
__ j) The concept of a conjugate class is only defined for a finite group.

*15.7 Find two Sylow 2-subgroups of S_4 and show that they are conjugate.

*15.8 Show that every group of order $(35)^3$ has a normal subgroup of order 125.

*15.9 Show that there are no simple groups of order $255 = (3)(5)(17)$.

*15.10 Show that there are no simple groups of order $p^r m$, where p is a prime and $m < p$.

*15.11 This exercise determines the conjugate classes of S_n for every integer $n \geq 1$.

a) Show that if $\sigma = (a_1, a_2, \ldots, a_m)$ is a cycle in S_n and τ is any element of S_n, then $\tau^{-1}\sigma\tau = (a_1\tau, a_2\tau, \ldots, a_m\tau)$.

b) Argue from (a) that any two cycles in S_n of the same length are conjugate.

c) Argue from (a) and (b) that a product of s disjoint cycles in S_n of lengths r_i for $i = 1, 2, \ldots, s$ is conjugate to every other product of s disjoint cycles of lengths r_i in S_n.

d) Show that the number of conjugate classes in S_n is $p(n)$, where $p(n)$ is the number of ways, neglecting the order of the summands, that n can be expressed as a sum of positive integers. The number $p(n)$ is the **number of partitions of** n.

e) Compute $p(n)$ for $n = 1, 2, 3, 4, 5, 6, 7$.

*15.12 Find the conjugate classes and the class equation for S_4. [*Hint:* Use Exercise 15.11.]

*15.13 Find the class equations for S_5 and S_6. [*Hint:* Use Exercise 15.11.]

*15.14 Show that the number of conjugate classes in S_n is also the number of different abelian groups (up to isomorphism) of order p^n, where p is a prime number. [*Hint:* Use Exercise 15.11.]

*15.15 Show that if $n > 2$, the center of S_n is the subgroup consisting of the identity permutation only. [*Hint:* Use Exercise 15.11.]

16 | Applications of the Sylow Theory

In this section we give several applications of the Sylow theorems. The student should find it intriguing to see how easily certain facts about groups of particular orders can be deduced. However, he should realize that we are working only with groups of finite order, and really making only a small dent in the general problem of determining the structure of all finite groups. If the order of a group has only a few factors, then the techniques illustrated in this section may be of some use in determining the structure of the group. This will be demonstrated further in Section 18, where we shall show how it is sometimes possible to describe all groups (up to isomorphism) of certain orders, even when some of the groups are not abelian. However, if the order of a finite group is highly composite, i.e., has a large number of factors, the problem is in general much harder.

*16.1 MORE APPLICATIONS TO p-GROUPS

Theorem 16.1 *Every group of prime power order (i.e., every finite p-group) is solvable.*

Proof. If G has order p^r, it is immediate from Theorem 15.3 that G has subgroups H_i of order p^i, each normal in a subgroup of order p^{i+1} for $1 \leq i < r$. Then

$$\{e\} = H_0 < H_1 < H_2 < \cdots < H_r = G$$

is a composition series, where the factor groups are of order p, and hence abelian and actually cyclic. Thus, G is solvable. ∎

We give an application of Theorem 15.2, although it is not one of the Sylow theorems.

Theorem 16.2 *Every group G of order a power of a prime p has a nontrivial center of at least p elements.*

Proof. Consider the class equation

$$n = c + n_1 + n_2 + \cdots + n_r$$

of G, as given in Theorem 15.2. Now n_i divides n for each i, so p divides each n_i and of course n. But then p must divide c, the order of the center. Since the center contains at least the identity element e, we have $c \geq 1$. Since p divides c, we have then that $c \geq p$. ∎

We turn now to a lemma on direct products which will be used in some of the following theorems.

Lemma 16.1 *Let G be a group containing normal subgroups H and K such that $H \cap K = \{e\}$ and $H \vee K = G$. Then G is isomorphic to $H \times K$.*

Proof. We show that the three conditions of Theorem 8.6 are satisfied. Since $H \cap K = \{e\}$ and $H \vee K = G$, we need show only that $hk = kh$ for $h \in H$ and $k \in K$. Consider the commutator $hkh^{-1}k^{-1}$. The grouping $(hkh^{-1})k^{-1}$ shows that the commutator is in K, since K is normal and $hkh^{-1} \in K$. Similarly, the grouping $h(kh^{-1}k^{-1})$ shows that the commutator is in H. Hence $hkh^{-1}k^{-1} \in (H \cap K)$, so $hkh^{-1}k^{-1} = e$, or $hk = kh$. Our lemma is proved. ∎

Theorem 16.3 *For a prime number p, every group G of order p^2 is abelian.*

Proof. If G is not cyclic, then every element except e must be of order p. Let a be such an element. Then the cyclic subgroup $\langle a \rangle$ of order p does not exhaust G. Also let $b \in G$ with $b \notin \langle a \rangle$. Then $\langle a \rangle \cap \langle b \rangle = \{e\}$, since an element c in $\langle a \rangle \cap \langle b \rangle$ with $c \neq e$ would generate both $\langle a \rangle$ and $\langle b \rangle$ giving $\langle a \rangle = \langle b \rangle$, contrary to construction. From Theorem 15.3, $\langle a \rangle$ is normal in some subgroup of order p^2 of G, that is, normal in all of G. Likewise $\langle b \rangle$ is normal in G. Now $\langle a \rangle \vee \langle b \rangle$ is a subgroup of G properly containing $\langle a \rangle$ and of order dividing p^2. Hence $\langle a \rangle \vee \langle b \rangle$ must be all of G. Thus the hypotheses of Lemma 16.1 are satisfied, and G is isomorphic to $\langle a \rangle \times \langle b \rangle$ and therefore abelian. ∎

*16.2 FURTHER APPLICATIONS

We turn now to a discussion of whether or not there exist simple groups of certain orders. We have seen that every group of prime order is simple. We also asserted that A_n is simple for $n \geq 5$ and that A_5 is the smallest simple group which is not of prime order. There was a famous conjecture of Burnside that every finite simple group of nonprime order must be of even order. It was a triumph when this was recently proved by Thompson and Feit (see Thompson-Feit [21]).

Theorem 16.4 *If p and q are distinct primes with $p < q$, then every group G of order pq has a single subgroup of order q and this subgroup is normal in G. Hence G is not simple. If q is not congruent to 1 modulo p, then G is abelian and cyclic.*

Proof. Theorems 15.3 and 15.4 tell us that G has a Sylow q-subgroup and that the number of such subgroups is congruent to 1 modulo q and divides pq, and therefore must divide p. Since $p < q$, the only possibility is the number 1. Thus there is only one Sylow q-subgroup H of G. This group H must be normal in G, for under an inner automorphism it would be carried into a group of the same order, hence itself. Thus G is not simple.

Likewise, there is a Sylow p-subgroup K of G, and the number of these divides q and is congruent to 1 modulo p. This number must be either 1 or q. If q is not congruent to 1 modulo p, then the number must be 1 and K is normal in G. Let us assume that $q \not\equiv 1 \pmod{p}$. Since every element in H other than e is of order q and every element in K other than e is of order p, we have $H \cap K = \{e\}$. Also $H \vee K$ must be a subgroup of G properly containing H and of order dividing pq. Hence $H \vee K = G$ and by Lemma 16.1, G is isomorphic to $H \times K$ or $\mathbf{Z}_q \times \mathbf{Z}_p$. Thus G is abelian and cyclic. ∎

We need another lemma for some of the counting arguments which follow.

Lemma 16.2 *If H and K are finite subgroups of a group G, then*

$$|HK| = \frac{(|H|)(|K|)}{|H \cap K|}.$$

Proof. Recall that $HK = \{hk \mid h \in H, k \in K\}$. Let $|H| = r$, $|K| = s$, and $|H \cap K| = t$. Clearly, HK has at most rs elements. However, it is possible for $h_1 k_1$ to equal $h_2 k_2$, for $h_1, h_2 \in H$ and $k_1, k_2 \in K$, that is, there may be some collapsing. If $h_1 k_1 = h_2 k_2$, then let

$$x = (h_2)^{-1} h_1 = k_2 (k_1)^{-1}.$$

Now $x = (h_2)^{-1} h_1$ shows that $x \in H$, and $x = k_2 (k_1)^{-1}$ shows that $x \in K$. Hence $x \in (H \cap K)$, and

$$h_2 = h_1 x^{-1} \qquad \text{and} \qquad k_2 = x k_1.$$

On the other hand, if for $y \in (H \cap K)$ we let $h_3 = h_1 y^{-1}$ and $k_3 = y k_1$, then clearly $h_3 k_3 = h_1 k_1$, with $h_3 \in H$ and $k_3 \in K$. Thus each element $hk \in HK$ can be represented in the form $h_i k_i$, for $h_i \in H$ and $k_i \in K$, as many times as there are elements of $H \cap K$, that is, t times. Therefore, the number of elements in HK is rs/t. ∎

The preceding lemma is another result which counts something, so don't underestimate it. The lemma will be used in the following way: a finite group can't have subgroups H and K that are too large with intersections that are too small, or the order of HK would have to exceed the order of G, which is impossible. For example, a group of order 24 can't have two subgroups of orders 12 and 8 with an intersection of order 2.

The remainder of this section consists of several examples illustrating techniques of proving that all groups of certain orders are abelian or that they have proper normal subgroups, i.e., that they are not simple. We will use one fact we mentioned before only in the exercises. *A subgroup H of index 2 in a finite group G is always normal,* for by counting, we see that there are only the left cosets H itself and the coset consisting of all elements in G not in H. The right cosets are the same. Thus every right coset is a left coset, and by Theorem 12.1 and the following discussion, H is normal in G.

Example 16.1 No group of order p^r for $r > 1$ is simple, where p is a prime. For by Theorem 15.3, such a group G contains a subgroup of order p^{r-1} normal in a subgroup of order p^r which must be all of G. Thus a group of order 16 is not simple; it has a normal subgroup of order 8. ‖

Example 16.2 Every group of order 15 is cyclic (hence abelian and not simple, since 15 is not a prime). This is because $15 = (5)(3)$, and 5 is not congruent to 1 modulo 3. By Theorem 16.4, we are done. ‖

Example 16.3 No group of order 20 is simple, for such a group G contains Sylow 5-subgroups in number congruent to 1 modulo 5 and a divisor of 20, hence only 1. This Sylow 5-subgroup is then normal, since all conjugates of it must be itself. ‖

Example 16.4 No group of order 30 is simple. We have seen that if there is only one Sylow p-subgroup for some prime p dividing 30, we are done. By Theorem 15.4, the possibilities for the number of Sylow 5-subgroups are 1 or 6, and those for Sylow 3-subgroups are 1 or 10. But if G has 6 Sylow 5-subgroups, then the intersection of any two is a subgroup of each of order dividing 5, and hence just $\{e\}$. Thus each contains 4 elements of order 5 which are in none of the others. Hence G must contain 24 elements of order 5. Similarly, if G has 10 Sylow 3-subgroups, it has at least 20 elements of order 3. The two types of Sylow subgroups together would require at least 44 elements in G. Thus there is a normal subgroup either of order 5 or of order 3. ‖

Example 16.5 No group of order 48 is simple. Indeed, we shall show that a group G of order 48 has a normal subgroup of either order 16 or order 8. By Theorem 15.4, G has either 1 or 3 Sylow 2-subgroups of order 16. If there is only one subgroup of order 16, it is normal in G, by a now familiar argument.

   Suppose that there are 3 subgroups of order 16, and let H and K be two of them. Then $H \cap K$ must be of order 8, for if $H \cap K$ were of order ≤ 4, then by Lemma 16.2, HK would have at least $(16)(16)/4$, or 64, elements, contradicting the fact that G has only 48 elements. Therefore, $H \cap K$ is normal in both H and K (being of index 2, or by Theorem 15.3). Hence the normalizer of $H \cap K$ contains both H and K and must have order a multiple > 1 of 16 and a divisor of 48, therefore 48. Thus $H \cap K$ must be normal in G. ‖

Example 16.6 No group of order 36 is simple. Such a group G has either 1 or 4 subgroups of order 9. If there is only 1 such subgroup, it is normal in G. If there are 4 such subgroups, let H and K be two of them. As in Example 16.5, $H \cap K$ must have at least 3 elements, or HK would have to have 81 elements, which is impossible. Thus the normalizer of $H \cap K$ has as order a multiple > 1 of 9 and a divisor of 36; hence the order must be either 18 or 36. If the order is 18, the normalizer is then of index 2, and therefore is normal in G. If the order is 36, then $H \cap K$ is normal in G. ‖

Example 16.7 Every group of order $255 = (3)(5)(17)$ is abelian (hence cyclic by Theorem 9.3 and not simple, since 255 is not a prime). By Theorem 15.4, such a group G has only one subgroup H of order 17. Then G/H has order 15 and is abelian by Example 16.2. By Theorem 12.6, we see that the commutator subgroup G' of G is contained in H. Thus as a subgroup of H, G' has either order 1 or 17. Theorem 15.4 also shows that G has either 1 or 85 subgroups of order 3 and either 1 or 51 subgroups of order 5. However, 85 subgroups of order 3 would require 170 elements of order 3, and 51 subgroups of order 5 would require 204 elements of order 5 in G; both together would then require 375 elements in G, which is impossible. Hence there is a subgroup K having either order 3 or order 5 and normal in G. Then G/K has either order $(5)(17)$ or order $(3)(17)$, and in either case, Theorem 16.4 shows that G/K is abelian. Thus $G' \leq K$ and has order either 3, 5, or 1. Since $G' \leq H$ showed that G' has order 17 or 1, we conclude that G' has order 1. Hence $G' = \{e\}$, and $G/G' \simeq G$ is abelian. Theorem 9.3 then shows that G is cyclic. ‖

EXERCISES

*16.1 By arguments similar to those used in the examples of this section, convince yourself that every group of order not a prime and less than 60 contains a proper normal subgroup and hence is not simple. You need not write out the details. (The hardest cases were discussed in the examples.)

†*16.2 Prove that every group of order $(5)(7)(47)$ is abelian and cyclic.

*16.3 Mark each of the following true or false.

— a) Every group of order 159 is cyclic.
— b) Every group of order 102 has a proper normal subgroup.
— c) Every solvable group is of prime power order.
— d) Every group of prime power order is solvable.
— e) It would become quite tedious to show that no group of nonprime order between 60 and 168 is simple by the methods illustrated in the text.
— f) One of the chief values of doing these numerical exercises is that it helps us to understand and to retain the theory.
— g) Every group of 125 elements has at least 5 elements that commute with every element in the group.
— h) Every group of order 42 has a normal subgroup of order 7.
— i) Every group of order 42 has a normal subgroup of order 8.
— j) The only simple groups are the groups $\mathbf{Z}_p$ and A_n, where p is a prime and $n \neq 4$.

*16.4 Prove that no group of order 96 is simple.

*16.5 Prove that no group of order 160 is simple.

17 | Free Groups

In this section and the next we discuss a portion of group theory which is of great interest not only in algebra but in topology as well. In fact, an excellent and readable discussion of free groups and presentations of groups is found in Crowell and Fox [45, Chapters III and IV].

*17.1 WORDS AND REDUCED WORDS

Let A be any (not necessarily finite) set of elements a_i for $i \in I$. We think of A as an **alphabet** and of the a_i as **letters** in the alphabet. Any symbol of the form a_i^n with $n \in \mathbf{Z}$ is a **syllable**, and a finite string w of syllables written in juxtaposition is a **word**. We also introduce the **empty word** 1 which has no syllables.

Example 17.1 Let $A = \{a_1, a_2, a_3\}$. Then

$$a_1 a_3^{-4} a_2^{\ 2} a_3, \qquad a_2^{\ 3} a_2^{-1} a_3 a_1^{\ 2} a_1^{-7}, \qquad \text{and} \qquad a_3^{\ 2}$$

are all words, if we follow the convention of understanding that $a_i^{\ 1}$ is the same as a_i. ‖

There are two natural types of modifications of certain words, the **elementary contractions**. The first type consists of replacing an occurrence of

$$a_i^{\ m} a_i^{\ n}$$

in a word by

$$a_i^{\ m+n}.$$

The second type consists of replacing an occurrence of

$$a_i^{\ 0}$$

in a word by

$$1,$$

that is, dropping it out of the word. By means of a finite number of elementary contractions, every word can be changed to a **reduced word**, one for which no more elementary contractions are possible. Note that these elementary contractions formally amount to the usual manipulations of integral exponents.

Example 17.2 The reduced form of the word $a_2^{\ 3} a_2^{-1} a_3 a_1^{\ 2} a_1^{-7}$ of Example 17.1 is $a_2^{\ 2} a_3 a_1^{-5}$. ‖

It should be said here once and for all that we are going to gloss over several points which some books spend pages "proving", usually by quite complicated induction arguments broken down into many cases. For example, suppose you are given a word and wish to find its reduced form. There may be a variety of elementary contractions which could be performed first. How do you know that the reduced word you end up with is the same no matter in which order you perform the elementary contractions? The student will probably say that this is obvious. Some authors spend considerable effort proving this. The author tends to agree here with the student. "Proofs" of this sort he regards as tedious, and they have never made him more comfortable about the situation. However, the author is the first to acknowledge that he is not a great mathematician. In deference to the fact that many mathematicians feel that these things do need considerable discussion, we shall mark an occasion when we just state such facts by the phrase, "It would seem obvious that," keeping the quotation marks.

*17.2   FREE GROUPS

Let the set of all reduced words formed from our alphabet A be $F[A]$. We now make $F[A]$ into a group in a natural way. For w_1 and w_2 in $F[A]$, define $w_1 \cdot w_2$ to be the reduced form of the word obtained by the juxtaposition $w_1 w_2$ of the two words.

Example 17.3 If

$$w_1 = a_2{}^3 a_1{}^{-5} a_3{}^2$$

and

$$w_2 = a_3{}^{-2} a_1{}^2 a_3 a_2{}^{-2},$$

then $w_1 \cdot w_2 = a_2{}^3 a_1{}^{-3} a_3 a_2{}^{-2}$. ‖

"It would seem obvious that" this operation of multiplication on $F[A]$ is well defined and associative. It is obvious that the empty word 1 acts as an identity element. "It would seem obvious that" given a reduced word $w \in F[A]$, if you form the word obtained by first writing the syllables of w in the opposite order and secondly replacing each $a_i{}^n$ by $a_i{}^{-n}$, then the resulting word w^{-1} is a reduced word also, and

$$w \cdot w^{-1} = w^{-1} \cdot w = 1.$$

Definition. The group $F[A]$ just described is the *free group generated by* A.

The student should look back at Theorem 9.1 and the definition preceding it to see that the present use of the term *generated* is consistent with the earlier use.

Starting with a group G and a generating set $\{a_i \mid i \in I\}$, one might ask if G is *free* on $\{a_i\}$, that is, if G is essentially the free group generated by $\{a_i\}$. We define precisely what this is to mean.

Definition. If G is a group with a set $A = \{a_i\}$ of generators, and if G is isomorphic to $F[A]$ under a map $\phi: G \to F[A]$ such that $a_i\phi = a_i$, then G is **free on** $\{a_i\}$ and the a_i are **free generators of** G. A group is **free** if it is free on some nonempty set $\{a_i\}$.

Example 17.4 The only example of a free group which has occurred before is **Z**, which is free on one generator. Clearly every free group is infinite. ‖

The student is referred to the literature for proofs of the next three theorems. We will not be using these results. They are stated simply to inform the student of these interesting facts.

Theorem 17.1 *If a group G is free on $\{a_i\}$ and also on $\{b_j\}$, then the sets $\{a_i\}$ and $\{b_j\}$ have the same number of elements, i.e., any two sets of free generators of a free group have the same cardinality.*

Definition. If G is free on $\{a_i\}$, the number of elements in $\{a_i\}$ is the **rank of the free group** G.

Actually, the next theorem is pretty evident from Theorem 17.1.

Theorem 17.2 *Two free groups are isomorphic if and only if they have the same rank.*

Theorem 17.3 *A proper subgroup of a free group is free.*

Example 17.5 Let $F[\{x, y\}]$ be the free group on $\{x, y\}$. Let

$$y_k = x^k y x^{-k}$$

for $k \geq 0$. The student will have no difficulty convincing himself that the y_k for $k \geq 0$ are free generators for the subgroup of $F[\{x, y\}]$ which they generate. This illustrates that although a subgroup of a free group is free, the rank of the subgroup may be much greater than the rank of the whole group! ‖

*17.3 HOMOMORPHISMS OF FREE GROUPS

Our work in this section will be concerned primarily with homomorphisms defined on a free group. The results here are simple and elegant.

Theorem 17.4 *Let G be generated by $\{a_i \mid i \in I\}$ and let G' be any group. If a'_i for $i \in I$ are any elements in G', not necessarily distinct, then there is at most one homomorphism $\phi: G \to G'$ such that $a_i\phi = a'_i$. If G is free on $\{a_i\}$, then there is exactly one such homomorphism.*

Proof. Let ϕ be a homomorphism from G into G' such that $a_i\phi = a'_i$. Now by Theorem 9.1, for any $x \in G$ we have

$$x = \prod_j a_{i_j}^{n_j}$$

for some finite product of the generators a_i, where the a_{i_j} appearing in the product need not be distinct. Then since ϕ is a homomorphism, we must have

$$x\phi = \prod_j (a_{i_j}{}^{n_j})\phi = \prod_j (a'_{i_j})^{n_j}.$$

Thus a homomorphism is completely determined by its values on elements of a generating set. This shows that there is at most one homomorphism such that $a_i\phi = a'_i$.

Now suppose G is free on $\{a_i\}$, that is, $G = F[\{a_i\}]$. For

$$x = \prod_j a_{i_j}{}^{n_j}$$

in G, define $\psi: G \to G'$ by

$$x\psi = \prod_j (a'_{i_j})^{n_j}.$$

This map is well defined, since $F[\{a_i\}]$ consists precisely of reduced words; no two different formal products in $F[\{a_i\}]$ are equal. Since the rules for computation involving exponents in G' are formally the same as those involving exponents in G, it is clear that $(xy)\psi = (x\psi)(y\psi)$ for any elements x and y in G, so ψ is indeed a homomorphism. ∎

Perhaps we should have proved the first part of this theorem earlier, rather than having relegated it to the exercises. Note that the theorem states that *a homomorphism of a group is completely determined if you know its value on each element of a generating set.* This is really very obvious and could have been mentioned immediately after the definition of a homomorphism. In particular, a homomorphism of a cyclic group is completely determined by its value on any single generator of the group.

Theorem 17.5 *Every group G' is a homomorphic image of a free group G.*

Proof. Let $G' = \{a'_i\}$, and let $\{a_i\}$ be a set with the same number of elements as G'. Let $G = F[\{a_i\}]$. Then by Theorem 17.4, there exists a homomorphism ψ mapping G into G' such that $a_i\psi = a'_i$. Clearly, the image of G under ψ is all of G'. ∎

*17.4 FREE ABELIAN GROUPS

"It would seem obvious that" *a free group is torsion free.* However $\mathbf{Z} \times \mathbf{Z}$ is torsion free but not free, since it is abelian but not cyclic. There is an important concept of a *free abelian group*, and $\mathbf{Z} \times \mathbf{Z}$ is an example of such a group. Let us form a new group, an abelianized version of a free group, by placing on the free group only the restriction that its elements are to commute. Theorem 12.6 and the discussion preceding it show us that this amounts to forming the factor group of the free group by its commutator subgroup.

Definition. A group G is a *free abelian group* if it is isomorphic to the factor group of some free group by its commutator subgroup. The images of the free generators of the free group under the canonical factor group homomorphism followed by the isomorphism form a **basis for G.**

Thus, if G is a free abelian group and is the factor group of $F[\{a_i\}]$, by renaming the image of a_i under the canonical homomorphism again "a_i", we can consider an element of G to be a finite product of the form

$$\prod_j a_{i_j}{}^{n_j},$$

where this time, in contrast to the situation for free groups, each a_{i_j} appears at most once. Thus

$$(a_{i_1}{}^{n_1}a_{i_2}{}^{n_2}\cdots a_{i_r}{}^{n_r})(a_{i_1}{}^{m_1}a_{i_2}{}^{m_2}\cdots a_{i_r}{}^{m_r}) = (a_{i_1}{}^{n_1+m_1}a_{i_2}{}^{n_2+m_2}\cdots a_{i_r}{}^{n_r+m_r}).$$

"It would seem obvious that" G can be viewed precisely as the set of all finite products of powers of the a_i, with no a_i appearing more than once.

In a free abelian group with a basis $\{a_i\}$, it is customary to use additive notation and to write $a_{i_1}{}^{n_1}a_{i_2}{}^{n_2}\cdots a_{i_r}{}^{n_r}$ as

$$"n_1a_{i_1} + n_2a_{i_2} + \cdots + n_ra_{i_r}",$$

where "na" denotes $a + a + \cdots + a$ for n summands if $n > 0$, $(-a) + (-a) + \cdots + (-a)$ for $|n|$ summands if $n < 0$, and where $0a = 0$. To avoid misunderstanding, note that in

$$0a = 0$$

the 0 on the left side of the equation is our usual number 0, but the 0 on the right is the identity of G, not necessarily the number 0.

There is a concept of a *basis* for a general (not necessarily free) abelian group (see Exercise 17.9). We shall use the term *basis* only for a free abelian group.

Theorems 17.1, 17.2, and 17.3 hold for free abelian groups as well as for free groups, the **rank of a free abelian group** being the number of elements in some basis for the group. Theorem 17.4 also holds for free abelian groups, with the obvious restriction that the image group G' must be abelian. The proof is about the same as that for free groups.

It is clear that

$$\mathbf{Z} \times \mathbf{Z} \times \cdots \times \mathbf{Z}$$

for m factors is a free abelian group of rank m. Lemma 9.2 thus states that *every torsion free, finitely generated abelian group is a free abelian group of finite rank m*, where m is the betti number of the group.

In contrast to Example 17.5 for free groups, it is true that for a free abelian group the rank of a subgroup is at most the rank of the entire group. It is very instructive for us to see this proved for the case of a group of finite rank. The technique employed is a bit different from any we have used before.

Theorem 17.6 *Let G be a free abelian group of finite rank, and let H be a subgroup of G. Then the rank of H is at most the rank of G.*

Proof. Let us first look at a particular basis $B = \{b_i\}$ for H, which exists for a free abelian group by the analog of Theorem 17.3. Let $\{a_1, \ldots, a_s\}$ be a basis for G. Then for each b_i, we have

$$b_i = \sum_{j=1}^{s} n_{ij}a_j.$$

Let b_1 be a particular b_i. It is easy to see that

$$\{b_1, b_i - q_i b_1 \mid i \neq 1\}$$

is again a basis for H for any integers q_i. Since $\{-b_i\}$ is also clearly a basis for H, we can assume, replacing b_i by $-b_i$ where necessary, that all n_{i1} are ≥ 0. If not all $n_{i1} = 0$, let us choose our b_1 so that $n_{11} > 0$ and n_{11} is minimal among the $n_{i1} \neq 0$. Let r_i and q_i be such that

$$n_{i1} = q_i n_{11} + r_i$$

for $i \neq 1$ and where $0 \leq r_i < n_{11}$ (see Lemma 6.1). But then

$$\{b_1, b_i - q_i b_1 \mid i \neq 1\}$$

is a basis for H such that either n_{11} is the only nonzero coefficient of a_1 when these basis elements are expressed in terms of the a_j, or there is a new coefficient r_{i_1} of a_1 for some $b_{i_1} - q_{i_1} b_1$ with $0 < r_{i_1} < n_{11}$. By repeating this process, in a finite number of steps we arrive at a basis

$$\{b_1^{(1)}, b_i^{(1)} \mid i \neq 1\},$$

where

$$b_1^{(1)} = \sum_{j=1}^{s} n_{1j}^{(1)} a_j,$$

with $n_{11}^{(1)} > 0$ and

$$b_i^{(1)} = \sum_{j=2}^{s} n_{ij}^{(1)} a_j \qquad \text{for} \quad i \neq 1,$$

that is, all coefficients of a_1 equal zero for $i \neq 1$. If our original $\{b_i\}$ was such that all $n_{i1} = 0$, we let $b_i^{(1)} = b_i$.

Now let us consider the subgroup $H^{(1)}$ of H having as basis $\{b_{(i)}^{(1)} \mid i \neq 1\}$. By repeating the above process with respect to the coefficient of a_2, we arrive at a basis

$$\{b_2^{(2)}, b_i^{(2)} \mid i \neq 1, 2\}$$

for this subgroup $H^{(1)}$ of H.

After t such steps, where $t \leq s$, we have constructed a basis

$$\{b_1^{(1)}, b_2^{(2)}, \ldots, b_t^{(t)}, b_i^{(t)} \mid i \neq 1, \ldots, t\}$$

of H, with the coefficients of a_j in the $b_i^{(t)}$ equal to 0 for all j if $i \neq 1, \ldots, t$, that is,

$$b_i^{(t)} = 0 \qquad \text{for} \quad i \neq 1, \ldots, t.$$

Thus

$$\{b_1^{(1)}, b_2^{(2)}, \ldots, b_t^{(t)}\}$$

is a basis for H which is therefore of rank $t \leq s.$ ∎

The student who was able to follow the above proof probably noticed that the method employed amounted to diagonalizing a matrix

$$\begin{pmatrix} n_{11} & n_{12} & \cdots & n_{1s} \\ n_{21} & n_{22} & \cdots & n_{2s} \\ n_{31} & n_{32} & \cdots & n_{3s} \\ \vdots & \vdots & \cdots & \vdots \end{pmatrix},$$

so that it becomes

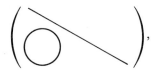

where all entries below the diagonal are zero.

EXERCISES

*17.1 Find the reduced form and the inverse of the reduced form of each of the following words.

a) $a^2 b^{-1} b^3 a^3 c^{-1} c^4 b^{-2}$ 　　　　　　　 b) $a^2 a^{-3} b^3 a^4 c^4 c^2 a^{-1}$

*17.2 Compute the products given in parts (a) and (b) of Exercise 17.1 in the case that $\{a, b, c\}$ is a set of generators forming a basis for a free abelian group. Find the inverses of these products.

*17.3 How many different homomorphisms are there of a free group of rank 2 into

a) Z_4? 　　　　b) Z_6? 　　　　c) S_3?

*17.4 How many different homomorphisms are there of a free group of rank 2 onto

a) Z_4? 　　　　b) Z_6? 　　　　c) S_3?

*17.5 How many different homomorphisms are there of a free abelian group of rank 2 into

a) Z_4? 　　　　b) Z_6? 　　　　c) S_3?

*17.6 How many different homomorphisms are there of a free abelian group of rank 2 onto

a) Z_4? 　　　　b) Z_6? 　　　　c) S_3?

†*17.7 Take one of the instances in this section in which the phrase "It would seem obvious that" was used and discuss your reaction in that instance.

*17.8 Mark each of the following true or false.

— a) Every proper subgroup of a free group is a free group.
— b) Every proper subgroup of every free abelian group is a free group.
— c) A homomorphic image of a free group is a free group.
— d) Every free abelian group has a basis.
— e) The free abelian groups of finite rank are precisely the finitely generated abelian groups.
— f) No free group is abelian.
— g) No free abelian group is free.
— h) This terminology drives me crazy.
— i) Any two free groups are isomorphic.
— j) Any two free abelian groups of the same rank are isomorphic.

---

*17.9 Let G be a finitely generated abelian group with identity 0. A finite set $\{b_1, \ldots, b_n\}$, where $b_i \in G$, is a **basis for** G if $\{b_1, \ldots, b_n\}$ generates G and $\sum_{i=1}^{n} m_i b_i = 0$ if and only if $m_i b_i = 0$, where $m_i \in \mathbf{Z}$.

a) Show that $\{2, 3\}$ is not a basis for $\mathbf{Z}_4$. Find a basis for $\mathbf{Z}_4$.
b) Show that both $\{1\}$ and $\{2, 3\}$ are bases for $\mathbf{Z}_6$. (This shows that for a finitely generated abelian group G with torsion, the number of elements in a basis may vary, i.e., it need not be an *invariant* of the group G.)
c) Is a basis for a free abelian group as we defined it in this section a basis in the sense in which it is used in this exercise?
d) Show that every finite abelian group has a basis $\{b_1, \ldots, b_n\}$, where the order of b_i divides the order of b_{i+1}. You may use any theorems in the text, even if they were not proved.

In present-day expositions of algebra, a frequently used technique (particularly by the disciples of N. Bourbaki) for introducing a new algebraic entity is the following:

1) Describe algebraic properties which this algebraic entity is to possess.

2) Prove that any two algebraic entities with these properties are isomorphic, i.e., that these properties *characterize* this entity.

3) Show that at least one such entity exists.

The next three exercises illustrate this technique for three algebraic entities, each of which the student has met before. So that we don't give away their identities, we use fictitious names for them in the first two exercises. The last part of these first two exercises asks the student to give the usual name for the entity.

*17.10 Let G be any group. An abelian group G^* is a **blip group of** G if there exists a fixed homomorphism ϕ of G onto G^* such that each homomorphism ψ of G into an abelian group G' can be factored as $\psi = \phi\theta$, where θ is a homomorphism of G^* into G' (see Fig. 17.1).

a) Show that any two blip groups of G are isomorphic. [*Hint:* Let G_1^* and G_2^* be two blip groups of G. Then each of the fixed homomorphisms $\phi_1: G \to G_1^*$ and $\phi_2: G \to G_2^*$ can be factored via the other blip group according to the definition of a blip group, that is, $\phi_1 = \phi_2\theta_1$ and $\phi_2 = \phi_1\theta_2$. Show that θ_1 is an isomorphism of G_2^* onto G_1^*. Apply Exercise 13.15.]

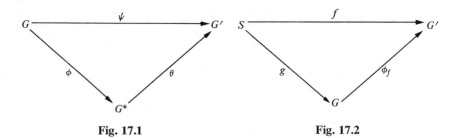

Fig. 17.1   **Fig. 17.2**

b) Show for every group G that a blip group G^* of G exists.

c) What concept that we have introduced before corresponds to this idea of a blip group of G?

*17.11 Let S be any set. A group G together with a fixed function $g: S \to G$, constitutes a **blop group on** S if for each group G' and map $f: S \to G'$ there exists a *unique* homomorphism ϕ_f of G into G' such that $f = g\phi_f$ (see Fig. 17.2).

a) Let S be a fixed set. Show that if both G_1, together with $g_1: S \to G_1$, and G_2, together with $g_2: S \to G_2$, are blop groups on S, then G_1 and G_2 are isomorphic. [*Hint:* Show that g_1 and g_2 are one-to-one maps and that Sg_1 and Sg_2 generate G_1 and G_2 respectively. Then proceed in a way analogous to that given by the hint of Exercise 17.10.]

b) Let S be a set. Show that a blop group on S exists. You may use any theorems of the text.

c) What concept that we have introduced before corresponds to this idea of a blop group on S?

*17.12 Characterize a free abelian group by properties in a fashion similar to that used in Exercise 17.11.

18 | Group Presentations

Following most of the literature on group presentations, in this section we let 1 be the identity of a group. The idea of a *group presentation* is to form a group by giving a set of generators for the group and certain equations or relations which you want the generators to satisfy. You want the group to be as "free as it possibly can be on the generators subject to these relations."

Example 18.1 Suppose G has generators x and y and is *free except for the relation* $xy = yx$, which we may express as $xyx^{-1}y^{-1} = 1$. Clearly, the condition $xy = yx$ is exactly what is needed to make G commutative, even though $xyx^{-1}y^{-1}$ is just one of the many possible commutators of $F[\{x, y\}]$. Thus G is free abelian on two generators and is isomorphic to $F[\{x, y\}]$ modulo its commutator subgroup. This commutator subgroup of $F[\{x, y\}]$ is the smallest normal subgroup containing $xyx^{-1}y^{-1}$, since any normal subgroup containing $xyx^{-1}y^{-1}$ gives rise to a factor group which is abelian, and thus contains the commutator subgroup by Theorem 12.6. ∥

The preceding example illustrates the general situation. Let $F[A]$ be a free group and suppose that you want to form a new group as much like $F[A]$ as it can be, subject to certain equations which you want satisfied. Any equation can be written in a form in which the right-hand side is 1. Thus we can consider the equations to be $r_i = 1$, where $r_i \in F[A]$. Clearly, if you require that $r_i = 1$, then you will have to have

$$x(r_i^n)x^{-1} = 1$$

for any $x \in F[A]$ and $n \in \mathbf{Z}$. Also any product of elements equal to 1 will again have to equal 1. Thus any finite product of the form

$$\prod_j x_j(r_{i_j}^{n_j})x_j^{-1},$$

where the r_{i_j} need not be distinct, will have to equal 1 in the new group. It is very easy to check that the set of all these finite products is a normal subgroup R of $F[A]$. Thus any group looking as much as possible like $F[A]$, subject to the requirements $r_i = 1$, also has $r = 1$ for every $r \in R$. But $F[A]/R$ looks like $F[A]$ (remember that we multiply cosets by choosing representatives) except that R has been collapsed to form the identity 1.

Thus the group we are after is (at least isomorphic to) $F[A]/R$. We can view this group as described by the generating set A and the set $\{r_i\}$.

> **Definition.** Let A be a set and let $\{r_i\} \subseteq F[A]$. Let R be the least normal subgroup of $F[A]$ containing the r_i. An isomorphism ϕ of $F[A]/R$ onto a group G is a **presentation of G**. The sets A and $\{r_i\}$ give a **group presentation**. The set A is the set of **generators for the presentation** and each r_i is a **relator**. Each $r \in R$ is a **consequence of** $\{r_i\}$. An equation $r_i = 1$ is a **relation**. A **finite presentation** is one in which both A and $\{r_i\}$ are finite sets.

This definition may seem complicated, but it really isn't. In Example 18.1, $\{x, y\}$ is our set of generators and $xyx^{-1}y^{-1}$ is the only relator. The equation $xyx^{-1}y^{-1} = 1$, or $xy = yx$, is a relation. This was an example of a finite presentation.

If a group presentation has generators x_j and relators r_i, we shall use the notations

$$\text{``}(x_j : r_i)\text{''} \qquad \text{or} \qquad \text{``}(x_j : r_i = 1)\text{''}$$

to denote the group presentation. We may refer to $F[\{x_j\}]/R$ as *the group with presentation* $(x_j : r_i)$.

*18.2   ISOMORPHIC PRESENTATIONS

Example 18.2   Consider the group presentation with

$$A = \{a\} \qquad \text{and} \qquad \{r_i\} = \{a^6\},$$

that is, the presentation

$$(a : a^6 = 1).$$

This group defined by one generator a, with the relation $a^6 = 1$, is clearly isomorphic to $\mathbf{Z}_6$.

Now consider the group defined by two generators a and b, with $a^2 = 1$, $b^3 = 1$, and $ab = ba$, that is, the group with presentation

$$(a, b : a^2, b^3, aba^{-1}b^{-1}).$$

The condition $a^2 = 1$ gives $a^{-1} = a$. Also $b^3 = 1$ gives $b^{-1} = b^2$. Thus every element in this group can be written as a product of nonnegative powers of a and b. The relation $aba^{-1}b^{-1} = 1$, that is, $ab = ba$, allows us to write first all the factors involving a and then the factors involving b. Hence every element of the group is equal to some $a^m b^n$. But then $a^2 = 1$ and $b^3 = 1$ show that there are just six distinct elements,

$$1, b, b^2, a, ab, ab^2.$$

Therefore this presentation also gives a group of order 6 which is abelian, and by Theorem 9.3, it must again be cyclic and isomorphic to $\mathbf{Z}_6$. ∥

The preceding example illustrates that different presentations may give isomorphic groups. When this happens, we have **isomorphic presentations**. To determine whether or not two presentations are isomorphic may be very hard. It has recently been shown (see Rabin [22]) that a number of such problems connected with this theory are not generally solvable, i.e., there is no *routine* and well-defined way of discovering a solution in all cases. These unsolvable problems include the problem of deciding whether or not two presentations are isomorphic, whether or not a group given by a presentation is finite, free, abelian, or trivial, and the famous *word problem* of determining whether or not a given word r is a consequence of a given set of words $\{r_i\}$.

The importance of this material is indicated by our Theorem 17.5, which guarantees that *every group has a presentation*.

Example 18.3 Let us show that

$$(x, y : y^2x = y, yx^2y = x)$$

is a presentation of the trivial group of one element. We need only show that x and y are consequences of the relators y^2xy^{-1} and yx^2yx^{-1}, or that $x = 1$ and $y = 1$ can be deduced from $y^2x = y$ and $yx^2y = x$. We illustrate both techniques.

As a consequence of y^2xy^{-1}, we get yx upon conjugation by y^{-1}. From yx we deduce $x^{-1}y^{-1}$, and then $(x^{-1}y^{-1})(yx^2yx^{-1})$ gives xyx^{-1}. Conjugating xyx^{-1} by x^{-1}, we get y. From y we get y^{-1}, and $y^{-1}(yx)$ is x.

Working with relations instead of relators, from $y^2x = y$ we deduce $yx = 1$ upon multiplication by y^{-1} on the left. Then substituting $yx = 1$ into $yx^2y = x$, that is, $(yx)(xy) = x$, we get $xy = x$. Then multiplying by x^{-1} on the left, we have $y = 1$. Substituting this in $yx = 1$, we get $x = 1$.

Both techniques amount to the same work, but it somehow seems more natural to most of us to work with relations. ‖

*18.3 APPLICATIONS

We conclude this section with two applications.

Example 18.4 Let us determine all groups of order 10 up to isomorphism. We know from Theorem 9.3 that every abelian group of order 10 is isomorphic to $\mathbf{Z}_{10}$. Suppose that G is nonabelian of order 10. By Sylow theory, G contains a normal subgroup H of order 5, and H must be cyclic. Let a be a generator of H. Then G/H is of order 2 and thus isomorphic to $\mathbf{Z}_2$. If $b \in G$ and $b \notin H$, we must then have $b^2 \in H$. Since every element of H except 1 has order 5, if b^2 were not equal to 1, then b^2 would have order 5, so b would have order 10. This would mean that G would be cyclic, contradicting our assumption that G is not abelian. Thus $b^2 = 1$. Finally, since H is a normal subgroup of G, $bHb^{-1} = H$, so in particular, $bab^{-1} \in H$.

Since conjugation by b is an automorphism of H, bab^{-1} must be another element of H of order 5, hence bab^{-1} equals a, a^2, a^3, or a^4. But $bab^{-1} = a$ would give $ba = ab$, and then clearly G would be abelian, since a and b generate G. Thus the possibilities for presentations of G are:

1) $(a, b : a^5 = 1, b^2 = 1, ba = a^2b)$,
2) $(a, b : a^5 = 1, b^2 = 1, ba = a^3b)$,
3) $(a, b : a^5 = 1, b^2 = 1, ba = a^4b)$.

Note that all three of these presentations can give groups of order at most 10, since the last relation $ba = a^ib$ enables us to express every product of a's and b's in G in the form "a^sb^t". Then $a^5 = 1$ and $b^2 = 1$ show that the set

$$S = \{a^0b^0, a^1b^0, a^2b^0, a^3b^0, a^4b^0, a^0b^1, a^1b^1, a^2b^1, a^3b^1, a^4b^1\}$$

includes all elements of G.

It is not yet clear that all these elements in S are distinct, so that we have in all three cases a group of order 10. For example, the group presentation

$$(a, b : a^5 = 1, b^2 = 1, ba = a^2b)$$

gives a group in which, using the associative law, we have

$$a = b^2a = (bb)a = b(ba) = b(a^2b) = (ba)(ab)$$
$$= (a^2b)(ab) = a^2(ba)b = a^2(a^2b)b = a^4b^2 = a^4.$$

Thus in this group, $a = a^4$, so $a^3 = 1$, which, together with $a^5 = 1$, yields $a^2 = 1$. But $a^2 = 1$, together with $a^3 = 1$, means that $a = 1$. Hence every element in the group with presentation

$$(a, b : a^5 = 1, b^2 = 1, ba = a^2b)$$

is equal to either 1 or b, that is, this group is isomorphic to $\mathbf{Z}_2$. A similar study of

$$(bb)a = b(ba)$$

for

$$(a, b : a^5 = 1, b^2 = 1, ba = a^3b)$$

shows that $a = a^4$ again, so this also yields a group isomorphic to $\mathbf{Z}_2$.

This leaves just

$$(a, b : a^5 = 1, b^2 = 1, ba = a^4b)$$

as a candidate for a nonabelian group of order 10. In this case, it can be shown that all elements of S are distinct, so this presentation does give a nonabelian group G of order 10. How can we show that all elements in S represent distinct elements of G? The easy way is to observe that we know that there is at least one nonabelian group of order 10, the dihedral group

D_5. Since G is the only remaining candidate, we must have $G \simeq D_5$. Another attack is as follows. Let us try to make S into a group by *defining* $(a^s b^t)(a^u b^v)$ to be $a^x b^y$, where x is the remainder of $s + u(4^t)$ when divided by 5, and y is the remainder of $t + v$ when divided by 2, in the sense of Lemma 6.1. In other words, we use the relation $ba = a^4 b$ as a guide in *defining* the product $(a^s b^t)(a^u b^v)$ of two elements of S. It is easy to see that $a^0 b^0$ acts as identity, and that given $a^u b^v$, we can determine t and s successively by letting

$$t \equiv -v \,(\text{mod } 2)$$

and then

$$s \equiv -u(4^t) \,(\text{mod } 5),$$

giving $a^s b^t$, which is a left inverse for $a^u b^v$. We will then have a group structure on S if and only if the associative law holds. In Exercise 18.7, we ask the student to carry out the straightforward computation for the associative law and to discover a condition for S to be a group under such a definition of multiplication. The criterion of the exercise in this case amounts to the valid congruence

$$4^2 \equiv 1 \,(\text{mod } 5).$$

Thus we do get a group of order 10. Note that

$$2^2 \not\equiv 1 \,(\text{mod } 5)$$

and

$$3^2 \not\equiv 1 \,(\text{mod } 5),$$

so Exercise 18.7 also shows that

$$(a, b : a^5 = 1, b^2 = 1, ba = a^2 b)$$

and

$$(a, b : a^5 = 1, b^2 = 1, ba = a^3 b)$$

do not give groups of order 10. ||

Example 18.5 Let us determine all groups of order 8 up to isomorphism. We know the three abelian ones:

$$\mathbf{Z}_8, \qquad \mathbf{Z}_2 \times \mathbf{Z}_4, \qquad \mathbf{Z}_2 \times \mathbf{Z}_2 \times \mathbf{Z}_2.$$

Using generators and relations, we shall give presentations of the nonabelian groups.

Let G be nonabelian of order 8. Since G is nonabelian, it has no elements of order 8, so every element but the identity is of order either 2 or 4. If every element were of order 2, then for $a, b \in G$, we would have $(ab)^2 = 1$, that is, $abab = 1$. Then since $a^2 = 1$ and $b^2 = 1$ also, we would have

$$ba = a^2 bab^2 = a(ab)^2 b = ab,$$

contrary to our assumption that G is not abelian. Thus G has an element of order 4.

Let $\langle a \rangle$ be a subgroup of G of order 4. If $b \notin \langle a \rangle$, the cosets $\langle a \rangle$ and $b\langle a \rangle$ exhaust all of G. Hence a and b are generators for G and $a^4 = 1$. Since $\langle a \rangle$ is normal in G (by Sylow theory, or because it is of index 2), $G/\langle a \rangle$ is isomorphic to $\mathbf{Z}_2$ and we have $b^2 \in \langle a \rangle$. If $b^2 = a$ or $b^2 = a^3$, then b would be of order 8. Hence $b^2 = 1$ or $b^2 = a^2$. Finally, since $\langle a \rangle$ is normal, we have $bab^{-1} \in \langle a \rangle$, and since $b\langle a \rangle b^{-1}$ is a subgroup conjugate to $\langle a \rangle$ and hence isomorphic to $\langle a \rangle$, we see that bab^{-1} must be an element of order 4. Thus $bab^{-1} = a$ or $bab^{-1} = a^3$. If bab^{-1} were equal to a, then ba would equal ab, which would make G abelian. Hence $bab^{-1} = a^3$ so $ba = a^3b$. Thus, we have two possibilities for G, namely

$$G_1 : (a, b : a^4 = 1, b^2 = 1, ba = a^3b)$$

and

$$G_2 : (a, b : a^4 = 1, b^2 = a^2, ba = a^3b).$$

Note that $a^{-1} = a^3$, and that b^{-1} is b in G_1 and b^3 in G_2. These facts, with the relation $ba = a^3b$, enable us to express every element in G_i in the form "$a^m b^n$", as in Examples 18.2 and 18.4. Since $a^4 = 1$ and either $b^2 = 1$ or $b^2 = a^2$, the possible elements in each group are

$$1, a, a^2, a^3, b, ab, a^2b, a^3b.$$

Thus G_1 and G_2 each have order at most 8. That G_1 is a group of order 8 can be seen from Exercise 18.7. An argument similar to that used in Exercise 18.7 shows that G_2 has order 8 also.

Since $ba = a^3b \neq ab$, we see that both G_1 and G_2 are nonabelian. That the two groups are not isomorphic follows from the fact that a computation shows that G_1 has only 2 elements of order 4, namely a and a^3. On the other hand, in G_2, all elements but 1 and a^2 are of order 4. We leave the computations of the tables for these groups for the student to perform in the third exercise at the end of this section. To illustrate, suppose you wish to compute $(a^2b)(a^3b)$. Using $ba = a^3b$ repeatedly, we get

$$(a^2b)(a^3b) = a^2(ba)a^2b = a^5(ba)ab = a^8(ba)b = a^{11}b^2.$$

Then for G_1, we have

$$a^{11}b^2 = a^{11} = a^3,$$

but if we are in G_2, we get

$$a^{11}b^2 = a^{13} = a.$$

The group G_1 is the **octic group** and is nothing more than our old friend the group D_4 of symmetries of the square. The group G_2 is the **quaternion group**; the reason for the name will be explained in Section 25.4. ∥

EXERCISES

*18.1 Give a presentation of Z_4 involving 1 generator; involving 2 generators; involving 3 generators.

*18.2 Give a presentation of S_3 involving 3 generators.

*18.3 Give the tables for both the octic group

$$(a, b : a^4 = 1, b^2 = 1, ba = a^3b)$$

and the quaternion group

$$(a, b : a^4 = 1, b^2 = a^2, ba = a^3b).$$

In both cases, write the elements in the order $1, a, a^2, a^3, b, ab, a^2b, a^3b$. (Note that you don't have to compute *every* product. You know that these presentations give groups of order 8, and once you have computed enough products, the rest are forced so that each row and each column of the table has each element exactly once.)

*18.4 Mark each of the following true or false.

— a) Every group has a presentation.

— b) Every group has many different presentations.

— c) Every group has two presentations which are not isomorphic.

— d) Every group has a finite presentation.

— e) Every group with a finite presentation is of finite order.

— f) Every cyclic group has a presentation with just one generator.

— g) Every conjugate of a relator is a consequence of the relator.

— h) Two presentations with the same number of generators are always isomorphic.

— i) In a presentation of an abelian group, the set of consequences of the relators contains the commutator subgroup of the free group on the generators.

— j) Every presentation of a free group has 1 as the only relator.

---

*18.5 Show that

$$(a, b : a^3 = 1, b^2 = 1, ba = a^2b)$$

gives a group of order 6. Show that it is nonabelian.

*18.6 Show that the presentation

$$(a, b : a^3 = 1, b^2 = 1, ba = a^2b)$$

of Exercise 18.5 gives (up to isomorphism) the only nonabelian group of order 6, and hence gives a group isomorphic to S_3.

*18.7 Let

$$S = \{a^ib^j \mid 0 \le i < m, 0 \le j < n\},$$

that is, S consists of all formal products a^ib^j starting with a^0b^0 and ending with $a^{m-1}b^{n-1}$. Let r be a positive integer, and define multiplication on S by

$$(a^sb^t)(a^ub^v) = a^xb^y,$$

where x is the remainder of $s + u(r^t)$ when divided by m, and y is the remainder of $t + v$ when divided by n, in the sense of Lemma 6.1.

a) Show that a necessary and sufficient condition for the associative law to hold and for S to be a group under this multiplication is that $r^n \equiv 1 \pmod{m}$.

b) Deduce from part (a) that the group presentation

$$(a, b : a^m = 1, b^n = 1, ba = a^r b)$$

gives a group of order mn if and only if $r^n \equiv 1 \pmod{m}$.

*18.8 Determine all groups of order 14 up to isomorphism. [*Hint:* Follow the outline of Example 18.4 and use Exercise 18.7 part (b).]

*18.9 Determine all groups of order 21 up to isomorphism. [*Hint:* Follow the outline of Example 18.4 and use Exercise 18.7 part (b). It may seem that there are two presentations giving nonabelian groups. Show that they are isomorphic.]

*18.10 Show that if $n = pq$, with p and q primes and $q > p$ and $q \equiv 1 \pmod{p}$, then there is exactly one nonabelian group (up to isomorphism) of order n. Assume (as will be proved later) that the $q - 1$ nonzero elements of $\mathbf{Z}_q$ form a cyclic group $\mathbf{Z}_q^*$ under multiplication modulo q. [*Hint:* The solutions of $x^p \equiv 1 \pmod{q}$ form a cyclic subgroup of $\mathbf{Z}_q^*$ with elements $1, r, r^2, \ldots, r^{p-1}$. In the group with presentation $(a, b : a^q = 1, b^p = 1, ba = a^r b)$, we have $bab^{-1} = a^r$, so $b^i a b^{-i} = a^{(r^i)}$. Thus, since b^i generates $\langle b \rangle$ for $j = 1, \ldots, p - 1$, this presentation is isomorphic to

$$(a, b^i : a^q = 1, (b^i)^p = 1, (b^i)a = a^{(r^i)}(b^i)),$$

so all the presentations $(a, b : a^q = 1, b^p = 1, ba = a^{(r^i)}b)$ are isomorphic.]

*18.11 Determine all groups (up to isomorphism) of order 12.

*18.12 Determine all groups (up to isomorphism) of order 30.

19 | Simplicial Complexes and Homology Groups

*19.1 MOTIVATION

In Section 10 we mentioned that topology concerns sets for which we have enough of an idea of when two points are close together to be able to define a continuous function. Two such sets, or *topological spaces*, are structurally the same if there is a one-to-one function mapping one onto the other such that both this function and its inverse are continuous. Naively, this means that one space can be stretched, twisted, and otherwise deformed, without being torn or cut, to look just like the other. Thus a big sphere is topologically the same structure as a small sphere, the boundary of a circle the same structure as the boundary of a square, etc. Two spaces which are structurally the same in this sense are **homeomorphic**. Hopefully the student recognizes that *the concept of homeomorphism is to topology as the concept of isomorphism* (*where sets have the same algebraic structure*) *is to algebra*.

The *main problem of topology* is to find useful, necessary and sufficient conditions, other than just the definition, for two spaces to be homeomorphic. Sufficient conditions are hard to come by in general. Necessary conditions are a dime a dozen, but some are very important and useful. A "nice" space has associated with it various kinds of groups, namely *homology groups*, *cohomology groups*, *homotopy groups*, and *cohomotopy groups*. If two spaces are homeomorphic, it can be shown that the groups of one are isomorphic to the corresponding groups associated with the other. Thus a necessary condition for spaces to be homeomorphic is that their groups be isomorphic. Some of these groups may reflect very interesting properties of the spaces. Moreover, a continuous mapping of one space into another gives rise to homomorphisms from the groups of one into the groups of the other. These group homomorphisms may reflect interesting properties of the mapping.

If the student could make neither head nor tail out of the preceding paragraphs, he need not worry. The above paragraphs were just intended as motivation for what follows. It is the purpose of this section to describe some groups, *homology groups*, which are associated with certain simple spaces, in our work, usually some subset of the familiar Euclidean 3-space $\mathbf{R}^3$.

A word about our function notation is necessary here. We have remarked that algebraists now tend to denote the image of an element a under a

157

mapping ϕ by "$a\phi$" rather than by "$\phi(a)$". Analysts, on the other hand, are keeping the classically conventional notation "$\phi(a)$". The topologist is caught squarely in the middle. So far he has been classical, i.e., on the analysts' side. We conform to this convention in our work here on topology so that the student may easily refer to texts on topology. But remember that now when we speak of a composition $\partial_{k-1}\partial_k$ of two functions, we mean first compute ∂_k and then ∂_{k-1} of the result.

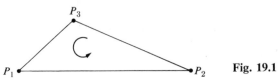

Fig. 19.1

*19.2 PRELIMINARY NOTIONS

First we introduce the idea of an *oriented n-simplex* in Euclidean 3-space $\mathbf{R}^3$ for $n = 0, 1, 2,$ and 3. An **oriented 0-simplex** is just a point P. An **oriented 1-simplex** is a *directed* line segment P_1P_2 joining the points P_1 and P_2 and imagined as traveled in the direction from P_1 to P_2. Thus $P_1P_2 \neq P_2P_1$. We will agree, however, that $P_1P_2 = -P_2P_1$. An **oriented 2-simplex** is a triangular region $P_1P_2P_3$, as in Fig. 19.1, together with a pre-scribed order of movement around the triangle, e.g., indicated by the arrow in Fig. 19.1 as the order $P_1P_2P_3$. The order $P_1P_2P_3$ is clearly the same order as $P_2P_3P_1$ and $P_3P_1P_2$, but the opposite order from $P_1P_3P_2$, $P_3P_2P_1$ and $P_2P_1P_3$. We will agree that

$$P_1P_2P_3 = P_2P_3P_1 = P_3P_1P_2 = -P_1P_3P_2 = -P_3P_2P_1 = -P_2P_1P_3.$$

Note that $P_iP_jP_k$ is equal to $P_1P_2P_3$ if

$$\begin{pmatrix} 1 & 2 & 3 \\ i & j & k \end{pmatrix}$$

is an even permutation, and is equal to $-P_1P_2P_3$ if the permutation is odd. The same could be said for an oriented 1-simplex P_1P_2. Note also that for $n = 0, 1, 2,$ an oriented n-simplex is an n-dimensional object.

The definition of an oriented 3-simplex should now be clear: An **oriented 3-simplex** is given by an ordered sequence $P_1P_2P_3P_4$ of four vertices of a solid tetrahedron, as in Fig. 19.2. We agree that $P_1P_2P_3P_4 = \pm P_iP_jP_rP_s$, depending on whether the permutation

$$\begin{pmatrix} 1 & 2 & 3 & 4 \\ i & j & r & s \end{pmatrix}$$

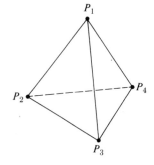

Fig. 19.2

is even or odd. Similar definitions hold for $n > 3$, but we shall stop here with dimensions that we can visualize. These simplexes are **oriented**, or have an **orientation**, meaning that we are concerned with the *order* of the vertices as well as with the actual points where the vertices are located. All our simplexes will be oriented, and we shall drop the adjective from now on.

We are now going to define the *boundary of an n-simplex* for $n = 0, 1, 2, 3$. The term *boundary* is intuitive. We define the **boundary of a 0-simplex** P to be the empty simplex which we denote this time by "0". The notation is

$$\text{``}\partial_0(P) = 0\text{''}.$$

The **boundary of a 1-simplex** P_1P_2 is defined by

$$\partial_1(P_1P_2) = P_2 - P_1,$$

that is, the formal difference of the end point and the beginning point. Likewise the **boundary of a 2-simplex** is defined by

$$\partial_2(P_1P_2P_3) = P_2P_3 - P_1P_3 + P_1P_2,$$

which we can remember by saying that it is the formal sum of terms which we obtain by dropping each P_i in succession from the 2-simplex $P_1P_2P_3$ and taking the sign to be $+$ if the first term is omitted, $-$ if the second is omitted, and $+$ if the third is omitted. Referring to Fig. 19.1, we see that this corresponds to going around what we naturally would call the boundary in the direction indicated by the orientation arrow. Note also that the equation $\partial_1(P_1P_2) = P_2 - P_1$ can be remembered in the same way. Thus we are led to the following definition of the **boundary of a 3-simplex**:

$$\partial_3(P_1P_2P_3P_4) = P_2P_3P_4 - P_1P_3P_4 + P_1P_2P_4 - P_1P_2P_3.$$

Similar definitions hold for the definition of ∂_n for $n > 3$. Each individual *summand* of the boundary of a simplex is a **face of the simplex**. Thus, $P_2P_3P_4$ is a face of $P_1P_2P_3P_4$, but $P_1P_3P_4$ is not a face. However, $P_1P_4P_3 = -P_1P_3P_4$ is a face of $P_1P_2P_3P_4$.

Suppose that you have a subset of $\mathbf{R}^3$ which is divided up "nicely" into simplexes, as for example the *surface S* of the tetrahedron in Fig. 19.2 which is split up into four 2-simplexes nicely fitted together. Thus on the surface of the tetrahedron, we have some 0-simplexes, or the vertices of the tetrahedron, some 1-simplexes, or the edges of the tetrahedron, and some 2-simplexes, or the triangles of the tetrahedron. In general, for a space to be divided up "nicely" into simplexes, we require that the following be true:

1) Each point of the space belongs to at least one simplex.

2) Each point of the space belongs to only a finite number of simplexes.

3) Two different (up to orientation) simplexes either have no points in common or one is (except possibly for orientation) a face of the other or a face of a face of the other, etc., or the set of points in common is (except possibly for orientation) a face or a face of a face etc. of each simplex.

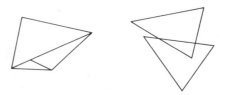

Fig. 19.3

Condition (3) excludes configurations like those shown in Fig. 19.3. A space divided up into simplexes according to these requirements is a **simplicial complex.**

*19.3   CHAINS, CYCLES, AND BOUNDARIES

Let us now describe some groups associated with a simplicial complex X. We shall illustrate each definition with the case of the *surface* S of our tetrahedron in Fig. 19.2.   The **group** $C_n(X)$ **of (oriented)** n**-chains** of X is the free abelian group generated by the (oriented) n-simplexes of X.   Thus every element of $C_n(X)$ is a finite sum of the form $\sum_i m_i \sigma_i$, where the σ_i are n-simplexes of X and $m_i \in \mathbf{Z}$. We accomplish addition of chains by taking the algebraic sum of the coefficients of each occurrence in the chains of a fixed simplex.

Example 19.1   For the surface S of our tetrahedron, every element of $C_2(S)$ is of the form

$$m_1 P_2 P_3 P_4 + m_2 P_1 P_3 P_4 + m_3 P_1 P_2 P_4 + m_4 P_1 P_2 P_3$$

for $m_i \in \mathbf{Z}$. As an illustration of addition, note that

$$(3P_2 P_3 P_4 - 5P_1 P_2 P_3) + (6P_2 P_3 P_4 - 4P_1 P_3 P_4)$$
$$= 9P_2 P_3 P_4 - 4P_1 P_3 P_4 - 5P_1 P_2 P_3.$$

An element of $C_1(S)$ is of the form

$$m_1 P_1 P_2 + m_2 P_1 P_3 + m_3 P_1 P_4 + m_4 P_2 P_3 + m_5 P_2 P_4 + m_6 P_3 P_4,$$

and an element of $C_0(S)$ is of the form

$$m_1 P_1 + m_2 P_2 + m_3 P_3 + m_4 P_4. \;\; \|$$

Now if σ is an n-simplex, $\partial_n(\sigma) \in C_{n-1}(X)$ for $n = 1, 2, 3$. Let us define $C_{-1}(X) = \{0\}$, the trivial group of one element, and then we will also have $\partial_0(\sigma) \in C_{-1}(X)$. Since $C_n(X)$ is free abelian, and since we can specify a homomorphism of such a group by giving its values on generators, we see that ∂_n gives a unique **boundary homomorphism,** which we denote again by "∂_n", mapping $C_n(X)$ into $C_{n-1}(X)$ for $n = 0, 1, 2, 3$.

Example 19.2   We have

$$\partial_n \left(\sum_i m_i \sigma_i \right) = \sum_i m_i \, \partial_n(\sigma_i).$$

For example,

$$\partial_1(3P_1P_2 - 4P_1P_3 + 5P_2P_4)$$
$$= 3\,\partial_1(P_1P_2) - 4\,\partial_1(P_1P_3) + 5\,\partial_1(P_2P_4)$$
$$= 3(P_2 - P_1) - 4(P_3 - P_1) + 5(P_4 - P_2)$$
$$= P_1 - 2P_2 - 4P_3 + 5P_4. \;\|$$

The student is reminded again that *any time you have a homomorphism, two things are of great interest, the kernel and the image.* The kernel of ∂_n consists of those n-chains with boundary 0. The elements of the kernel are **n-cycles**. The usual notation for the kernel of ∂_n, that is, the **group of n-cycles**, is "$Z_n(X)$".

Example 19.3 If $z = P_1P_2 + P_2P_3 + P_3P_1$, then

$$\partial_1(z) = (P_2 - P_1) + (P_3 - P_2) + (P_1 - P_3) = 0.$$

Thus z is a 1-cycle. However, if we let $c = P_1P_2 + 2P_2P_3 + P_3P_1$, then

$$\partial_1(c) = (P_2 - P_1) + 2(P_3 - P_2) + (P_1 - P_3) = -P_2 + P_3 \neq 0.$$

Thus $c \notin Z_1(X)$. $\;\|$

Note that $z = P_1P_2 + P_2P_3 + P_3P_1$ of Example 19.3 corresponds to one circuit or *cycle* around a triangle with vertices P_1, P_2, and P_3.

The image of ∂_n, the **group of $(n - 1)$-boundaries**, consists exactly of those $(n - 1)$-chains which are boundaries of n-chains. This group is denoted by "$B_{n-1}(X)$".

Example 19.4 Referring to Example 19.3, we see that if

$$P_1P_2 + 2P_2P_3 + P_3P_1$$

is a 1-chain in $C_1(X)$, then $P_3 - P_2$ is a 0-boundary. Note that $P_3 - P_2$ *bounds P_2P_3.* $\;\|$

Let us now compute $Z_n(X)$ and $B_n(X)$ for a more complicated example. In topology, if a group is the trivial group consisting just of the identity 0, one usually denotes it by "0" rather than "$\{0\}$". We shall follow this convention.

Example 19.5 Let us compute for $n = 0, 1, 2$ the groups $Z_n(S)$ and $B_n(S)$ for the *surface S* of the tetrahedron of Fig. 19.2.

First, for the easier cases, since the highest dimensional simplex for the surface is a 2-simplex, we have $C_3(S) = 0$, so

$$B_2(S) = \partial_3(C_3(S)) = 0.$$

Also since $C_{-1}(S) = 0$ by our definition, we see that

$$Z_0(S) = C_0(S).$$

Thus $Z_0(S)$ is free abelian on four generators, P_1, P_2, P_3, and P_4. It is easily seen that the image of a group under a homomorphism is generated by the images of generators of the original group. Thus, since $C_1(S)$ is generated by $P_1P_2, P_1P_3, P_1P_4, P_2P_3, P_2P_4$, and P_3P_4, we see that $B_0(S)$ is generated by

$$P_2 - P_1, \ P_3 - P_1, \ P_4 - P_1, \ P_3 - P_2, \ P_4 - P_2, \ P_4 - P_3.$$

However, $B_0(S)$ is not free abelian on these generators. For example, $P_3 - P_2 = (P_3 - P_1) - (P_2 - P_1)$. It is easy to see that $B_0(S)$ is free abelian on $P_2 - P_1, P_3 - P_1$, and $P_4 - P_1$.

Now let us go after the tougher group $Z_1(S)$. An element c of $C_1(S)$ is a formal sum of integral multiples of edges P_iP_j. It is clear that $\partial_1(c) = 0$ if and only if each vertex which is the beginning point of a total (counting multiplicity) of r edges of c is also the end point of exactly r edges. Thus

$$
\begin{aligned}
z_1 &= P_2P_3 + P_3P_4 + P_4P_2, \\
z_2 &= P_1P_4 + P_4P_3 + P_3P_1, \\
z_3 &= P_1P_2 + P_2P_4 + P_4P_1, \\
z_4 &= P_1P_3 + P_3P_2 + P_2P_1
\end{aligned}
$$

are all 1-cycles. These are exactly the boundaries of the individual 2-simplexes. We claim that the z_i generate $Z_1(S)$. Let $z \in Z_1(S)$, and choose a particular vertex, say P_1; let us work on edges having P_1 as an end point. These edges are P_1P_2, P_1P_3, and P_1P_4. Let the coefficient of P_1P_j in z be m_j. Then

$$z + m_2z_4 - m_4z_2$$

is again a cycle but does not contain the edges P_1P_2 or P_1P_4. Thus the only edge having P_1 as a vertex in the cycle $z + m_2z_4 - m_4z_2$ is possibly P_1P_3, but this edge could not appear with a nonzero coefficient as it would contribute a nonzero multiple of the vertex P_1 to the boundary, contradicting the fact that a cycle has boundary 0. Thus $z + m_2z_4 - m_4z_2$ consists of the edges of the 2-simplex $P_2P_3P_4$. Since in a 1-cycle each of P_2, P_3, and P_4 must serve the same number of times as a beginning and an end point of edges in the cycle, counting multiplicity, we see that

$$z + m_2z_4 - m_4z_2 = rz_1$$

for some integer r. Thus $Z_1(S)$ is generated by the z_i, actually by any three of the z_i. Since the z_i are the individual boundaries of the 2-simplexes, as we observed, we see that

$$Z_1(S) = B_1(S).$$

The student should see geometrically what this computation means in terms of Fig. 19.2.

Finally, we describe $Z_2(S)$. Now $C_2(S)$ is generated by the simplexes $P_2P_3P_4, P_3P_1P_4, P_1P_2P_4$, and $P_2P_1P_3$. If $P_2P_3P_4$ has coefficient r_1 and $P_3P_1P_4$ has coefficient r_2 in a 2-cycle, then the common edge P_3P_4 has coefficient $r_1 - r_2$ in its boundary. Thus we must have $r_1 = r_2$, and by a similar argument, in a cycle each one of the four 2-simplexes appears with the same coefficient. Thus $Z_2(S)$ is generated by

$$P_2P_3P_4 + P_3P_1P_4 + P_1P_2P_4 + P_2P_1P_3,$$

that is, $Z_2(S)$ is infinite cyclic. Again, the student should interpret this computation geometrically in terms of Fig. 19.2. ‖

*19.4 $\partial^2 = 0$ AND HOMOLOGY GROUPS

We now come to one of the most important equations in all of mathematics. We shall state it only for $n = 1, 2,$ and 3, but it holds for all $n > 0$.

Theorem 19.1 *Let X be a simplicial complex, and let $C_n(X)$ be the n-chains of X for $n = 0, 1, 2, 3$. Then the composite homomorphism $\partial_{n-1}\partial_n$ mapping $C_n(X)$ into $C_{n-2}(X)$ maps everything into 0 for $n = 1, 2, 3$. That is, for each $c \in C_n(X)$ we have $\partial_{n-1}(\partial_n(c)) = 0$. We use the nota-tion "$\partial_{n-1}\partial_n = 0$", or more briefly, "$\partial^2 = 0$".*

Proof. Since a homomorphism is completely determined by its values on generators, it is enough to check that for an n-simplex σ, we have $\partial_{n-1}(\partial_n(\sigma)) = 0$. For $n = 1$ this is obvious, since ∂_0 maps everything into 0. For $n = 2$,

$$\begin{aligned}
\partial_1(\partial_2(P_1P_2P_3)) &= \partial_1(P_2P_3 - P_1P_3 + P_1P_2) \\
&= (P_3 - P_2) - (P_3 - P_1) + (P_2 - P_1) \\
&= 0.
\end{aligned}$$

The case $n = 3$ will make an excellent exercise for the student in the defini-tion of the boundary operator (see Exercise 19.2). ∎

Corollary. *For $n = 0, 1, 2,$ and 3, $B_n(X)$ is a subgroup of $Z_n(X)$.*

Proof. For $n = 0, 1,$ and 2, we have $B_n(X) = \partial_{n+1}(C_{n+1}(X))$. Then if $b \in B_n(X)$, we must have $b = \partial_{n+1}(c)$ for some $c \in C_{n+1}(X)$. Thus

$$\partial_n(b) = \partial_n(\partial_{n+1}(c)) = 0,$$

so $b \in Z_n(X)$.

For $n = 3$, since we are not concerned with simplexes of dimension greater than 3, $B_3(X) = 0$. ∎

Definition. The factor group $H_n(X) = Z_n(X)/B_n(X)$ is the **n-dimensional homology group** of X.

Example 19.6 Let us calculate $H_n(S)$ for $n = 0, 1, 2,$ and 3 and where S is the *surface* of the tetrahedron in Fig. 19.2.

We found $Z_n(S)$ and $B_n(S)$ in Example 19.5. Now $C_3(S) = 0$, so $Z_3(S)$ and $B_3(S)$ are both 0 and hence

$$H_3(S) = 0.$$

Also, $Z_2(S)$ is infinite cyclic and we saw that $B_2(S) = 0$. Thus $H_2(S)$ is infinite cyclic, that is,

$$H_2(S) \simeq \mathbf{Z}.$$

We saw that $Z_1(S) = B_1(S)$, so the factor group $Z_1(S)/B_1(S)$ is the trivial group of one element, that is,

$$H_1(S) = 0.$$

Finally, $Z_0(S)$ was free abelian on P_1, P_2, P_3, and P_4, while $B_0(S)$ was generated by $P_2 - P_1$, $P_3 - P_1$, $P_4 - P_1$, $P_3 - P_2$, $P_4 - P_2$, and $P_4 - P_3$. We claim that every coset of $Z_0(S)/B_0(S)$ contains exactly one element of the form rP_1. Let $z \in Z_0(S)$, and suppose that the coefficient of P_2 in z is s_2, of P_3 is s_3, and of P_4 is s_4. Then

$$z - [s_2(P_2 - P_1) + s_3(P_3 - P_1) + s_4(P_4 - P_1)] = rP_1$$

for some r, so $z \in [rP_1 + B_0(S)]$, that is, any coset does contain an element of the form rP_1. If the coset also contains $r'P_1$, then $r'P_1 \in [rP_1 + B_0(S)]$, so $(r' - r)P_1$ is in $B_0(S)$. Clearly, the only multiple of P_1 which is a boundary is zero, so $r = r'$ and the coset contains exactly one element of the form rP_1. We may then choose the rP_1 as representatives of the cosets in computing $H_0(S)$. Thus $H_0(S)$ is infinite cyclic, that is,

$$H_0(S) \simeq \mathbf{Z}. \;\|$$

These definitions and computations probably seem very complicated to the student. The ideas are very natural, but we admit that they are a bit messy to write down. However, the arguments used in these calculations are typical for homology theory, i.e., if you can understand them, you will understand all our others. Furthermore, we can make them *geometrically*, looking at the picture of the space. The next section will be devoted to further computations of homology groups of certain simple but important spaces.

EXERCISES

*19.1 Assume that $c = 2P_1P_3P_4 - 4P_3P_4P_6 + 3P_3P_2P_4 + P_1P_6P_4$ is a 2-chain of a certain simplicial complex X.

a) Compute $\partial_2(c)$.   b) Is c a 2-cycle?   c) Is $\partial_2(c)$ a 1-cycle?

†*19.2 Compute $\partial_2(\partial_3(P_1P_2P_3P_4))$ and show that it is 0, completing the proof of Theorem 19.1.

*19.3 Describe $C_i(P)$, $Z_i(P)$, $B_i(P)$, and $H_i(P)$ for the space consisting of just the 0-simplex P. (This is really a trivial problem.)

*19.4 Describe $C_i(X)$, $Z_i(X)$, $B_i(X)$, and $H_i(X)$ for the space X consisting of two distinct 0-simplexes, P and P'. (*Note:* The line segment joining the two points is *not* part of the space.)

*19.5 Describe $C_i(X)$, $Z_i(X)$, $B_i(X)$, and $H_i(X)$ for the space X consisting of the 1-simplex P_1P_2.

*19.6 Mark each of the following true or false.

— a) Every boundary is a cycle.
— b) Every cycle is a boundary.
— c) $C_n(X)$ is always a free abelian group.
— d) $B_n(X)$ is always a free abelian group.
— e) $Z_n(X)$ is always a free abelian group.
— f) $H_n(X)$ is always abelian.
— g) The boundary of a 3-simplex is a 2-simplex.
— h) The boundary of a 2-simplex is a 1-chain.
— i) The boundary of a 3-cycle is a 2-chain.
— j) If $Z_n(X) = B_n(X)$, then $H_n(X)$ is the trivial group of one element.

---

*19.7 Define the following concepts so as to generalize naturally the definitions in the text given for dimensions 0, 1, 2, and 3.

a) An oriented n-simplex
b) The boundary of an oriented n-simplex
c) A face of an oriented n-simplex

*19.8 Continuing the idea of Exercise 19.7, what would be an easy way to answer a question asking you to define $C_n(X)$, $\partial_n: C_n(X) \to C_{n-1}(X)$, $Z_n(X)$, and $B_n(X)$ for a simplicial complex X perhaps containing some simplexes of dimension greater than 3?

*19.9 Following the ideas of Exercises 19.7 and 19.8, prove that $\partial^2 = 0$ in general, i.e., that $\partial_{n-1}(\partial_n(c)) = 0$ for every $c \in C_n(X)$, where n may be greater than 3.

*19.10 Let X be a simplicial complex. For an (oriented) n-simplex σ of X, the **coboundary** $\delta^{(n)}(\sigma)$ **of** σ is the $(n + 1)$-chain $\sum \tau$, where the sum is taken over all $(n + 1)$-simplexes τ which have σ as a face. That is, the simplexes τ appearing in the sum are precisely those that have σ as a *summand* of $\partial_{n+1}(\tau)$. *Orientation is important here*. Thus P_2 is a face of P_1P_2, but P_1 is not. However, P_1 is a face of P_2P_1. Let X be the simplicial complex consisting of the *solid tetrahedron* of Fig. 19.2.

a) Compute $\delta^{(0)}(P_1)$ and $\delta^{(0)}(P_4)$.
b) Compute $\delta^{(1)}(P_3P_2)$.
c) Compute $\delta^{(2)}(P_3P_2P_4)$.

*19.11 Following the idea of Exercise 19.10, let X be a simplicial complex, and let the group $C^{(n)}(X)$ of **n-cochains** be the same as the group $C_n(X)$.

a) Define $\delta^{(n)}: C^{(n)}(X) \to C^{(n+1)}(X)$ in a way analogous to the way we defined $\partial_n: C_n(X) \to C_{n-1}(X)$.
b) Show that $\delta^2 = 0$, that is, that $\delta^{(n+1)}(\delta^{(n)}(c)) = 0$ for each $c \in C^{(n)}(X)$.

*19.12 Following the ideas of Exercises 19.10 and 19.11, define the *group* $Z^{(n)}(X)$ *of n-cocycles* of X, the *group* $B^{(n)}(X)$ *of n-coboundaries* of X, and show that $B^{(n)}(X) \le Z^{(n)}(X)$.

*19.13 Following the ideas of Exercises 19.10, 19.11, and 19.12, define the *n-dimensional cohomology group* $H^{(n)}(X)$ of X. Compute $H^{(n)}(S)$ for the *surface S* of the tetrahedron of Fig. 19.2.

20 | Computations of Homology Groups

*20.1 TRIANGULATIONS

Suppose you wish to calculate homology groups for the surface of a sphere. The first thing you probably will say, if you are alert, is that the surface of a sphere is not a simplicial complex, since this surface is curved and a triangle is a plane surface. Remember that two spaces are topologically the same if one can be obtained from the other by bending, twisting, etc. Imagine our 3-simplex, the tetrahedron, to have a rubber surface and to be filled with air. If the rubber surface is flexible, like the rubber of a balloon, it will promptly deform itself into a sphere and the four faces of the tetrahedron will then appear as "triangles" drawn on the surface of the sphere. This illustrates what is meant by a *triangulation* of a space. The term *triangulation* need not refer to a division into 2-simplexes only, but is also used for a division into n-simplexes for any $n \geq 0$. If a space is divided up into pieces in such a way that near each point the space can be deformed to look like a part of some Euclidean space $\mathbf{R}^n$ and the pieces into which the space was divided appear after this deformation as part of a simplicial complex, then the original division of the space is a **triangulation of the space**. The homology groups of the space are then defined formally just as in the last section.

*20.2 INVARIANCE PROPERTIES

There are two very important *invariance properties* of homology groups, the proofs of which require quite a lot of machinery, but which are easy for us to explain roughly. First, the homology groups of a space are defined in terms of a triangulation, but actually they are the same (i.e., isomorphic) groups no matter how the space is triangulated. For example, a square region can be triangulated in many ways, two of which are shown in Fig. 20.1. The homology groups are the same no matter which triangulation is used to compute them. *This is not obvious!*

For the second invariance property, if one triangulated space is homeomorphic to another, e.g., can be deformed into the other without being torn or cut, the homology groups of the two spaces are the same (i.e., isomorphic)

166

Fig. 20.1

in each dimension n. *This is again not obvious.* We shall use both of these facts without proof.

Example 20.1 The homology groups of the surface of a sphere are the same as those for the surface of our tetrahedron in Example 19.6, since the two spaces are homeomorphic. ‖

Two important types of spaces in topology are the spheres and the cells. Let us introduce them and the usual notations. The *n*-**sphere** S^n is the set of all points a distance of 1 unit from the origin in $(n + 1)$-dimensional Euclidean space $\mathbf{R}^{n+1}$. Thus the 2-sphere S^2 is what is usually called the *surface* of a sphere in $\mathbf{R}^3$, S^1 is the rim of a circle, and S^0 is two points. Of course, the choice of 1 for the distance from the origin is not important. A 2-sphere of radius 10 is homeomorphic to one of radius 1 and homeomorphic to the surface of an ellipsoid for that matter. The *n*-**cell** or *n*-**ball** E^n is the set of all points in $\mathbf{R}^n$ a distance ≤ 1 from the origin. Thus E^3 is what you usually think of as a solid sphere, E^2 is a circular region, and E^1 is a line segment.

Example 20.2 The above remarks and the computations of Example 19.6 show that $H_2(S^2)$ and $H_0(S^2)$ are both isomorphic to $\mathbf{Z}$, and $H_1(S^2) = 0$. ‖

*20.3 CONNECTED AND CONTRACTIBLE SPACES

There is a very nice interpretation of $H_0(X)$ for a space X with a triangulation. A space is **connected** if any two points in it can be joined by a path (a concept which we won't define) lying totally in the space. If a space is not connected, then it is split up into a number of pieces, each of which is connected but no two of which can be joined by a path in the space. These pieces are the **connected components of the space**.

> **Theorem 20.1** *If a space X is triangulated into a finite number of simplexes, then $H_0(X)$ is isomorphic to $\mathbf{Z} \times \mathbf{Z} \times \cdots \times \mathbf{Z}$, and the betti number m of factors $\mathbf{Z}$ is the number of connected components of X.*

Proof. Now $C_0(X)$ is the free abelian group generated by the finite number of vertices P_i in the triangulation of X. Also, $B_0(X)$ is generated by expressions of the form

$$P_{i_2} - P_{i_1},$$

where $P_{i_1}P_{i_2}$ is an edge in the triangulation. Fix P_{i_1}. Any vertex P_{i_r} in the same connected component of X as P_{i_1} can be joined to P_{i_1} by a finite sequence

$$P_{i_1}P_{i_2}, P_{i_2}P_{i_3}, \ldots, P_{i_{r-1}}P_{i_r}$$

of edges. Then

$$P_{i_r} = P_{i_1} + (P_{i_2} - P_{i_1}) + (P_{i_3} - P_{i_2}) + \cdots + (P_{i_r} - P_{i_{r-1}}),$$

showing that $P_{i_r} \in [P_{i_1} + B_0(X)]$. Clearly, if P_{i_s} is not in the same connected component with P_{i_1}, then $P_{i_s} \notin [P_{i_1} + B_0(X)]$, since no edge joins the two components. Thus, if we select one vertex from each connected component, each coset of $H_0(X)$ contains exactly one representative which is an integral multiple of one of the selected vertices. The theorem follows at once. ∎

Example 20.3 We have at once that

$$H_0(S^n) \simeq \mathbf{Z}$$

for $n > 0$, since S^n is connected for $n > 0$. However,

$$H_0(S^0) \simeq \mathbf{Z} \times \mathbf{Z}$$

(see Exercise 19.4). Also,

$$H_0(E^n) \simeq \mathbf{Z}$$

for $n \geq 1$. ‖

A space is **contractible** if it can be compressed to a point without being torn or cut, *but always kept within the space it originally occupied.* We just state the next theorem.

Theorem 20.2 *If X is a contractible space triangulated into a finite number of simplexes, then $H_n(X) = 0$ for $n \geq 1$.*

Example 20.4 It is a fact that S^2 is not contractible. It is not too easy to prove this fact. The student will, however, probably be willing to take it as self-evident that you can't compress the "surface of a sphere" to a point without tearing it, *keeping it always within the original space S^2 that it occupied.* It is not fair to compress it all to the "center of the sphere." We saw that $H_2(S^2) \neq 0$ but is isomorphic to $\mathbf{Z}$.

Suppose, however, we consider $H_2(E^3)$, where we can regard E^3 as our solid tetrahedron of Fig. 19.2, for it is homeomorphic to E^3. The surface S of this tetrahedron is homeomorphic to S^2. The simplexes here for E^3 are the same as they are for S (or S^2), which we examined in Examples 19.5 and 19.6, except for the whole 3-simplex σ which now appears. Remember that a generator of $Z_2(S)$, and hence of $Z_2(E^3)$, was exactly the entire boundary of σ. Viewed in E^3, this is $\partial_3(\sigma)$, an element of $B_2(E^3)$, so $Z_2(E^3) = B_2(E^3)$ and $H_2(E^3) = 0$. Since E^3 is obviously contractible, this is consistent with Theorem 20.2. ‖

In general E^n is contractible for $n \geq 1$, so we have by Theorem 20.2,

$$H_i(E^n) = 0$$

for $i > 0$.

*20.4 FURTHER COMPUTATIONS

We have seen a nice interpretation for $H_0(X)$ in Theorem 20.1. As the preceding examples illustrate, the 1-cycles in a triangulated space are generated by closed curves of the space formed by edges of the triangulation. The 2-cycles can be thought of as generated by 2-spheres or other closed 2-dimensional surfaces in the space. Forming the factor group

$$H_1(X) = Z_1(X)/B_1(X)$$

amounts roughly to counting the closed curves that appear in the space which are not there simply because they appear as the boundary of a 2-dimensional piece (i.e., collection of 2-simplexes) of the space. Similarly, forming $H_2(X) = Z_2(X)/B_2(X)$ amounts roughly to counting the closed 2-dimensional surfaces in the space which can't be "filled in solid" within the space, i.e., are not boundaries of some collection of 3-simplexes. Thus for $H_1(S^2)$, every closed curve drawn on the surface of the 2-sphere bounds a 2-dimensional piece of the sphere, so $H_1(S^2) = 0$. However, the only possible closed 2-dimensional surface, S^2 itself, can't be "filled in solid" *within the whole space S^2 itself,* so $H_2(S^2)$ is free abelian on one generator.

Example 20.5 According to the reasoning above, one would expect $H_1(S^1)$ to be free abelian on one generator, i.e., isomorphic to **Z**, since the circle itself is not the boundary of a 2-dimensional part of S^1. You see, there *is* no 2-dimensional part of S^1. We compute and see whether this is indeed so.

A triangulation of S^1 is given in Fig. 20.2. Now $C_1(S^1)$ is generated by P_1P_2, P_2P_3, and P_3P_1. If a 1-chain is a cycle so that its boundary is zero, then it must contain P_1P_2 and P_2P_3 the same number of times, otherwise its boundary would contain a nonzero multiple of P_2. A similar argument holds for any two edges. Thus $Z_1(S^1)$ is generated by $P_1P_2 + P_2P_3 + P_3P_1$. Since $B_1(S^1) = \partial_2(C_2(S^1)) = 0$, there being no 2-simplexes, we see that $H_1(S^1)$ is free abelian on one generator, that is,

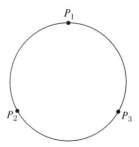

Fig. 20.2

$$H_1(S^1) \simeq \textbf{Z}. \; \|$$

*It can be proved that for $n > 0$, $H_n(S^n)$ and $H_0(S^n)$ are isomorphic to **Z**, while $H_i(S^n) = 0$ for $0 < i < n$.*

To conform to topological terminology, we shall call an element of $H_n(X)$,

that is, a coset of $B_n(X)$ in $Z_n(X)$, an "**homology class.**" Cycles in the same homology class are **homologous.**

Example 20.6 Let us compute the homology groups of a plane annular region X between two concentric circles. A triangulation is indicated in Fig. 20.3. Of course, since X is connected, it follows that

$$H_0(X) \simeq \mathbf{Z}.$$

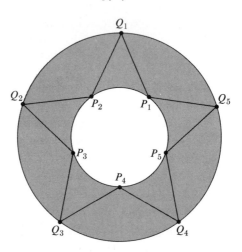

Fig. 20.3

If z is any 1-cycle, and if P_1P_2 has coefficient r in z, then $z - r\,\partial_2(P_1P_2Q_1)$ is a cycle without P_1P_2 homologous to z. By continuing this argument, we find that there is a 1-cycle homologous to z containing no edge on the inner circle of the annulus. Adjusting further by multiples of $\partial_2(Q_iP_iQ_j)$, we arrive at z' containing no edge Q_iP_i either. But then if Q_5P_1 appears in z' with nonzero coefficient, P_1 appears with nonzero coefficient in $\partial_1(z')$, contradicting the fact that z' is a cycle. Thus z is homologous to a cycle made up of edges only on the outer circle. By a familiar argument, such a cycle must be of the form

$$n(Q_1Q_2 + Q_2Q_3 + Q_3Q_4 + Q_4Q_5 + Q_5Q_1).$$

It is then clear that

$$H_1(X) \simeq \mathbf{Z}.$$

We showed that we could "push" any 1-cycle to the outer circle. Of course, we could have pushed it to the inner circle equally well.

For $H_2(X)$, note that $Z_2(X) = 0$, since every 2-simplex has in its boundary an edge on either the inner or the outer circle of the annulus which appears in no other 2-simplex. The boundary of any nonzero 2-chain must then contain some nonzero multiples of these edges. Hence

$$H_2(X) = 0. \;\|$$

Example 20.7 We shall compute the homology groups of the torus surface X which looks like the surface of a doughnut, as in Fig. 20.4. To visualize a triangulation of the torus, imagine that you cut it on the circle marked a, then cut it all around the circle marked b, and flatten it out as in Fig. 20.5. Then draw the triangles. To recover the torus from Fig. 20.5, join the left edge b to the right edge b in such a way that the arrows are going in the same direction. This gives a cylinder with circle a at each end. Then bend the cylinder around and join the circles a, again keeping the arrows going the same way around the circles.

Since the torus is connected, $H_0(X) \simeq \mathbf{Z}$.

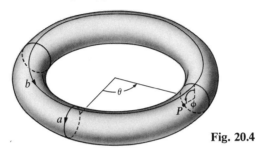

Fig. 20.4

For $H_1(X)$, let z be a 1-cycle. By changing z by a multiple of the boundary of the triangle numbered 1 in Fig. 20.5, you can get a homologous cycle not containing the side $/$ of triangle 1. Then by changing this new 1-cycle by a suitable multiple of the boundary of triangle 2, you can further eliminate the side $|$ of 2. Continuing, we can then eliminate $/$ of 3, $|$ of 4, $/$ of 5, $—$ of 6, $/$ of 7, $|$ of 8, $/$ of 9, $|$ of 10, $/$ of 11, $—$ of 12, $/$ of 13, $|$ of 14, $/$ of 15, $|$ of 16, and $/$ of 17. The resulting cycle, homologous to z, can then only contain the edges shown in Fig. 20.6. But such a cycle couldn't contain, with non-zero coefficient, any of the edges we have numbered in Fig. 20.6, or it would not have boundary 0. Thus z is homologous to a 1-cycle having edges only on the circle a or the circle b (refer to Fig. 20.4). By a now hopefully familiar argument, every edge on circle a must appear the same number of times, and the same is also true for edges on circle b; however, an edge on circle b need

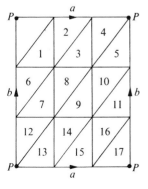

Fig. 20.5

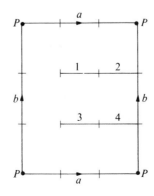

Fig. 20.6

not appear the same number of times as an edge appears on a. Furthermore, if a 2-chain is to have a boundary just containing a and b, all the triangles oriented counterclockwise must appear with the same coefficient so that the inner edges will cancel out. The boundary of such a 2-chain is 0. Thus every homology class (coset) contains one and only one element

$$ra + sb,$$

where r and s are integers. Hence $H_1(X)$ is free abelian on two generators, represented by the two circles a and b. Therefore,

$$H_1(X) \simeq \mathbf{Z} \times \mathbf{Z}.$$

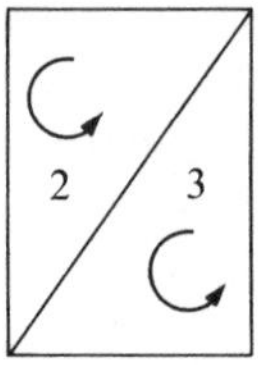

Fig. 20.7

Finally, for $H_2(X)$, a 2-cycle must contain the triangle numbered 2 of Fig. 20.5 with counterclockwise orientation the same number of times as it contains the triangle numbered 3, also with counterclockwise orientation, in order for the common edge / of these triangles not to be in the boundary. These orientations are illustrated in Fig. 20.7. The same holds true for any two adjacent triangles, and thus every triangle with the counterclockwise orientation must appear the same number of times in a 2-cycle. Clearly, any multiple of the formal sum of all the 2-simplexes, all with counterclockwise orientation, is a 2-cycle. Thus $Z_2(X)$ is infinite cyclic, isomorphic to $\mathbf{Z}$. Also $B_2(X) = 0$, there being no 3-simplexes, so

$$H_2(X) \simeq \mathbf{Z}. \;\|$$

EXERCISES

In these exercises, you need not write out in detail your computations or arguments.

*20.1 Compute the homology groups of the space consisting of two tangent 1-spheres, i.e., a figure eight.

*20.2 Compute the homology groups of the space consisting of two tangent 2-spheres.

*20.3 Compute the homology groups of the space consisting of a 2-sphere with an annular ring (as in Fig. 20.3) which does not touch the 2-sphere.

*20.4 Compute the homology groups of the space consisting of a 2-sphere with an annular ring whose inner circle is a great circle of the 2-sphere.

*20.5 Compute the homology groups of the space consisting of a circle touching a 2-sphere at one point.

*20.6 Compute the homology groups of the surface consisting of a 2-sphere with a handle (see Fig. 20.8).

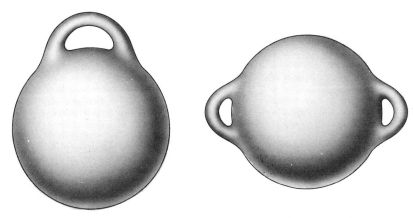

Fig. 20.8   Fig. 20.9

*20.7 Mark each of the following true or false.

— a) Homeomorphic simplicial complexes have isomorphic homology groups.
— b) If two simplicial complexes have isomorphic homology groups, then the spaces are homeomorphic.
— c) S^n is homeomorphic to E^n.
— d) $H_n(X)$ is trivial for $n > 0$ if X is a connected space with a finite triangulation.
— e) $H_n(X)$ is trivial for $n > 0$ if X is a contractible space with a finite triangulation.
— f) $H_n(S^n) = 0$ for $n > 0$.
— g) $H_n(E^n) = 0$ for $n > 0$.
— h) A torus is homeomorphic to S^2.
— i) A torus is homeomorphic to E^2.
— j) A torus is homeomorphic to a sphere with a handle on it (see Fig. 20.8).

*20.8 Compute the homology groups of the space consisting of two torus surfaces having no points in common.

*20.9 Compute the homology groups of the space consisting of two stacked torus surfaces, stacked as one would stack two inner tubes.

*20.10 Compute the homology groups of the space consisting of a torus tangent to a 2-sphere at all points of a great circle of the 2-sphere, i.e., a balloon wearing an inner tube.

*20.11 Compute the homology groups of the surface consisting of a 2-sphere with two handles (see Fig. 20.9).

*20.12 Compute the homology groups of the surface consisting of a 2-sphere with n handles (generalizing Exercises 20.6 and 20.11).

21 | More Homology Computations and Applications

*21.1 ONE-SIDED SURFACES

Thus far all the homology groups we have found have been free abelian, so that there were no nonzero elements of finite order. This can be shown always to be the case for the homology groups of a *closed surface* (i.e., a surface like S^2 which has no boundary) which has two sides. Our next example is of a *one-sided closed surface*, the *Klein bottle*. Here the 1-dimensional homology group will have a nontrivial torsion subgroup reflecting the twist in the surface.

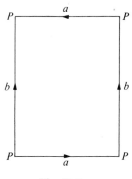

Fig. 21.1

Example 21.1 Let us calculate the homology groups of the Klein bottle X. Figure 21.1 represents the Klein bottle cut apart, just as Fig. 20.5 represents the torus cut apart. The only difference is that the arrows on the top and bottom edge a of the rectangle are in *opposite* directions this time. To recover a Klein bottle from Fig. 21.1, again bend the rectangle joining the edges labeled b so that the directions of the arrows match up. This gives a cylinder which is shown somewhat deformed, with one end pushed a little way inside the cylinder, in Fig. 21.2. Such deformations are legitimate in topology. Now the circles a must be joined so that the arrows go around the same way. *This can't be done in* $\mathbf{R}^3$. You must imagine that you are in $\mathbf{R}^4$, so that you can bend the neck of the bottle around and "through" the side without intersecting the side, as shown in Fig. 21.3. With a little thought, you can see that this resulting surface really has only one side. That is, if you start at any place and begin to paint "one side," you will wind up painting the whole thing. There is no concept of an *inside* of a Klein bottle.

We can calculate the homology groups of the Klein bottle much as we calculated the homology groups of the torus in Example 20.7, by splitting Fig. 21.1 into triangles exactly as we did for the torus. Of course,

$$H_0(X) \simeq \mathbf{Z},$$

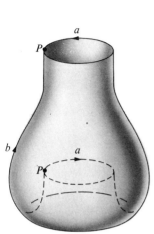

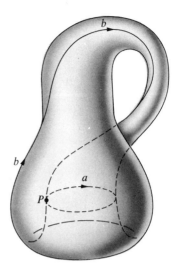

Fig. 21.2   Fig. 21.3

since X is connected. As we found for the torus, if we triangulate the Klein bottle by dividing Fig. 21.1 into triangles, every 1-cycle is homologous to a cycle of the form

$$ra + sb$$

for r and s integers. If a 2-chain is to have a boundary containing just a and b, again all the triangles oriented counterclockwise must appear with the same coefficient so that the inner edges will cancel each other. In the case of the torus, the boundary of such a 2-chain was 0. Here, however, it is $k(2a)$, where k is the number of times each triangle appears. Thus $H_1(X)$ is an abelian group with generators the homology classes of a and b and the relations $a + b = b + a$ and $2a = 0$. Therefore,

$$H_1(X) \simeq \mathbf{Z}_2 \times \mathbf{Z},$$

a group with torsion coefficient 2 and betti number 1. Our argument above regarding 2-chains shows that there are no 2-cycles this time, so

$$H_2(X) = 0. \; \|$$

A torsion coefficient does not have to be present in some homology group of a one-sided surface *with boundary*. Mostly for the sake of completeness, we give this standard example of the *Möbius strip*.

Example 21.2 Let X be the Möbius strip, which we can form by taking a rectangle of paper and joining the two ends marked a with a half twist so that the arrows match up, as indicated in Fig. 21.4. Note that the Möbius strip is a surface with a boundary, and the boundary is just one closed curve

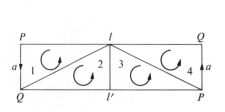
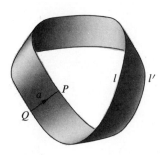

Fig. 21.4

(homeomorphic to a circle) made up of l and l'. It is clear that the Möbius strip, like the Klein bottle, has just one side, in the sense that if you were asked to color only one side of it, you would wind up coloring the whole thing.

Of course, since X is connected,

$$H_0(X) \simeq \mathbf{Z}.$$

Let z be any 1-cycle. By subtracting in succession suitable multiples of the triangles numbered 2, 3, and 4 in Fig. 21.4, we can eliminate edges / of triangle 2, | of triangle 3, and \ of triangle 4. Thus z is homologous to a cycle z' having edges on only l, l', and a, and as before, both edges on l' must appear the same number of times. But if c is a 2-chain consisting of the formal sum of the triangles oriented as shown in Fig. 21.4, we see that $\partial_2(c)$ consists of the edges on l and l' plus $2a$. Since both edges on l' must appear in z' the same number of times, by subtracting a suitable multiple of $\partial_2(c)$, we see that z is homologous to a cycle with edges just lying on l and a. By a familiar argument, all these edges properly oriented must appear the same number of times in this new cycle, and thus the homology class containing their formal sum is a generator for $H_1(X)$. Therefore,

$$H_1(X) \simeq \mathbf{Z}.$$

This generating cycle starts at P and goes around the strip, then cuts across it at Q via a, and arrives at its starting point.

If z'' were a 2-cycle, it would have to contain the triangles 1, 2, 3, and 4 of Fig. 21.4 an equal number r of times with the indicated orientation. But then $\partial_2(z'')$ would be $r(2a + l + l') \neq 0$. Thus $Z_2(X) = 0$, so

$$H_2(X) = 0. \ \|$$

*21.2   THE EULER CHARACTERISTIC

Let us turn from the computation of homology groups to a few interesting facts and applications. Let X be a finite simplicial complex (or triangulated space) consisting of simplexes of dimension 3 and less. Let n_0 be the total number of vertices in the triangulation, n_1 the number of edges, n_2 the number of 2-simplexes, and n_3 the number of 3-simplexes. The number

$$n_0 - n_1 + n_2 - n_3 = \sum_{i=0}^{3} (-1)^i n_i$$

can be shown to be the same no matter how the space X is triangulated. This number is the **Euler characteristic** $\chi(X)$ of the space. We just state the following fascinating theorem.

Theorem 21.1 *Let X be a finite simplicial complex (or triangulated space) of dimension ≤ 3. Let $\chi(X)$ be the Euler characteristic of the space X, and let β_j be the betti number of $H_j(X)$. Then*

$$\chi(X) = \sum_{j=0}^{3} (-1)^j \beta_j.$$

This theorem holds also for X of dimension greater than 3, with the obvious extension of the definition of the Euler characteristic to dimension greater than 3.

Example 21.3 Consider the solid tetrahedron E^3 of Fig. 19.2. Here $n_0 = 4$, $n_1 = 6$, $n_2 = 4$, and $n_3 = 1$, so

$$\chi(E^3) = 4 - 6 + 4 - 1 = 1.$$

Remember that we saw that $H_3(E^3) = H_2(E^3) = H_1(E^3) = 0$ and $H_0(E^3) \simeq \mathbf{Z}$. Thus $\beta_3 = \beta_2 = \beta_1 = 0$ and $\beta_0 = 1$, so

$$\sum_{j=0}^{3} (-1)^j \beta_j = 1 = \chi(E^3).$$

For the surface S^2 of the tetrahedron in Fig. 19.2, we have $n_0 = 4$, $n_1 = 6$, $n_2 = 4$, and $n_3 = 0$, so

$$\chi(S^2) = 4 - 6 + 4 = 2.$$

Also $H_3(S^2) = H_1(S^2) = 0$, and $H_2(S^2)$ and $H_0(S^2)$ are both isomorphic to $\mathbf{Z}$. Thus $\beta_3 = \beta_1 = 0$ and $\beta_2 = \beta_0 = 1$, so

$$\sum_{j=0}^{3} (-1)^j \beta_j = 2 = \chi(S^2).$$

Finally, for S^1 in Fig. 20.2, $n_0 = 3$, $n_1 = 3$, and $n_2 = n_3 = 0$, so

$$\chi(S^1) = 3 - 3 = 0.$$

Here $H_1(S^1)$ and $H_0(S^1)$ are both isomorphic to $\mathbf{Z}$, and

$$H_3(S^1) = H_2(S^1) = 0.$$

Thus $\beta_0 = \beta_1 = 1$ and $\beta_2 = \beta_3 = 0$, giving

$$\sum_{j=0}^{3} (-1)^j \beta_j = 0 = \chi(S^1). \; \|$$

*21.3 MAPPINGS OF SPACES

A continuous function f mapping a space X into a space Y gives rise to a homomorphism f_{*n} mapping $H_n(X)$ into $H_n(Y)$ for $n \geq 0$. The demonstration of the existence of this homomorphism takes more machinery than we wish to develop here, but let us attempt to describe how these homomorphisms can be computed in certain cases. The following is true:

If $z \in Z_n(X)$, and if $f(z)$, regarded as the result of picking up z and setting it down in Y in the naively obvious way, should be an n-cycle in Y, then

$$f_{*n}(z + B_n(X)) = f(z) + B_n(Y).$$

That is, if z represents a homology class in $H_n(X)$ and $f(z)$ is an n-cycle in Y, then $f(z)$ represents the image homology class under f_{*n} of the homology class containing z.

Let's illustrate this and attempt to show just what we mean here by $f(z)$.

Example 21.4 Consider the unit circle

$$S^1 = \{(x, y) \mid x^2 + y^2 = 1\}$$

in $\mathbf{R}^2$. Any point in S^1 has coordinates $(\cos \theta, \sin \theta)$, as indicated in Fig. 21.5. Let $f : S^1 \to S^1$ be given by

$$f((\cos \theta, \sin \theta)) = (\cos 3\theta, \sin 3\theta).$$

Obviously, this function f is continuous. Now f should induce

$$f_{*1}: H_1(S^1) \to H_1(S^1).$$

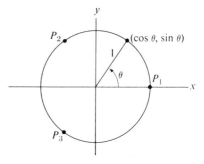

Fig. 21.5

Here $H_1(S^1)$ is isomorphic to $\mathbf{Z}$ and has as generator the homology class of $z = P_1P_2 + P_2P_3 + P_3P_1$, as seen in Example 20.5. Now if P_1, P_2, and P_3 are evenly spaced about the circle, then f maps each of the arcs P_1P_2, P_2P_3, and P_3P_1 onto the whole perimeter of the circle, that is,

$$f(P_1P_2) = f(P_2P_3) = f(P_3P_1) = P_1P_2 + P_2P_3 + P_3P_1.$$

Thus,

$$f_{*1}(z + B_1(S^1)) = 3(P_1P_2 + P_2P_3 + P_3P_1) + B_1(S^1)$$
$$= 3z + B_1(S^1),$$

that is, f_{*1} maps a generator of $H_1(S^1)$ onto three times itself. This obviously reflects the fact that f winds S^1 around itself three times. ‖

Example 21.4 illustrates our previous assertion that the homomorphisms of homology groups associated with a continuous mapping f may mirror important properties of the mapping.

Finally we use these concepts to indicate a proof of the famous *Brouwer Fixed-Point Theorem*. This theorem states that a continuous map f of E^n into itself has a *fixed point*, i.e., there is some $x \in E^n$ such that $f(x) = x$. Let us see what this means for E^2, a circular region. Imagine that you have a thin sheet of rubber stretched out on a table to form a circular disk. Mark with a pencil the outside boundary of the rubber circle on the table. Then stretch, compress, bend, twist, and fold the rubber in any fashion without tearing it, but keep it always within the penciled circle on the table. When you finish, some point on the rubber will be over exactly the same point on the table at which it first started.

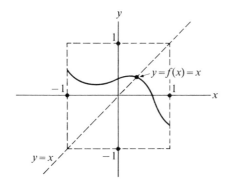

Fig. 21.6

The proof we outline is good for any $n > 1$. For $n = 1$, looking at the graph of a function $f: E^1 \rightarrow E^1$, we find that the theorem simply states that any continuous path joining the left and right sides of a square must cross the diagonal somewhere, as indicated in Fig. 21.6. The student should visualize the construction of our proof with E^3 having boundary S^2 and E^2 having boundary S^1. The proof contains a figure illustrating the construction for the case of E^2.

Theorem 21.2 (Brouwer Fixed-Point Theorem). *A continuous map f of E^n into itself has a fixed point for $n \geq 1$.*

Proof. The case $n = 1$ was considered above. Let f be a map of E^n into E^n for $n > 1$. We shall assume that f has no fixed point and shall derive a contradiction.

If $f(x) \neq x$ for all $x \in E^n$, we can consider the line segment from $f(x)$ to x. Let us extend this line segment *in the direction from $f(x)$ to x* until it goes through the boundary S^{n-1} of E^n at some point y. This defines for us a function g with $g(x) = y$, as illustrated in Fig. 21.7. Note that for y on the boundary, we have $g(y) = y$. Now since f is continuous, it is pretty obvious that g is also continuous. (A continuous function is roughly one that maps points that are sufficiently close to-gether into points that are close together. If x_1 and x_2 are sufficiently close together, then $f(x_1)$ and $f(x_2)$ are sufficiently close together so that the line segment joining $f(x_1)$ and x_1 is so close to the line segment joining $f(x_2)$ and x_2 that $y_1 = g(x_1)$ is close to $y_2 = g(x_2)$.) Then g is a continuous mapping of E^n into S^{n-1}, and thus induces a homomorphism

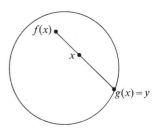

Fig. 21.7

$$g_{*(n-1)}: H_{n-1}(E^n) \to H_{n-1}(S^{n-1}).$$

Now we said that $H_{n-1}(E^n) = 0$, for $n > 1$, since E^n is contractible, and we checked it for $n = 2$ and $n = 3$. Since $g_{*(n-1)}$ is a homomorphism, we must have $g_{*(n-1)}(0) = 0$. But an $(n-1)$-cycle representing the homology class 0 of $H_{n-1}(E^n)$ is the whole complex S^{n-1} with proper orientation of simplexes, and $g(S^{n-1}) = S^{n-1}$, since $g(y) = y$ for all $y \in S^{n-1}$. Thus

$$g_{*(n-1)}(0) = S^{n-1} + B_{n-1}(S^{n-1}).$$

which is a generator $\neq 0$ of $H_{n-1}(S^{n-1})$, a contradiction. ∎

We find the preceding proof very satisfying aesthetically, and hope you agree.

EXERCISES

*21.1 Verify by direct calculation that both triangulations of the square region X in Fig. 20.1 give the same value for the Euler characteristic $\chi(X)$.

*21.2 Illustrate Theorem 21.1, as we did in Example 21.3, for each of the following spaces.

a) The annular region of Example 20.6
b) The torus of Example 20.7
c) The Klein bottle of Example 21.1

*21.3 Will the Brouwer Fixed-Point Theorem hold for a continuous map of a square region into itself? Why? Will it hold for a continuous map of a space consisting of two disjoint 2-cells into itself? Why?

*21.4 Compute the homology groups of the space consisting of a 2-sphere touching a Klein bottle at one point.

*21.5 Compute the homology groups of the space consisting of two Klein bottles with no points in common.

*21.6 Mark each of the following true or false.

— a) Every homology group of a contractible space is the trivial group of one
element.

— b) A continuous map from a simplicial complex X into a simplicial complex Y
induces a homomorphism of $H_n(X)$ into $H_n(Y)$.

— c) All homology groups are abelian.

— d) All homology groups are free abelian.

— e) All 0-dimensional homology groups are free abelian.

— f) If a space X has n-simplexes but none of dimension greater than n and
$H_n(X) \neq 0$, then $H_n(X)$ is free abelian.

— g) The boundary of an n-chain is an $(n - 1)$-chain.

— h) The boundary of an n-chain is an $(n - 1)$-cycle.

— i) The n-boundaries form a subgroup of the n-cycles.

— j) The n-dimensional homology group of a simplicial complex is always a
subgroup of the group of n-chains.

---

*21.7 Find the Euler characteristic of a 2-sphere with n handles (see Exercise 20.12).

*21.8 We can form the topological *real projective plane* X, using Fig. 21.8, by joining the semicircles a so that the directions of the arrows match up. *This can't be
done in Euclidean 3-space* $\mathbf{R}^3$. One must go to $\mathbf{R}^4$. Triangulate this space X, starting
with the form exhibited in Fig. 21.8, and compute its homology groups.

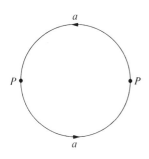

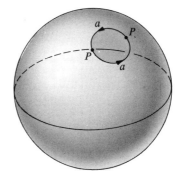

Fig. 21.8   Fig. 21.9

*21.9 The circular disk shown in Fig. 21.8 can be deformed topologically to appear
as a 2-sphere with a hole in it, as shown in Fig. 21.9. We form the real projective
plane from this configuration by sewing up the hole in such a way that only diametrically opposite points on the rim of the hole are sewn together. This can't
be done in $\mathbf{R}^3$.

Extending this idea, a 2-sphere with q holes in it, which are then sewn up by
bringing together diametrically opposite points on the rims of the holes, gives a
2-sphere with q cross caps. Find the homology groups of a 2-sphere with q cross
caps. (To see a triangulation, view the space as the disk in Fig. 21.8 but with
$q - 1$ holes in it to be sewn up as described above. Then triangulate this disk with
these holes.)

Comment: It can be shown that every sufficiently nice closed surface, namely a *closed 2-manifold*, is homeomorphic to a 2-sphere with some number $h \geq 0$ of handles if the surface is two sided, and is homeomorphic to a 2-sphere with $q > 0$ cross caps if it is one sided. The number h or q, as the case may be, is the **genus of the surface**.

Fig. 21.10

*21.10 Every point P on a regular torus X can be described by means of two angles θ and ϕ, as shown in Fig. 21.10. That is, we can associate *coordinates* (θ, ϕ) with P. For each of the mappings f of the torus X onto itself given below, describe the induced map f_{*n} of $H_n(X)$ into $H_n(X)$ for $n = 0, 1$, and 2, by finding the images of the generators for $H_n(X)$ described in Example 20.7. Interpret these group homomorphisms geometrically as we did in Example 21.4.

a) $f: X \to X$ given by $f((\theta, \phi)) = (2\theta, \phi)$
b) $f: X \to X$ given by $f((\theta, \phi)) = (\theta, 2\phi)$
c) $f: X \to X$ given by $f((\theta, \phi)) = (2\theta, 2\phi)$

*21.11 With reference to Exercise 21.10, the torus X can be mapped onto its circle b (which is homeomorphic to S^1) by a variety of maps. For each such map $f: X \to b$ given below, describe the homomorphism $f_{*n}: H_n(X) \to H_n(b)$ for $n = 0, 1$, and 2, by describing the image of generators of $H_n(X)$ as in Exercise 21.10.

a) $f: X \to b$ given by $f((\theta, \phi)) = (\theta, 0)$
b) $f: X \to b$ given by $f((\theta, \phi)) = (2\theta, 0)$

*21.12 Repeat Exercise 21.11, but view the map f as a map of the torus X into itself, inducing maps $f_{*n}: H_n(X) \to H_n(X)$.

*21.13 Consider the map f of the Klein bottle in Fig. 21.1 given by mapping a point Q of the rectangle in Fig. 21.1 onto the point of b directly opposite it. Note that b is topologically a 1-sphere. Compute the induced maps $f_{*n}: H_n(X) \to H_n(b)$ for $n = 0, 1$, and 2, by describing images of generators of $H_n(X)$.

*21.14 Repeat Exercise 21.13 with the map f sending each point Q of the Klein bottle onto the point of the 1-sphere a directly above it in Fig. 21.1.

22 | Homological Algebra

*22.1 CHAIN COMPLEXES AND MAPPINGS

The subject of algebraic topology is responsible for a great surge in a new direction in algebra. You see, if you have a simplicial complex X, then you naturally get chain groups $C_k(X)$ and maps ∂_k, as indicated in the diagram

$$C_n(X) \xrightarrow{\partial_n} C_{n-1}(X) \xrightarrow{\partial_{n-1}} \cdots \xrightarrow{\partial_2} C_1(X) \xrightarrow{\partial_1} C_0(X) \xrightarrow{\partial_0} 0,$$

with $\partial_{k-1}\partial_k = 0$. You then abstract the purely algebraic portion of this situation and consider any sequence of abelian groups A_k and homomorphisms $\partial_k \colon A_k \to A_{k-1}$ such that $\partial_{k-1}\partial_k = 0$ for $k \geq 1$. So that you don't always have to require $k \geq 1$ in $\partial_{k-1}\partial_k = 0$, it is convenient to consider "doubly infinite" sequences of groups A_k for *all* $k \in \mathbf{Z}$. Often $A_k = 0$ for $k < 0$ and $k > n$ in applications. The study of such sequences and maps of such sequences is a topic of *homological algebra*.

Definition. A *chain complex* $\langle A, \partial \rangle$ is a doubly infinite sequence

$$A = \{\ldots, A_2, A_1, A_0, A_{-1}, A_{-2}, \ldots\}$$

of abelian groups A_k, together with a collection $\partial = \{\partial_k \mid k \in \mathbf{Z}\}$ of homomorphisms such that $\partial_k \colon A_k \to A_{k-1}$ and $\partial_{k-1}\partial_k = 0$.

As a convenience similar to our notation in group theory, we shall be sloppy and let "A" denote the chain complex $\langle A, \partial \rangle$. We can now imitate in a completely algebraic setting our constructions and definitions of Section 19.

Theorem 22.1 *If A is a chain complex, then the image of ∂_k is a subgroup of the kernel of ∂_{k-1}.*

Proof. Consider

$$A_k \xrightarrow{\partial_k} A_{k-1} \xrightarrow{\partial_{k-1}} A_{k-2}.$$

Now $\partial_{k-1}\partial_k = 0$, since A is a chain complex. That is, $\partial_{k-1}(\partial_k(A_k)) = 0$. This tells us at once that $\partial_k(A_k)$ is contained in the kernel of ∂_{k-1}, which is what we wished to prove. ∎

Definition. If A is a chain complex, then the kernel $Z_k(A)$ of ∂_k is the **group of k-cycles**, and the image $B_k(A) = \partial_{k+1}(A_{k+1})$ is the **group of k-boundaries**. The factor group $H_k(A) = Z_k(A)/B_k(A)$ is the **kth homology group of A**.

We stated in the last section that for simplicial complexes X and Y, a continuous mapping f from X into Y induces a homomorphism of $H_k(X)$ into $H_k(Y)$. This mapping of the homology groups arises in the following way. For suitable triangulations of X and Y, the mapping f gives rise to a homomorphism f_k of $C_k(X)$ into $C_k(Y)$ which has the important property that *it commutes with ∂_k*, that is,

$$\partial_k f_k = f_{k-1}\partial_k.$$

Let us turn to the purely algebraic situation and see how this induces a map of the homology groups.

Theorem 22.2 (Fundamental Lemma). *Let A and A' with collections ∂ and ∂' of homomorphisms be chain complexes, and suppose that there is a collection f of homomorphisms $f_k\colon A_k \to A'_k$ as indicated in the diagram*

$$\cdots \xrightarrow{\partial_{k+2}} A_{k+1} \xrightarrow{\partial_{k+1}} A_k \xrightarrow{\partial_k} A_{k-1} \xrightarrow{\partial_{k-1}} \cdots$$

$$\left\downarrow f_{k+1}\right. \qquad \left\downarrow f_k\right. \qquad \left\downarrow f_{k-1}\right.$$

$$\cdots \xrightarrow{\partial'_{k+2}} A'_{k+1} \xrightarrow{\partial'_{k+1}} A'_k \xrightarrow{\partial'_k} A'_{k-1} \xrightarrow{\partial'_{k-1}} \cdots$$

Suppose furthermore that every square is commutative, that is,

$$f_{k-1}\partial_k = \partial'_k f_k$$

for all k. Then f_k induces a natural homomorphism $f_{*k}\colon H_k(A) \to H_k(A')$.

Proof. Let $z \in Z_k(A)$. Now

$$\partial'_k(f_k(z)) = f_{k-1}(\partial_k(z)) = f_{k-1}(0) = 0,$$

so $f_k(z) \in Z_k(A')$. Let us attempt to define $f_{*k}\colon H_k(A) \to H_k(A')$ by

$$f_{*k}(z + B_k(A)) = f_k(z) + B_k(A'). \tag{1}$$

We must first show that f_{*k} is well defined, i.e., independent of our choice of a representative of $z + B_k(A)$. Suppose that $z_1 \in (z + B_k(A))$. Then $(z_1 - z) \in B_k(A)$, so there exists $c \in A_{k+1}$ such that $z_1 - z = \partial_{k+1}(c)$. But then

$$f_k(z_1) - f_k(z) = f_k(z_1 - z) = f_k(\partial_{k+1}(c)) = \partial'_{k+1}(f_{k+1}(c)),$$

and this last term is an element of $\partial'_{k+1}(A'_{k+1}) = B_k(A')$. Hence

$$f_k(z_1) \in (f_k(z) + B_k(A')).$$

Thus two representatives of the same coset in $H_k(A) = Z_k(A)/B_k(A)$ are mapped into representatives of just one coset in $H_k(A') = Z_k(A')/B_k(A')$. This shows that $f_{*k} \colon H_k(A) \to H_k(A')$ is well defined by equation (1).

Now we compute f_{*k} by taking f_k of representatives of cosets, and we define the group operation of a factor group by applying the group operation of the original group to representatives of cosets. It follows at once from the fact that the action of f_k on $Z_k(A)$ is a homomorphism of $Z_k(A)$ into $Z_k(A')$ that f_{*k} is a homomorphism of $H_k(A)$ into $H_k(A')$. ▮

If the collections of maps f, ∂, and ∂' have the property, given in Theorem 22.2, that the squares are commutative, then f **commutes with** ∂.

After another definition, we shall give a seemingly trivial but very important illustration of Theorem 22.2.

Definition. A chain complex $\langle A', \partial' \rangle$ is a **subcomplex of a chain complex** $\langle A, \partial \rangle$, if, for all k, A'_k is a subgroup of A_k and $\partial'_k(c) = \partial_k(c)$ for every $c \in A'_k$, that is, ∂'_k and ∂_k have the same effect on elements of the subgroup A'_k of A_k.

Example 22.1 Let A be a chain complex, and let A' be a subcomplex of A. Let i be the collection of injection mappings $i_k \colon A'_k \to A_k$ given by $i_k(c) = c$ for $c \in A'_k$. It is obvious that i commutes with ∂. Thus we have induced homomorphisms $i_{*k} \colon H_k(A') \to H_k(A)$. One might naturally suspect that i_{*k} must be an isomorphic mapping of $H_k(A')$ into $H_k(A)$. *This need not be so!* For example, let us consider the 2-sphere S^2 as a subcomplex of the 3-cell E^3. This gives rise to $i_2 \colon C_2(S^2) \to C_2(E^3)$ and induces

$$i_{*2} \colon H_2(S^2) \to H_2(E^3).$$

But we have seen that $H_2(S^2) \simeq \mathbf{Z}$, while $H_2(E^3) = 0$. Thus i_{*2} can't possibly be an isomorphic mapping. ‖

*22.2   RELATIVE HOMOLOGY

Suppose that A' is a subcomplex of the chain complex A. The topological situation from which this arises is the consideration of a *simplicial subcomplex* Y (in the obvious sense) of a simplicial complex X. We can then naturally consider $C_k(Y)$ a subgroup of $C_k(X)$, just as in the algebraic situation where we have A'_k a subgroup of A_k. Clearly, we would have

$$\partial_k\bigl(C_k(Y)\bigr) \leq C_{k-1}(Y).$$

Let us deal now with the algebraic situation and remember that it can be applied to our topological situation at any time.

If A' is a subcomplex of the chain complex A, we can form the collection A/A' of factor groups A_k/A'_k. We claim that A/A' again gives rise to a chain complex in a natural way, and we must exhibit a collection $\bar{\partial}$ of homomorphisms

$$\bar{\partial}_k \colon (A_k/A'_k) \to (A_{k-1}/A'_{k-1})$$

such that $\bar{\partial}_{k-1}\bar{\partial}_k = 0$. The definition of $\bar{\partial}_k$ to attempt is obvious, namely define

$$\bar{\partial}_k(c + A'_k) = \partial_k(c) + A'_{k-1}$$

for $c \in A_k$. We have to show three things: that $\bar{\partial}_k$ is well defined, that it is a homomorphism, and that $\bar{\partial}_{k-1}\bar{\partial}_k = 0$.

First, to show that $\bar{\partial}_k$ is well defined, let c_1 also be in $c + A'_k$. Then $(c_1 - c) \in A'_k$, so $\partial_k(c_1 - c) \in A'_{k-1}$. Thus

$$\partial_k(c_1) \in \big(\partial_k(c) + A'_{k-1}\big)$$

also. This shows that $\bar{\partial}_k$ is well defined.

The equation

$$
\begin{aligned}
\bar{\partial}_k[(c_1 + A'_k) + (c_2 + A'_k)] &= \bar{\partial}_k[(c_1 + c_2) + A'_k] \\
&= \partial_k(c_1 + c_2) + A'_{k-1} \\
&= \big(\partial_k(c_1) + \partial_k(c_2)\big) + A'_{k-1} \\
&= \bar{\partial}_k(c_1 + A'_k) + \bar{\partial}_k(c_2 + A'_k)
\end{aligned}
$$

shows that $\bar{\partial}_k$ is a homomorphism.

Finally,

$$
\begin{aligned}
\bar{\partial}_{k-1}\big(\bar{\partial}_k(c + A'_k)\big) &= \bar{\partial}_{k-1}\big(\partial_k(c) + A'_{k-1}\big) \\
&= \partial_{k-1}\big(\partial_k(c)\big) + A'_{k-2} = 0 + A'_{k-2},
\end{aligned}
$$

so $\bar{\partial}_{k-1}\bar{\partial}_k = 0$.

The preceding arguments are typical routine computations to the homo-logical algebraist, just as addition and multiplication of integers are routine to you. We gave them in great detail. One has to be a little careful to keep track of *dimension*, i.e., to keep track of subscripts. Actually, the expert in homological algebra usually doesn't write most of these indices, but he always knows precisely with which group he is working. We gave all the indices so that you could keep track of exactly which groups were under consideration. Let us summarize the above work in a theorem.

Theorem 22.3 *If A' is a subcomplex of the chain complex A, then the collection A/A' of factor groups A_k/A'_k, together with the collection $\bar{\partial}$ of homomorphisms $\bar{\partial}_k$ defined by*

$$\bar{\partial}_k(c + A'_k) = \partial_k(c) + A'_{k-1}$$

for $c \in A_k$, is a chain complex.

Since A/A' is a chain complex, we can then form the homology groups $H_k(A/A')$.

Definition. The homology group $H_k(A/A')$ is the **kth relative homology group of A modulo A'**.

In our topological situation where Y is a subcomplex of a simplicial complex X, we shall conform to the usual notation of topologists and denote the kth relative homology group arising from the subcomplex $C(Y)$ of the chain complex $C(X)$ by "$H_k(X, Y)$". All the chains of Y are thus "set equal to 0." Geometrically, this corresponds to shrinking Y to a point.

Example 22.2 Let X be the simplicial complex consisting of the edges (excluding the inside) of the triangle in Fig. 22.1, and let Y be the subcomplex consisting of the edge P_2P_3. We have seen that $H_1(X) \simeq H_1(S^1) \simeq \mathbf{Z}$. Shrinking P_2P_3 to a point collapses the rim of the triangle, as shown in Fig. 22.2. The result is still topologically the same as S^1. Thus, we would expect again to have $H_1(X, Y) \simeq \mathbf{Z}$.

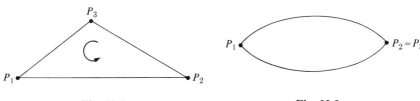

Fig. 22.1   Fig. 22.2

Generators for $C_1(X)$ are P_1P_2, P_2P_3, and P_3P_1. Since $P_2P_3 \in C_1(Y)$, we see that generators of $C_1(X)/C_1(Y)$ are

$$P_1P_2 + C_1(Y) \qquad \text{and} \qquad P_3P_1 + C_1(Y).$$

To find $Z_1(X, Y)$ we compute

$$\bar{\partial}_1(nP_1P_2 + mP_3P_1 + C_1(Y)) = \partial_1(nP_1P_2) + \partial_1(mP_3P_1) + C_0(Y)$$
$$= n(P_2 - P_1) + m(P_1 - P_3) + C_0(Y)$$
$$= (m - n)P_1 + C_0(Y),$$

since $P_2, P_3 \in C_0(Y)$. Thus for a cycle, we must have $m = n$, so a generator of $Z_1(X, Y)$ is $(P_1P_2 + P_3P_1) + C_1(Y)$. Since $B_1(X, Y) = 0$, we see that indeed

$$H_1(X, Y) \simeq \mathbf{Z}.$$

Since $P_1 + C_0(Y)$ generates $Z_0(X, Y)$ and

$$\bar{\partial}_1(P_2P_1 + C_1(Y)) = (P_1 - P_2) + C_0(Y) = P_1 + C_0(Y),$$

we see that $H_0(X, Y) = 0$. This is characteristic of relative homology groups of dimension 0 for connected simplicial complexes. ∥

Example 22.3 Let us consider S^1 as a subcomplex (the boundary) of E^2 and compute $H_2(E^2, S^1)$. Remember that E^2 is a circular disk, so S^1 can be indeed thought of as its boundary (see Fig. 22.3). You can demonstrate the shrinking of S^1 to a point by putting a drawstring around the edge of a circular piece of cloth and then drawing the string so that the rim of the circle comes in to one point. The resulting space is then a closed bag or S^2. Thus, while $H_2(E^2) = 0$, since E^2 is a contractible space, we would expect

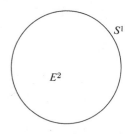

$$H_2(E^2, S^1) \simeq \mathbf{Z}.$$

Fig. 22.3

For purposes of computation, we can regard E^2 topologically as the triangular region of Fig. 22.1 and S^1 as the rim of the triangle. Then $C_2(E^2, S^1)$ is generated by $P_1P_2P_3 + C_2(S^1)$, and

$$\bar{\partial}_2(P_1P_2P_3 + C_2(S^1)) = \partial_2(P_1P_2P_3) + C_1(S^1)$$
$$= (P_2P_3 - P_1P_3 + P_1P_2) + C_1(S^1).$$

But $(P_2P_3 - P_1P_3 + P_1P_2) \in C_1(S^1)$, so we have

$$\bar{\partial}_2(P_1P_2P_3 + C_2(S^1)) = 0.$$

Hence $P_1P_2P_3 + C_2(S^1)$ is an element of $Z_2(E^2, S^1)$. Since

$$B_2(E^2, S^1) = 0,$$

we see that

$$H_2(E^2, S^1) \simeq \mathbf{Z},$$

as we expected. ‖

*22.3   THE EXACT HOMOLOGY SEQUENCE OF A PAIR

We now describe the *exact homology sequence of a pair* and give an application. We shall not carry out all the details of the computations. The computations are routine and straightforward. We shall give all the necessary definitions, and shall let the student supply the details in the exercises.

Lemma 22.1 *Let A' be a subcomplex of a chain complex A. Let j be the collection of natural homomorphisms $j_k\colon A_k \to (A_k/A'_k)$. Then*

$$j_{k-1}\partial_k = \bar{\partial}_k j_k,$$

that is, j commutes with ∂.

Proof. We leave this easy computation to the exercises (see Exercise 22.12). ∎

Theorem 22.4 *The map j_k of Lemma 22.1 induces a natural homomorphism*

$$j_{*k}\colon H_k(A) \to H_k(A/A').$$

Proof. This is immediate from Lemma 22.1 and Theorem 22.2. ∎

Let A' be a subcomplex of the chain complex A. Let $h \in H_k(A/A')$. Then $h = z + B_k(A/A')$ for $z \in Z_k(A/A')$, and in turn $z = c + A'_k$ for some $c \in A_k$. (Note that we arrive at c from h by two successive choices of representatives.) Now $\bar{\partial}_k(z) = 0$, which implies that $\partial_k(c) \in A'_{k-1}$. This, together with $\partial_{k-1}\partial_k = 0$, gives us $\partial_k(c) \in Z_{k-1}(A')$. Define

$$\partial_{*k} \colon H_k(A/A') \to H_{k-1}(A')$$

by

$$\partial_{*k}(h) = \partial_k(c) + B_{k-1}(A').$$

This definition of ∂_{*k} looks very complicated. Think of it as follows. Start with an element of $H_k(A/A')$. Now such an element is represented by a relative k-cycle modulo A'. To say it is a relative k-cycle modulo A' is to say that its boundary is in A'_{k-1}. Since its boundary is in A'_{k-1} and is a boundary of something in A_k, this boundary must be a $(k-1)$-cycle in A'_{k-1}. Thus starting with $h \in H_k(A/A')$, we have arrived at a $(k-1)$-cycle representing a homology class in $H_{k-1}(A')$.

Lemma 22.2 *The map* $\partial_{*k} \colon H_k(A/A') \to H_{k-1}(A')$*, which we have just defined, is well defined, and is a homomorphism of* $H_k(A/A')$ *into* $H_{k-1}(A')$*.*

Proof. We leave this proof to the exercises (see Exercise 22.13). ∎

Let i_{*k} be the map of Example 22.1. We now can construct the following diagram.

$$\cdots \xrightarrow{\partial_{*k+1}} H_k(A') \xrightarrow{i_{*k}} H_k(A) \xrightarrow{j_{*k}} H_k(A/A')$$

$$\xrightarrow{\partial_{*k}} H_{k-1}(A') \xrightarrow{i_{*k-1}} H_{k-1}(A) \xrightarrow{j_{*k-1}} H_{k-1}(A/A') \xrightarrow{\partial_{*k-1}} \cdots \qquad (1)$$

Lemma 22.3 *The groups in diagram (1), together with the given maps, form a chain complex.*

Proof. You need only check that a sequence of two consecutive maps always gives 0. We leave this for the exercises (see Exercise 22.14). ∎

Since diagram (1) gives a chain complex, we could (horrors!) ask for the homology groups of this chain complex. We have been aiming at this question, the answer to which is actually quite easy. *All the homology groups of this chain complex are* 0. You may think that such a chain complex is uninteresting. Far from it. Such a chain complex even has a special name.

Definition. A sequence of groups A_k and homomorphisms ∂_k forming a chain complex is an ***exact sequence*** if all the homology groups of the chain complex are 0, that is, if for all k we have that the image of ∂_k is equal to the kernel of ∂_{k-1}.

Exact sequences are of great importance in topology. We shall give some elementary properties of them in the exercises.

Theorem 22.5 *The groups and maps of the chain complex in diagram (1) form an exact sequence.*

Proof. We leave this proof to the exercises (see Exercise 22.15). ∎

Definition. The exact sequence in diagram (1) is the **exact homology sequence of the pair** (A, A').

Example 22.4 Let us now give an application of Theorem 22.5 to topology. We have stated without proof that $H_n(S^n) \simeq \mathbf{Z}$ and $H_0(S^n) \simeq \mathbf{Z}$, but that $H_k(S^n) = 0$ for $k \neq 0, n$. We have also stated without proof that $H_k(E^n) = 0$ for $k \neq 0$, since E^n is contractible. Let us assume the result for E^n and now *derive* from this the result for S^n.

We can view S^n as a subcomplex of the simplicial complex E^{n+1}. For example, E^{n+1} is topologically equivalent to an $(n+1)$-simplex, and S^n is topologically equivalent to its boundary. Let us form the exact homology sequence of the pair (E^{n+1}, S^n). We have

$$\underbrace{H_{n+1}(S^n)}_{=\,0} \xrightarrow{i_{*n+1}} \underbrace{H_{n+1}(E^{n+1})}_{=\,0} \xrightarrow{j_{*n+1}} \underbrace{H_{n+1}(E^{n+1}, S^n)}_{\simeq\,\mathbf{Z}} \xrightarrow{\partial_{*n+1}}$$

$$\underbrace{H_n(S^n)}_{=\,?} \xrightarrow{i_{*n}} \underbrace{H_n(E^{n+1})}_{=\,0} \xrightarrow{j_{*n}} \underbrace{H_n(E^{n+1}, S^n)}_{=\,0} \xrightarrow{\partial_{*n}} \cdots \xrightarrow{j_{*k+1}}$$

$$\underbrace{H_{k+1}(E^{n+1}, S^n)}_{=\,0} \xrightarrow{\partial_{*k+1}} \underbrace{H_k(S^n)}_{=\,?} \xrightarrow{i_{*k}} \underbrace{H_k(E^{n+1})}_{=\,0} \xrightarrow{j_{*k}} \cdots \qquad (2)$$

for $1 \leq k < n$. The fact that E^{n+1} is contractible gives $H_k(E^{n+1}) = 0$ for $k \geq 1$. We have indicated this on diagram (2). Viewing E^{n+1} as an $(n+1)$-simplex and S^n as its boundary, we see that $C_k(E^{n+1}) \leq C_k(S^n)$ for $k \leq n$. Therefore $H_k(E^{n+1}, S^n) = 0$ for $k \leq n$. We also indicated this on diagram (2). Just as in Example 22.3, one sees that $H_{n+1}(E^{n+1}, S^n) \simeq \mathbf{Z}$, with a generating homology class containing as representative

$$P_1 P_2 \cdots P_{n+2} + C_{n+1}(S^n).$$

For $1 \leq k < n$, the exact sequence in the last row of diagram (2) tells us that $H_k(S^n) = 0$, for from $H_k(E^{n+1}) = 0$, we see that

$$(\text{kernel } i_{*k}) = H_k(S^n).$$

But from $H_{k+1}(E^{n+1}, S^n) = 0$, we see that (image ∂_{*k+1}) = 0. From exactness, (kernel i_{*k}) = (image ∂_{*k+1}), so $H_k(S^n) = 0$ for $1 \leq k < n$.

The following chain of reasoning leads to $H_n(S^n) \simeq \mathbf{Z}$. Refer to diagram (2) above.

a) Since $H_{n+1}(E^{n+1}) = 0$, we have (image j_{*n+1}) $= 0$.
b) Hence (kernel ∂_{*n+1}) $=$ (image j_{*n+1}) $= 0$ by exactness, that is, ∂_{*n+1} is an isomorphic mapping.
c) Therefore (image ∂_{*n+1}) $\simeq \mathbf{Z}$.
d) Since $H_n(E^{n+1}) = 0$, we have (kernel i_{*n}) $= H_n(S^n)$.
e) By exactness, (image ∂_{*n+1}) $=$ (kernel i_{*n}), so $H_n(S^n) \simeq \mathbf{Z}$.

Thus we see that $H_n(S^n) \simeq \mathbf{Z}$ and $H_k(S^n) = 0$ for $1 \le k < n$.

Since S^n is connected, $H_0(S^n) \simeq \mathbf{Z}$. This fact could also be deduced from the exact sequence

$$\underbrace{H_1(E^{n+1}, S^n)}_{= 0} \xrightarrow{\partial_{*1}} \underbrace{H_0(S^n)}_{\simeq \mathbf{Z}} \xrightarrow{i_{*0}} \underbrace{H_0(E^{n+1})}_{\simeq \mathbf{Z}} \xrightarrow{j_{*0}} \underbrace{H_0(E^{n+1}, S^n).}_{= 0} \parallel$$

EXERCISES

*22.1 Let A and B be additive groups, and suppose that the sequence

$$0 \to A \xrightarrow{f} B \to 0$$

is exact. Show that $A \simeq B$.

*22.2 Let A, B, and C be additive groups and suppose that the sequence

$$0 \to A \xrightarrow{i} B \xrightarrow{j} C \to 0$$

is exact. Show that

a) j maps B onto C,
b) i is an isomorphism of A into B,
c) C is isomorphic to $B/i(A)$.

*22.3 Let A, B, C, and D be additive groups and let

$$A \xrightarrow{i} B \xrightarrow{j} C \xrightarrow{k} D$$

be an exact sequence. Show that the following three conditions are equivalent:

1) i is onto B;
2) j maps all of B onto 0;
3) k is a one-to-one map.

*22.4 Show that if

$$A \xrightarrow{g} B \xrightarrow{h} C \xrightarrow{i} D \xrightarrow{j} E \xrightarrow{k} F$$

is an exact sequence of additive groups, then the following are equivalent:

1) h and j both map everything onto 0;
2) i is an isomorphism of C onto D;
3) g is onto B and k is one to one.

*22.5 Theorem 22.2 and Theorem 22.3 are closely connected with Exercise 13.16. Show the connection.

*22.6 In a computation analogous to Examples 22.2 and 22.3 of the text, find the relative homology groups $H_n(X, a)$ for the torus X with subcomplex a, as shown in Fig. 20.4 and Fig. 20.5. (Since we can regard these relative homology groups as the homology groups of the space obtained from X by shrinking a to a point, these should be the homology groups of the *pinched torus*.)

*22.7 For the simplicial complex X and subcomplex a of Exercise 22.6, form the exact homology sequence of the pair (X, a) and verify by direct computation that this sequence is exact.

*22.8 Repeat Exercise 22.6 with X the Klein bottle of Fig. 21.1 and Fig. 21.3. (This should give the homology groups of the *pinched Klein bottle*.)

*22.9 For the simplicial complex X and subcomplex a of Exercise 22.8, form the exact homology sequence of the pair (X, a) and verify by direct computation that this sequence is exact.

*22.10 Find the relative homology groups $H_n(X, Y)$, where X is the annular region of Fig. 20.3 and Y is the subcomplex consisting of the two boundary circles.

*22.11 For the simplicial complex X and subcomplex Y of Exercise 22.10, form the exact homology sequence of the pair (X, Y) and verify by direct computation that this sequence is exact.

*22.12 Prove Lemma 22.1.

*22.13 Prove Lemma 22.2.

*22.14 Prove Lemma 22.3.

*22.15 Prove Theorem 22.5 by means of the following steps.

a) Show (image i_{*k}) $\subseteq$ (kernel j_{*k}).
b) Show (kernel j_{*k}) $\subseteq$ (image i_{*k}).
c) Show (image j_{*k}) $\subseteq$ (kernel ∂_{*k}).
d) Show (kernel ∂_{*k}) $\subseteq$ (image j_{*k}).
e) Show (image ∂_{*k}) $\subseteq$ (kernel i_{*k-1}).
f) Show (kernel i_{*k-1}) $\subseteq$ (image ∂_{*k}).

*22.16 Let $\langle A, \partial \rangle$ and $\langle A', \partial' \rangle$ be chain complexes, and let f and g be collections of homomorphisms $f_k: A_k \to A'_k$ and $g_k: A_k \to A'_k$ such that both f and g commute with ∂. An **algebraic homotopy** between f and g is a collection D of homomorphisms $D_k: A_k \to A'_{k+1}$ such that for all $c \in A_k$, we have

$$f_k(c) - g_k(c) = \partial'_{k+1}(D_k(c)) + D_{k-1}(\partial_k(c)).$$

(One abbreviates this condition by $f - g = \partial D + D\partial$.) Show that if there exists an algebraic homotopy between f and g, that is, if f and g are **homotopic**, then f_{*k} and g_{*k} are the same homomorphism of $H_k(A)$ into $H_k(A')$.

PART II | RINGS AND FIELDS

23 | Rings

23.1 DEFINITION AND BASIC PROPERTIES

All our work thus far has been concerned with sets on which a single binary operation has been defined. Our familiar examples of sets of numbers show that a study of sets on which two binary operations have been defined should be of great importance. The most general system of this nature that we shall study is a ring.

Definition. A *ring* $\langle R, +, \cdot \rangle$ is a set R together with two binary operations $+$ and $\cdot$ of addition and multiplication defined on R such that the following axioms are satisfied:

$\mathcal{R}_1$. $\langle R, + \rangle$ is an abelian group.
$\mathcal{R}_2$. Multiplication is associative.
$\mathcal{R}_3$. For all $a, b, c \in R$, the *left distributive law*, $a(b + c) = (ab) + (ac)$, and the *right distributive law*, $(a + b)c = (ac) + (bc)$, hold.

Example 23.1 The student is well aware that axioms $\mathcal{R}_1$, $\mathcal{R}_2$, and $\mathcal{R}_3$ for a ring hold in any subset of the complex numbers which is a group under addition and on which multiplication is closed. For example, $\langle Z, +, \cdot \rangle$, $\langle Q, +, \cdot \rangle$, $\langle R, +, \cdot \rangle$, and $\langle C, +, \cdot \rangle$ are all rings. ‖

We shall observe the usual convention that multiplication is performed before addition, so the left distributive law, for example, becomes

$$a(b + c) = ab + ac,$$

without the parentheses on the right side of the equation. Also, as a convenience analogous to our notation in group theory, we shall somewhat incorrectly refer to *a ring R* in place of *a ring* $\langle R, +, \cdot \rangle$, provided that no confusion will result. In particular, from now on Z will always be $\langle Z, +, \cdot \rangle$, and Q, R, and C will also be the obvious rings. We may on occasion refer to $\langle R, + \rangle$ as *the additive group of the ring R*.

Example 23.2 Consider the cyclic group $\langle Z_n, + \rangle$. If one defines for $a, b \in Z_n$ the product ab to be the remainder of the usual product of integers when divided by n, it can be shown that $\langle Z_n, +, \cdot \rangle$ is a ring. We shall feel free to use this fact. For example, in $\mathbf{Z}_{10}$ we have $(3)(7) = 1$. This operation on $\mathbf{Z}_n$ is **multiplication modulo** n. We do not check the ring axioms here, for they will later follow directly from some of the theory we have to develop anyway. ‖

From now on, $\mathbf{Z}_n$ will always be $\langle \mathbf{Z}_n, +, \cdot \rangle$. Continuing matters of notation, we shall always let 0 be the additive identity of a ring. The additive inverse of an element a of a ring is $-a$. We shall frequently have occasion to refer to a sum

$$a + a + \cdots + a$$

having n summands. We shall let this sum be $n \cdot a$. *However, $n \cdot a$ is not to be construed as a multiplication of n and a in the ring, for the integer n may not be in the ring at all.* If $n < 0$, we let

$$n \cdot a = (-a) + (-a) + \cdots + (-a)$$

for $|n|$ summands. Finally, we define

$$0 \cdot a = 0$$

for $0 \in \mathbf{Z}$ on the left side of the equation and $0 \in R$ on the right side. Actually the equation $0a = 0$ holds also for $0 \in R$ on both sides. The following theorem proves this and various other easy but important facts. Note the strong use of the distributive laws in the proof of this theorem. *These distributive laws are the only means we have of relating additive concepts to multiplicative concepts in a ring.*

Theorem 23.1 *If R is a ring with additive identity 0, then for any $a, b \in R$ we have*

1) $0a = a0 = 0$,
2) $a(-b) = (-a)b = -(ab)$,
3) $(-a)(-b) = ab$.

Proof. For (1), note that

$$a0 = a(0 + 0) = a0 + a0.$$

Then by the cancellation law for the additive group $\langle R, + \rangle$, we have $0 = a0$. Likewise,

$$0a = (0 + 0)a = 0a + 0a$$

implies that $0a = 0$. This proves (1).

In order to understand the proof of (2), you must remember that, by *definition*, $-(ab)$ is the element which when added to ab gives 0. Thus to show that $a(-b) = -(ab)$, you must show precisely that $a(-b) + ab = 0$. By the left distributive law,

$$a(-b) + ab = a(-b + b) = a0 = 0,$$

since $a0 = 0$ by (1). Likewise,

$$(-a)b + ab = (-a + a)b = 0b = 0.$$

For (3), note that

$$(-a)(-b) = -(a(-b))$$

by (2). Again by (2),

$$-(a(-b)) = -(-(ab)),$$

and $-(-(ab))$ is the element which when added to $-(ab)$ gives 0. This is ab by definition of $-(ab)$ and by the uniqueness of an inverse in a group. Thus, $(-a)(-b) = ab$. ∎

It is important that the student *understand* the preceding proof. If he can't follow the logic and use of definitions, he will come to grief later. (Actually he has probably already come to grief.) The theorem allows us to use our usual rules for signs.

Hopefully the student is beginning to realize that in the study of any sort of mathematical structure, an idea of basic importance is the concept of two systems being structurally identical, i.e., one being just like the other except for its name. In algebra this concept is always called "*isomorphism*." The concept of two rings being just alike except for names of elements leads us, just as it did for groups, to the following definition.

Definition. An *isomorphism* ϕ *of a ring* R *with a ring* R' is a one-to-one function mapping R onto R' such that for all $a, b \in R$,

1) $(a + b)\phi = a\phi + b\phi$,
2) $(ab)\phi = (a\phi)(b\phi)$.

The rings R and R' are then *isomorphic*.

Example 23.3 As abelian groups, $\langle \mathbf{Z}, + \rangle$ and $\langle 2\mathbf{Z}, + \rangle$ are isomorphic under the map $\phi: \mathbf{Z} \to 2\mathbf{Z}$, with $x\phi = 2x$ for $x \in \mathbf{Z}$. Note that $2\mathbf{Z}$ is closed under the usual multiplication and that $\langle 2\mathbf{Z}, +, \cdot \rangle$ is a ring. Here ϕ is *not* a ring isomorphism, for $(xy)\phi = 2xy$, while $(x\phi)(y\phi) = 2x2y = 4xy$. ‖

From now on, $n\mathbf{Z}$ will always be the ring $\langle n\mathbf{Z}, +, \cdot \rangle$.

23.2 MULTIPLICATIVE QUESTIONS; FIELDS

All the rings we have seen so far have a multiplication which is commutative. Many of them, such as $\mathbf{Z}$, $\mathbf{Q}$, and $\mathbf{R}$, also have a multiplicative identity 1. However, $2\mathbf{Z}$ does not have an identity element for multiplication. There are many rings in which multiplication is not commutative. The student familiar with a little matrix theory will see that the $n \times n$ matrices with entries in $\mathbf{Z}$ (or $\mathbf{Q}$, $\mathbf{R}$, or $\mathbf{C}$) form a ring under matrix addition and multiplication in which multiplication is not commutative if $n \geq 2$. These rings of matrices do have an identity element for multiplication. We treat them in more detail in Section 25.

It is evident that $\{0\}$, with $0 + 0 = 0$ and $(0)(0) = 0$, gives a ring. Here 0 acts as multiplicative as well as additive identity. By Theorem 23.1, this is the only case in which 0 could act as a multiplicative identity, for if $0a = a$, we can deduce that $a = 0$. We shall exclude this trivial case when speaking of a multiplicative identity in a ring, i.e., whenever we speak of a multiplicative identity, we will assume that it is nonzero.

Definition. A ring in which the multiplication is commutative is a *commutative ring*. A ring R with a multiplicative identity 1 such that $1x = x1 = x$ for all $x \in R$ is a *ring with unity*. A multiplicative identity in a ring is *unity*.

Theorem 23.2 *If R is a ring with unity, then this unity 1 is the only multiplicative identity.*

Proof. We proceed exactly as we did for groups. Let 1 and $1'$ both be multiplicative identities in a ring R, and set up a competition. Regarding 1 as identity, we have

$$(1)(1') = 1'.$$

Regarding $1'$ as identity, we have

$$(1)(1') = 1.$$

Thus $1 = 1'$. ∎

If $R_1, R_2, \ldots, R_n$ are rings, we can form the set $R_1 \times R_2 \times \cdots \times R_n$ of all ordered n-tuples $(r_1, r_2, \ldots, r_n)$, where $r_i \in R_i$. Defining addition and multiplication of n-tuples by components (just as for groups), we see at once from the ring axioms in each component that the set of all these n-tuples forms a ring under addition and multiplication by components. We shall usually use additive terminology and notation, so that

$$R_1 + R_2 + \cdots + R_n$$

is the **direct sum** of these rings R_i. Clearly, such a direct sum is commutative or has unity if and only if each R_i is commutative or has unity, respectively.

In a ring R with unity, the set R^* of nonzero elements, if closed under the ring multiplication, will be a multiplicative group if inverses exist. A **multiplicative inverse** of an element a in a ring R with unity 1 is an element $a^{-1} \in R$ such that $aa^{-1} = a^{-1}a = 1$. Precisely as for groups, a multiplicative inverse for an element a in R is unique, if it exists at all (see Exercise 23.9). Theorem 23.1 shows that it would be hopeless to have a multiplicative inverse for 0 unless one wishes to regard the set $\{0\}$, where $0 + 0 = 0$ and $(0)(0) = 0$, as a ring with 0 as both additive and multiplicative identity. We have agreed to exclude this trivial case when speaking of a ring with unity. We are thus led to discuss the existence of multiplicative inverses for nonzero elements in a ring with unity. The student is probably getting tired of definitions, but we must define our terms.

Definition. Let R be a ring with unity. An element u in R is a *unit of R* if it has a multiplicative inverse in R. If every nonzero element of R is a unit, then R is a *skew field* or *division ring*. A *field* is a commutative division ring.

Example 23.4 **Z** is not a field, since 2, for example, has no multiplicative inverse, so 2 is not a unit in **Z**. The only units in **Z** are 1 and -1. Clearly, **Q** and **R** are fields. ‖

One has the natural concepts of a subring of a ring and a subfield of a field. A **subring of a ring** is a subset of the ring which is a ring under induced operations from the whole ring; a **subfield** is defined similarly for a subset of a field. In fact, let us say here once and for all that if we have a set, together with a certain specified type of algebraic structure on the set, the resulting conglomeration being a **glob** (group, ring, field, integral domain, vector space, etc.), then any subset of this set, together with a natural induced algebraic structure *which yields an algebraic structure of the same type*, is a **subglob**. If K and L are globs, we shall let "$K \leq L$" denote that K is a subglob of L, and "$K < L$" denote that $K \leq L$ but $K \neq L$.

Finally, we wish to caution the student not to confuse our use of the words *unit* and *unity*. Unity is the multiplicative identity, while a unit is any element having a multiplicative inverse. Thus the multiplicative identity or unity is a unit, but not every unit is unity. For example, -1 is a unit in **Z**, but -1 is not unity, that is, $-1 \neq 1$.

EXERCISES

23.1 State for which of the following sets the indicated operations of addition and multiplication are defined, i.e., closed, and give a ring structure. If a ring is not formed, tell why this is the case.

a) $n\mathbf{Z}$ with the usual addition and multiplication
b) $\mathbf{Z}^+$ with the usual addition and multiplication
c) $\mathbf{Z} + \mathbf{Z}$ with addition and multiplication by components
d) $2\mathbf{Z} + \mathbf{Z}$ with addition and multiplication by components
e) $\{a + b\sqrt{2} \mid a, b \in \mathbf{Z}\}$ with the usual addition and multiplication
f) $\{a + b\sqrt{2} \mid a, b \in \mathbf{Q}\}$ with the usual addition and multiplication
g) The set of all pure imaginary complex numbers ri for $r \in \mathbf{R}$ with the usual addition and multiplication

23.2 For each part of Exercise 23.1 which gives a ring, state whether or not the ring is commutative, has unity, and is a field.

23.3 Describe all units in each of the following rings.

a) $\mathbf{Z}$ b) $\mathbf{Z} + \mathbf{Z}$
c) $\mathbf{Z}_5$ d) $\mathbf{Q}$
e) $\mathbf{Z} + \mathbf{Q} + \mathbf{Z}$ f) $\mathbf{Z}_4$

†**23.4** Show that if U is the collection of all units in a ring $\langle R, +, \cdot \rangle$ with unity, then $\langle U, \cdot \rangle$ is a group. (*Warning:* Be sure to show that U is closed under multiplication.)

23.5 Mark each of the following true or false.

— a) Every field is also a ring.

— b) Every ring has a multiplicative identity.

— c) Every ring with unity has at least 2 units.

— d) Every ring with unity has at most 2 units.

— e) It is possible for a subset of some field to be a ring but not a subfield, under the induced operations.

— f) The distributive laws for a ring are not very important.

— g) Multiplication in a field is commutative.

— h) The nonzero elements of a field form a group under the multiplication in the field.

— i) Addition in every ring is commutative.

— j) Every element in a ring has an additive inverse.

23.6 Show that the rings $2\mathbf{Z}$ and $3\mathbf{Z}$ are not isomorphic. Show that the fields $\mathbf{R}$ and $\mathbf{C}$ are not isomorphic.

23.7 Give an example of a ring with unity 1 which has a subring with unity $1' \neq 1$.

23.8 Show that the unity element in a subfield of a field must be the unity of the whole field, in contrast to Exercise 23.7 for rings.

23.9 Show that the multiplicative inverse of a unit in a ring with unity is unique.

23.10 Show that a subset S of a ring R gives a subring of R if and only if the following hold:

1) $0 \in S$;
2) $(a - b) \in S$ for all $a, b \in S$;
3) $ab \in S$ for all $a, b \in S$.

23.11 a) Show that an intersection of subrings of a ring R is again a subring of R.
b) Show that an intersection of subfields of a field F is again a subfield of F.

23.12 Let R be a ring, and let a be a fixed element of R. Let $I_a = \{x \in R \mid ax = 0\}$. Show that I_a is a subring of R.

23.13 Let R be a ring, and let a be a fixed element of R. Let R_a be the subring of R which is the intersection of all subrings of R containing a (see Exercise 23.11). The ring R_a is the **subring of R generated by** a. Show that the abelian group $\langle R_a, + \rangle$ is generated (in the sense of Section 9) by $\{a^n \mid n \in \mathbf{Z}^+\}$.

23.14 Consider $\langle S, +, \cdot \rangle$, where S is a set and $+$ and $\cdot$ are binary operations on S such that

1) $\langle S, + \rangle$ is a group,
2) $\langle S^*, \cdot \rangle$ is a group, where S^* consists of all elements of S except the additive identity,
3) $a(b + c) = (ab) + (ac)$ and $(a + b)c = (ac) + (bc)$ for all $a, b, c \in S$.

Show that $\langle S, +, \cdot \rangle$ is a division ring. [*Hint:* Apply the distributive laws to $(1 + 1)(a + b)$ to prove the commutativity of addition.]

23.15 A ring R is a **Boolean ring** if $a^2 = a$ for all $a \in R$. Show that every Boolean ring is commutative.

23.16 (For students having some knowledge of the laws of set theory.) For a set S let $\mathcal{P}(S)$ be the collection of all subsets of S. Define binary operations $+$ and $\cdot$ on $\mathcal{P}(S)$ by

$$A + B = (A \cup B) - (A \cap B) = \{x \mid x \in A \text{ or } x \in B \text{ but } x \notin (A \cap B)\}$$

and

$$A \cdot B = A \cap B$$

for $A, B \in \mathcal{P}(S)$.

a) Give the tables for $+$ and $\cdot$ for $\mathcal{P}(S)$, where $S = \{a, b\}$. [*Hint:* $\mathcal{P}(S)$ has 4 elements.]

b) Show that for *any* set S, $\langle \mathcal{P}(S), +, \cdot \rangle$ is a Boolean ring (see Exercise 23.15).

24 | Integral Domains

While a careful treatment of polynomials is not given until Section 30, for purposes of motivation we shall make naive use of them in this section.

24.1 DIVISORS OF ZERO AND CANCELLATION

One of the most important algebraic properties of our usual number system is that a product of two numbers can only be zero if at least one of the factors is zero. The student uses this fact constantly, perhaps without realizing it. Suppose, for example, you are asked to "solve the equation

$$x^2 - 5x + 6 = 0."$$

The first thing you do is to factor the left side:

$$x^2 - 5x + 6 = (x - 2)(x - 3).$$

Then you conclude that the only possible values for x are 2 and 3. Why? It is because if x is replaced by any number a, the product $(a - 2)(a - 3)$ of the resulting numbers is zero if and only if either $a - 2 = 0$ or $a - 3 = 0$.

Example 24.1 Let us solve the equation $x^2 - 5x + 6 = 0$ in $\mathbf{Z}_{12}$. Now $x^2 - 5x + 6 = (x - 2)(x - 3)$ is still valid if we think of x as standing for any number in $\mathbf{Z}_{12}$. But in $\mathbf{Z}_{12}$, not only is $0a = a0 = 0$ for all $a \in \mathbf{Z}_{12}$, but also,

$$(2)(6) = (6)(2) = (3)(4) = (4)(3) = (3)(8) = (8)(3)$$
$$= (4)(6) = (6)(4) = (4)(9) = (9)(4) = (6)(6) = (6)(8)$$
$$= (8)(6) = (6)(10) = (10)(6) = (8)(9) = (9)(8) = 0.$$

Thus our equation has not only 2 and 3 as solutions, but also 6 and 11, for $(6 - 2)(6 - 3) = (4)(3) = 0$ and $(11 - 2)(11 - 3) = (9)(8) = 0$ in $\mathbf{Z}_{12}$. ‖

These ideas are of such importance that we formalize them in a definition.

Definition. If a and b are two nonzero elements of a ring R such that $ab = 0$, then a and b are ***divisors of zero*** (or ***zero divisors***). In particular, a is a ***left divisor of zero*** and b is a ***right divisor of zero***.

In a commutative ring, every left divisor of zero is also a right divisor of zero and conversely. Thus there is no distinction between left and right divisors of zero in a commutative ring.

202

Example 24.1 shows that in $\mathbf{Z}_{12}$ the elements 2, 3, 4, 6, 8, 9, and 10 are all divisors of zero. Note that these are exactly the numbers in $\mathbf{Z}_{12}$ which are not relatively prime to 12, that is, whose gcd with 12 is not 1. Our next theorem shows that this is an example of a general situation.

Theorem 24.1 *In the ring $\mathbf{Z}_n$, the divisors of zero are precisely those elements which are not relatively prime to n.*

Proof. Let $m \in \mathbf{Z}_n$, where $m \neq 0$, and let the gcd of m and n be $d \neq 1$. Then

$$m\left(\frac{n}{d}\right) = \left(\frac{m}{d}\right)n,$$

and $(m/d)n$ gives 0 as a multiple of n. Thus, $m(n/d) = 0$ in $\mathbf{Z}_n$, while neither m nor n/d is 0, so m is a divisor of zero.

On the other hand, suppose $m \in \mathbf{Z}_n$ is relatively prime to n. If for $s \in \mathbf{Z}_n$ we have $ms = 0$, then n divides the product ms of m and s as elements in the ring $\mathbf{Z}$. Since n has no factors >1 in common with m, it must be that n divides s, so $s = 0$ in $\mathbf{Z}_n$. ∎

Corollary. *If p is a prime, then $\mathbf{Z}_p$ has no divisors of zero.*

Proof. This is immediate from Theorem 24.1. ∎

Another indication of the importance of the concept of zero divisors is shown in the theorem which follows. Let R be a ring, and let $a, b, c \in R$. The **cancellation laws** hold in R if $ab = ac$, with $a \neq 0$, implies $b = c$, and $ba = ca$, with $a \neq 0$, implies $b = c$. These are multiplicative cancellation laws. Of course, the additive cancellation laws hold in R, since $\langle R, + \rangle$ is a group.

Theorem 24.2 *The cancellation laws hold in a ring R if and only if R has no left or right divisors of zero.*

Proof. Let R be a ring in which the cancellation laws hold, and suppose $ab = 0$ for some $a, b \in R$. We must show that either a or b is 0. If $a \neq 0$, then $ab = a0$ implies that $b = 0$ by the cancellation laws. Similarly, $b \neq 0$ implies that $a = 0$, so there can be no left or right divisors of zero if the cancellation laws hold.

Conversely, suppose that R has no left or right divisors of zero, and suppose that $ab = ac$ with $a \neq 0$. Then

$$ab - ac = a(b - c) = 0.$$

Since $a \neq 0$, and since R has no left divisors of zero, we must have $b - c = 0$, so $b = c$. A similar argument shows that $ba = ca$ with $a \neq 0$ implies $b = c$. ∎

Suppose that R is a ring with no divisors of zero. Then an equation $ax = b$, with $a \neq 0$, in R can have at most one solution x in R, for if

$ax_1 = b$ and $ax_2 = b$, then $ax_1 = ax_2$, and by Theorem 24.2, $x_1 = x_2$, since R has no divisors of zero. If R has unity 1 and a is a unit in R with multiplicative inverse a^{-1}, then clearly the solution x of $ax = b$ is $a^{-1}b$. In the case that R is commutative, in particular if R is a field, it is customary to denote $a^{-1}b$ and ba^{-1} (they are equal by commutativity) by the formal quotient "b/a". This quotient notation must not be used in the event that R is not commutative, for then one does not know whether "b/a" denotes $a^{-1}b$ or ba^{-1}. In a field F it is usual to *define* a **quotient** b/a, where $a \neq 0$, as the solution x in F of the equation $ax = b$. This definition is consistent with our preceding remarks, and we shall be using this quotient notation when we work in a field. In particular, the multiplicative inverse of a non-zero element a of a field is $1/a$.

24.2 INTEGRAL DOMAINS

Definition. An *integral domain* D is a commutative ring with unity containing no divisors of zero.

Thus, if the coefficients of a polynomial are from an integral domain, one can solve a polynomial equation in which the polynomial can be factored into linear factors in the usual fashion by setting each factor equal to zero.

In our hierarchy of algebraic structures, an integral domain belongs between a commutative ring with unity and a field, as we shall now show. Theorem 24.2 shows that the cancellation laws for multiplication hold in an integral domain. We have seen that $\mathbf{Z}$ and $\mathbf{Z}_p$ for any prime p are integral domains, but $\mathbf{Z}_n$ is not an integral domain if n is not prime.

Theorem 24.3 *Every field F is an integral domain.*

Proof. Let $a, b \in F$, and suppose that $a \neq 0$. Then if $ab = 0$, we have

$$\left(\frac{1}{a}\right)(ab) = \left(\frac{1}{a}\right)0 = 0.$$

But then

$$0 = \left(\frac{1}{a}\right)(ab) = \left[\left(\frac{1}{a}\right)a\right]b = 1b = b.$$

We have shown that $ab = 0$ with $a \neq 0$ implies that $b = 0$ in F, so there are no divisors of zero in F. Of course, F is a commutative ring with unity, so our theorem is proved. ∎

Thus far the only fields we have seen are $\mathbf{Q}$, $\mathbf{R}$, and $\mathbf{C}$. The corollary of the next theorem will give us some fields of finite order!

Theorem 24.4 *Every finite integral domain is a field.*

Proof. Let

$$0, 1, a_1, \ldots, a_n$$

be all the elements of a finite integral domain D. We need to show that for $a \in D$, where $a \neq 0$, there exists $b \in D$ such that $ab = 1$. Now consider

$$a1, aa_1, \ldots, aa_n.$$

We claim that all these elements of D are distinct, for $aa_i = aa_j$ implies that $a_i = a_j$, by the cancellation laws which hold in an integral domain. Also, since D has no zero divisors, none of these elements is zero. Hence by counting, we find that $a1, aa_1, \ldots, aa_n$ are the elements $1, a_1, \ldots, a_n$ in some order, so that either $a1 = 1$, that is, $a = 1$, or $aa_i = 1$ for some i. Thus a has a multiplicative inverse. ∎

Corollary. *If p is a prime, then $\mathbf{Z}_p$ is a field.*

Proof. This follows immediately from the fact that $\mathbf{Z}_p$ is an integral domain and from Theorem 24.4. ∎

24.3 THE CHARACTERISTIC OF A RING

Let R be any ring. We might ask whether or not there is a positive integer n such that $n \cdot a = 0$ for all $a \in R$, where $n \cdot a$ means $a + a + \cdots + a$ for n summands, as explained before. For example, the integer m has this property for the ring $\mathbf{Z}_m$.

Definition. If for a ring R a positive integer n exists such that $n \cdot a = 0$ for all $a \in R$, then the least such positive integer is the **characteristic of the ring** R. If no such positive integer exists, then R is of **characteristic** 0.

We shall be using the concept of a characteristic chiefly for fields.

Example 24.2 The ring $\mathbf{Z}_n$ is of characteristic n, while $\mathbf{Z}$, $\mathbf{Q}$, $\mathbf{R}$, and $\mathbf{C}$ all have characteristic 0. ‖

Theorem 24.5 *If R is a ring with unity 1, then R has characteristic $n > 0$ if and only if n is the smallest positive integer such that $n \cdot 1 = 0$.*

Proof. By definition, if R has characteristic $n > 0$, then $n \cdot a = 0$ for all $a \in R$, so in particular $n \cdot 1 = 0$.

Conversely, suppose that n is a positive integer such that $n \cdot 1 = 0$. Then for any $a \in R$, we have

$$n \cdot a = a + a + \cdots + a = a(1 + 1 + \cdots + 1) = a(n \cdot 1) = a0 = 0.$$

Our theorem follows directly. ∎

24.4 FERMAT'S THEOREM

We conclude this section with some elegant applications to number theory. It is easy to see that *for any field, the nonzero elements form a group under the field multiplication.* In particular, for $\mathbf{Z}_p$, the elements

$$1, 2, 3, \ldots, p - 1$$

form a group of order $p - 1$ under multiplication modulo p. Since the order of any element in a group divides the order of the group, we see for $a \neq 0$ and $a \in \mathbf{Z}_p$ that $a^{p-1} = 1$ in $\mathbf{Z}_p$. We shall see in detail later, for multiplication as well as addition, that $a \in \mathbf{Z}_p$ can be viewed as representing the coset $a + p\mathbf{Z}$, and that the product of cosets can be computed by multiplication modulo p of representatives in a fashion analogous to the way we compute sums. The collection $\mathbf{Z}/p\mathbf{Z}$ of these cosets becomes a ring isomorphic to $\mathbf{Z}_p$. Let us assume this now. This gives us immediately the so-called *Little Theorem of Fermat.*

Theorem 24.6 (Fermat). *If $a \in \mathbf{Z}$ and p is a prime not dividing a, then p divides $a^{p-1} - 1$, that is, $a^{p-1} \equiv 1 \pmod{p}$ for $a \not\equiv 0 \pmod{p}$.*

Corollary. *If $a \in \mathbf{Z}$, then $a^p \equiv a \pmod{p}$ for any prime p.*

Proof. This follows from Theorem 24.6, if $a \not\equiv 0 \pmod{p}$. If $a \equiv 0 \pmod{p}$, then both sides reduce to 0 modulo p. ∎

This corollary will be of great importance to us later in our work with finite fields.

Example 24.3 Let us compute the remainder of 8^{103} when divided by 13. Using Fermat's theorem, we have

$$8^{103} \equiv (8^{12})^8 (8^7) \equiv (1^8)(8^7) \equiv 8^7 \equiv (-5)^7$$
$$\equiv (25)^3(-5) \equiv (-1)^3(-5) \equiv 5 \pmod{13}. \;\|$$

*24.5 EULER'S GENERALIZATION

Euler gave a generalization of Fermat's theorem. His generalization will follow at once from our next theorem.

Theorem 24.7 *The set G_n of nonzero elements of $\mathbf{Z}_n$ which are not zero divisors forms a group under multiplication modulo n.*

Proof. First we must show that G_n is closed under multiplication modulo n. Let $a, b \in G_n$. If $ab \notin G_n$, then there would exist $c \neq 0$ in $\mathbf{Z}_n$ such that $(ab)c = 0$. Now $(ab)c = 0$ implies that $a(bc) = 0$. Since $b \in G_n$ and $c \neq 0$, we have $bc \neq 0$ by definition of G_n. But then $a(bc) = 0$ would imply that $a \notin G_n$ contrary to assumption. *Note that we have shown that for any ring the set of elements which are not divisors of 0 is closed under multiplication.* No structure of $\mathbf{Z}_n$ other than ring structure has been involved so far.

We now show that G_n is a group. Of course, multiplication modulo n is associative, and $1 \in G_n$. It remains to show that for $a \in G_n$, there is

$b \in G_n$ such that $ab = 1$. Let

$$1, a_1, \ldots, a_r$$

be the elements of G_n. The elements

$$a1, aa_1, \ldots, aa_r$$

are all different, for if $aa_i = aa_j$, then $a(a_i - a_j) = 0$, and since $a \in G_n$ and thus is not a divisor of 0, we must have $a_i - a_j = 0$ or $a_i = a_j$. Therefore by counting, we find that either $a1 = 1$, or some aa_i must be 1, so a has a multiplicative inverse. ∎

Note that the only property of $\mathbf{Z}_n$ used in this last theorem, other than the fact that it was a ring with unity, was that it was finite. In both Theorem 24.7 and Theorem 24.4 we have (in essentially the same construction) employed a counting argument. *Counting arguments are often simple, but are among the most powerful tools of all mathematics.*

Let $\varphi(n)$ be defined as the number of positive integers less than or equal to n and relatively prime to n. For example, if $n = 12$, the positive integers less than or equal to 12 and relatively prime to 12 are 1, 5, 7, and 11, so $\varphi(12) = 4$. By Theorem 24.1, $\varphi(n)$ is the number of elements of $\mathbf{Z}_n$ which are not divisors of zero. This function $\varphi \colon \mathbf{Z}^+ \to \mathbf{Z}^+$ is the **Euler phi-function**. We can now describe Euler's generalization of Fermat's theorem.

Theorem 24.8 (Euler). *If a is an integer relatively prime to n, then $a^{\varphi(n)} - 1$ is divisible by n, that is, $a^{\varphi(n)} \equiv 1 \pmod{n}$.*

Proof. If a is relatively prime to n, then the coset $a + n\mathbf{Z}$ of $n\mathbf{Z}$ containing a contains an integer $b < n$ and relatively prime to n. Using the fact (which will be proved later) that multiplication of these cosets by multiplication modulo n of representatives is well defined, we have

$$a^{\varphi(n)} \equiv b^{\varphi(n)} \pmod{n}.$$

But by Theorems 24.1 and 24.7, b can be viewed as an element of the multiplicative group G_n of order $\varphi(n)$ consisting of the $\varphi(n)$ elements of $\mathbf{Z}_n$ relatively prime to n. Thus

$$b^{\varphi(n)} \equiv 1 \pmod{n},$$

and our theorem follows. ∎

EXERCISES

24.1 Find all solutions of the equation $x^3 - 2x^2 - 3x = 0$ in $\mathbf{Z}_{12}$.

24.2 Solve the equation $3x = 2$ in the field $\mathbf{Z}_7$; in the field $\mathbf{Z}_{23}$.

24.3 Find the characteristic of each of the following rings.

a) $2\mathbf{Z}$ b) $\mathbf{Z} + \mathbf{Z}$

c) $\mathbf{Z}_3 + 3\mathbf{Z}$ d) $\mathbf{Z}_3 + \mathbf{Z}_3$

e) $\mathbf{Z}_3 + \mathbf{Z}_4$ f) $\mathbf{Z}_6 + \mathbf{Z}_{15}$

24.4 Using Fermat's theorem, find the remainder of 3^{47} when divided by 23.

†24.5 a) Show that 1 and $p - 1$ are the only elements of the field $\mathbf{Z}_p$ which are their own multiplicative inverse. [*Hint:* Consider the equation $x^2 - 1 = 0$.]

b) From part (a), deduce the half of *Wilson's theorem* which states that if p is a prime, then $(p - 1)! \equiv -1 \pmod{p}$. (The other half states that if

$$(n - 1)! \equiv -1 \pmod{n},$$

then n is a prime.)

24.6 Mark each of the following true or false.

— a) $n\mathbf{Z}$ has zero divisors if n is not prime.

— b) Every field is an integral domain.

— c) The characteristic of $n\mathbf{Z}$ is n.

— d) As a ring, $\mathbf{Z}$ is isomorphic to $n\mathbf{Z}$ for all $n \geq 1$.

— e) The cancellation law holds in any ring which is isomorphic to an integral domain.

— f) Every integral domain of characteristic 0 is infinite.

— g) The direct sum of two integral domains is again an integral domain.

— h) A divisor of zero in a commutative ring with unity can have no multiplicative inverse.

— i) $n\mathbf{Z}$ is a subdomain of $\mathbf{Z}$.

— j) $\mathbf{Z}$ is a subfield of $\mathbf{Q}$.

24.7 Find all solutions of the equation $x^2 + 2x + 2 = 0$ in $\mathbf{Z}_6$; of the equation $x^2 + 2x + 4 = 0$ in $\mathbf{Z}_6$.

24.8 Show that an intersection of subdomains of an integral domain D is again a subdomain of D.

24.9 Show that a finite ring R with unity and no divisors of zero is a division ring. (It is actually a field, although this is not easy to prove. See Theorem 25.5.) [*Note:* In your proof, to show that $a \neq 0$ is a unit, you must show that a "left multiplicative inverse" of $a \neq 0$ in R is also a "right multiplicative inverse."]

24.10 Show that the characteristic of a subdomain of an integral domain D is equal to the characteristic of D.

24.11 Show that if D is an integral domain, then $\{n \cdot 1 \mid n \in \mathbf{Z}\}$ is a subdomain of D contained in every subdomain of D.

24.12 Show that the characteristic of an integral domain D must be either 0 or a prime p. [*Hint:* If the characteristic of D is mn, consider $(m \cdot 1)(n \cdot 1)$ in D.]

24.13 Use Fermat's theorem to show that for any positive integer n, $n^{37} - n$ is divisible by 383838. [*Hint:* $383838 = (37)(19)(13)(7)(3)(2)$.]

***24.14** Give the group multiplication table for the multiplicative group of those elements of $\mathbf{Z}_{12}$ relatively prime to 12. To which group of order 4 is it isomorphic?

***24.15** Make a table of the values of $\varphi(n)$ for $n \leq 30$.

***24.16** Use Euler's generalization of Fermat's theorem to find the remainder of 7^{1000} when divided by 24.

25 | Some Noncommutative Examples

We shall be doing almost nothing with noncommutative rings and skew fields. So that the student may realize that there are many important noncommutative rings occurring very naturally in algebra, in this section we give several examples of such rings.

*25.1 MATRICES OVER A FIELD

Let F be any field (say $\mathbf{Q}$, $\mathbf{R}$, or $\mathbf{C}$), and consider the set $M_2(F)$ of all 2×2 square arrays

$$(a_{ij}) = \begin{pmatrix} a_{11} & a_{12} \\ a_{21} & a_{22} \end{pmatrix},$$

where the a_{ij} are all in F. The first subscript i of a_{ij} indicates the *row* in which a_{ij} occurs in the square array, and the second subscript j indicates the *column*. Thus a_{12} is the element of F in the first row and second column of the square array. Such a square array is a **2 × 2 matrix over** F. The set $M_n(F)$ of all $n \times n$ **matrices over** F is similarly defined.

We define matrix addition on $M_2(F)$ by

$$\begin{pmatrix} a_{11} & a_{12} \\ a_{21} & a_{22} \end{pmatrix} + \begin{pmatrix} b_{11} & b_{12} \\ b_{21} & b_{22} \end{pmatrix} = \begin{pmatrix} a_{11} + b_{11} & a_{12} + b_{12} \\ a_{21} + b_{21} & a_{22} + b_{22} \end{pmatrix},$$

that is, by adding corresponding entries. After a few moments of thought, it is clear from the field axioms for F that $\langle M_2(F), + \rangle$ is an abelian group, with additive identity

$$\begin{pmatrix} 0 & 0 \\ 0 & 0 \end{pmatrix},$$

and with

$$-\begin{pmatrix} a_{11} & a_{12} \\ a_{21} & a_{22} \end{pmatrix} = \begin{pmatrix} -a_{11} & -a_{12} \\ -a_{21} & -a_{22} \end{pmatrix}.$$

Matrix multiplication on $M_2(F)$ is defined by

$$\begin{pmatrix} a_{11} & a_{12} \\ a_{21} & a_{22} \end{pmatrix} \begin{pmatrix} b_{11} & b_{12} \\ b_{21} & b_{22} \end{pmatrix} = \begin{pmatrix} a_{11}b_{11} + a_{12}b_{21} & a_{11}b_{12} + a_{12}b_{22} \\ a_{21}b_{11} + a_{22}b_{21} & a_{21}b_{12} + a_{22}b_{22} \end{pmatrix}.$$

This multiplication looks difficult, and is best remembered by

$$(a_{ij})(b_{ij}) = (c_{ij}),$$

where

$$c_{rs} = \sum_{i=1}^{2} a_{ri} b_{is}.$$

With the analogous definition for a matrix multiplication in which the sum goes from $i = 1$ to n and the obvious analogous definition of matrix addition, everything we have done is valid for the set $M_n(F)$ of all $n \times n$ matrices over F.

Example 25.1 In $M_2(Q)$,

$$\begin{pmatrix} 2 & 1 \\ -3 & 4 \end{pmatrix} + \begin{pmatrix} -1 & 0 \\ 2 & -5 \end{pmatrix} = \begin{pmatrix} 1 & 1 \\ -1 & -1 \end{pmatrix}$$

and

$$\begin{pmatrix} 2 & 1 \\ -3 & 4 \end{pmatrix} \begin{pmatrix} -1 & 0 \\ 2 & -5 \end{pmatrix} = \begin{pmatrix} 0 & -5 \\ 11 & -20 \end{pmatrix}. \; \|$$

To show that $\langle M_n(F), +, \cdot \rangle$ is a ring, it remains to prove the associative and distributive laws. We illustrate with the associative law for matrix multiplication in $M_n(F)$. Using field properties of F and the definition of matrix multiplication in $M_n(F)$, if d_{rs} is an entry in $(a_{ij})[(b_{ij})(c_{ij})]$, we have

$$d_{rs} = \sum_{k=1}^{n} a_{rk} \left(\sum_{j=1}^{n} b_{kj} c_{js} \right) = \sum_{j=1}^{n} \left(\sum_{k=1}^{n} a_{rk} b_{kj} \right) c_{js} = e_{rs},$$

where e_{rs} is the entry in the rth row and sth column of $[(a_{ij})(b_{ij})](c_{ij})$. The distributive laws are similarly proved. We consider the following theorem to have been proved.

Theorem 25.1 *If F is a field, then the set $M_n(F)$ of all $n \times n$ matrices with entries from F forms a ring under matrix addition and multiplication.*

These rings of matrices are considered in linear algebra. In this context, they can be viewed as corresponding to certain functions, and matrix multiplication, when viewed in this light, can be shown to be just function composition. Since function composition is always associative, this gives another, more elegant, demonstration of the associative law.

Note that $M_1(F)$ is isomorphic to F under the map $\phi: F \to M_1(F)$ given by $a\phi = (a)$ for $a \in F$. Remember that this section is supposed to be concerned with noncommutative rings. It is true that $M_n(F)$ is noncommutative if $n \geq 2$. Example 25.2 illustrates this for $M_2(F)$.

Example 25.2 Since every field F contains elements 0 and 1, $M_2(F)$ always has among its elements

$$\begin{pmatrix} 0 & 1 \\ 0 & 0 \end{pmatrix} \quad \text{and} \quad \begin{pmatrix} 0 & 0 \\ 0 & 1 \end{pmatrix}.$$

The definition of matrix multiplication shows that

$$\begin{pmatrix} 0 & 1 \\ 0 & 0 \end{pmatrix}\begin{pmatrix} 0 & 0 \\ 0 & 1 \end{pmatrix} = \begin{pmatrix} 0 & 1 \\ 0 & 0 \end{pmatrix},$$

while

$$\begin{pmatrix} 0 & 0 \\ 0 & 1 \end{pmatrix}\begin{pmatrix} 0 & 1 \\ 0 & 0 \end{pmatrix} = \begin{pmatrix} 0 & 0 \\ 0 & 0 \end{pmatrix}.$$

Thus $M_2(F)$ is noncommutative. Since

$$\begin{pmatrix} 0 & 0 \\ 0 & 0 \end{pmatrix}$$

is the additive identity, this example also shows that there exist divisors of zero in $M_2(F)$. The same is true of $M_n(F)$ for $n \geq 2$. We leave as an exercise the demonstration that

$$\begin{pmatrix} 1 & 0 \\ 0 & 1 \end{pmatrix}$$

is unity in $M_2(F)$ (see Exercise 25.2). ‖

*25.2   RINGS OF ENDOMORPHISMS

Let A be any abelian group. A homomorphism of A into itself is an **endomorphism of** A. Let the set of all endomorphisms of A be $Hom(A)$. Since the composition of two homomorphisms of A into itself is again such a homomorphism, we define multiplication on $Hom(A)$ by function composition, and thus multiplication is associative.

To define addition, for $\phi, \psi \in Hom(A)$, we have to describe the value of $(\phi + \psi)$ on each $a \in A$. Define

$$a(\phi + \psi) = (a\phi) + (a\psi).$$

Since

$$(a + b)(\phi + \psi) = (a + b)\phi + (a + b)\psi = (a\phi + b\phi) + (a\psi + b\psi)$$
$$= (a\phi + a\psi) + (b\phi + b\psi) = a(\phi + \psi) + b(\phi + \psi),$$

we see that $\phi + \psi$ is again in $Hom(A)$.

Since A is commutative, we have

$$a(\phi + \psi) = (a\phi) + (a\psi) = (a\psi) + (a\phi) = a(\psi + \phi)$$

for all $a \in A$, so $\phi + \psi = \psi + \phi$ and addition in $Hom(A)$ is commutative. The associativity of addition follows from

$$a[\phi + (\psi + \theta)] = a\phi + a(\psi + \theta) = a\phi + (a\psi + a\theta)$$
$$= (a\phi + a\psi) + a\theta = a(\phi + \psi) + a\theta = a[(\phi + \psi) + \theta].$$

If e is the additive identity of A, then the homomorphism 0 defined by

$$a0 = e$$

for $a \in A$ is clearly an additive identity in $Hom(A)$. Finally, for

$$\phi \in Hom(A),$$

$-\phi$ defined by

$$a(-\phi) = -(a\phi)$$

is in $Hom(A)$, since

$$(a + b)(-\phi) = -((a + b)\phi) = -(a\phi + b\phi)$$
$$= -(a\phi) + (-(b\phi)) = a(-\phi) + b(-\phi).$$

Clearly, $\phi + (-\phi) = 0$. Thus $\langle Hom(A), + \rangle$ is an abelian group.

Note that we have not yet used the fact that our functions are *homomorphisms* except to show that $\phi + \psi$ and $-\phi$ are again *homomorphisms*. Thus the set A^A of *all functions* from A into A is an abelian group under exactly the same definition of addition, and, of course, function composition again gives a nice associative multiplication in A^A. However, we do need the fact that these functions in $Hom(A)$ are homomorphisms now to prove the right distributive law in $Hom(A)$. Except for this right distributive law, $\langle A^A, +, \cdot \rangle$ satisfies all the axioms for a ring. Let ϕ, ψ, and θ be in $Hom(A)$, and let $a \in A$. Then

$$a[(\phi + \psi)\theta] = [a(\phi + \psi)]\theta = (a\phi + a\psi)\theta.$$

Since θ is a *homomorphism*,

$$(a\phi + a\psi)\theta = (a\phi)\theta + (a\psi)\theta = a(\phi\theta) + a(\psi\theta) = a(\phi\theta + \psi\theta).$$

Thus $(\phi + \psi)\theta = \phi\theta + \psi\theta$. The left distributive law causes no trouble, even in A^A, and follows from

$$a[\phi(\psi + \theta)] = a\phi(\psi + \theta) = (a\phi)\psi + (a\phi)\theta = a(\phi\psi) + a(\phi\theta) = a(\phi\psi + \phi\theta).$$

Thus we have proved the following theorem.

Theorem 25.2 *The set $Hom(A)$ of all endomorphisms of an abelian group A forms a ring under homomorphism addition and homomorphism multiplication (function composition).*

Again, to show applicability to this section, we should give an example showing that $Hom(A)$ need not be commutative. Since function composition is in general not commutative, this seems reasonable to expect. However, $Hom(A)$ may be commutative in some cases. In fact, $Hom(\langle \mathbf{Z}, + \rangle)$ is commutative. We ask the student to show this in an exercise (see Exercise 25.10).

Example 25.3 Consider the free abelian group $\langle \mathbf{Z} \times \mathbf{Z}, + \rangle$ discussed in Part I. We can specify an endomorphism of this free abelian group by giving its values on the generators $(1, 0)$ and $(0, 1)$ of the group. Define

$$\phi \in Hom(\langle \mathbf{Z} \times \mathbf{Z}, + \rangle)$$

by

$$(1, 0)\phi = (1, 0) \quad \text{and} \quad (0, 1)\phi = (1, 0).$$

Define ψ by

$$(1, 0)\psi = (0, 0) \quad \text{and} \quad (0, 1)\psi = (0, 1).$$

Naively, ϕ maps everything onto the first factor of $\mathbf{Z} \times \mathbf{Z}$, and ψ collapses the first factor. Thus

$$(n, m)(\phi\psi) = (n + m, 0)\psi = (0, 0),$$

while

$$(n, m)(\psi\phi) = (0, m)\phi = (m, 0).$$

Hence $\phi\psi \neq \psi\phi$. ∥

*25.3   GROUP RINGS AND GROUP ALGEBRAS

Let $G = \{g_i \mid i \in I\}$ be any multiplicative group, and let R be any commutative ring with unity. Let $R(G)$ be the set of all *formal sums*

$$\sum_{i \in I} a_i g_i$$

for $a_i \in R$ and $g_i \in G$, *where all but a finite number of the a_i are* 0. Define the sum of two elements of $R(G)$ by

$$\left(\sum_{i \in I} a_i g_i\right) + \left(\sum_{i \in I} b_i g_i\right) = \sum_{i \in I} (a_i + b_i) g_i.$$

It is clear that $(a_i + b_i) = 0$ except for a finite number of indices i, so $\sum_{i \in I} (a_i + b_i) g_i$ is again in $R(G)$. It is immediate that $\langle R(G), + \rangle$ is an abelian group with additive identity $\sum_{i \in I} 0 g_i$.

Multiplication of two elements of $R(G)$ is defined by the use of the multiplications in G and R as follows:

$$\left(\sum_{i \in I} a_i g_i\right)\left(\sum_{i \in I} b_i g_i\right) = \sum_{i \in I} \left(\sum_{g_j g_k = g_i} a_j b_k\right) g_i.$$

Naively, we formally distribute the sum $\sum_{i \in I} a_i g_i$ over the sum $\sum_{i \in I} b_i g_i$ and rename a term $a_j g_j b_k g_k$ by "$a_j b_k g_i$", where $g_j g_k = g_i$ in G. Since a_i and b_i are 0 for all but a finite number of i, the sum $\sum_{g_j g_k = g_i} a_j b_k$ contains only a finite number of nonzero summands $a_j b_k \in R$ and may thus be viewed as an element of R. Clearly, again at most a finite number of such sums $\sum_{g_j g_k = g_i} a_j b_k$ are nonzero. Thus multiplication is closed on $R(G)$.

The distributive laws follow at once from the definition of addition and the formal way we used distributivity to define multiplication. For the associativity of multiplication,

$$\left(\sum_{i \in I} a_i g_i\right)\left[\left(\sum_{i \in I} b_i g_i\right)\left(\sum_{i \in I} c_i g_i\right)\right] = \left(\sum_{i \in I} a_i g_i\right)\left[\sum_{i \in I}\left(\sum_{g_j g_k = g_i} b_j c_k\right) g_i\right]$$

$$= \sum_{i \in I}\left(\sum_{g_h g_j g_k = g_i} a_h b_j c_k\right) g_i$$

$$= \left[\sum_{i \in I}\left(\sum_{g_h g_j = g_i} a_h b_j\right) g_i\right]\left(\sum_{i \in I} c_i g_i\right)$$

$$= \left[\left(\sum_{i \in I} a_i g_i\right)\left(\sum_{i \in I} b_i g_i\right)\right]\left(\sum_{i \in I} c_i g_i\right).$$

Thus we have proved the following theorem.

Theorem 25.3 *If G is any multiplicative group, then $\langle R(G), +, \cdot \rangle$ is a ring.*

If we rename the element $\sum_{i \in I} a_i g_i$ of $R(G)$, where $a_i = 0$ for $i \neq j$ and $a_j = 1$, by "g_j", we see that $\langle R(G), \cdot \rangle$ can be considered to contain G naturally as a multiplicative subsystem. Thus, if G is not abelian, $R(G)$ will not be a commutative ring.

Definition. The ring $R(G)$ defined above is the **group ring of G over R**. If F is a field, then $F(G)$ is the **group algebra of G over F**.

+	0	a	e	$e + a$
0	0	a	e	$e + a$
a	a	0	$e + a$	e
e	e	$e + a$	0	a
$e + a$	$e + a$	e	a	0

$\cdot$	0	a	e	$e + a$
0	0	0	0	0
a	0	e	a	$e + a$
e	0	a	e	$e + a$
$e + a$	0	$e + a$	$e + a$	0

Fig. 25.1

Example 25.4 Let us give the addition and multiplication tables for the group algebra $\mathbf{Z}_2(G)$, where $G = \{e, a\}$ is cyclic of order 2. The elements of $\mathbf{Z}_2(G)$ are

$$0e + 0a, \qquad 0e + 1a, \qquad 1e + 0a, \qquad \text{and} \qquad 1e + 1a.$$

If we denote these elements in the obvious, natural way by

$$\text{"0"}, \quad \text{"}a\text{"}, \quad \text{"}e\text{"}, \quad \text{and} \quad \text{"}e + a\text{"},$$

respectively, we get the tables in Fig. 25.1.

For example, to see that $(e + a)(e + a) = 0$, we have

$$(1e + 1a)(1e + 1a) = (1 + 1)e + (1 + 1)a = 0e + 0a.$$

This example shows that a group algebra may have zero divisors. Indeed, this is usually the case. ‖

*25.4   THE QUATERNIONS

The student has as yet seen no example of a skew field. The *quaternions* of Hamilton are the standard example of a skew field; let us describe them.

Let the set Q be $\mathbf{R} \times \mathbf{R} \times \mathbf{R} \times \mathbf{R}$. Now $\langle \mathbf{R} \times \mathbf{R} \times \mathbf{R} \times \mathbf{R}, + \rangle$ is a group under addition by components, the direct product of $\mathbf{R}$ under addition with itself four times. This gives the operation of addition on Q. Let us rename certain elements of Q. We shall let

$$1 = (1, 0, 0, 0), \qquad i = (0, 1, 0, 0),$$
$$j = (0, 0, 1, 0), \qquad \text{and} \qquad k = (0, 0, 0, 1).$$

We furthermore agree to let

$$a_1 = (a_1, 0, 0, 0), \qquad a_2 i = (0, a_2, 0, 0),$$
$$a_3 j = (0, 0, a_3, 0), \qquad \text{and} \qquad a_4 k = (0, 0, 0, a_4).$$

In view of our definition of addition, we then have

$$(a_1, a_2, a_3, a_4) = a_1 + a_2 i + a_3 j + a_4 k.$$

Thus

$$(a_1 + a_2 i + a_3 j + a_4 k) + (b_1 + b_2 i + b_3 j + b_4 k)$$
$$= (a_1 + b_1) + (a_2 + b_2)i + (a_3 + b_3)j + (a_4 + b_4)k.$$

To define multiplication on Q, we start by defining

$$1a = a1 = a \qquad \text{for} \quad a \in Q,$$
$$i^2 = j^2 = k^2 = -1,$$

and

$$ij = k, \quad jk = i, \quad ki = j, \quad ji = -k, \quad kj = -i, \quad \text{and} \quad ik = -j.$$

The student should note the similarity with the so-called *cross product of vectors*. These formulas are easy to remember if you think of the sequence

$$i, j, k, i, j, k.$$

The product from left to right of two adjacent elements is the next one to the right. The product from right to left of two adjacent elements is the negative of the next one to the left. One then defines a product to be what it must be to make the distributive laws hold, namely

$$(a_1 + a_2i + a_3j + a_4k)(b_1 + b_2i + b_3j + b_4k)$$
$$= (a_1b_1 - a_2b_2 - a_3b_3 - a_4b_4) + (a_1b_2 + a_2b_1 + a_3b_4 - a_4b_3)i$$
$$+ (a_1b_3 - a_2b_4 + a_3b_1 + a_4b_2)j$$
$$+ (a_1b_4 + a_2b_3 - a_3b_2 + a_4b_1)k.$$

Verification that Q is a skew field is now a tedious chore, some of which is assigned in an exercise. Since $ij = k$ and $ji = -k$, we see that multiplication is not commutative, so Q is definitely not a field. The only axiom which cannot be verified mechanically is the existence of a multiplicative inverse for $a = a_1 + a_2i + a_3j + a_4k$, with not all $a_i = 0$. The student can check that

$$(a_1 + a_2i + a_3j + a_4k)(a_1 - a_2i - a_3j - a_4k)$$
$$= a_1{}^2 + a_2{}^2 + a_3{}^2 + a_4{}^2.$$

If we let

$$|a|^2 = a_1{}^2 + a_2{}^2 + a_3{}^2 + a_4{}^2 \quad \text{and} \quad \bar{a} = a_1 - a_2i - a_3j - a_4k,$$

we see that

$$\frac{\bar{a}}{|a|^2} = \frac{a_1}{|a|^2} - \left(\frac{a_2}{|a|^2}\right)i - \left(\frac{a_3}{|a|^2}\right)j - \left(\frac{a_4}{|a|^2}\right)k$$

is a multiplicative inverse for a. We consider that we have demonstrated the following theorem.

Theorem 25.4 *The quaternions Q form a skew field under addition and multiplication.*

Note that $G = \{\pm 1, \pm i, \pm j, \pm k\}$ is a group of order 8 under quaternion multiplication. In terms of generators and relations, this group is generated by i and j, where

$$i^4 = 1, \quad j^2 = i^2 \quad \text{and} \quad ji = i^3j.$$

Since we saw in Example 18.5 that G_2, with presentation

$$(a, b : a^4 = 1, b^2 = a^2, ba = a^3b),$$

is a group of order 8, we must have $G_2 \simeq G$. This explains why the group G_2 of Example 18.5 was called the "*quaternion group.*"

Algebra is not as rich in (strictly) skew fields as it is in fields. For example, there are no finite skew fields (which are not fields). This is the content of a famous theorem of Wedderburn which we state without proof.

Theorem 25.5 (Wedderburn). *A finite division ring is a field.*

Proof. See the literature. ∎

EXERCISES

*25.1 Compute

$$\begin{pmatrix} 3 & -4 \\ 1 & 5 \end{pmatrix} + \begin{pmatrix} 4 & 17 \\ 5 & -3 \end{pmatrix}$$

and

$$\begin{pmatrix} 3 & -4 \\ 1 & 5 \end{pmatrix} \begin{pmatrix} 4 & 17 \\ 5 & -3 \end{pmatrix}$$

in $M_2(Q)$.

*25.2 Show that

$$\begin{pmatrix} 1 & 0 \\ 0 & 1 \end{pmatrix}$$

is unity in $M_2(F)$. Describe the unity element in $M_n(F)$.

*25.3 Let ϕ be the element of $Hom(\langle \mathbf{Z} \times \mathbf{Z}, + \rangle)$ given in Example 25.3. This example showed that ϕ is a left divisor of zero. Show that ϕ is also a right divisor of zero.

*25.4 Let $G = \{e, a, b\}$ be a cyclic group of order 3 with identity element e. Write each of the following elements in the group algebra $\mathbf{Z}_5(G)$ in the form

$$re + sa + tb \qquad \text{for} \quad r, s, t \in \mathbf{Z}_5.$$

a) $(2e + 3a + 0b) + (4e + 2a + 3b)$
b) $(2e + 3a + 0b)(4e + 2a + 3b)$
c) $(3e + 3a + 3b)^4$

*25.5 Write the following elements of Q in the form $a_1 + a_2i + a_3j + a_4k$ for $a_i \in \mathbf{R}$.

a) $(i + 3j)(4 + 2j - k)$
b) $i^2j^3kji^5$
c) $(i + j)^{-1}$
d) $[(1 + 3i)(4j + 3k)]^{-1}$

*25.6 Mark each of the following true or false.

— a) $M_n(F)$ has no divisors of zero for any n.
— b) Every nonzero element of $M_2(\mathbf{Z}_2)$ is a unit.
— c) $Hom(A)$ is always a ring with unity $\neq 0$ for every abelian group A.
— d) $Hom(A)$ is never a ring with unity $\neq 0$ for any abelian group A.
— e) The subset $Iso(A)$ of $Hom(A)$, consisting of the isomorphisms of A onto A, forms a subring of $Hom(A)$ for every abelian group A.
— f) $R(\langle \mathbf{Z}, + \rangle)$ is isomorphic to $\langle \mathbf{Z}, +, \cdot \rangle$ for every commutative ring R with unity.
— g) The group ring $R(G)$ of an abelian group G is a commutative ring for any commutative ring R with unity.
— h) The quaternions are a field.
— i) $\langle Q^*, \cdot \rangle$ is a group where Q^* is the set of nonzero quaternions.
— j) No subring of Q is a field.

*25.7 Show that the matrix

$$\begin{pmatrix} 0 & 0 \\ 0 & 1 \end{pmatrix}$$

in $M_2(F)$ is not only a left divisor of zero, as was shown in Example 25.2, but is also a right divisor of zero.

*25.8 Prove the left distributive law in $M_2(F)$.

*25.9 Show that $M_2(F)$ has at least 6 units for every field F. Exhibit these units. [*Hint:* F has at least two elements, 0 and 1.]

*25.10 Show that $Hom(\langle \mathbf{Z}, + \rangle)$ is naturally isomorphic to $\langle \mathbf{Z}, +, \cdot \rangle$ and that $Hom(\langle \mathbf{Z}_n, + \rangle)$ is naturally isomorphic to $\langle \mathbf{Z}_n, +, \cdot \rangle$.

*25.11 Show that $Hom(\langle \mathbf{Z}_2 \times \mathbf{Z}_2, + \rangle)$ is not isomorphic to $\langle \mathbf{Z}_2 \times \mathbf{Z}_2, +, \cdot \rangle$.

*25.12 Referring to the group S_3 given in Example 4.1, compute the product

$$(0\rho_0 + 1\rho_1 + 0\rho_2 + 0\mu_1 + 1\mu_2 + 1\mu_3)(1\rho_0 + 1\rho_1 + 0\rho_2 + 1\mu_1 + 0\mu_2 + 1\mu_3)$$

in the group algebra $\mathbf{Z}_2(S_3)$.

*25.13 If $G = \{e\}$, the group of one element, show that $R(G)$ is isomorphic to R for any ring R.

*25.14 Find two subsets of $\mathbb{Q}$ different from $\mathbf{C}$ and from each other, each of which is a field isomorphic to $\mathbf{C}$ under the induced addition and multiplication from $\mathbb{Q}$.

*25.15 Show by an example that a polynomial equation of degree n may have more than n solutions in a skew field. [*Hint:* Consider $n = 2$ and the skew field $\mathbb{Q}$.]

*25.16 Prove the associative law for multiplication in $\mathbb{Q}$. (This should cure you of wanting to verify any other skew field axioms for $\mathbb{Q}$.)

*25.17 Find the center of the group $\langle \mathbb{Q}^*, \cdot \rangle$, where $\mathbb{Q}^*$ is the set of nonzero quaternions.

26 | The Field of Quotients of an Integral Domain

If an integral domain is such that every nonzero element has a multiplicative inverse, then it is a field. However, many integral domains, such as the integers $\mathbf{Z}$, do not form a field. This dilemma is not too serious. It is the purpose of this section to show that every integral domain can be regarded as being contained in a certain field, *a field of quotients of the integral domain.* This field will be a minimal field containing the integral domain in a sense which we shall describe. For example, the integers are contained in the field $\mathbf{Q}$, whose elements can all be expressed as quotients of integers. Our construction of a field of quotients of an integral domain is exactly the same as the construction of the rational numbers from the integers, which the student may have seen in a course in foundations or advanced calculus. To follow this construction through is such a good exercise in the use of definitions and the concept of isomorphism that we discuss it in some detail, although to write out, or to read, every last detail would be tedious. We can be motivated at every step by the way $\mathbf{Q}$ can be formed from $\mathbf{Z}$. Recall that the different representations of a rational number as a quotient of integers constituted our motivation for the discussion of equivalence relations in Article 0.3.

26.1 THE CONSTRUCTION

Let D be an integral domain which we desire to enlarge to a field of quotients F. A coarse outline of the steps we take is as follows:

1) Define what the elements of F are to be.
2) Define the binary operations of addition and multiplication on F.
3) Check all the field axioms to show that F is a field under these operations.
4) Show that F can be viewed as containing D as an integral subdomain.

Steps (1), (2), and (4) are very interesting, and step (3) is largely a boring chore. We proceed with the construction.

STEP 1. Let D be a given integral domain, and form the Cartesian product

$$D \times D = \{(a, b) \mid a, b \in D\}.$$

We are going to think of an ordered pair (a, b) as representing a *formal quotient* a/b, that is, if $D = \mathbf{Z}$, the pair $(2, 3)$ will eventually represent the

219

number $\frac{2}{3}$ for us. The pair $(2, 0)$ represents no element of $\mathbf{Q}$, and in the general case also, we cut the set $D \times D$ down a bit. Let S be the subset of $D \times D$ given by

$$S = \{(a, b) \mid a, b \in D, b \neq 0\}.$$

Now S is still not going to be our field as is indicated by the fact that, with $D = \mathbf{Z}$, *different* pairs of integers such as $(2, 3)$ and $(4, 6)$ can represent the *same* rational number. We next define when two elements of S represent the same element of F, or as we shall say, when two elements of S are *equivalent*.

> **Definition.** Two elements (a, b) and (c, d) in S are **equivalent**, denoted by "$(a, b) \sim (c, d)$", if and only if $ad = bc$.

Observe that this definition is reasonable, since the criterion for $(a, b) \sim (c, d)$ is an equation $ad = bc$ involving elements in D and concerning the known multiplication in D. Note also that for $D = \mathbf{Z}$, it gives us our usual definition of *equality*, for example, $\frac{2}{3} = \frac{4}{6}$, since $(2)(6) = (3)(4)$. The rational number which we usually denote by "$\frac{2}{3}$" can be thought of as the collection of *all* quotients of integers which reduce to, or are equivalent to, $\frac{2}{3}$.

> **Lemma 26.1** *The relation $\sim$ between elements of the set S as just described is an equivalence relation.*

Proof. We must check the three properties of an equivalence relation.

Reflexive: $(a, b) \sim (a, b)$ since $ab = ba$, for multiplication in D is commutative.

Symmetric: If $(a, b) \sim (c, d)$, then $ad = bc$. Since multiplication in D is commutative, we deduce that $cb = da$, and consequently $(c, d) \sim (a, b)$.

Transitive: If $(a, b) \sim (c, d)$ and $(c, d) \sim (r, s)$, then $ad = bc$ and $cs = dr$. Using these relations and the fact that multiplication in D is commutative, we have

$$asd = sad = sbc = bcs = bdr = brd.$$

Now $d \neq 0$, and D is an integral domain, so cancellation is valid; this is a crucial step in the argument. Hence from $asd = brd$ we obtain $as = br$, so that $(a, b) \sim (r, s)$. ∎

It is worth comparing the preceding proof with the demonstration in Example 0.1. The steps are identical.

We now know, in view of Theorem 0.1, that $\sim$ gives a partition of S into equivalence classes. To avoid long bars over extended expressions, we shall let $[(a, b)]$, rather than $\overline{(a, b)}$, be the equivalence class of (a, b) in S under the relation $\sim$. We now finish step (1) by defining F to be the set of all equivalence classes $[(a, b)]$ for $(a, b) \in S$.

STEP 2. The next lemma serves to define addition and multiplication in F. The student should check to see that if $D = \mathbf{Z}$ and $[(a, b)]$ is viewed as $(a/b) \in \mathbf{Q}$, these definitions applied to $\mathbf{Q}$ give the usual operations.

Lemma 26.2 *For $[(a, b)]$ and $[(c, d)]$ in F, the equations*

$$[(a, b)] + [(c, d)] = [(ad + bc), bd]$$

and

$$[(a, b)][(c, d)] = [(ad, + bc, bd)]$$

give well-defined operations of addition and multiplication on F.

Proof. Note first that if $[(a, b)]$ and $[(c, d)]$ are in F, then (a, b) and (c, d) are in S, so $b \neq 0$ and $d \neq 0$. Since D is an integral domain, $bd \neq 0$, so both $(ad + bc, bd)$ and (ac, bd) are in S. (Note the crucial use here of the fact that D has no divisors of zero.) This shows that the right-hand sides of the defining equations are at least in F.

It remains for us to show that these operations of addition and multiplication are well defined. That is, they were defined by means of representatives in S of elements of F. We must show that if different representatives in S are chosen, the same element of F will result. To this end, suppose that $(a_1, b_1) \in [(a, b)]$ and $(c_1, d_1) \in [(c, d)]$. We must show that

$$(a_1 d_1 + b_1 c_1, b_1 d_1) \in [(ad + bc, bd)]$$

and

$$(a_1 c_1, b_1 d_1) \in [(ac, bd)].$$

Now $(a_1, b_1) \in [(a, b)]$ means that $(a_1, b_1) \sim (a, b)$, that is,

$$a_1 b = b_1 a.$$

Similarly, $(c_1, d_1) \in [(c, d)]$ implies that

$$c_1 d = d_1 c.$$

Multiplying the first equation by $d_1 d$, the second by $b_1 b$, and adding the resulting equations, we obtain the following equation in D:

$$a_1 b d_1 d + c_1 d b_1 b = b_1 a d_1 d + d_1 c b_1 b.$$

Using various axioms for an integral domain, we see that

$$(a_1 d_1 + b_1 c_1) b d = b_1 d_1 (ad + bc),$$

so

$$(a_1 d_1 + b_1 c_1, b_1 d_1) \sim (ad + bc, bd),$$

giving $(a_1 d_1 + b_1 c_1, b_1 d_1) \in [(ad + bc, bd)]$. This takes care of addition in F. For multiplication in F, on multiplying the equations $a_1 b = b_1 a$ and $c_1 d = d_1 c$, we obtain

$$a_1 b c_1 d = b_1 a d_1 c,$$

so, using axioms of D, we get

$$a_1 c_1 bd = b_1 d_1 ac,$$

which implies that

$$(a_1 c_1, b_1 d_1) \sim (ac, bd).$$

Thus $(a_1 c_1, b_1 d_1) \in [(ac, bd)]$, which completes the proof. ∎

The student should be sure he *understands* the meaning of the last lemma and the necessity for proving it. This completes our step (2).

STEP 3. Step (3) is pretty boring, but it is good for the student to work through a few of these details. The reason for this is that he cannot work through them unless he *understands* what we have done. Thus working through them will contribute to his understanding of this construction. We outline the things that must be proved, and prove a couple of them. The rest are left to the exercises.

a) Addition in F is commutative.

Proof. Now $[(a, b)] + [(c, d)]$ is by definition $[(ad + bc, bd)]$. Also $[(c, d)] + [(a, b)]$ is by definition $[(cb + da, db)]$. We need to show that $(ad + bc, bd) \sim (cb + da, db)$. This is clear, since $ad + bc = cb + da$ and $bd = db$, by the axioms of D. ∎

b) Addition is associative.
c) $[(0, 1)]$ is an identity for addition in F.
d) $[(-a, b)]$ is an additive inverse for $[(a, b)]$ in F.
e) Multiplication in F is associative.
f) Multiplication in F is commutative.
g) The distributive laws hold in F.
h) $[(1, 1)]$ is a multiplicative identity in F.
i) If $[(a, b)] \in F$ is not the additive identity, then $a \neq 0$ in D and $[(b, a)]$ is a multiplicative inverse for $[(a, b)]$.

Proof. Let $[(a, b)] \in F$. If $a = 0$, then

$$a1 = b0 = 0,$$

so

$$(a, b) \sim (0, 1),$$

that is, $[(a, b)] = [(0, 1)]$. But $[(0, 1)]$ is the additive identity by (c). Thus, if $[(a, b)]$ is not the additive identity in F, we have $a \neq 0$, so it makes sense to talk about $[(b, a)]$ in F. Now $[(a, b)][(b, a)] = [(ab, ba)]$. But in D we have $ab = ba$, or $(ab)1 = (ba)1$, so

$$(ab, ba) \sim (1, 1).$$

Thus

$$[(a, b)][(b, a)] = [(1, 1)],$$

and $[(1, 1)]$ is the multiplicative identity by (h). ∎

This completes step (3).

STEP 4. It remains for us to show that F can be regarded as containing D. To do this, we show that there is an isomorphism i of D with a subdomain of F. Then if we rename the image of D under i by the names of the elements of D, we will be done. The next lemma gives us this isomorphism.

Lemma 26.3 *The map* $i: D \rightarrow F$ *given by* $ai = [(a, 1)]$ *is an isomorphism of* D *with a subdomain of* F.

Proof. For a and b in D, we have

$$(a + b)i = [(a + b, 1)].$$

Also,

$$(ai) + (bi) = [(a, 1)] + [(b, 1)] = [(a1 + 1b, 1)] = [(a + b, 1)],$$

so $(a + b)i = (ai) + (bi)$. Furthermore,

$$(ab)i = [(ab, 1)],$$

while

$$(ai)(bi) = [(a, 1)][(b, 1)] = [(ab, 1)],$$

so $(ab)i = (ai)(bi)$.

It remains for us to show only that i is one to one. If $ai = bi$, then

$$[(a, 1)] = [(b, 1)],$$

so $(a, 1) \sim (b, 1)$ giving $a1 = 1b$, that is,

$$a = b.$$

Thus i is an isomorphism of D with Di, and, of course, Di is then a subdomain of F. ∎

Since $[(a, b)] = [(a, 1)][(1, b)] = [(a, 1)]/[(b, 1)] = ai/bi$ clearly holds in F, we have now proved the following theorem.

Theorem 26.1 *Any integral domain* D *can be enlarged to (or embedded in) a field* F *such that every element of* F *can be expressed as a quotient of two elements of* D. *(Such a field* F *is a* **field of quotients** *of* D.)

26.2   UNIQUENESS

We said in the beginning that F could be regarded in some sense as a minimal field containing D. This is pretty obvious, since every field containing D must contain all elements a/b for every $a, b \in D$ with $b \neq 0$. The next theorem actually gives a generalization of this idea, and is perhaps an unnecessarily fancy way of achieving our aim. Will you let us strive for a bit of elegance and generality just this once? Also, this theorem is the first real *mapping extension theorem* we have seen. If you get some idea of the concept now, perhaps the idea of our main *Isomorphism Extension Theorem* in Section 41 will come easier to you.

Theorem 26.2 *Let D and D′ be isomorphic integral domains, and let* $\phi: D \to D'$ *be an isomorphism of D onto D′. Let F be a field of quotients of D, and let F′ be any field containing D′. Then there is an isomorphism* ψ *of F into F′, that is, an isomorphism of F with a subfield of F′, such that* $a\psi = a\phi$ *for* $a \in D$. *We say that* ϕ *is* **extended to an isomorphism** ψ **of all of F.**

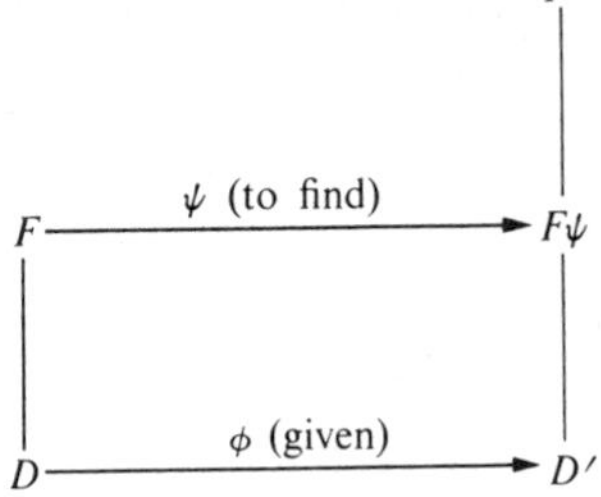

Proof. The lattice and mapping diagram in Fig. 26.1 may help you to visualize the situation for this theorem.

Every $x \in F$ is a quotient a/b of some two elements a and b, $b \neq 0$, of D. Let us attempt to define ψ by

$$(a/b)\psi = (a\phi)/(b\phi).$$

Fig. 26.1

We must first show that this map ψ is sensible and well defined. Since ϕ is an isomorphism, for $b \neq 0$ we have $b\phi \neq 0$, so our definition of $(a/b)\psi$ as $(a\phi)/(b\phi)$ makes sense. If $a/b = c/d$ in F, then $ad = bc$ in D, so $(ad)\phi = (bc)\phi$. But since ϕ is an isomorphism,

$$(ad)\phi = (a\phi)(d\phi), \quad \text{and} \quad (bc)\phi = (b\phi)(c\phi).$$

Thus

$$(a\phi)/(b\phi) = (c\phi)/(d\phi)$$

in F', so ψ is well defined.

The equations

$$(xy)\psi = (x\psi)(y\psi)$$

and

$$(x + y)\psi = x\psi + y\psi$$

follow easily from the definition of ψ and from the fact that ϕ is an isomorphism.

If $(a/b)\psi = (c/d)\psi$, we have

$$(a\phi)/(b\phi) = (c\phi)/(d\phi),$$

so

$$(a\phi)(d\phi) = (b\phi)(c\phi).$$

Since ϕ is an isomorphism, we then deduce that $(ad)\phi = (bc)\phi$, so $ad = bc$, since ϕ is one to one. This implies that $a/b = c/d$. Thus ψ is one to one.

Finally, for $a \in D$, we have

$$a\psi = (a/1)\psi = (a\phi)/(1\phi) = a\phi,$$

since 1ϕ is the identity of D'. Thus $a\phi = a\psi$ for $a \in D$, that is, ψ extends ϕ. ∎

Corollary 1 *Every field F' containing an integral domain D contains a field of quotients of D.*

Proof. This follows from Theorem 26.2 if we put $D' = D$ and put ϕ equal to the identity map. ∎

Corollary 2 *Any two fields of quotients of an integral domain D are isomorphic.*

Proof. Suppose in Theorem 26.2 that F' is a field of quotients of D', so that every element $x' \in F'$ can be expressed in the form a'/b' for $a', b' \in D'$. The fact that ϕ is onto D' tells us that given $a', b' \in D'$, there exist $a, b \in D$ with $a\phi = a'$ and $b\phi = b'$. Then $(a/b)\psi = a'/b'$, that is, ψ is onto F'. Putting $D' = D$ and ϕ equal to the identity map, we then have this corollary. ∎

EXERCISES

26.1 Describe the field F of quotients of the integral subdomain

$$D = \{n + mi \mid n, m \in \mathbf{Z}\}$$

of **C**. "Describe" means give the elements of **C** which comprise the field of quotients of D in **C**.

26.2 Describe (see Exercise 26.1) the field F of quotients of the integral subdomain $D = \{n + m\sqrt{2} \mid n, m \in \mathbf{Z}\}$ of **R**.

26.3 Show by an example that a field F' of quotients of a proper subdomain D' of an integral domain D may also be a field of quotients for D.

†**26.4** Prove part (g) of step (3). You may assume any preceding part of step (3).

26.5 Mark each of the following true or false.

— a) **Q** is a field of quotients of **Z**.
— b) **R** is a field of quotients of **Z**.
— c) **R** is a field of quotients of **R**.
— d) **C** is a field of quotients of **R**.
— e) If D is a field, then any field of quotients of D is isomorphic to D.
— f) The fact that D has no divisors of zero was used strongly several times in the construction of a field F of quotients of the integral domain D.
— g) Every element of an integral domain D is a unit in a field F of quotients of D.
— h) Every nonzero element of an integral domain D is a unit in a field F of quotients of D.
— i) A field of quotients F' of a subdomain D' of an integral domain D can be regarded as a subfield of some field of quotients of D.
— j) Every permutation of every set S is an extension of itself (as a mapping).

26.6 Prove part (b) of step (3). You may assume any preceding part of step (3).

26.7 Prove part (c) of step (3). You may assume any preceding part of step (3).

26.8 Prove part (d) of step (3). You may assume any preceding part of step (3).

26.9 Prove part (e) of step (3). You may assume any preceding part of step (3).

26.10 Prove part (f) of step (3). You may assume any preceding part of step (3).

26.11 Let R be a commutative ring, and let $T \neq \{0\}$ be a nonempty subset of R closed under multiplication and containing no divisors of zero. Starting with $R \times T$ and otherwise exactly following the construction in this section, one can show that the ring R can be enlarged to a *partial ring of quotients* $Q(R, T)$. Think about this for fifteen minutes or so; look back over the construction and see why things still work. In particular, show the following:

a) $Q(R, T)$ has unity even if R does not.

b) In $Q(R, T)$, every nonzero element of T is a unit.

26.12 Prove from Exercise 26.11 that every commutative ring containing an element a which is not a divisor of zero can be enlarged to a commutative ring with unity.

26.13 With reference to Exercise 26.11, how many elements are there in the ring $Q(\mathbf{Z}_4, \{1, 3\})$?

26.14 With reference to Exercise 26.11, describe the ring $Q(\mathbf{Z}, \{2^n \mid n \in \mathbf{Z}^+\})$ by describing a subring of $\mathbf{R}$ to which it is isomorphic.

26.15 With reference to Exercise 26.11, describe the ring $Q(3\mathbf{Z}, \{6^n \mid n \in \mathbf{Z}^+\})$ by describing a subring of $\mathbf{R}$ to which it is isomorphic.

26.16 With reference to Exercise 26.11, suppose we drop the condition that T have no divisors of zero and just require that nonempty $T \neq \{0\}$ be closed under multiplication. The attempt to enlarge R to a commutative ring with unity in which every nonzero element of T is a unit must fail if T contains an element a which is a divisor of zero, for a divisor of zero cannot also be a unit. Try to discover where a construction parallel to that in the text but starting with $R \times T$ first runs into trouble. In particular, for $R = \mathbf{Z}_6$ and $T = \{1, 2, 4\}$, illustrate the first difficulty encountered. [*Hint:* It is in step (1).]

27 | Our Basic Goal

This section is of an expository nature and is designed to give the student the proper perspective on the unstarred portion of the rest of this text. There are no exercises.

In the following two sections, we shall be concerned with topics in ring theory which are analogous to the material on factor groups and homomorphisms for group theory. However, our aim in developing these analogous concepts is quite different from our aims in group theory. In group theory, we used (partly in starred sections) the concepts of factor groups and homomorphisms to study the structure of a given group and to determine the types of group structures of certain orders which could exist. With apologies to the professional mathematician, we explain now to the student, in terms with which he is familiar and which he thinks he understands, our purpose in developing this analogous machinery for rings.

Every single bit of the unstarred material in the rest of this text is aimed at finding and studying solutions of polynomial equations such as

$$x^2 + x - 6 = 0.$$

Let us take a moment to talk about this aim in the light of mathematical history.

We start way back with the Pythagorean school of mathematics of about 525 B.C. The Pythagoreans asserted with an almost fanatical fervor that all distances are **commensurable**, that is, that given distances a and b, there should exist a unit of distance u and integers n and m such that $a = (n)(u)$ and $b = (m)(u)$. In terms of numbers, then, thinking of u as being 1 unit of distance, they maintained that all numbers are integers. This idea of commensurability can be rephrased according to our ideas as an assertion that all numbers are rational, for if a and b are rational numbers, then each is an integral multiple of the reciprocal of the least common multiple of their denominators. For example, if $a = \frac{7}{12}$ and $b = \frac{19}{15}$, then $a = (35)(\frac{1}{60})$ and $b = (76)(\frac{1}{60})$.

The Pythagoreans knew, of course, what is now called the *Pythagorean theorem*, i.e., that for a right triangle with legs of lengths a and b and a hypotenuse of length c, one has

$$a^2 + b^2 = c^2.$$

They also had to grant the existence of a hypotenuse of a right triangle having two legs of equal lengths, say one unit each. The hypotenuse of such a right triangle would, as we know, have to have a length of $\sqrt{2}$. Imagine then their consternation, dismay, and even fury when one of their society, according to some stories it was Pythagoras himself, came up with the embarrassing fact which is stated in our terminology in the following theorem.

Theorem 27.1 *The equation* $x^2 = 2$ *has no solutions in rational numbers. Thus* $\sqrt{2}$ *is not a rational number.*

Proof. Suppose that m/n for $m,\ n \in \mathbf{Z}$ is a rational number such that $(m/n)^2 = 2$. Then

$$m^2 = 2n^2,$$

where both m^2 and $2n^2$ are integers. Since m^2 and $2n^2$ are the same integer, and since 2 is a factor of $2n^2$, we see that 2 must be one of the factors of m^2. But as a square, m^2 has as factors the factors of m repeated twice. Thus m^2 must have two factors 2. Then $2n^2$ must have two factors 2, so n^2 must have 2 as a factor. But then as a square, n^2 must have two factors 2. Therefore $2n^2$ has three factors 2. But then m^2 must have three factors 2 and hence four factors 2 as a square. But then $2n^2$ must have at least four factors 2 and hence ... By bouncing back and forth from m^2 to $2n^2$ in this fashion, you get yourself into an impossible situation. We have shown by contradiction that $2 \neq (m/n)^2$ for $m,\ n \in \mathbf{Z}$. This argument can be made much more concise and elegant, but we are in a reckless mood. ∎

Thus the Pythagoreans ran right into the question of a solution of a polynomial equation, $x^2 - 2 = 0$. We refer the student to Shanks [35, Chapter 3], for a lively and totally delightful account of this Pythagorean dilemma and its significance in mathematics.

In our motivation of the definition of a group, we commented on the necessity of having negative numbers, so that equations such as $x + 2 = 0$ might have solutions. The introduction of negative numbers caused a certain amount of consternation in some philosophical circles. One can visualize 1 apple, 2 apples, and even $\frac{13}{11}$ apples, but how can you point to anything and say that it is -17 apples? Finally, consideration of the equation $x^2 + 1 = 0$ led to the introduction of the number i. The very name of an "imaginary number" given to i shows how this number was regarded. Even today many students are led by this name to regard i with some degree of suspicion. The negative numbers were introduced to the student at such an early stage in his mathematical development that he accepts them without question.

So much for history. We reiterate:

The unstarred portion of the rest of this text is devoted to finding and studying solutions of polynomial equations.

The student first met polynomials in high school freshman algebra. The first problem there was to learn how to add, multiply, and factor polynomials. Then, in both freshman algebra and in the second course in algebra in high school, considerable emphasis was placed on solving polynomial equations. These topics are exactly those with which we shall be concerned. The difference is that while in high school only polynomials with real number coefficients were considered, *we shall be doing our work for polynomials with coefficients from any arbitrary given field.*

As we said above, the next two sections deal with factor rings and ring homomorphisms, material formally analogous to our work on factor groups and group homomorphisms. Polynomials won't even be mentioned. Then we shall introduce polynomials and demonstrate how the idea of solving a polynomial equation can be phrased in terms of the language of homomorphisms. Don't be scared by the terminology at that point. Remember always:

We shall just be doing high school algebra in a more general context.

And then, with amazing ease, elegance, beauty, and sophistication, as a bright star suddenly appears in your possibly drab mathematical life, we shall achieve our

Basic Goal: *To show that given any polynomial equation of degree* ≥ 1, *where the coefficients of the polynomial may be from any field, there exists a solution of the equation.*

If you feel that all this fuss is ridiculous, just think back in history. *This is the culmination of more than 2000 years of mathematical endeavor in working with polynomial equations.* After achieving our *Basic Goal*, we shall spend the rest of our time studying the nature of these solutions of polynomial equations. We emphasize again that you need have no fear in approaching this material. *We shall be dealing with familiar topics of high school algebra. This work should seem much more natural to you than group theory.*

In conclusion, we remark that the machinery of factor rings and ring homomorphisms is not really necessary in order for us to achieve our *Basic Goal.* For a direct demonstration, see Artin [26, p. 29]. However, factor rings and ring homomorphisms are fundamental ideas which the student should grasp, and our *Basic Goal* will follow very easily once we have mastered them. We will further use these concepts effectively in the study of properties of solutions of polynomial equations.

28 | Quotient Rings and Ideals

28.1 INTRODUCTION

This section begins a study of rings which is analogous to the material on groups, contained in Sections 11, 12, and 13, concerning factor groups and homomorphisms. Since $\langle R, + \rangle$ is a group for every ring R, indeed $\langle R, + \rangle$ is an abelian group, the additive part of this theory is already done. We shall have to concern ourselves only with its multiplicative aspects. To make the analogy with the situation for groups as clear as possible, we shall develop this theory according to the same plan that we used for groups. This will give the student another chance to master it.

Let R be a ring. We are concerned with studying a partition of R into disjoint subsets or cells such that these cells themselves can be viewed as elements in a ring where *both* addition and multiplication of the cells are the induced operations from R. That is, we wish to define *both* operations by choosing representatives from the cells, adding or multiplying these representatives in R, and *defining* the sum or product of the cells to be the cell in which the sum or product of the representatives is found. *Both* these operations must be *well defined*, i.e., independent of the choices of representatives from the cells.

Theorem 28.1 (Analog of Theorem 11.1). *If a ring R can be partitioned into cells with both the induced operations described above well defined, and if the cells form a ring under these induced operations, then the cell containing the additive identity 0 of R must be an additive subgroup N of the additive group $\langle R, + \rangle$. Furthermore, N must have the additional property that for all $r \in R$ and $n \in N$, both $rn \in N$ and $nr \in N$. We express this last condition as $rN \subseteq N$ and $Nr \subseteq N$.*

Proof. Viewing $\langle R, + \rangle$ as a group under addition, we know from Theorem 11.1 that N must be an additive subgroup of $\langle R, + \rangle$. Indeed, we know from work following Theorem 11.1 that N must be a normal additive subgroup of $\langle R, + \rangle$, but since addition in R is commutative, every additive subgroup of $\langle R, + \rangle$ is normal in $\langle R, + \rangle$.

We need only show that for $r \in R$ we have $rN \subseteq N$ and $Nr \subseteq N$. Let $r \in R$. Now r is in some cell $A \subseteq R$, and by our hypothesis that the induced cell multiplication is well defined, we can compute the products AN and NA by choosing any representatives. Let us choose $r \in A$ and $0 \in N$. Then

230

AN and NA are the cells containing $r0 = 0r = 0$, that is, $AN = NA = N$. Therefore, for all possible representatives $n \in N$, we have $rn \in N$ and $nr \in N$. ∎

28.2 CRITERIA FOR THE EXISTENCE OF A COSET RING

Theorem 28.2 (Analog of Theorem 11.2). *If a ring R can be partitioned into cells with both the induced operations well defined and giving a ring, then the cells must be precisely the left (and also the right) cosets with respect to addition of the additive subgroup $\langle N, + \rangle$ of $\langle R, + \rangle$, where N is the cell containing 0.*

Proof. This is immediate from Theorem 11.2 if we consider just $\langle R, + \rangle$, an additive group, and forget about multiplication. ∎

Of course, Theorem 11.3 holds here also, i.e., these cosets are disjoint. Since addition is commutative, the left coset $r + N$ is the same as the right coset $N + r$.

Lemma 28.1 (Analog of Lemma 12.1). *If $\langle N, + \rangle$ is an additive subgroup of $\langle R, + \rangle$ for a ring R, and if the induced operations of addition and multiplication on cosets $r + N$ for $r \in R$ are well defined, i.e., independent of the choice of representatives, then the collection of these cosets $r + N$ is a ring under these induced coset operations.*

Proof. Lemma 12.1 shows us that the cosets form a group under the induced addition. The ring axioms involving multiplication follow at once, since we also compute products by choosing representatives, and since the ring axioms hold in R. For example, the left distributive law

$$(r_1 + N)[(r_2 + N) + (r_3 + N)]$$
$$= [(r_1 + N)(r_2 + N)] + [(r_1 + N)(r_3 + N)]$$

follows when we choose $r_i \in (r_i + N)$ as representatives and observe that the required

$$r_1(r_2 + r_3) \in [(r_1 r_2 + r_1 r_3) + N]$$

is true, since $r_1(r_2 + r_3) = (r_1 r_2 + r_1 r_3)$ by the left distributive law in R. The other distributive law and the associative law for multiplication follow similarly. ∎

Theorem 28.3 (Analog of Theorem 12.1). *If $\langle N, + \rangle$ is an additive subgroup of the additive group $\langle R, + \rangle$ of a ring R, then the operations of induced addition and multiplication are both well defined on the cosets $r + N$ for $r \in R$ if and only if $rN \subseteq N$ and $Nr \subseteq N$ for all $r \in R$.*

Proof. The fact that if the operations are well defined, then $rN \subseteq N$ and $Nr \subseteq N$ follows immediately from Lemma 28.1 and Theorem 28.1.

Suppose that $\langle N, + \rangle$ is an additive subgroup of $\langle R, + \rangle$ such that for all $r \in R$ we have $rN \subseteq N$ and $Nr \subseteq N$. Coset addition is well defined, by Theorem 12.1 applied to the subgroup $\langle N, + \rangle$ of $\langle R, + \rangle$. For multiplication, we must show that a product $(r_1 + N)(r_2 + N)$, computed by choosing representatives, is well defined. It is no loss of generality to take as two representatives of $r_1 + N$ the elements r_1 itself and $r_1 + n_1$ for $n_1 \in N$. Similarly, let r_2 and $r_2 + n_2$ be two elements of $r_2 + N$. We must show that $(r_1 + n_1)(r_2 + n_2)$ is in the same coset $r_1 r_2 + N$ as $r_1 r_2$. Now

$$(r_1 + n_1)(r_2 + n_2) = r_1 r_2 + n_1 n_2 + n_1 r_2 + r_1 n_2,$$

by the distributive laws in R. The assumption that $rN \subseteq N$ and $Nr \subseteq N$ implies that $r_1 n_2 \in N$ and $n_1 r_2 \in N$. Viewing n_1 as an element of R, we have $n_1 n_2 \in n_1 N$ and $n_1 N \subseteq N$, so $n_1 n_2 \in N$ also. Then since $\langle N, + \rangle$ is an additive subgroup of $\langle R, + \rangle$, $(n_1 r_2 + r_1 n_2 + n_1 n_2) \in N$, so $(r_1 + n_1)(r_2 + n_2) \in (r_1 r_2 + N)$. ∎

It is now clear that these special additive subgroups $\langle N, + \rangle$ of a ring R having the property that $rN \subseteq N$ and $Nr \subseteq N$ for all $r \in R$ are going to play a role of basic importance in ring theory, analogous to the role of a normal subgroup in a group. Note that the conditions $rN \subseteq N$ and $Nr \subseteq N$ imply in particular, for $r \in N$, that N is closed under the multiplication of R. Thus, we can consider N with the induced addition and multiplication from R to be a subring of R. Not every subring N of every ring R satisfies the conditions $rN \subseteq N$ and $Nr \subseteq N$ however. For example, $\mathbf{Q} \leq \mathbf{R}$, but $\pi \mathbf{Q} \not\subseteq \mathbf{Q}$.

28.3 IDEALS AND QUOTIENT RINGS

Definition (*Analog of the definition of a normal subgroup*). A subring N of a ring R satisfying $rN \subseteq N$ and $Nr \subseteq N$ for all $r \in R$ is an *ideal* (or *two-sided ideal*) *of* R. A subring N of R satisfying $rN \subseteq N$ for all $r \in R$ is a *left ideal of* R, and one satisfying $Nr \subseteq N$ for all $r \in R$ is a *right ideal of* R.

When we refer to an *ideal*, it will always be assumed to be a two-sided ideal. Left or right ideals which are not two sided will not concern us. Let us reiterate for emphasis:

An ideal is to a ring as a normal subgroup is to a group.

Definition (*Analog of the definition of a factor group*). If N is an ideal in a ring R, then the ring of cosets $r + N$ under the induced operations is the *quotient ring*, or *factor ring*, or *residue class ring of R modulo N*, and is denoted by "R/N". The cosets are *residue classes modulo N*.

Example 28.1 Consider the ring $\mathbf{Z}$. The only additive subgroups of $\langle \mathbf{Z}, + \rangle$ are the subgroups $n\mathbf{Z}$, as we have seen. Clearly, if r is any integer and

$m \in n\mathbf{Z}$, then $rm = mr$ is again a multiple of n, that is, if $m = ns$, then $rm = mr = n(sr)$, and $n(sr) \in n\mathbf{Z}$. Thus, $n\mathbf{Z}$ is an ideal, and the cosets $a + n\mathbf{Z}$ of $n\mathbf{Z}$ form a ring $\mathbf{Z}/n\mathbf{Z}$ under the induced operations of addition and multiplication. ‖

In Example 23.2, we defined for a and b in $\mathbf{Z}_n$ the product ab modulo n to be the remainder when divided by n of the usual product of a and b in $\mathbf{Z}$. The map $\phi: \mathbf{Z}_n \rightarrow \mathbf{Z}/n\mathbf{Z}$ given by

$$a\phi = a + n\mathbf{Z}$$

is clearly a one-to-one onto map such that $(a + b)\phi = a\phi + b\phi$ and $(ab)\phi = (a\phi)(b\phi)$. Thus $\mathbf{Z}_n$ under addition and multiplication modulo n may be viewed as $\mathbf{Z}/n\mathbf{Z}$ renamed, and consequently is a ring under this addition and multiplication. We used this fact in the preceding sections with the comment that it would be justified in a general situation later. *This is the justification.* We shall often identify $\mathbf{Z}/n\mathbf{Z}$ with $\mathbf{Z}_n$ by means of this isomorphism ϕ, which is a natural or canonical isomorphism.

Example 28.2 As was shown in the corollary of Theorem 24.4, $\mathbf{Z}_p$, which is isomorphic to $\mathbf{Z}/p\mathbf{Z}$, is a field for p a prime. *Thus a quotient ring of an integral domain may be a field.* We shall say more about this in the next section. ‖

Example 28.3 The subset $N = \{0, 3\}$ of $\mathbf{Z}_6$ is easily seen to be an ideal of $\mathbf{Z}_6$, and $\mathbf{Z}_6/N$ has three elements, $0 + N$, $1 + N$, and $2 + N$. These obviously add and multiply in such a fashion as to show that $\mathbf{Z}_6/N \simeq \mathbf{Z}_3$ under the correspondence

$$(0 + N) \leftrightarrow 0, \qquad (1 + N) \leftrightarrow 1, \qquad (2 + N) \leftrightarrow 2. \; ‖$$

Example 28.4 The ring $\mathbf{Z} + \mathbf{Z}$ is not an integral domain, for

$$(0, 1)(1, 0) = (0, 0),$$

showing that $(0, 1)$ and $(1, 0)$ are zero divisors. If $N = \{(0, n) \mid n \in \mathbf{Z}\}$, it is clear that N is an ideal of $\mathbf{Z} + \mathbf{Z}$, and that $(\mathbf{Z} + \mathbf{Z})/N$ is isomorphic to $\mathbf{Z}$ under the correspondence $[(m, 0) + N] \leftrightarrow m$, the residue classes being of the form $(m, 0) + N$, where $m \in \mathbf{Z}$. *Thus a quotient ring of a ring may be an integral domain, even though the original ring is not.* We shall say more about this also in the next section. ‖

The preceding examples may indicate to the student the tremendous importance of the concepts of ideals and quotient rings. Every ring R has two ideals, R itself and $\{0\}$. For these **improper ideals**, the factor rings are R/R, which has only one element, and $R/\{0\}$, which is clearly isomorphic to R. These are uninteresting cases. Just as for a subgroup of a group, a **proper ideal** of a ring R is an ideal N of R such that $N \neq R$ and $N \neq \{0\}$.

While quotient rings of rings and integral domains may be of great interest, as the above examples indicate, the corollary of the next and last theorem of this section shows that a quotient ring of a field is really not of much additional use.

Theorem 28.4 *If R is a ring with unity, and N is an ideal of R containing a unit, then $N = R$.*

Proof. Let N be an ideal of R, and suppose that $u \in N$ for some unit u in R. Then the condition $rN \subseteq N$ for all $r \in R$ implies, if we take $r = u^{-1}$ and $u \in N$, that $1 = u^{-1}u$ is in N. But then $rN \subseteq N$ for all $r \in R$ implies that $r1 = r$ is in N for all $r \in R$, so $N = R$. ∎

Corollary. *A field contains no proper ideals.*

Proof. Since every nonzero element of a field is a unit, it follows at once from Theorem 28.4 that an ideal of a field F is either $\{0\}$ or all of F. ∎

EXERCISES

28.1 Find all ideals N of $\mathbf{Z}_{12}$. In each case compute $\mathbf{Z}_{12}/N$, that is, find a known ring to which the quotient ring is isomorphic.

28.2 Give addition and multiplication tables for $2\mathbf{Z}/8\mathbf{Z}$. Are $2\mathbf{Z}/8\mathbf{Z}$ and $\mathbf{Z}_4$ isomorphic rings?

28.3 Find a subring of the ring $\mathbf{Z} + \mathbf{Z}$ which is not an ideal of $\mathbf{Z} + \mathbf{Z}$.

†**28.4** Prove that a quotient ring of a ring R modulo an ideal N is commutative if and only if $(rs - sr) \in N$ for all $r, s \in R$.

28.5 Mark each of the following true or false.

___ a) $\mathbf{Q}$ is an ideal in $\mathbf{R}$.
___ b) Every ideal in a ring is a subring of the ring.
___ c) Every subring of every ring is an ideal of the ring.
___ d) Every quotient ring of every commutative ring is again a commutative ring.
___ e) The rings $\mathbf{Z}/4\mathbf{Z}$ and $\mathbf{Z}_4$ are isomorphic.
___ f) An ideal N in a ring with unity R is all of R if and only if $1 \in N$.
___ g) The concept of an ideal is to the concept of a ring as the concept of a normal subgroup is to the concept of a group.
___ h) $\mathbf{Z}_4$ is an ideal of $4\mathbf{Z}$.
___ i) If a ring R has zero divisors, then every quotient ring of R has zero divisors.
___ j) $\mathbf{Z}$ is an ideal in $\mathbf{Q}$.

28.6 Show that a factor ring of a field is either the trivial ring of one element or is isomorphic to the field.

28.7 Show that if R is a ring with unity and N is an ideal of R such that $N \neq R$, then R/N is a ring with unity $\neq 0$.

28.8 Let R be a commutative ring and let $a \in R$. Show that $I_a = \{x \in R \mid ax = 0\}$ is an ideal of R.

28.9 Show that an intersection of ideals of a ring R is again an ideal of R.

28.10 Determine all ideals of $\mathbf{Z} + \mathbf{Z}$.

28.11 An element a of a ring R is **nilpotent** if $a^n = 0$ for some $n \in \mathbf{Z}^+$. Show that the collection of all nilpotent elements in a commutative ring R is an ideal, the **radical of** R.

28.12 Referring to the definition given in Exercise 28.11, find the radical of the ring $\mathbf{Z}_{12}$ and observe that it is one of the ideals of $\mathbf{Z}_{12}$ found in Exercise 28.1. What is the radical of $\mathbf{Z}$? of $\mathbf{Z}_{32}$?

28.13 Referring to Exercise 28.11, show that if N is the radical of a commutative ring R, then R/N has as radical the trivial ideal $\{0 + N\}$.

28.14 Let R be a commutative ring and N an ideal of R. Referring to Exercise 28.11, show that if every element of N is nilpotent and the radical of R/N is R/N, then the radical of R is R.

28.15 Let R be a commutative ring and N an ideal of R. Show that the set $\sqrt{N}$ of all $a \in R$, such that $a^n \in N$ for some $n \in \mathbf{Z}^+$, is an ideal of R, the **radical of** N. Is this terminology consistent with that in Exercise 28.11?

28.16 Referring to Exercise 28.15, show by examples that for proper ideals N of a commutative ring R,

a) $\sqrt{N}$ need not equal N,   b) $\sqrt{N}$ may equal N.

28.17 What is the relation of the ideal $\sqrt{N}$ of Exercise 28.15 to the radical of R/N (see Exercise 28.11)? Word your answer carefully.

There is a sort of *arithmetic* of ideals in a commutative ring. The next three exercises define sum, product, and quotient of ideals.

28.18 If A and B are ideals of a ring R, the **sum** $A + B$ of A and B is defined by

$$A + B = \{a + b \mid a \in A, b \in B\}.$$

a) Show that $A + B$ is an ideal.   b) Show that $A \subseteq A + B$ and $B \subseteq A + B$.

28.19 Let A and B be ideals of a ring R. The **product** AB of A and B is defined by

$$AB = \left\{ \sum_{i=1}^{n} a_i b_i \mid a_i \in A, b_i \in B, n \in \mathbf{Z}^+ \right\}.$$

a) Show that AB is an ideal in R.   b) Show that $AB \subseteq (A \cap B)$.

28.20 Let A and B be ideals of a commutative ring R. The **quotient** $A : B$ **of** A **by** B is defined by

$$A : B = \{r \in R \mid rb \in A \text{ for all } b \in B\}.$$

Show that $A : B$ is an ideal of R.

*28.21 Show that for a field F, the set of all matrices of the form

$$\begin{pmatrix} a & b \\ 0 & 0 \end{pmatrix}$$

for $a, b \in F$ is a right ideal but not a left ideal of $M_2(F)$.

*28.22 Show that the matrix ring $M_2(\mathbf{Z}_2)$ is a **simple ring**, that is, $M_2(\mathbf{Z}_2)$ has no proper ideals.

29 | Homomorphisms of Rings

29.1 DEFINITION AND ELEMENTARY PROPERTIES

We keep our discussion of homomorphisms of rings parallel to that in Section 13 for homomorphisms of groups.

Definition (*Analog of the definition of a group homomorphism*). A map ϕ of a ring R into a ring R' is a **homomorphism** if

$$(a + b)\phi = a\phi + b\phi$$

and

$$(ab)\phi = (a\phi)(b\phi)$$

for all elements a and b in R.

Theorem 29.1 (Analog of Theorem 13.1). *If N is an ideal of a ring R, then the canonical map $\gamma: R \rightarrow R/N$ given by $a\gamma = a + N$ for $a \in R$ is a homomorphism.*

Proof. Theorem 13.1 applied to $\langle R, + \rangle$ as an additive group with $\langle N, + \rangle$ a normal subgroup shows that

$$(a + b)\gamma = a\gamma + b\gamma.$$

Also,

$$(ab)\gamma = ab + N = (a + N)(b + N) = (a\gamma)(b\gamma). \quad \blacksquare$$

Definition (*Analog of the definition of the kernel of a group homomorphism*). The **kernel of a homomorphism** ϕ of a ring R into a ring R' is the set of all elements of R mapped onto the additive identity $0'$ of R' by ϕ.

Theorem 29.2 (Analog of Theorem 13.2). *Let ϕ be a homomorphism of a ring R into a ring R'. If 0 is the additive identity in R, then $0\phi = 0'$ is the additive identity in R', and if $a \in R$, then $(-a)\phi = -(a\phi)$. If S is a subring of R, then $S\phi$ is a subring of R', and S an ideal of R implies that $S\phi$ is an ideal of $R\phi$. Going the other way, if S' is a subring of R', then $S'\phi^{-1}$ is a subring of R, and S' an ideal of $R\phi$ implies that $S'\phi^{-1}$ is an ideal of R. Finally, if R has unity 1 and $1\phi \neq 0'$, then $1\phi = 1'$ is unity for $R\phi$. Loosely, subrings correspond to subrings, ideals to ideals, and rings with unity to rings with unity under a ring homomorphism.*

236

Proof. Let ϕ be a homomorphism of a ring R into a ring R'. Since, in particular, ϕ can be viewed as a group homomorphism of $\langle R, + \rangle$ into $\langle R', +' \rangle$, Theorem 13.2 tells us that $0\phi = 0'$ is the additive identity of R' and that $(-a)\phi = -(a\phi)$.

Theorem 13.2 also tells us that if S is a subring of R, then, considering the additive group $\langle S, + \rangle$, we find that $\langle S\phi, +' \rangle$ is a subgroup of $\langle R', +' \rangle$. If $s_1\phi$ and $s_2\phi$ are two elements of $S\phi$, then

$$(s_1\phi)(s_2\phi) = (s_1 s_2)\phi$$

and $(s_1 s_2)\phi \in S\phi$, so multiplication is closed on $S\phi$, which is then a subring of R'. If S is an ideal of R, then for $s \in S$ and $r \in R$, we have

$$(r\phi)(s\phi) = (rs)\phi$$

and $(rs)\phi \in S\phi$. Also,

$$(s\phi)(r\phi) = (sr)\phi$$

and $(sr)\phi \in S\phi$. Thus $S\phi$ is an ideal in $R\phi$.

Going the other way, Theorem 13.2 again shows that if S' is a subring of R', then $\langle S'\phi^{-1}, + \rangle$ is a subgroup of $\langle R, + \rangle$. If $a\phi \in S'$ and $b\phi \in S'$, then

$$(ab)\phi = (a\phi)(b\phi)$$

and $[(a\phi)(b\phi)] \in S'$, so multiplication is closed on $S'\phi^{-1}$, which is then a subring of R. If S' is an ideal of $R\phi$, then for $a \in S'\phi^{-1}$ and any $r \in R$ we have

$$(ra)\phi = (r\phi)(a\phi)$$

and $[(r\phi)(a\phi)] \in S'$, so $ra \in S'\phi^{-1}$. Similarly, $ar \in S'\phi^{-1}$. Thus $S'\phi^{-1}$ is an ideal of R.

Finally, if R has unity 1, then for all $r \in R$,

$$r\phi = (1r)\phi = (r1)\phi = (1\phi)(r\phi) = (r\phi)(1\phi),$$

so $1' = 1\phi$ is a multiplicative identity for $R\phi$. If $1' \neq 0'$, then $1'$ is unity for $R\phi$. ∎

Theorem 29.2 shows in particular that for a homomorphism $\phi: R \to R'$, the kernel $K = \{0'\}\phi^{-1}$ is an ideal of R.

Theorem 29.3 (Fundamental Homomorphism Theorem; Analog of Theorem 13.3). *Let ϕ be a homomorphism of a ring R into a ring R' with kernel K. Then $R\phi$ is a ring and there is a canonical isomorphism of $R\phi$ with R/K.*

Proof. Theorem 29.2 shows that $R\phi$ is a ring. Let $(a + K) \in R/K$, and define the map $\psi: R/K \to R\phi$ by

$$(a + K)\psi = a\phi.$$

Theorem 13.3 shows that ψ is well defined, one to one and onto with

$$[(a + K) + (b + K)]\psi = (a + K)\psi + (b + K)\psi.$$

Now

$$[(a + K)(b + K)]\psi = (ab + K)\psi = (ab)\phi = (a\phi)(b\phi)$$
$$= [(a + K)\psi][(b + K)\psi].$$

Thus ψ is a ring isomorphism.

Again ψ is canonical in the sense that if $\gamma: R \to R/K$ is the canonical map, then $\phi = \gamma\psi$. ∎

29.2 MAXIMAL AND PRIME IDEALS

We now take up the question of when a quotient ring of a ring is a field and when it is an integral domain. The analogy with groups in Section 13 can be stretched a bit further to cover the case in which the quotient ring is a field.

> **Definition** (*Analog of the definition of a maximal normal subgroup*). A **maximal ideal of a ring** R is an ideal M different from R such that there is no proper ideal N of R properly containing M.

> **Theorem 29.4** (Analog of Theorem 13.4). *Let R be a commutative ring with unity. Then M is a maximal ideal of R if and only if R/M is a field.*

Proof. Suppose M is a maximal ideal in R. It is easy to see that if R is a commutative ring with unity, then R/M is also a commutative ring with unity if $M \neq R$, which is the case if M is maximal. Let $(a + M) \in R/M$, with $a \notin M$, so that $a + M$ is not the additive identity of R/M. We must show that $a + M$ has a multiplicative inverse in R/M. Let

$$N = \{ra + m \mid r \in R, m \in M\}.$$

Then $\langle N, + \rangle$ is a group, for

$$(r_1 a + m_1) + (r_2 a + m_2) = (r_1 + r_2)a + (m_1 + m_2),$$

and the latter is clearly in N, while also,

$$0 = 0a + 0 \quad \text{and} \quad -(ra + m) = (-r)a + (-m).$$

Now

$$r_1(ra + m) = (r_1 r)a + r_1 m$$

shows that $r_1(ra + m) \in N$ for $r_1 \in R$, and since R is a commutative ring, $(ra + m)r_1 \in N$ also. Thus N is an ideal. But

$$a = 1a + 0$$

shows that $a \in N$, and for $m \in M$,

$$m = 0a + m$$

shows that $M \subseteq N$. Hence N is an ideal of R properly containing M, since $a \in N$ and $a \notin M$. Since M is maximal, we must have $N = R$. In particular, $1 \in N$. Then by definition of N, there is $b \in R$ and $m \in M$ such that $1 = ba + m$. Therefore,

$$1 + M = ba + M = (b + M)(a + M),$$

so $b + M$ is a multiplicative inverse of $a + M$.

Conversely, suppose that R/M is a field. By Theorem 29.2, if N is any ideal of R such that $M \subset N \subset R$ and γ is the canonical homomorphism of R onto R/M, then $N\gamma$ is an ideal of R/M with $\{(0 + M)\} \subset N\gamma \subset R/M$, contrary to the corollary of Theorem 28.4 which states that the field R/M contains no proper ideals. Hence if R/M is a field, M is maximal. ∎

Corollary. *A commutative ring with unity is a field if and only if it has no proper ideals.*

Proof. The corollary of Theorem 28.4 shows that a field has no proper ideals.

Conversely, if a commutative ring R with unity has no proper ideals, then $\{0\}$ is a maximal ideal and $R/\{0\}$, which is isomorphic to R, is a field by Theorem 29.4. ∎

We now turn to the question of the characterization of the ideals $N \neq R$ for a commutative ring R with unity such that R/N is an integral domain. The answer here is rather obvious. The factor ring R/N will be an integral domain if and only if $(a + N)(b + N) = N$ implies that either

$$a + N = N \quad \text{or} \quad b + N = N.$$

This is exactly the statement that R/N has no divisors of zero, since the coset N plays the roll of zero in R/N. Looking at representatives, we see that this condition amounts to saying that $ab \in N$ implies that either $a \in N$ or $b \in N$.

Example 29.1 The ideals of $\mathbf{Z}$ are of the form $n\mathbf{Z}$. We have seen that $\mathbf{Z}/n\mathbf{Z} \simeq \mathbf{Z}_n$ and that $\mathbf{Z}_n$ is an integral domain if and only if n is a prime. Thus the ideals $n\mathbf{Z}$ such that $\mathbf{Z}/n\mathbf{Z}$ is an integral domain are of the form $p\mathbf{Z}$, where p is a prime. Of course, $\mathbf{Z}/p\mathbf{Z}$ is actually a field, so that $p\mathbf{Z}$ is a maximal ideal of $\mathbf{Z}$. Note that for a product rs of integers to be in $p\mathbf{Z}$, the prime p must divide either r or s. The role of prime integers in this example makes the use of the word *prime* in the next definition more reasonable. ‖

Definition. *An ideal $N \neq R$ in a commutative ring R is a **prime ideal** if $ab \in N$ implies that either $a \in N$ or $b \in N$ for all $a, b \in R$.*

Our remarks preceding Example 29.1 constitute a proof of the following theorem.

Theorem 29.5 *Let R be a commutative ring with unity, and let $N \neq R$ be an ideal in R. Then R/N is an integral domain if and only if N is a prime ideal in R.*

Corollary. *Every maximal ideal in a commutative ring R with unity is a prime ideal.*

Proof. If M is maximal in R, then R/M is a field, hence an integral domain, and therefore M is a prime ideal by Theorem 29.5. ∎

The material that has just been presented regarding maximal and prime ideals is very important and we shall be using it quite a lot. The student should keep the main ideas well in mind, even though he may have become lost in the fascinating proof of Theorem 29.4. That is, he must know and understand the definitions of maximal and prime ideals, and must remember the following facts which we have demonstrated.

For a commutative ring R with unity:

1) *An ideal M of R is maximal if and only if R/M is a field.*
2) *An ideal N of R is prime if and only if R/N is an integral domain.*
3) *Every maximal ideal of R is a prime ideal.*

29.3 PRIME FIELDS

Let R be any ring with unity 1. Recall that by $n \cdot 1$ we mean $1 + 1 + \cdots + 1$ for n summands for $n > 0$, and $(-1) + (-1) + \cdots + (-1)$ for $|n|$ summands for $n < 0$, while $n \cdot 1 = 0$ for $n = 0$.

Theorem 29.6 *If R is a ring with unity 1, then the map $\phi\colon \mathbf{Z} \to R$ given by*

$$n\phi = n \cdot 1$$

for $n \in \mathbf{Z}$ is a homomorphism of $\mathbf{Z}$ into R.

Proof. It is obvious that

$$(n + m)\phi = (n + m) \cdot 1 = (n \cdot 1) + (m \cdot 1) = n\phi + m\phi.$$

The distributive laws in R show that

$$\underbrace{(1 + 1 + \cdots + 1)}_{n \text{ summands}}\underbrace{(1 + 1 + \cdots + 1)}_{m \text{ summands}} = \underbrace{(1 + 1 + \cdots + 1)}_{nm \text{ summands}}.$$

Thus $(n \cdot 1)(m \cdot 1) = (nm) \cdot 1$ for $n, m > 0$. Similar arguments with the distributive laws shows that for all $n, m \in \mathbf{Z}$, we have

$$(n \cdot 1)(m \cdot 1) = (nm) \cdot 1.$$

Thus

$$(nm)\phi = (nm) \cdot 1 = (n \cdot 1)(m \cdot 1) = (n\phi)(m\phi). \quad ∎$$

Corollary. *If R is a ring with unity and characteristic n > 1, then R contains a subring isomorphic to $\mathbf{Z}_n$. If R has characteristic 0, then R contains a subring isomorphic to $\mathbf{Z}$.*

Proof. The map $\phi \colon \mathbf{Z} \to R$ given by $m\phi = m \cdot 1$ for $m \in \mathbf{Z}$ is a homomorphism by Theorem 29.6. The kernel must be an ideal in $\mathbf{Z}$. All ideals in $\mathbf{Z}$ are of the form $s\mathbf{Z}$ for some $s \in \mathbf{Z}$. By Theorem 24.5, it is clear that if R has characteristic $n > 0$, then the kernel of ϕ is $n\mathbf{Z}$. Then the image $\mathbf{Z}\phi \leq R$ is isomorphic to $\mathbf{Z}/n\mathbf{Z} \simeq \mathbf{Z}_n$. If the characteristic of R is 0, then $m \cdot 1 \neq 0$ for all $m \neq 0$, so the kernel of ϕ is $\{0\}$. Thus, the image $\mathbf{Z}\phi \leq R$ is isomorphic to $\mathbf{Z}$. ∎

Theorem 29.7 *A field F is either of prime characteristic p and contains a subfield isomorphic to $\mathbf{Z}_p$ or of characteristic 0 and contains a subfield isomorphic to $\mathbf{Q}$.*

Proof. If the characteristic of F is not 0, the above corollary shows that F contains a subring isomorphic to $\mathbf{Z}_n$. Then n must be a prime p, or F would have zero divisors. If F is of characteristic 0, then F must contain a subring isomorphic to $\mathbf{Z}$. In this case the corollaries of Theorem 26.2 show that F must contain a field of quotients of this subring, and that this field of quotients must be isomorphic to $\mathbf{Q}$. ∎

Thus every field contains either a subfield isomorphic to $\mathbf{Z}_p$ for some prime p or a subfield isomorphic to $\mathbf{Q}$. These fields $\mathbf{Z}_p$ and $\mathbf{Q}$ are the fundamental building blocks on which all fields rest.

Definition. The fields $\mathbf{Z}_p$ and $\mathbf{Q}$ are *prime fields*.

EXERCISES

29.1 Describe all ring homomorphisms of $\mathbf{Z}$ into $\mathbf{Z}$. [*Hint:* By group theory, a homomorphism of a ring R is determined by its values on an additive generating set of the group $\langle R, + \rangle$. Describe the homomorphisms by giving their values on the generator 1 of $\langle \mathbf{Z}, + \rangle$.]

29.2 Describe all ring homomorphisms of $\mathbf{Z} + \mathbf{Z}$ into $\mathbf{Z}$. (See the hint of Exercise 29.1.)

29.3 Find all prime ideals and all maximal ideals of $\mathbf{Z}_{12}$.

29.4 Find a maximal ideal of $\mathbf{Z} + \mathbf{Z}$. Find a prime ideal of $\mathbf{Z} + \mathbf{Z}$ which is not maximal. Find a proper ideal of $\mathbf{Z} + \mathbf{Z}$ which is not prime.

†**29.5** Prove directly from the definitions of maximal and prime ideals that every maximal ideal of a commutative ring R with unity is a prime ideal. [*Hint:* Suppose M is maximal in R, $ab \in M$, and $a \notin M$. Consider the ideal of R generated by a and M.]

29.6 Mark each of the following true or false.

___ a) The concept of a ring homomorphism is closely connected with the idea of a factor ring.

— b) A homomorphism is to a ring as an isomorphism is to a group.
— c) A ring homomorphism is a one-to-one map if and only if the kernel is zero.
— d) In a sense, a field is to the theory of commutative rings with unity as a simple group is to the theory of groups.
— e) The kernel of a ring homomorphism is an ideal of the whole ring.
— f) Every prime ideal of every commutative ring with unity is a maximal ideal.
— g) Every maximal ideal of every commutative ring with unity is a prime ideal.
— h) $\mathbf{Q}$ is its own prime subfield.
— i) The prime subfield of $\mathbf{C}$ is $\mathbf{R}$.
— j) Every field contains a prime field as a subfield.

29.7 Describe all ring homomorphisms of $\mathbf{Z} + \mathbf{Z}$ into $\mathbf{Z} + \mathbf{Z}$. (See the hint of Exercise 29.1.)

29.8 Show that each homomorphism of a field is either an isomorphism or maps everything onto 0.

29.9 Show that if R, R', and R'' are rings, and if $\phi: R \to R'$ and $\psi: R' \to R''$ are homomorphisms, then the composite function $\phi\psi: R \to R''$ is a homomorphism. (Use Exercise 13.13.)

29.10 Show that N is a maximal ideal in a ring R if and only if R/N is a simple ring, i.e., has no proper ideals. (Compare with Theorem 13.4.)

29.11 The corollary of Theorem 29.6 tells us that every ring with unity contains a subring isomorphic to either $\mathbf{Z}$ or some $\mathbf{Z}_n$. Is it possible that a ring with unity may simultaneously contain two subrings isomorphic to $\mathbf{Z}_n$ and $\mathbf{Z}_m$ for $n \neq m$? Is it possible that a ring with unity may simultaneously contain two subrings isomorphic to the fields $\mathbf{Z}_p$ and $\mathbf{Z}_q$ for two different primes p and q? If it is impossible, prove it. If it is possible, give illustrations. (This is related to Exercise 23.7.)

29.12 Following the idea of Exercise 29.11, is it possible for an integral domain to contain two subrings isomorphic to $\mathbf{Z}_p$ and $\mathbf{Z}_q$ for $p \neq q$ and p and q both prime. Give reasons or an illustration. (This is related to Exercise 23.8.)

29.13 Let R and R' be rings and let N and N' be ideals of R and R', respectively. Let ϕ be a homomorphism of R into R'. Show that ϕ induces a natural homomorphism $\phi_*: R/N \to R'/N'$ if $N\phi \subseteq N'$. (Use Exercise 13.16.)

29.14 Let ϕ be a homomorphism of a ring R with unity onto a ring R'. Let u be a unit in R. Show that $u\phi$ is a unit in R' if and only if u is not in the kernel of ϕ.

*29.15 (First Isomorphism Theorem for Rings) Let M and N be ideals of a ring R such that $M \leq N$. Show that there is a natural isomorphism of R/N with $(R/M)(N/M)$. (Use Theorem 13.6.)

*29.16 (Second Isomorphism Theorem for Rings) Let M and N be ideals of a ring R and let

$$M + N = \{m + n \mid m \in M, n \in N\}.$$

Show that $M + N$ is an ideal of R, and that $(M + N)/N$ is naturally isomorphic to $M/(M \cap N)$. (Use Exercise 13.17.)

*29.17 Show that $\phi: \mathbf{C} \rightarrow M_2(\mathbf{R})$ given by

$$(a + bi)\phi = \begin{pmatrix} a & b \\ -b & a \end{pmatrix}$$

for $a, b \in \mathbf{R}$ is an isomorphism of $\mathbf{C}$ into $M_2(\mathbf{R})$.

*29.18 Let R be a ring with unity and let $Hom(\langle R, + \rangle)$ be the ring of endomorphisms of $\langle R, + \rangle$ as described in Section 25.2. Let $a \in R$, and let $\rho_a: R \rightarrow R$ be given by

$$x\rho_a = xa$$

for $x \in R$.

a) Show that ρ_a is an endomorphism of $\langle R, + \rangle$.
b) Show that $R' = \{\rho_a \,|\, a \in R\}$ is a subring of $Hom(\langle R, + \rangle)$.
c) Prove the analog of Cayley's theorem for R by showing that R' of (b) is isomorphic to R.

30 | Rings of Polynomials

30.1 POLYNOMIALS IN AN INDETERMINATE

The reader probably has a pretty workable idea of what constitutes a *polynomial in x with coefficients in a ring R*. He knows how to add and multiply such objects, he has done this for years, and he knows what is meant by the *degree of a polynomial*. We suggest that this fortunate man proceed at once to the statement of Theorem 30.1, skip the proof as trivial, and start reading after the proof. Oh yes, we should say that we are considering x an *indeterminate* rather than a variable.

Our problem is twofold: to explain what a polynomial is, and to explain what x is. If we just define a polynomial with coefficients in a ring R to be a *finite formal sum*

$$\sum_{i=0}^{n} a_i x^i = a_0 + a_1 x + \cdots + a_n x^n,$$

where $a_i \in R$, we get ourselves into a bit of trouble. For surely $0 + a_1 x$ and $0 + a_1 x + 0x^2$ are different as formal sums, but we want to regard them as the same polynomial. Perhaps the easy way out is to define a polynomial as an *infinite formal sum*

$$\sum_{i=0}^{\infty} a_i x^i = a_0 + a_1 x + \cdots + a_n x^n + \cdots,$$

where $a_i = 0$ for all but a finite number of values of i. Now there is no problem of having more than one formal sum represent what we wish to consider a single polynomial.

This brings us to the question of what x is. The *elegant* thing to do is to throw x away altogether. A polynomial $a_0 + a_1 x + \cdots + a_n x^n + \cdots$ is completely determined by its sequence of coefficients

$$a_0, a_1, \ldots, a_n, \ldots$$

which doesn't involve the problem child x. A polynomial could be *elegantly defined* as such a sequence. Why carry x around when you don't even need it? Mathematicians have simply become used to x, that is why. It has seniority, tenure, or whatever it is that things have in the WFMN (World

Federation of Mathematical Notations). If your ring has unity 1 and you must have x, the *elegant* thing to do is to define x to be the sequence

$$0, 1, 0, 0, 0, \ldots$$

But then x would probably lodge a protest with the Grievance Committee at such treatment. Since its role is that of excess baggage anyway, let's not fuss too much about what it really is. Let's agree to just call it an "**indeterminate**." You have to admit, that is a pretty good term for something you find hard to analyze. Perhaps *indeterminable* would be better. One thing is sure, it isn't 0 or 2 or any other number. *Thus from now on, we shall never be writing expressions such as "$x = 0$" or "$x = 2$".*

We are now about ready to define a polynomial with coefficients in a ring R as an infinite formal sum

$$\sum_{i=0}^{\infty} a_i x^i = a_0 + a_1 x + \cdots + a_n x^n + \cdots,$$

where $a_i \in R$ and $a_i = 0$ for all but a finite number of values of i. But with $R = \mathbf{Z}, 2 + x^2$ would not be a polynomial! We would always have to write

$$2 + 0x + 1x^2 + 0x^3 + 0x^4 + \cdots$$

Thus we shall change the notation a bit after the definition.

Definition. Let R be a ring. A *polynomial $f(x)$ with coefficients in R* is an infinite formal sum

$$\sum_{i=0}^{\infty} a_i x^i = a_0 + a_1 x + \cdots + a_n x^n + \cdots,$$

where $a_i \in R$ and $a_i = 0$ for all but a finite number of values of i. The a_i are *coefficients of $f(x)$*. If for some $i > 0$ it is true that $a_i \neq 0$, the largest such value of i is the *degree of $f(x)$*. If no such $i > 0$ exists, then $f(x)$ is of *degree zero*.

Let us agree that if $f(x) = a_0 + a_1 x + \cdots + a_n x^n + \cdots$ has $a_i = 0$ for $i > n$, then we may denote $f(x)$ by "$a_0 + a_1 x + \cdots + a_n x^n$". Also, if some $a_i = 1$, we may drop this from the formal sum, so that we shall consider, for example, $2 + x$ to be the polynomial $2 + 1x$ with coefficients in $\mathbf{Z}$. Finally, we shall agree that we may omit altogether from the formal sum any term $0x^i$, or a_0, if $a_0 = 0$ but not all $a_i = 0$. Thus $0, 2, x$, and $2 + x^2$ are all polynomials with coefficients in $\mathbf{Z}$. An element of R is a **constant polynomial**.

Addition and multiplication of polynomials with coefficients in a ring R are defined in a way formally familiar to the student. If

$$f(x) = a_0 + a_1 x + \cdots + a_n x^n + \cdots$$

and

$$g(x) = b_0 + b_1 x + \cdots + b_n x^n + \cdots,$$

then for polynomial addition, we have

$$f(x) + g(x) = c_0 + c_1x + \cdots + c_nx^n + \cdots,$$

where $c_n = a_n + b_n$, and for polynomial multiplication, we have

$$f(x)g(x) = d_0 + d_1x + \cdots + d_nx^n + \cdots,$$

where $d_n = \sum_{i=0}^{n} a_ib_{n-i}$. It is clear that again both c_i and d_i are 0 for all but a finite number of values of i, so these definitions make sense. Note that $\sum_{i=0}^{n} a_ib_{n-i}$ need not equal $\sum_{i=0}^{n} b_ia_{n-i}$ if R is not commutative. With these definitions of addition and multiplication, we have the following theorem.

> **Theorem 30.1** *The set $R[x]$ of all polynomials in an indeterminate x with coefficients in a ring R is a ring under polynomial addition and multiplication. If R is commutative, then so is $R[x]$, and if R has unity 1, then 1 is also unity for $R[x]$.*

Proof. That $\langle R[x], + \rangle$ is an abelian group is obvious. The associative law for multiplication and the distributive laws are straightforward, but a bit cumbersome, computations. We illustrate by proving the associative law. Applying ring axioms to $a_i, b_j, c_k \in R$, we obtain

$$\left[\left(\sum_{i=0}^{\infty} a_ix^i\right)\left(\sum_{j=0}^{\infty} b_jx^j\right)\right]\left(\sum_{k=0}^{\infty} c_kx^k\right) = \left[\sum_{n=0}^{\infty}\left(\sum_{i=0}^{n} a_ib_{n-i}\right)x^n\right]\left(\sum_{k=0}^{\infty} c_kx^k\right)$$

$$= \sum_{s=0}^{\infty}\left[\sum_{n=0}^{s}\left(\sum_{i=0}^{n} a_ib_{n-i}\right)c_{s-n}\right]x^s$$

$$= \sum_{s=0}^{\infty}\left(\sum_{i+j+k=s} a_ib_jc_k\right)x^s$$

$$= \sum_{s=0}^{\infty}\left[\sum_{m=0}^{s} a_{s-m}\left(\sum_{j=0}^{m} b_jc_{m-j}\right)\right]x^s$$

$$= \left(\sum_{i=0}^{\infty} a_ix^i\right)\left[\sum_{m=0}^{\infty}\left(\sum_{j=0}^{m} b_jc_{m-j}\right)x^m\right]$$

$$= \left(\sum_{i=0}^{\infty} a_ix^i\right)\left[\left(\sum_{j=0}^{\infty} b_jx^j\right)\left(\sum_{k=0}^{\infty} c_kx^k\right)\right].$$

Whew!! The distributive laws are similarly proved.

The comments prior to the statement of the theorem show that $R[x]$ is a commutative ring if R is commutative, and a unity 1 in R is obviously also unity for $R[x]$, in view of the definition of multiplication in $R[x]$. ∎

Thus $\mathbf{Z}[x]$ is the ring of polynomials in the indeterminate x with integral coefficients, $\mathbf{Q}[x]$ the ring of polynomials in x with rational coefficients, etc.

Example 30.1 In $\mathbf{Z}_2[x]$, we have

$$(x + 1)^2 = (x + 1)(x + 1) = x^2 + (1 + 1)x + 1 = x^2 + 1.$$

Still working in $\mathbf{Z}_2[x]$, we obtain

$$(x + 1) + (x + 1) = (1 + 1)x + (1 + 1) = 0x + 0 = 0. \;\|$$

If R is a ring and x and y are two indeterminates, then we can form the ring $(R[x])[y]$, that is, the ring of polynomials in y with coefficients which are polynomials in x. It is pretty obvious, but somewhat tedious to prove carefully, that $(R[x])[y]$ is naturally isomorphic to $(R[y])[x]$. Naively, every polynomial in y with coefficients which are polynomials in x can be naturally "rewritten" as a polynomial in x with coefficients which are polynomials in y. We shall identify these rings by means of this natural isomorphism, and shall consider this ring $R[x, y]$ the **ring of polynomials in two indeterminates x and y with coefficients in R**. The **ring $R[x_1, \ldots, x_n]$ of polynomials in the n indeterminates x_i with coefficients in R** is similarly defined.

We leave as an exercise the easy proof that if D is an integral domain, then so is $D[x]$. In particular, if F is a field, then $F[x]$ is an integral domain. Note that $F[x]$ is not a field, for x is not a unit in $F[x]$. That is, there is no polynomial $f(x) \in F[x]$ such that $xf(x) = 1$. By Theorem 26.1, one can construct the field of quotients $F(x)$ of $F[x]$. Any element in $F(x)$ can be represented as a quotient $f(x)/g(x)$ of two polynomials in $F[x]$ with $g(x) \neq 0$. We similarly define $F(x_1, \ldots, x_n)$ to be the field of quotients of $F[x_1, \ldots, x_n]$. This field $F(x_1, \ldots, x_n)$ is the **field of rational functions in n indeterminates over F**. These fields play a very important role in algebraic geometry.

30.2 THE EVALUATION HOMOMORPHISMS

We are now ready to proceed to show, as we promised in Section 27, how the machinery of homomorphisms and factor rings can be used to study what the student has always referred to as "solving a polynomial equation." Let E and F be fields, with F a subfield of E, that is, $F \leq E$. The next theorem asserts the existence of very important homomorphisms of $F[x]$ into E. *These homomorphisms will be the fundamental tools for all the rest of our work.* If you really comprehend them and the homomorphism theory in the preceding section, you are in pretty good shape for the rest of the course.

Theorem 30.2 (The Evaluation Homomorphisms for Field Theory). *Let F be a subfield of a field E, let α be any element of E, and let x be an indeterminate. The map $\phi_\alpha: F[x] \to E$ defined by*

$$(a_0 + a_1 x + \cdots + a_n x^n)\phi_\alpha = a_0 + a_1 \alpha + \cdots + a_n \alpha^n$$

*for $(a_0 + a_1x + \cdots + a_nx^n) \in F[x]$ is a homomorphism of $F[x]$ into E.
Also, $x\phi_\alpha = \alpha$ and ϕ_α maps F isomorphically by the identity map, that is,
$a\phi_\alpha = a$ for $a \in F$. The homomorphism ϕ_α is **evaluation at** α.*

Proof. The lattice and mapping diagram in Fig. 30.1 may help you to
visualize this situation. The dashed lines indicate an element of the set.
The theorem is really an immediate consequence of our definitions of addition
and multiplication in $F[x]$. It is clear that the map ϕ_α is well defined, i.e.,
independent of our representation of $f(x) \in F[x]$ as a finite sum

$$a_0 + a_1x + \cdots + a_nx^n.$$

Such a finite sum representing $f(x)$ can only be changed by insertion or
deletion of terms $0x^i$, which clearly does not affect the value of $(f(x))\phi_\alpha$.

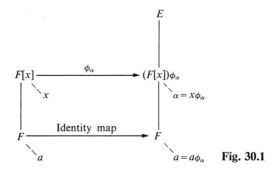

                                                           Fig. 30.1

If $f(x) = a_0 + a_1x + \cdots + a_nx^n$, $g(x) = b_0 + b_1x + \cdots + b_mx^m$,
and $h(x) = f(x) + g(x) = c_0 + c_1x + \cdots + c_rx^r$, then

$$(f(x) + g(x))\phi_\alpha = (h(x))\phi_\alpha = c_0 + c_1\alpha + \cdots + c_r\alpha^r,$$

while

$$(f(x))\phi_\alpha + (g(x))\phi_\alpha$$
$$= (a_0 + a_1\alpha + \cdots + a_n\alpha^n) + (b_0 + b_1\alpha + \cdots + b_m\alpha^m).$$

Since by definition of polynomial addition we have $c_i = a_i + b_i$, we see that

$$(f(x) + g(x))\phi_\alpha = (f(x))\phi_\alpha + (g(x))\phi_\alpha.$$

Turning to multiplication, we see that if

$$f(x)g(x) = d_0 + d_1x + \cdots + d_sx^s,$$

then

$$(f(x)g(x))\phi_\alpha = d_0 + d_1\alpha + \cdots + d_s\alpha^s,$$

while

$$[(f(x))\phi_\alpha][(g(x))\phi_\alpha]$$
$$= (a_0 + a_1\alpha + \cdots + a_n\alpha^n)(b_0 + b_1\alpha + \cdots + b_m\alpha^m).$$

Since by definition of polynomial multiplication, $d_j = \sum_{i=0}^{j} a_i b_{j-i}$, we see that

$$(f(x)g(x))\phi_\alpha = [(f(x))\phi_\alpha][(g(x))\phi_\alpha].$$

Thus ϕ_α is a homomorphism.

The very definition of ϕ_α applied to a constant polynomial $a \in F[x]$, where $a \in F$, gives $a\phi_\alpha = a$, so ϕ_α maps F isomorphically by the identity map. Again by definition of ϕ_α, we have $x\phi_\alpha = (1x)\phi_\alpha = 1\alpha = \alpha$. ∎

We point out the fact that this theorem is valid with the identical proof if F and E are merely commutative rings with unity rather than fields. However, we shall be interested only in the case in which they are fields.

It is hard to overemphasize the importance of this simple theorem for us. It is the very foundation for all of our further work in field theory. It is so simple that it could justifiably be called an *observation* rather than a theorem. Perhaps we shouldn't have tried to write out the proof. The polynomial notation makes it look so complicated that you may be fooled into thinking that it is a difficult theorem.

Example 30.2 Let F be Q and E be R in Theorem 30.2, and consider the evaluation homomorphism $\phi_0 \colon Q[x] \to R$. Here

$$(a_0 + a_1 x + \cdots + a_n x^n)\phi_0 = a_0 + a_1 0 + \cdots + a_n 0^n = a_0.$$

Thus every polynomial is mapped onto its constant term. Note that the kernel of ϕ_0 is the ideal N of all polynomials with constant term 0. By Theorem 29.3, the image $(Q[x])\phi_0 = Q$ is naturally isomorphic to $Q[x]/N$. A coset element of $Q[x]/N$ consists precisely of all polynomials having a given fixed constant term. ‖

Example 30.3 Let F be Q and E be R in Theorem 30.2, and consider the evaluation homomorphism $\phi_2 \colon Q[x] \to R$. Here

$$(a_0 + a_1 x + \cdots + a_n x^n)\phi_2 = a_0 + a_1 2 + \cdots + a_n 2^n.$$

Note that

$$(x^2 + x - 6)\phi_2 = 2^2 + 2 - 6 = 0.$$

Thus $x^2 + x - 6$ is in the kernel N of ϕ_2. Of course,

$$x^2 + x - 6 = (x - 2)(x + 3),$$

and, if you like, the reason that $(x^2 + x - 6)\phi_2 = 0$ is that $(x - 2)\phi_2 = 2 - 2 = 0$. We shall see later that N is precisely the ideal of all polynomials of the form $(x - 2)f(x)$ for $f(x) \in Q[x]$. Here the image of $Q[x]$ under ϕ_2 is again just Q, and again by Theorem 29.3, $Q[x]/N$ is naturally isomorphic to Q. ‖

Example 30.4 Let F be $\mathbf{Q}$ and E be $\mathbf{C}$ in Theorem 30.2, and consider the evaluation homomorphism $\phi_i \colon \mathbf{Q}[x] \to \mathbf{C}$. Here

$$(a_0 + a_1 x + \cdots + a_n x^n)\phi_i = a_0 + a_1 i + \cdots + a_n i^n$$

and $x\phi_i = i$. Note that

$$(x^2 + 1)\phi_i = i^2 + 1 = 0,$$

so $x^2 + 1$ is in the kernel N of ϕ_i. We shall see later that N is precisely the ideal of all polynomials of the form $(x^2 + 1)f(x)$ for $f(x) \in \mathbf{Q}[x]$. By Theorem 29.3, the image $(\mathbf{Q}[x])\phi_i$ is naturally isomorphic to $\mathbf{Q}[x]/N$. We shall see later that this ring $(\mathbf{Q}[x])\phi_i$ consists of all complex numbers of the form $q_1 + q_2 i$ for $q_1, q_2 \in \mathbf{Q}$, and is a subfield of $\mathbf{C}$. ‖

Example 30.5 Let F be $\mathbf{Q}$ and let E be $\mathbf{R}$ in Theorem 30.2, and consider the evaluation homomorphism $\phi_\pi \colon \mathbf{Q}[x] \to \mathbf{R}$. Here

$$(a_0 + a_1 x + \cdots + a_n x^n)\phi_\pi = a_0 + a_1 \pi + \cdots + a_n \pi^n.$$

It can be proved that $a_0 + a_1 \pi + \cdots + a_n \pi^n = 0$ if and only if $a_i = 0$ for $i = 0, 1, \ldots, n$. Thus, the kernel of ϕ_π is $\{0\}$, and ϕ_π is an isomorphic mapping. This shows that all *formal polynomials in π with rational coefficients* form a ring isomorphic to $\mathbf{Q}[x]$ in a natural way with $x\phi_\pi = \pi$. ‖

30.3 THE NEW APPROACH

We now complete the connection between our new ideas and the classical concept of "solving a polynomial equation." Rather than speak of *solving a polynomial equation*, we shall refer to *finding a zero of a polynomial*.

Definition. Let F be a subfield of a field E, and let α be an element of E. Let $f(x) = a_0 + a_1 x + \cdots + a_n x^n$ be in $F[x]$, and let $\phi_\alpha \colon F[x] \to E$ be the evaluation homomorphism of Theorem 30.2. Let "$f(\alpha)$" denote

$$(f(x))\phi_\alpha = a_0 + a_1 \alpha + \cdots + a_n \alpha^n.$$

If $f(\alpha) = 0$, then α is a **zero of** $f(x)$.

In terms of this definition, we can rephrase the classical problem of "finding all real solutions of the polynomial equation $r^2 + r - 6 = 0$" by letting $F = \mathbf{Q}$ and $E = \mathbf{R}$ and *finding all $\alpha \in \mathbf{R}$ such that*

$$(x^2 + x - 6)\phi_\alpha = 0,$$

that is, finding all zeros of $x^2 + x - 6$ in $\mathbf{R}$. Both problems have the same answer, since

$$\{\alpha \in \mathbf{R} \mid (x^2 + x - 6)\phi_\alpha = 0\} = \{r \in \mathbf{R} \mid r^2 + r - 6 = 0\} = \{2, -3\}.$$

It probably seems to the student that we have merely succeeded in making a simple problem seem quite complicated. *What we have done is to phrase the problem in the language of mappings, and we can now use all the mapping machinery that we have developed for its solution.* Remember our *Basic Goal,* which can now be rephrased as follows.

Basic Goal: *To show that for a field F, every nonconstant polynomial f(x) ∈ F[x] has a zero.*

Let us cheat and take a little peek ahead and see how our *Basic Goal* might be accomplished. If $f(x)$ has no zero in F, we have to *manufacture* in some fashion a field E containing F such that there is an α in E with $f(\alpha) = (f(x))\phi_\alpha = 0$. How can we build E? Now E is to contain the image $(F[x])\phi_\alpha$ of $F[x]$ under our evaluation homomorphism ϕ_α. Remember that by Theorem 29.3 $(F[x])\phi_\alpha$ is isomorphic to $F[x]/(\text{kernel of }\phi)_\alpha$. This suggests that we try to form E by forming a factor ring $F[x]/N$ for some suitable ideal N in $F[x]$. Of course to make $F[x]/N$ a field, N must be a *maximal ideal* of $F[x]$, by Theorem 29.4. Thus our task in the next section, as our final step toward the achievement of our *Basic Goal,* will be to examine the nature of ideals in $F[x]$. This is how all the machinery we have developed concerning factor rings and homomorphisms comes into play. There is a saying: "Don't bring a cannon on stage unless you intend to fire it." We brought the cannon onto the stage in Sections 28 and 29, and we are now priming it for firing.

EXERCISES

30.1 Find the sum and product of $f(x) = 2x^3 + 4x^2 + 3x + 2$ and $g(x) = 3x^4 + 2x + 4$, given that $f(x), g(x) \in Z_5[x]$.

30.2 Consider the element

$$f(x, y) = (3x^3 + 2x)y^3 + (x^2 - 6x + 1)y^2 + (x^4 - 2x)y + (x^4 - 3x^2 + 2)$$

of $(Q[x])[y]$. Write $f(x, y)$ as it would appear if viewed as an element of $(Q[y])[x]$.

30.3 Let $F = E = Z_7$ in Theorem 30.2. Evaluate each of the following for the indicated evaluation homomorphism $\phi_a: Z_7[x] \to Z_7$.

a) $(x^2 + 3)\phi_2$ b) $(2x^3 - x^2 + 3x + 2)\phi_0$
c) $[(x^4 + 2x)(x^3 - 3x^2 + 3)]\phi_3$ d) $[(x^3 + 2)(4x^2 + 3)(x^7 + 3x^2 + 1)]\phi_5$

30.4 Consider the evaluation homomorphism $\phi_5: Q[x] \to R$. Find 6 elements in the kernel of ϕ_5.

30.5 Find all zeros in Z_5 of $(x^5 + 3x^3 + x^2 + 2x) \in Z_5[x]$. [*Hint:* There are only five candidates. Try them.]

†**30.6** Prove that if D is an integral domain, then $D[x]$ is an integral domain.

30.7 Mark each of the following true or false.

— a) The polynomial $(a_nx^n + \cdots + a_1x + a_0) \in R[x]$ is zero if and only if $a_i = 0, i = 0, 1, \ldots, n$.
— b) If R is a commutative ring, then $R[x]$ is commutative.
— c) If D is an integral domain, then $D[x]$ is an integral domain.

— d) If R is a ring containing divisors of zero, then $R[x]$ has divisors of zero.

— e) If R is a ring and $f(x)$ and $g(x)$ in $R[x]$ are of degrees 3 and 4, respectively, then $f(x)g(x)$ may be of degree 8 in $R[x]$.

— f) If R is any ring and $f(x)$ and $g(x)$ in $R[x]$ are of degrees 3 and 4, respectively, then $f(x)g(x)$ is always of degree 7.

— g) If F is a subfield of a field E and $a \in E$ is a zero of $f(x) \in F[x]$, then a is a zero of $h(x) = f(x)g(x)$ for all $g(x) \in F[x]$.

— h) The evaluation homomorphism ϕ_α of Theorem 30.2 is an extension of the injection map $i: F \to E$ to $F[x]$, where $(a)i = a$ for $a \in F$.

— i) If F is a subfield of a field E and $f(x) \in F[x]$, then the set of all zeros of $f(x)$ in E is an ideal of E.

— j) If F is a subfield of a field E and $\alpha \in E$, then the set of all $f(x) \in F[x]$ such that $f(\alpha) = 0$ is an ideal of $F[x]$.

30.8 Prove the left distributive law for $R[x]$, where R is a ring and x is an indeterminate.

30.9 Let $\phi_a: Z_5[x] \to Z_5$ be an evaluation homomorphism as in Theorem 30.2. Use Fermat's theorem to evaluate $(x^{231} + 3x^{117} - 2x^{53} + 1)\phi_3$.

30.10 Use Fermat's theorem to find all zeros in Z_5 of

$$2x^{219} + 3x^{74} + 2x^{57} + 3x^{44}.$$

30.11 Let D be an integral domain and x an indeterminate.

a) Describe the units in $D[x]$.

b) Find the units in $Z[x]$.

c) Find the units in $Z_7[x]$.

30.12 Let F be a subfield of a field E.

a) Define an *evaluation homomorphism*

$$\phi_{\alpha_1}, \ldots, {}_{\alpha_n}: F[x_1, \ldots, x_n] \to E \qquad \text{for} \quad \alpha_i \in E,$$

stating the analog of Theorem 30.2.

b) With $E = F = Q$, compute $(x_1{}^2 x_2{}^3 + 3x_1{}^4 x_2)\phi_{-3,2}$.

c) Define the concept of a *zero of a polynomial* $f(x_1, \ldots, x_n) \in F[x_1, \ldots, x_n]$ in a way analogous to the definition in the text of a zero of $f(x)$.

30.13 Let F be a field. Show that all polynomials with constant term $a_0 = 0$ form an ideal $\langle x \rangle$ of $F[x]$.

30.14 Let F be a field, and let $\langle x \rangle$ be the ideal in $F[x]$ defined in Exercise 30.13. Show that $F[x]/\langle x \rangle$ is a field isomorphic to F by

a) showing directly that each residue class in $F[x]/\langle x \rangle$ contains exactly one element of F which may be chosen as a representative for computing in $F[x]/\langle x \rangle$,

b) considering the evaluation homomorphism $\phi_0: F[x] \to F$ as defined in Theorem 30.2, showing that $\langle x \rangle$ is the kernel of ϕ_0 and applying Theorem 29.3.

30.15 Let R be a ring, and let R^R be the set of all functions mapping R into R. For $\phi, \psi \in R^R$, define the sum $\phi + \psi$ by

$$r(\phi + \psi) = (r\phi) + (r\psi)$$

and the product $\phi \cdot \psi$ by

$$r(\phi \cdot \psi) = (r\phi)(r\psi)$$

for $r \in R$. Note that $\cdot$ is *not* function composition. Show that $\langle R^R, +, \cdot \rangle$ is a ring.

30.16 Referring to Exercise 30.15, let F be a field. An element ϕ of F^F is a **polynomial function on** F, if there exists $f(x) \in F[x]$ such that $a\phi = f(a)$ for all $a \in F$.

a) Show that the set P_F of all polynomial functions on F forms a subring of F^F.

b) Show that the ring P_F is not necessarily isomorphic to $F[x]$. [*Hint:* Show that if F is a finite field, P_F and $F[x]$ don't even have the same number of elements.]

30.17 Refer to Exercises 30.15 and 30.16 for the following questions.

a) How many elements are there in $\mathbf{Z}_2^{\mathbf{Z}_2}$? in $\mathbf{Z}_3^{\mathbf{Z}_3}$?

b) Classify $\langle \mathbf{Z}_2^{\mathbf{Z}_2}, + \rangle$ and $\langle \mathbf{Z}_3^{\mathbf{Z}_3}, + \rangle$ by the Fundamental Theorem of Finitely Generated Abelian Groups.

c) Show that if F is a finite field, then $F^F = P_F$. [*Hint:* Of course, $P_F \subseteq F^F$. Let F have as elements $a_1, \ldots, a_n$. Note that if

$$f_i(x) = c(x - a_1) \cdots (x - a_{i-1})(x - a_{i+1}) \cdots (x - a_n),$$

then $f_i(a_j) = 0$ for $j \neq i$, and the value $f_i(a_i)$ can be controlled by the choice of $c \in F$. Use this to show that every function on F is a polynomial function.]

31 | Factorization of Polynomials Over a Field

31.1 THE DIVISION ALGORITHM IN $F[x]$

We indicated at the end of the last section that the final step for the achievement of our *Basic Goal* would be the study of ideal structure in polynomial rings $F[x]$, where F is a field. Of particular interest is the characterization of maximal ideals of $F[x]$ to determine when $F[x]/N$ is a field. It may seem odd that we have to study factorization of polynomials to do this, but we can make this seem reasonable in two ways. First, our main purpose is to study zeros of polynomials, and suppose that $f(x) \in F[x]$ factors in $F[x]$, so that $f(x) = g(x)h(x)$ for $g(x), h(x) \in F[x]$. If $F \le E$ and $\alpha \in E$, then $f(\alpha) = 0$ if and only if either $g(\alpha) = 0$ or $h(\alpha) = 0$. Thus the attempt to find a zero of $f(x)$ is reduced to the problem of finding a zero of a factor of $f(x)$. Also, we wish to find maximal ideals of $F[x]$. Now any maximal ideal M is also a prime ideal. Therefore, if $f(x) = g(x)h(x)$ and $f(x) \in M$ for a maximal ideal M, then we must have either $g(x) \in M$ or $h(x) \in M$. These two considerations thus suggest the study of factorization of polynomials in $F[x]$.

The following theorem is the basic tool for our work in this section. The student should note the similarity with Lemma 6.1, *the division algorithm for* **Z**, the importance of which has been amply demonstrated. Division algorithms will be treated in a more general setting in starred Section 33.

Theorem 31.1 (Division Algorithm for $F[x]$). *Let*

$$f(x) = a_n x^n + a_{n-1} x^{n-1} + \cdots + a_0$$

and

$$g(x) = b_m x^m + b_{m-1} x^{m-1} + \cdots + b_0$$

be two elements of $F[x]$, with a_n and b_m both nonzero elements of F and $m > 0$. Then there are unique polynomials $q(x)$ and $r(x)$ in $F[x]$ such that $f(x) = g(x)q(x) + r(x)$, with the degree of $r(x)$ less than $m = $ degree $g(x)$.

Proof. Consider the set $S = \{f(x) - g(x)s(x) \mid s(x) \in F[x]\}$. Let $r(x)$ be an element of minimal degree in S. Then

$$f(x) = g(x)q(x) + r(x)$$

for some $q(x) \in F[x]$. We must show that the degree of $r(x)$ is less than m. Suppose that

$$r(x) = c_t x^t + c_{t-1} x^{t-1} + \cdots + c_0,$$

with $c_j \in F$ and $c_t \neq 0$ if $t \neq 0$. If $t \geq m$, then

$$f(x) - q(x)g(x) - (c_t/b_m)x^{t-m}g(x) = r(x) - (c_t/b_m)x^{t-m}g(x), \qquad (1)$$

and the latter is of the form

$$r(x) - (c_t x^t + \text{terms of lower degree}),$$

which is a polynomial of degree lower than t, the degree of $r(x)$. However, the polynomial in (1) can be written in the form

$$f(x) - g(x)[q(x) + (c_t/b_m)x^{t-m}],$$

so it is in S, contradicting the fact that $r(x)$ was selected to have minimal degree in S. Thus the degree of $r(x)$ is less than $m = $ degree $g(x)$.

For uniqueness, if

$$f(x) = g(x)q_1(x) + r_1(x)$$

and

$$f(x) = g(x)q_2(x) + r_2(x),$$

then subtracting we have

$$g(x)[q_1(x) - q_2(x)] = r_2(x) - r_1(x).$$

Since the degree of $r_2(x) - r_1(x)$ is less than the degree of $g(x)$, this can only hold if $q_1(x) - q_2(x) = 0$ or $q_1(x) = q_2(x)$. Then we must have $r_2(x) - r_1(x) = 0$ or $r_1(x) = r_2(x)$. ∎

You can compute the polynomials $q(x)$ and $r(x)$ of Theorem 31.1 by *long division* just as you divided polynomials in $\mathbf{R}[x]$ in high school. After the following corollaries, we give some computational examples.

You have no doubt seen in high school algebra the first corollary for the special case $\mathbf{R}[x]$. We phrase our proof in terms of the mapping (homomorphism) approach described in Section 30.

Corollary 1 *An element $a \in F$ is a zero of $f(x) \in F[x]$ if and only if $x - a$ is a factor of $f(x)$ in $F[x]$.*

Proof. Suppose that for $a \in F$ we have $f(a) = 0$. By Theorem 31.1, there exist $q(x), r(x) \in F[x]$ such that

$$f(x) = (x - a)q(x) + r(x),$$

where the degree of $r(x)$ is less than 1. Then we must have $r(x) = c$ for $c \in F$, so

$$f(x) = (x - a)q(x) + c.$$

Applying our evaluation homomorphism $\phi_a: F[x] \to F$ of Theorem 30.2, we find
$$0 = f(a) = 0q(a) + c,$$
so it must be that $c = 0$. Then $f(x) = (x - a)q(x)$, so $x - a$ is a factor of $f(x)$.

Conversely, if $x - a$ is a factor of $f(x)$ in $F[x]$, where $a \in F$, then applying our evaluation homomorphism ϕ_a to $f(x) = (x - a)q(x)$, we have $f(a) = 0q(a) = 0$. ∎

The next corollary should certainly also look familiar.

Corollary 2 *A nonzero polynomial $f(x) \in F[x]$ of degree n can have at most n zeros in a field F.*

Proof. The above corollary shows that if $a_1 \in F$ is a zero of $f(x)$, then
$$f(x) = (x - a_1)q_1(x),$$
where of course the degree of $q_1(x)$ is $n - 1$. A zero $a_2 \in F$ of $q_1(x)$ then results in a factorization
$$f(x) = (x - a_1)(x - a_2)q_2(x).$$
Continuing this process, we arrive at
$$f(x) = (x - a_1) \cdots (x - a_r)q_r(x),$$
where $q_r(x)$ has no further zeros in F. Clearly, $r \leq n$. Also, if $b \neq a_i$ for $i = 1, \ldots, r$ and $b \in F$, then
$$f(b) = (b - a_1) \cdots (b - a_r)q_r(b) \neq 0,$$
since F has no divisors of zero and none of $b - a_i$ or $q_r(b)$ are zero by construction. Hence the a_i for $i = 1, \ldots, r \leq n$ are all the zeros in F of $f(x)$. ∎

Example 31.1 Let us work with polynomials in $\mathbf{Z}_5[x]$ and "divide
$$f(x) = x^4 - 3x^3 + 2x^2 + 4x - 1$$
by $g(x) = x^2 - 2x + 3$" to find $q(x)$ and $r(x)$ of Theorem 31.1. The long division should be easy for you to follow, but remember that we are in $\mathbf{Z}_5[x]$, so, for example, $4x - (-3x) = 2x$.

$$
\begin{array}{r}
x^2 - x - 3 \\
x^2 - 2x + 3 \overline{\smash{\big)}\ x^4 - 3x^3 + 2x^2 + 4x - 1} \\
\underline{x^4 - 2x^3 + 3x^2} \\
- x^3 - x^2 + 4x \\
\underline{- x^3 + 2x^2 - 3x} \\
- 3x^2 + 2x - 1 \\
\underline{- 3x^2 + x - 4} \\
x + 3
\end{array}
$$

Thus

$$x^4 - 3x^3 + 2x^2 + 4x - 1 = (x^2 - 2x + 3)(x^2 - x - 3) + (x + 3),$$

so $q(x) = x^2 - x - 3$, and $r(x) = x + 3$. ∥

Example 31.2 Working again in $\mathbf{Z}_5[x]$, note that 1 is a zero of

$$(x^4 + 3x^3 + 2x + 4) \in \mathbf{Z}_5[x].$$

Thus by Corollary 1 of Theorem 31.1, we should be able to factor $x^4 + 3x^3 + 2x + 4$ into $(x - 1)q(x)$ in $\mathbf{Z}_5[x]$. Let us try again by long division.

$$
\begin{array}{r}
x^3 + 4x^2 + 4x\ + 1 \\
x - 1 \overline{\smash{\big)}\ x^4 + 3x^3 + 2x + 4} \\
\underline{x^4 - x^3} \\
4x^3 \\
\underline{4x^3 - 4x^2} \\
4x^2 + 2x \\
\underline{4x^2 - 4x} \\
x + 4 \\
\underline{x - 1} \\
0
\end{array}
$$

Thus $x^4 + 3x^3 + 2x + 4 = (x - 1)(x^3 + 4x^2 + 4x + 1)$ in $\mathbf{Z}_5[x]$. Since 1 is seen to be a zero of $x^3 + 4x^2 + 4x + 1$ also, we can divide this polynomial by $x - 1$ and get

$$
\begin{array}{r}
x^2 + 4 \\
x - 1 \overline{\smash{\big)}\ x^3 + 4x^2 + 4x + 1} \\
\underline{x^3 - x^2} \\
0\ + 4x + 1 \\
\underline{4x - 4} \\
0
\end{array}
$$

Since $x^2 + 4$ still has 1 as a zero, we can divide again by $x - 1$ and get

$$
\begin{array}{r}
x + 1 \\
x - 1 \overline{\smash{\big)}\ x^2 + 4} \\
\underline{x^2 - x} \\
x + 4 \\
\underline{x - 1} \\
0
\end{array}
$$

Thus $x^4 + 3x^3 + 2x + 4 = (x - 1)^3(x + 1)$ in $\mathbf{Z}_5[x]$. ∥

The technique of *synthetic division* is undoubtedly valid in $F[x]$, if you can remember how it goes.

31.2 IRREDUCIBLE POLYNOMIALS

Our next definition singles out a type of polynomial in $F[x]$ which will be of utmost importance to us. Again, the student is probably already familiar with the concept. You see, we really *are* doing high school algebra in a more general setting.

> **Definition.** A nonconstant polynomial $f(x) \in F[x]$ is **irreducible over F** or is an **irreducible polynomial in $F[x]$**, if $f(x)$ cannot be expressed as a product $g(x)h(x)$ of two polynomials $g(x)$ and $h(x)$ in $F[x]$ both of lower degree than the degree of $f(x)$.

Note that the preceding definition concerns the concept *irreducible over F* and not just the concept *irreducible*. A polynomial $f(x)$ may be irreducible over F, but may not be irreducible if viewed over a larger field E containing F. We illustrate this.

Example 31.3 Theorem 27.1 showed that $x^2 - 2$ viewed in $\mathbf{Q}[x]$ has no zeros in $\mathbf{Q}$. This shows that $x^2 - 2$ is irreducible over $\mathbf{Q}$, for a factorization $x^2 - 2 = (ax + b)(cx + d)$ for $a, b, c, d \in \mathbf{Q}$ would clearly give rise to zeros of $x^2 - 2$ in $\mathbf{Q}$. However, $x^2 - 2$ viewed in $\mathbf{R}[x]$ is not irreducible over $\mathbf{R}$, for $x^2 - 2$ factors in $\mathbf{R}[x]$ into $(x - \sqrt{2})(x + \sqrt{2})$. ‖

It is worth while for the student to remember that *the units in $F[x]$ are precisely the nonzero elements of F.* Thus we could have defined an irreducible polynomial $f(x)$ as a nonconstant polynomial such that in any factorization $f(x) = g(x)h(x)$ in $F[x]$, either $g(x)$ or $h(x)$ is a unit. This viewpoint will be elaborated in the next starred sections, which deal with factorization in rings more general than $F[x]$.

Example 31.4 Let us show that $f(x) = x^3 + 3x + 2$ viewed in $\mathbf{Z}_5[x]$ is irreducible over $\mathbf{Z}_5$. If $x^3 + 3x + 2$ factored in $\mathbf{Z}_5[x]$ into polynomials of lower degree, then there would exist at least one linear factor of $f(x)$ of the form $x - a$ for some $a \in \mathbf{Z}_5$. But then $f(a)$ would be zero, by Corollary 1 of Theorem 31.1. However, $f(0) = 2$, $f(1) = 1$, $f(-1) = -2$, $f(2) = 1$, and $f(-2) = -2$, showing that $f(x)$ has no zeros in $\mathbf{Z}_5$. Thus $f(x)$ is irreducible over $\mathbf{Z}_5$. This test for irreducibility by finding zeros works nicely for quadratic and cubic polynomials over a finite field with a small number of elements. ‖

Irreducible polynomials will play a very important role in our work from now on. The problem of determining whether or not a given $f(x) \in F[x]$ is irreducible over F may be difficult. We now give some criteria for irreducibility which are useful in certain cases. One technique for determining irreducibility of quadratic and cubic polynomials was illustrated in Examples 31.3 and 31.4. We formalize it in a theorem.

Theorem 31.2 Let $f(x) \in F[x]$, and let $f(x)$ be of degree 2 or 3. Then $f(x)$ is reducible over F if and only if it has a zero in F.

Proof. If $f(x)$ is reducible so that $f(x) = g(x)h(x)$, where the degree of $g(x)$ and the degree of $h(x)$ are both less than the degree of $f(x)$, then since $f(x)$ is either quadratic or cubic, either $g(x)$ or $h(x)$ is of degree 1. If, say, $g(x)$ is of degree 1, then except for a possible factor in F, $g(x)$ is of the form $x - a$. Then $g(a) = 0$, which implies that $f(a) = 0$, so $f(x)$ has a zero in F.

Conversely, Corollary 1 of Theorem 31.1 shows that if $f(a) = 0$ for $a \in F$, then $x - a$ is a factor of $f(x)$, so $f(x)$ is reducible. ∎

We turn to some conditions for irreducibility over $\mathbf{Q}$ of polynomials in $\mathbf{Q}[x]$. The most important condition that we shall give is contained in the next theorem. We shall not prove this theorem here; it is proved in a more general situation in starred Section 32 (see the corollary of Lemma 32.6).

Theorem 31.3 If $f(x) \in \mathbf{Z}[x]$, then $f(x)$ factors into a product of two polynomials of lower degree in $\mathbf{Q}[x]$ if and only if it has such a factorization with polynomials of the same degree in $\mathbf{Z}[x]$.

Proof. See the corollary of Lemma 32.6. ∎

Corollary. If $f(x) = x^n + a_{n-1}x^{n-1} + \cdots + a_0$ is in $\mathbf{Z}[x]$ with $a_0 \neq 0$, and if $f(x)$ has a zero in $\mathbf{Q}$, then it has a zero m in $\mathbf{Z}$, and m must divide a_0.

Proof. If $f(x)$ has a zero a in $\mathbf{Q}$, then $f(x)$ has a linear factor $x - a$ in $\mathbf{Q}[x]$ by Corollary 1 of Theorem 31.1. But then by Theorem 31.3, $f(x)$ has a factorization with a linear factor in $\mathbf{Z}[x]$, so for some $m \in \mathbf{Z}$ we must have

$$f(x) = (x - m)(x^{n-1} + \cdots + a_0/m).$$

Thus a_0/m is in $\mathbf{Z}$, so m divides a_0. ∎

Example 31.5 This corollary of Theorem 31.3 gives us another proof of the irreducibility of $x^2 - 2$ over $\mathbf{Q}$, for $x^2 - 2$ factors nontrivially in $\mathbf{Q}[x]$ if and only if it has a zero in $\mathbf{Q}$ by Theorem 31.2. By the corollary of Theorem 31.3, it has a zero in $\mathbf{Q}$ if and only if it has a zero in $\mathbf{Z}$, and moreover the only possibilities are the divisors ± 1 and ± 2 of 2. A check shows that none of these numbers is a zero of $x^2 - 2$. ‖

Example 31.6 Let us use Theorem 31.3 to show that

$$f(x) = x^4 - 2x^2 + 8x + 1$$

viewed in $\mathbf{Q}[x]$ is irreducible over $\mathbf{Q}$. If $f(x)$ has a linear factor in $\mathbf{Q}[x]$, then it has a zero in $\mathbf{Z}$, and by the corollary of Theorem 31.3, this zero would have to be a divisor in $\mathbf{Z}$ of 1, that is, either ± 1. But $f(1) = 8$, and $f(-1) = -8$, so such a factorization is impossible.

If $f(x)$ factors into two quadratic factors in $\mathbf{Q}[x]$, then by Theorem 31.3 it has a factorization

$$(x^2 + ax + b)(x^2 + cx + d)$$

in $\mathbf{Z}[x]$. Equating coefficients of powers of x, we find that we must have

$$bd = 1, \qquad ad + bc = 8, \qquad ac + b + d = -2, \qquad \text{and} \qquad a + c = 0$$

for integers $a, b, c, d \in \mathbf{Z}$. From $bd = 1$, we see that either $b = d = 1$ or $b = d = -1$. If $b = d = 1$, then $ac + b + d = -2$ implies that $ac = -4$, which, with $a + c = 0$, gives $a = -c = \pm 2$. If $b = d = -1$, then $ac + b + d = -2$ implies that $ac = 0$, so $a = -c = 0$. No combination of the numbers arising from these possibilities can make $ad + bc$ as large as 8. Thus a factorization into two quadratic polynomials is also impossible, and $f(x)$ is irreducible over $\mathbf{Q}$. ‖

We conclude our irreducibility criteria with the famous Eisenstein test for irreducibility. An additional and very useful criterion is given in Exercise 31.18.

Theorem 31.4 (Eisenstein). *Let $p \in \mathbf{Z}$ be a prime. Suppose that $f(x) = a_n x^n + \cdots + a_0$ is in $\mathbf{Z}[x]$, and $a_n \not\equiv 0 \pmod{p}$, but $a_i \equiv 0 \pmod{p}$ for $i < n$, with $a_0 \not\equiv 0 \pmod{p^2}$. Then $f(x)$ is irreducible over $\mathbf{Q}$.*

Proof. By Theorem 31.3 we need only show that $f(x)$ does not factor into polynomials of lower degree in $\mathbf{Z}[x]$. If

$$f(x) = (b_r x^r + \cdots + b_0)(c_s x^s + \cdots + c_0)$$

is a factorization in $\mathbf{Z}[x]$, with $b_r \neq 0$, $c_s \neq 0$ and r and s both less than n, then $a_0 \not\equiv 0 \pmod{p^2}$ implies that not both b_0 and c_0 are congruent to 0 modulo p. Suppose that $b_0 \not\equiv 0 \pmod{p}$ and $c_0 \equiv 0 \pmod{p}$. Now $a_n \not\equiv 0 \pmod{p}$ implies that b_r and c_s are not congruent to 0 modulo p, since $a_n = b_r c_s$. Let m be the smallest value of k such that $c_k \not\equiv 0 \pmod{p}$. Then

$$a_m = b_0 c_m + b_1 c_{m-1} + \cdots + b_{m-i} c_i$$

for some i, $0 \leq i < m$. Now b_0 and c_m neither congruent to 0 modulo p and $c_{m-1}, \ldots, c_i$ all congruent to 0 modulo p implies that $a_m \not\equiv 0$ modulo p, so $m = n$. Consequently, $s = n$, contradicting our assumption that $s < n$, that is, that our factorization was nontrivial. ∎

Note that if we take $p = 2$, the Eisenstein criterion gives us still another proof of the irreducibility of $x^2 - 2$ over $\mathbf{Q}$.

Example 31.7 Taking $p = 3$, we see by Theorem 31.4 that

$$25x^5 - 9x^4 + 3x^2 - 12$$

is irreducible over $\mathbf{Q}$. ‖

Corollary. *The cyclotomic polynomial*

$$\Phi_p(x) = \frac{x^p - 1}{x - 1} = x^{p-1} + x^{p-2} + \cdots + x + 1$$

is irreducible over **Q** *for any prime p.*

Proof. Again by Theorem 31.3, we need only consider factorizations in **Z**[x]. Let

$$g(x) = \Phi_p(x + 1) = \frac{(x + 1)^p - 1}{(x + 1) - 1} = \frac{x^p + \binom{p}{1} x^{p-1} + \cdots + px}{x}.$$

Then

$$g(x) = x^{p-1} + \binom{p}{1} x^{p-2} + \cdots + p$$

satisfies the Eisenstein criterion for the prime p, and is thus irreducible over **Q**. But clearly if $\Phi_p(x) = h(x)r(x)$ were a nontrivial factorization of $\Phi_p(x)$ in **Z**[x], then

$$\Phi_p(x + 1) = g(x) = h(x + 1)r(x + 1)$$

would give a nontrivial factorization of $g(x)$ in **Z**[x]. Thus $\Phi_p(x)$ must also be irreducible over **Q**. ∎

31.3 IDEAL STRUCTURE IN $F[x]$

We give the next definition for a general commutative ring R with unity, although we are only interested in the case $R = F[x]$, where F is a field. It is clear that for a commutative ring R with unity and $a \in R$, the set $\{ra \mid r \in R\}$ is an ideal in R which contains the element a.

Definition. If R is a commutative ring with unity and $a \in R$, the ideal $\{ra \mid r \in R\}$ of all multiples of a is the *principal ideal generated by* a and is denoted by "$\langle a \rangle$". An ideal N of R is a *principal ideal* if $N = \langle a \rangle$ for some $a \in R$.

Example 31.8 The ideal $\langle x \rangle$ in $F[x]$ clearly consists of all polynomials in $F[x]$ having zero constant term. ‖

The next theorem is another easy but very important application of Theorem 31.1, which is our basic tool for this whole section. You might be prepared to prove this theorem if your instructor ever asks for proofs. The proof of this theorem is to the division algorithm in $F[x]$ as the proof that a subgroup of a cyclic group is cyclic is to the division algorithm in **Z**.

Theorem 31.5 *If F is a field, every ideal in F[x] is principal.*

Proof. Let N be an ideal of $F[x]$. If $N = \{0\}$, then $N = \langle 0 \rangle$. Suppose that $N \neq \{0\}$, and let $g(x)$ be a nonzero element of N of minimal degree. If the degree of $g(x)$ is zero, then $g(x) \in F$ and is a unit, so $N = F[x] = \langle 1 \rangle$ by

Theorem 28.4, so N is principal. If the degree of $g(x)$ is ≥ 1, let $f(x)$ be any element of N. Then by Theorem 31.1, $f(x) = g(x)q(x) + r(x)$, where $(\text{degree } r(x)) < (\text{degree } g(x))$. Now $f(x) \in N$ and $g(x) \in N$ implies that $f(x) - g(x)q(x) = r(x)$ is in N by definition of an ideal. Since $g(x)$ was a nonzero element of minimal degree in N, we must have $r(x) = 0$. Thus $f(x) = g(x)q(x)$ and $N = \langle g(x) \rangle$. ∎

We can now achieve our aim of characterizing the maximal ideals of $F[x]$.

Theorem 31.6 *An ideal $\langle p(x) \rangle \neq \{0\}$ of $F[x]$ is maximal if and only if $p(x)$ is irreducible over F.*

Proof. Suppose that $\langle p(x) \rangle \neq \{0\}$ is a maximal ideal of $F[x]$. Then $\langle p(x) \rangle \neq F[x]$, so $p(x) \notin F$. Let $p(x) = f(x)g(x)$ be a factorization of $p(x)$ in $F[x]$. Since $\langle p(x) \rangle$ is a maximal ideal and hence also a prime ideal, $(f(x)g(x)) \in \langle p(x) \rangle$ implies that $f(x) \in \langle p(x) \rangle$, or $g(x) \in \langle p(x) \rangle$, that is, either $f(x)$ or $g(x)$ has $p(x)$ as a factor. But then we can't have the degrees of both $f(x)$ and $g(x)$ less than the degree of $p(x)$. This shows that $p(x)$ is irreducible over F.

Conversely, if $p(x)$ is irreducible over F, suppose that N is an ideal such that $\langle p(x) \rangle \subseteq N \subseteq F[x]$. Now N is a principal ideal by Theorem 31.5, so $N = \langle g(x) \rangle$ for some $g(x) \in N$. Then $p(x) \in N$ implies that $p(x) = g(x)q(x)$ for some $q(x) \in F[x]$. But $p(x)$ is irreducible, which implies that either $g(x)$ or $q(x)$ is of degree 0. If $g(x)$ is of degree 0, that is, a nonzero constant in F, then $g(x)$ is a unit in $F[x]$, so $\langle g(x) \rangle = N = F[x]$. If $q(x)$ is of degree 0, then $q(x) = c$, where $c \in F$, and $g(x) = (1/c)p(x)$ is in $\langle p(x) \rangle$, so $N = \langle p(x) \rangle$. Thus $\langle p(x) \rangle \subset N \subset F[x]$ is impossible, so $\langle p(x) \rangle$ is maximal. ∎

Example 31.9 Example 31.4 shows that $x^3 + 3x + 2$ is irreducible in $\mathbf{Z}_5[x]$. Thus $\mathbf{Z}_5[x]/\langle x^3 + 3x + 2 \rangle$ is a field. Similarly, Theorem 27.1 shows that $x^2 - 2$ is irreducible in $\mathbf{Q}[x]$, so $\mathbf{Q}[x]/\langle x^2 - 2 \rangle$ is a field. We shall examine such fields in more detail later. ‖

31.4 UNIQUENESS OF FACTORIZATION IN $F[x]$

We shall demonstrate that polynomials in $F[x]$ can be factored into a product of irreducible polynomials in $F[x]$ in an essentially unique way, as the student would guess from his high school work in $\mathbf{R}[x]$. We shall consider unique factorization in more general situations in the starred section which follows. Since the result for $F[x]$ can be obtained cheaply at this point, we decided to include this special case here. We shall be dealing exclusively with fields in the remaining unstarred portion of this text, so this special case is all that we shall need for our work.

For $f(x), g(x) \in F[x]$ we say that $g(x)$ **divides** $f(x)$ **in** $F[x]$ if there exists $q(x) \in F[x]$ such that $f(x) = g(x)q(x)$.

Theorem 31.7 *Let $p(x)$ be an irreducible polynomial in $F[x]$. If $p(x)$ divides $r(x)s(x)$ for $r(x)$, $s(x) \in F[x]$, then either $p(x)$ divides $r(x)$ or $p(x)$ divides $s(x)$.*

Proof. Suppose that $p(x)$ divides $r(x)s(x)$. Then $r(x)s(x) \in \langle p(x) \rangle$, which is maximal by Theorem 31.6. Therefore, $\langle p(x) \rangle$ is a prime ideal by the corollary to Theorem 29.5. Hence $r(x)s(x) \in \langle p(x) \rangle$ implies that either $r(x) \in \langle p(x) \rangle$, giving $p(x)$ divides $r(x)$, or $s(x) \in \langle p(x) \rangle$, giving $p(x)$ divides $s(x)$. ∎

Corollary. *If $p(x)$ is irreducible in $F[x]$ and $p(x)$ divides the product $r_1(x) \cdots r_n(x)$ for $r_i(x) \in F[x]$, then $p(x)$ divides $r_i(x)$ for at least one i.*

Proof. Using mathematical induction, we find that this is immediate from Theorem 31.7. ∎

Theorem 31.8 *If F is a field, then every nonconstant polynomial $f(x) \in F[x]$ can be factored in $F[x]$ into a product of irreducible polynomials, the irreducible polynomials being unique except for order and for unit (i.e., nonzero constant) factors in F.*

Proof. Let $f(x) \in F[x]$ be a nonconstant polynomial. If $f(x)$ is not irreducible, then $f(x) = g(x)h(x)$, with the degree of $g(x)$ and the degree of $h(x)$ both less than the degree of $f(x)$. If $g(x)$ and $h(x)$ are both irreducible, we stop here. If not, at least one of them factors into polynomials of lower degree. Continuing this process (an induction argument really), we arrive at a factorization

$$f(x) = p_1(x)p_2(x) \cdots p_r(x),$$

where $p_i(x)$ is irreducible.

It remains for us to show uniqueness. Suppose that

$$f(x) = p_1(x)p_2(x) \cdots p_r(x) = q_1(x)q_2(x) \cdots q_s(x)$$

are two factorizations of $f(x)$ into irreducible polynomials. Then by the corollary to Theorem 31.7, $p_1(x)$ divides some $q_j(x)$, let us assume $q_1(x)$. Since $q_1(x)$ is irreducible,

$$q_1(x) = u_1 p_1(x),$$

where $u_1 \neq 0$, but u_1 is in F and thus is a unit. Then substituting $u_1 p_1(x)$ for $q_1(x)$ and canceling, we get

$$p_2(x) \cdots p_r(x) = u_1 q_2(x) \cdots q_s(x).$$

By a similar argument, say $q_2(x) = u_2 p_2(x)$, so

$$p_3(x) \cdots p_r(x) = u_1 u_2 q_3(x) \cdots q_s(x).$$

Continuing in this manner, we eventually arrive at

$$1 = u_1u_2 \cdots u_r q_{r+1}(x) \cdots q_s(x).$$

Clearly, this is only possible if $s = r$, so that this equation is actually $1 = u_1u_2 \cdots u_r$. Thus the irreducible factors $p_i(x)$ and $q_j(x)$ were the same except possibly for order and unit factors. $\blacksquare$

Example 31.10 Example 31.2 shows that the factorization of $x^4 + 3x^3 + 2x + 4$ in $Z_5[x]$ is $(x - 1)^3(x + 1)$. These irreducible factors in $Z_5[x]$ are only defined up to units in $Z_5[x]$, that is, nonzero constants in Z_5. For example, $(x - 1)^3(x + 1) = (x - 1)^2(2x - 2)(3x + 3)$. $\|$

EXERCISES

31.1 Let $f(x) = x^6 + 3x^5 + 4x^2 - 3x + 2$ and $g(x) = x^2 + 2x - 3$ be in $Z_7[x]$. Find $q(x)$ and $r(x)$ in $Z_7[x]$ such that $f(x) = g(x)q(x) + r(x)$, with $(\text{degree } r(x)) < 2$.

31.2 Let $f(x) = x^6 + 3x^5 + 4x^2 - 3x + 2$ and $g(x) = 3x^2 + 2x - 3$ be in $Z_7[x]$. Find $q(x)$ and $r(x)$ in $Z_7[x]$ such that $f(x) = g(x)q(x) + r(x)$, with $(\text{degree } r(x)) < 2$.

31.3 The polynomial $x^4 + 4$ can be factored into linear factors in $Z_5[x]$. Find this factorization.

31.4 Is $x^3 + 2x + 3$ an irreducible polynomial of $Z_5[x]$? Why? Express it as a product of irreducible polynomials of $Z_5[x]$.

31.5 Show that $f(x) = x^2 + 8x - 2$ is irreducible over Q. Is $f(x)$ irreducible over R? over C?

31.6 Repeat Exercise 31.5 with $g(x) = x^2 + 6x + 12$ in place of $f(x)$.

31.7 a) Show that $x^3 + 3x^2 - 8$ is irreducible over Q.
b) Show that $x^4 - 22x^2 + 1$ is irreducible over Q.

†31.8 Prove that if F is a field, every proper prime ideal of $F[x]$ is maximal.

31.9 Mark each of the following true or false.

___ a) $x - 2$ is irreducible over Q.
___ b) $3x - 6$ is irreducible over Q.
___ c) $x^2 - 3$ is irreducible over Q.
___ d) $x^2 + 3$ is irreducible over Z_7.
___ e) If F is a field, the units of $F[x]$ are precisely the nonzero elements of F.
___ f) If F is a field, the units of $F(x)$ are precisely the nonzero elements of F.
___ g) A polynomial $f(x)$ of degree n with coefficients in a field F can have at most n zeros in F.
___ h) A polynomial $f(x)$ of degree n with coefficients in a field F can have at most n zeros in any given field E such that $F \leq E$.
___ i) Every ideal of $F[x]$ is principal.
___ j) Every principal ideal in $F[x]$ is a maximal ideal.

31.10 Determine which of the following polynomials in $\mathbf{Z}[x]$ satisfy an Eisenstein criterion for irreducibility over $\mathbf{Q}$.

a) $x^2 - 12$ b) $8x^3 + 6x^2 - 9x + 24$
c) $4x^{10} - 9x^3 + 24x - 18$ d) $2x^{10} - 25x^3 + 10x^2 - 30$

31.11 Is $2x^3 + x^2 + 2x + 2$ an irreducible polynomial in $\mathbf{Z}_5[x]$? Why? Express it as a product of irreducible polynomials in $\mathbf{Z}_5[x]$.

31.12 Find all zeros of $6x^4 + 17x^3 + 7x^2 + x - 10$ in $\mathbf{Q}$. (This is a tedious high school algebra problem. *You* might use a bit of analytic geometry and calculus and make a graph, or use Newton's method to see which are the best candidates for zeros.)

31.13 If F is a field and $a \neq 0$ is a zero of $f(x) = a_0 + a_1x + \cdots + a_nx^n$ in $F[x]$, show that $1/a$ is a zero of $a_n + a_{n-1}x + \cdots + a_0x^n$.

31.14 Find all irreducible polynomials of degree 2 or 3 in $\mathbf{Z}_2[x]$ and $\mathbf{Z}_3[x]$.

31.15 Is $\mathbf{Q}[x]/\langle x^2 - 5x + 6 \rangle$ a field? Why? What about $\mathbf{Q}[x]/\langle x^2 - 6x + 6 \rangle$?

31.16 Let F be a field and $f(x), g(x) \in F[x]$. Show that $f(x)$ divides $g(x)$ if and only if $g(x) \in \langle f(x) \rangle$.

31.17 Let F be a field and let $f(x), g(x) \in F[x]$. Show that

$$N = \{r(x)f(x) + s(x)g(x) \mid r(x), s(x) \in F[x]\}$$

is an ideal of $F[x]$. Show that if $f(x)$ and $g(x)$ have different degrees and $N \neq F[x]$, then $f(x)$ and $g(x)$ cannot both be irreducible over F.

31.18 Let $\sigma_m: \mathbf{Z} \to \mathbf{Z}_m$ be the natural homomorphism given by $a\sigma_m = $ (the remainder of a when divided by m) for $a \in \mathbf{Z}$.

a) Show that $\overline{\sigma_m}: \mathbf{Z}[x] \to \mathbf{Z}_m[x]$ given by

$$(a_0 + a_1x + \cdots + a_nx^n)\overline{\sigma_m} = a_0\sigma_m + (a_1\sigma_m)x + \cdots + (a_n\sigma_m)x^n$$

is a homomorphism of $\mathbf{Z}[x]$ onto $\mathbf{Z}_m[x]$.

b) Show that if $f(x) \in \mathbf{Z}[x]$ has degree n and $(f(x))\overline{\sigma_m}$ does not factor in $\mathbf{Z}_m[x]$ into two polynomials of degree less than n, then $f(x)$ is irreducible in $\mathbf{Q}[x]$.

c) Use part (b) to show that $x^3 + 17x + 36$ is irreducible in $\mathbf{Q}[x]$. [*Hint:* Try a prime value of m which simplifies the coefficients.]

31.19 Let F be a field, and let S be any subset of $F \times F \times \cdots \times F$ for n factors. Show that the set of all $f(x_1, \ldots, x_n) \in F[x_1, \ldots, x_n]$ which are zero at every $(a_1, \ldots, a_n) \in S$ (see Exercise 30.12) forms an ideal in $F[x_1, \ldots, x_n]$. This is of importance in algebraic geometry.

32 | Unique Factorization Domains

*32.1 INTRODUCTION

This section and the next two sections are concerned with integral domains and the problem of factorization of elements in an integral domain. The integral domain $\mathbf{Z}$ is our standard example of an integral domain in which there is unique factorization into primes (irreducibles). Section 31.4 showed that for a field F, $F[x]$ is also such an integral domain with unique factorization. In order to discuss analogous ideas in an arbitrary integral domain, we shall give several definitions, some of which are repetitions of earlier ones. It is nice to have them all in one place for reference.

Definition. Let D be an integral domain and $a, b, \in D$. If there exists $c \in D$ such that $b = ac$, then a **divides** b (or a **is a factor of** b), denoted by "$a \mid b$".

Definition. An element u of an integral domain D is a **unit of** D if u divides 1, that is, if u has a multiplicative inverse in D. Two elements $a, b \in D$ are **associates in** D if $a = bu$, where u is a unit in D.

In Exercise 32.7, we ask you to show that this criterion for a and b to be associates is an equivalence relation on D.

Example 32.1 The only units in $\mathbf{Z}$ are 1 and -1. Thus the only associates of 26 in $\mathbf{Z}$ are 26 and -26. ‖

It should be noted that while our condition $a = bu$ for elements $a, b \in D$ to be associates is not formally symmetric, if $a = bu$, then $b = au^{-1}$, where u^{-1} exists and is a unit of D, since u is a unit of D.

Definition. A nonzero element p which is not a unit of an integral domain D is an **irreducible** of D if in any factorization $p = ab$ in D either a or b is a unit.

It is easily seen that an associate of an irreducible is again an irreducible.

Definition. An integral domain D is a **unique factorization domain** (abbreviated UFD) if the following conditions are satisfied:

1) Every element of D which is neither zero nor a unit can be factored into a product of a finite number of irreducibles.

266

2) If $p_1 \cdots p_r$ and $q_1 \cdots q_s$ are two factorizations of the same element of D into irreducibles, then $r = s$ and the q_j can be renumbered so that p_i and q_i are associates.

Example 32.2 Theorem 31.8 shows that for a field F, $F[x]$ is a UFD. Also the student knows that **Z** is a UFD; we have made frequent use of this fact, although we have never proved it. For example, in **Z** we have

$$24 = (2)(2)(3)(2) = (-2)(-3)(2)(2).$$

Here 2 and -2 are associates as are 3 and -3. Thus except for order and associates, the irreducible factors in these two factorizations of 24 are the same. ‖

After just one more definition we can describe what we wish to achieve in this section.

Definition. An integral domain D is a ***principal ideal domain*** (abbreviated PID) if every ideal in D is a principal ideal.

Our purpose in this section is to prove two exceedingly important results:
(A) *Every* PID *is a* UFD.
(B) *If D is a* UFD, *then $D[x]$ is a* UFD.

Note that the fact that $F[x]$ is a UFD, where F is a field (by Theorem 31.8), illustrates both (A) and (B). For by Theorem 31.5, $F[x]$ is a PID. Also, since F has no nonzero elements which are not units, F satisfies (in a vacuous sense which bothers some students) our definition for a UFD. Thus (B) would give another proof that $F[x]$ is a UFD, except for the fact that we shall actually use Theorem 31.8 quite often in proving (B). In the following section, we shall study properties of a certain special class of UFD's, the *Euclidean domains*.

Let us proceed to prove (A) and (B). The author has always disliked the proof of (B). Let's put it off as long as possible and prove (A) first.

*32.2 EVERY PID IS A UFD

The steps leading up to Theorem 31.8 and its proof indicate the way for our proof of (A). Much of the material will be repetitive. We inefficiently handled the special case of $F[x]$ separately in Theorem 31.8, since it was easy and was the only case we shall need for our field theory in general.

To prove that an integral domain D is a UFD, it is necessary to show that both conditions (1) and (2) of the definition of a UFD are satisfied. For our special case of $F[x]$ in Theorem 31.8, (1) was very easy and resulted from an argument that in a factorization of a polynomial of degree >0 into a product of two nonconstant polynomials, the degree of each factor was less than the degree of the original polynomial. Thus you couldn't keep on factoring indefinitely without running into unit factors, i.e., polynomials of degree zero. For the general case of a PID, it is harder to show that this is so. We now turn to this problem. We shall need one more set-theoretic concept.

Definition. If $\{A_i \mid i \in I\}$ is a collection of sets, then the **union** $\bigcup_{i \in I} A_i$ *of the sets* A_i is the set of all x such that $x \in A_i$ for at least one $i \in I$.

Lemma 32.1 (Ascending chain condition for a PID). *Let* D *be a* PID. *If* $N_1 \subseteq N_2 \subseteq \cdots$ *is a monotonic ascending chain of ideals* N_i, *then there exists a positive integer* r *such that* $N_r = N_s$ *for all* $s \geq r$. *Therefore, every strictly ascending chain* $K_1 \subset K_2 \subset \cdots$ *of ideals in a* PID *is of finite length, i.e., the* **ascending chain condition** (ACC) *holds for ideals in a* PID.

Proof. Let $N_1 \subseteq N_2 \subseteq \cdots$ be a monotonic ascending chain of ideals N_i in D. Let $N = \bigcup_i N_i$. Clearly, $N \subseteq D$. We claim that N is an ideal in D. Let $a, b \in N$. Then there are ideals N_{i_1} and N_{i_2}, with $a \in N_{i_1}$ and $b \in N_{i_2}$. Now either $N_{i_1} \subseteq N_{i_2}$ or $N_{i_2} \subseteq N_{i_1}$; let us assume that $N_{i_1} \subseteq N_{i_2}$, so both a and b are in N_{i_2}. This implies that $a \pm b$ and ab are in N_{i_2}, so $a \pm b$ and ab are in N. Taking $a = 0$, we see that $b \in N$ implies $-b \in N$. Clearly, $0 \in N$. Thus N is a subring of D. For $a \in N$ and $d \in D$, we must have $a \in N_{i_1}$ for some N_{i_1}. Then since N_{i_1} is an ideal, $da = ad$ is in N_{i_1}. Therefore, $da \in \bigcup_i N_i$, that is, $da \in N$. Hence N is an ideal.

Now as an ideal in D which is a PID, $N = \langle c \rangle$ for some $c \in D$. Since $N = \bigcup_i N_i$, we must have $c \in N_r$ for some $r \in \mathbf{Z}^+$. For $s \geq r$, we have

$$\langle c \rangle \subseteq N_r \subseteq N_s \subseteq N = \langle c \rangle.$$

Thus $N_r = N_s$ for $s \geq r$. ∎

We can now prove condition (1) of the definition of a UFD for an integral domain which is a PID.

Theorem 32.1 *Let* D *be a* PID. *Every element which is neither zero nor a unit in* D *is a product of irreducibles.*

Proof. Let $a \in D$, where a is neither zero nor a unit. We first show that a has at least one irreducible factor. If a is an irreducible, we are done. If a is not an irreducible, then $a = a_1 b_1$, where neither a_1 nor b_1 is a unit. Now

$$\langle a \rangle \subset \langle a_1 \rangle,$$

for $\langle a \rangle \subseteq \langle a_1 \rangle$ is obvious, and if $\langle a \rangle = \langle a_1 \rangle$, we would have $a_1 = va$ for some $v \in D$. But then we would have

$$a = a_1 b_1 = vab_1 = vb_1 a,$$

so $1 = vb_1$ and b_1 would be a unit, contrary to construction. Continuing this procedure then, starting now with a_1, we arrive at a strictly ascending chain of ideals

$$\langle a \rangle \subset \langle a_1 \rangle \subset \langle a_2 \rangle \subset \cdots$$

By Lemma 32.1, this chain terminates with some $\langle a_r \rangle$, and a_r must then be irreducible. Thus a has an irreducible factor a_r.

By what we have just proved, for an element a which is neither zero nor a unit in D, either a is irreducible or $a = p_1 c_1$ for p_1 an irreducible, and c_1 not a unit. By an argument similar to the one just made, in the latter case we can conclude that $\langle a \rangle \subset \langle c_1 \rangle$. If c_1 is not irreducible, then $c_1 = p_2 c_2$ for an irreducible p_2 with c_2 not a unit. Continuing, we get a strictly ascending chain of ideals

$$\langle a \rangle \subset \langle c_1 \rangle \subset \langle c_2 \rangle \subset \cdots$$

This chain must terminate, by Lemma 32.1, with some $c_r = q_r$ which is an irreducible. Then $a = p_1 p_2 \cdots p_r q_r$. ∎

This completes our demonstration of condition (1) of the definition of a UFD. Let us turn to condition (2). Our arguments here are parallel to those leading to Theorem 31.8. The results we encounter along the way are of some interest in themselves.

Lemma 32.2 (Generalization of Theorem 31.6). *An ideal $\langle p \rangle$ in a PID is maximal if and only if p is an irreducible.*

Proof. Let $\langle p \rangle$ be a maximal ideal of D, a PID. Suppose that $p = ab$ in D. Clearly, $\langle p \rangle \subseteq \langle a \rangle$. Suppose that $\langle a \rangle = \langle p \rangle$. Then there exists $v \in D$ such that $a = vp$, so

$$a = vab = avb.$$

Thus $1 = vb$, so b must be a unit. If $\langle a \rangle \neq \langle p \rangle$, then we must have $\langle a \rangle = \langle 1 \rangle = D$, since $\langle p \rangle$ is maximal. But then there exists $u \in D$ such that $au = 1$, so a is a unit. Thus, if $p = ab$, either a or b must be a unit. Hence p is an irreducible of D.

Conversely, suppose that p is an irreducible in D. Then if $\langle p \rangle \subseteq \langle a \rangle$, we must have $p = ab$. Now if a is a unit, then $\langle a \rangle = \langle 1 \rangle = D$. If a is not a unit, then b must be a unit, so there exists $u \in D$ such that $bu = 1$. Then $pu = abu = a$, so $\langle a \rangle \subseteq \langle p \rangle$, and we have $\langle a \rangle = \langle p \rangle$. Thus $\langle p \rangle \subseteq \langle a \rangle$ implies that either $\langle a \rangle = D$ or $\langle a \rangle = \langle p \rangle$, and $\langle p \rangle \neq D$ or p would be a unit. Hence $\langle p \rangle$ is a maximal ideal. ∎

Lemma 32.3 (Generalization of Theorem 31.7). *In a PID, if an irreducible p divides ab, then either $p \mid a$ or $p \mid b$.*

Proof. Let D be a PID and suppose that for an irreducible p in D we have $p \mid ab$. Then $(ab) \in \langle p \rangle$. Since every maximal ideal in D is a prime ideal by the corollary to Theorem 29.5, $(ab) \in \langle p \rangle$ implies that either $a \in \langle p \rangle$ or $b \in \langle p \rangle$, giving either $p \mid a$ or $p \mid b$. ∎

Corollary. *If p is an irreducible in a PID and p divides the product $a_1 a_2 \cdots a_n$ for $a_i \in D$, then $p \mid a_i$ for at least one i.*

Proof. This is immediate from Lemma 32.3 if we use mathematical induction. ∎

Definition. A nonzero nonunit element p of an integral domain D with the property that $p \mid ab$ implies either $p \mid a$ or $p \mid b$ is a **prime**.

Lemma 32.3 focused our attention on the defining property of a prime. In exercise 32.5, we ask you to show that a prime in an integral domain is always an irreducible, and that in a UFD, an irreducible is also a prime. Thus the concepts of prime and irreducible coincide in a UFD. Example 34.3 will exhibit an integral domain containing some irreducibles which are not primes, so the concepts do not coincide in every domain. The defining property of a prime is precisely what is needed to establish the uniqueness property (2) in the definition of a UFD. We now complete the proof of (A) by demonstrating the property (2) for a PID.

Theorem 32.2 (Generalization of Theorem 31.8). *Every PID is a UFD.*

Proof. Theorem 32.1 shows that if D is a PID, then each $a \in D$, where a is neither zero nor a unit, has a factorization

$$a = p_1 p_2 \cdots p_r$$

into irreducibles. It remains for us to show uniqueness. Let

$$a = q_1 q_2 \cdots q_s$$

be another such factorization into irreducibles. Then we have $p_1 \mid (q_1 q_2 \cdots q_s)$ which implies that $p_1 \mid q_{j_1}$ for some j_1 by the corollary of Lemma 32.3. By changing the order of the q_j if necessary, we can assume that $j_1 = 1$ or $p_1 \mid q_1$. Then $q_1 = p_1 u_1$, and since p_1 is an irreducible, u_1 is a unit, so p_1 and q_1 are associates. We have then

$$p_1 p_2 \cdots p_r = p_1 u_1 q_2 \cdots q_s,$$

so by the cancellation law in D,

$$p_2 \cdots p_r = u_1 q_2 \cdots q_s.$$

Continuing this process, starting with p_2, etc., we finally arrive at

$$1 = u_1 u_2 \cdots u_r q_{r+1} \cdots q_s.$$

Since the q_j are irreducibles, we must have $r = s$. ∎

Many algebra texts start by proving the following corollary of Theorem 32.2. We have assumed that the student was familiar with this corollary and used it freely in our other work.

Corollary (Fundamental Theorem of Arithmetic). *The integral domain* **Z** *is a UFD.*

Proof. We have seen that all ideals in **Z** are of the form $n\mathbf{Z} = \langle n \rangle$ for $n \in \mathbf{Z}$. Thus **Z** is a PID, and Theorem 32.2 applies. ∎

It is worth noting that the proof that **Z** is a PID was really way back in the corollary of Theorem 6.2. We proved Theorem 6.2 by using the division algorithm for $\mathbf{Z}$ exactly as we proved, in Theorem 31.5, that $F[x]$ is a PID by using the division algorithm for $F[x]$. In Section 33 we shall examine this parallel more closely.

*32.3   IF D IS A UFD, THEN $D[x]$ IS A UFD

We proceed now with the proof of (B), our second main result for this section. The reason we don't like the proofs here is because they are all so naively obvious, but so messy to write down. The idea of the argument is as follows. Let D be a UFD. We can form a field of quotients F of D. Then $F[x]$ is a UFD by Theorem 31.8, and we shall show that we can recover a factorization for $f(x) \in D[x]$ from its factorization in $F[x]$. It will be necessary to compare the irreducibles in $F[x]$ with those in $D[x]$, of course. This approach, which we prefer as more intuitive than some more efficient modern ones, is essentially due to Gauss.

Definition.   Let D be a UFD. A nonconstant polynomial

$$f(x) = a_0 + a_1 x + \cdots + a_n x^n$$

in $D[x]$ is ***primitive*** if the only common divisors of all the a_i are units of D.

Example 32.3   In $\mathbf{Z}[x]$, $4x^2 + 3x + 2$ is primitive, but $4x^2 + 6x + 2$ is not, since 2, a nonunit in $\mathbf{Z}$, is a common divisor of 4, 6, and 2. ‖

Clearly, every nonconstant irreducible in $D[x]$ must be a primitive polynomial.

Lemma 32.4   *If D is a UFD, then for every nonconstant $f(x) \in D[x]$ we have $f(x) = (c)g(x)$, where $c \in D$, $g(x) \in D[x]$, and $g(x)$ is primitive. The element c is unique up to a unit factor in D and is the* **content** *of $f(x)$. Also $g(x)$ is unique up to a unit factor in D.*

Proof.   Let $f(x) \in D[x]$ be given, where $f(x)$ is a nonconstant polynomial. Since D is a UFD, each coefficient of $f(x)$ can be factored into a finite product of irreducibles in D, uniquely up to order and associates.   Imagine each coefficient of $f(x)$ to be so factored. If p is a particular irreducible dividing every coefficient in $f(x)$, replace every occurrence of every associate of p in a factorization of a coefficient by pu for some unit u. Continuing this procedure for another irreducible q appearing in the factorization of some coefficient of $f(x)$, etc., we arrive eventually at a factorization of the coefficients of $f(x)$ in which each irreducible p_i appearing in the factorization of one coefficient and dividing all coefficients actually appears in the factorization of all coefficients, but no other associate of p_i appears in the factorization of any coefficient. Let $c = \prod_i p_i^{\nu_i}$, where the product is taken over all irreducibles p_i appearing in the factorizations of all the coefficients in this adjusted factorization, and where ν_i is the greatest integer such that $p_i^{\nu_i}$

divides all the coefficients. Clearly, we have $f(x) = (c)g(x)$, where $c \in D$, $g(x) \in D[x]$, and $g(x)$ is primitive by construction.

For uniqueness, if also $f(x) = (d)h(x)$ for $d \in D$, $h(x) \in D[x]$, and $h(x)$ primitive, then each irreducible factor of c must divide d and conversely. By setting $(c)g(x) = (d)h(x)$ and canceling irreducible factors of c into d, we arrive at $(u)g(x) = (v)h(x)$ for a unit $u \in D$. But then obviously we must also have v a unit of D or we would be able to cancel irreducible factors of v into u. Thus u and v are both units, so c is unique up to a unit factor. From $f(x) = (c)g(x)$, we see that the primitive polynomial $g(x)$ is also unique up to a unit factor. ∎

The preceding lemma illustrates what we meant when we said that the proofs here are obvious but messy to write down carefully. Surely this lemma falls into the "intuitively obvious" category.

Example 32.4 In $\mathbf{Z}[x]$,

$$4x^2 + 6x - 8 = (2)(2x^2 + 3x - 4),$$

where $2x^2 + 3x - 4$ is primitive. ‖

Lemma 32.5 (Gauss). *If D is a UFD, then a product of two primitive polynomials in $D[x]$ is again primitive.*

Proof. Let

$$f(x) = a_0 + a_1 x + \cdots + a_n x^n$$

and

$$g(x) = b_0 + b_1 x + \cdots + b_m x^m$$

be primitive in $D[x]$, and let $h(x) = f(x)g(x)$. Let p be an irreducible in D. Then p does not divide all a_i and p does not divide all b_j, since $f(x)$ and $g(x)$ are primitive. Let a_r be the first coefficient of $f(x)$ not divisible by p, that is, $p \mid a_i$ for $i < r$, but $p \nmid a_r$ (that is, p does not divide a_r). Similarly, let $p \mid b_j$ for $j < s$, but $p \nmid b_s$. The coefficient of x^{r+s} in $h(x) = f(x)g(x)$ is

$$c_{r+s} = (a_0 b_{r+s} + \cdots + a_{r-1} b_{s+1}) + a_r b_s + (a_{r+1} b_{s-1} + \cdots + a_{r+s} b_0).$$

Now $p \mid a_i$ for $i < r$ implies that

$$p \mid (a_0 b_{r+s} + \cdots + a_{r-1} b_{s+1}),$$

and also $p \mid b_j$ for $j < s$ implies that

$$p \mid (a_{r+1} b_{s-1} + \cdots + a_{r+s} b_0).$$

But p does not divide a_r or b_s, so p does not divide $a_r b_s$, and consequently p does not divide c_{r+s}. This shows that given any irreducible $p \in D$, there is some coefficient of $f(x)g(x)$ not divisible by p. Thus $f(x)g(x)$ is primitive. ∎

Corollary. *If D is a UFD, then a finite product of primitive polynomials in $D[x]$ is again primitive.*

Proof. This follows from Lemma 32.5 by induction. ∎

Now let D be a UFD and let F be a field of quotients of D. By Theorem 31.8, $F[x]$ is a UFD. As we said earlier, we shall show that $D[x]$ is a UFD by carrying a factorization in $F[x]$ of $f(x) \in D[x]$ back into one in $D[x]$. The next lemma relates the nonconstant irreducibles of $D[x]$ to those of $F[x]$. This is the last important step.

> **Lemma 32.6** *Let D be a UFD and let F be a field of quotients of D. Let $f(x) \in D[x]$, where $(degree\ f(x)) > 0$. If $f(x)$ is an irreducible in $D[x]$, then $f(x)$ is also an irreducible in $F[x]$. Also, if $f(x)$ is primitive in $D[x]$ and irreducible in $F[x]$, then $f(x)$ is irreducible in $D[x]$.*

Proof. Suppose that a nonconstant $f(x) \in D[x]$ factors into polynomials of lower degree in $F[x]$, that is, $f(x) = r(x)s(x)$ for $r(x), s(x) \in F[x]$. Then since F is a field of quotients of D, each coefficient in $r(x)$ and $s(x)$ is of the form a/b for some $a, b \in D$. By clearing denominators, we can get

$$(d)f(x) = (e)r_1(x)s_1(x)$$

for $d, e \in D$, and $r_1(x), s_1(x) \in D[x]$, where the degrees of $r_1(x)$ and $s_1(x)$ are the degrees of $r(x)$ and $s(x)$, respectively. By Lemma 32.4, $f(x) = (c)g(x)$, $r_1(x) = (c_1)r_2(x)$, and $s_1(x) = (c_2)s_2(x)$ for primitive polynomials $g(x)$, $r_2(x)$, and $s_2(x)$, and $c, c_1, c_2 \in D$. Then

$$(dc)g(x) = (ec_1c_2)r_2(x)s_2(x),$$

and by Lemma 32.5, $r_2(x)s_2(x)$ is primitive. By the uniqueness part of Lemma 32.4, $ec_1c_2 = dcu$ for some unit u in D. But then

$$(dc)g(x) = (dcu)r_2(x)s_2(x),$$

so

$$f(x) = (c)g(x) = (cu)r_2(x)s_2(x).$$

We have shown that if $f(x)$ factors nontrivially in $F[x]$, then $f(x)$ factors nontrivially into polynomials of the same degree in $D[x]$. Thus, if $f(x)$ is irreducible in $D[x]$, it must be irreducible in $F[x]$.

It is obvious that a nonconstant $f(x) \in D[x]$ which is primitive in $D[x]$ and irreducible in $F[x]$ is also irreducible in $D[x]$, since $D[x] \subseteq F[x]$. ∎

Lemma 32.6 shows that *if D is a UFD the irreducibles in $D[x]$ are precisely the irreducibles in D, together with the nonconstant primitive polynomials which are irreducible in $F[x]$, where F is a field of quotients of $D[x]$.*

The preceding lemma is really very important in its own right. This is indicated by the following corollary, a special case of which was our Theorem 31.3 of the preceding section. (We admit that it does not seem very sensible to call a special case of a corollary of a lemma a theorem. The label which you assign to a result depends somewhat on the context in which it appears.)

Corollary. *If D is a* UFD *and F is a field of quotients of D, then a non-constant $f(x) \in D[x]$ factors into a product of two polynomials of lower degree in $F[x]$ if and only if it has a factorization into polynomials of the same degree in $D[x]$.*

Proof. It was shown in the proof of Lemma 32.6 that if $f(x)$ factors into a product of two polynomials of lower degree in $F[x]$, then it has a factorization into polynomials of the same degree in $D[x]$ (see the next to last sentence of the first paragraph of the proof).

The converse is obvious, since $D[x] \subseteq F[x]$. ∎

We are now prepared to prove our main theorem. We shall repeat the construction of the proof of Lemma 32.6 again; it is the core of this whole argument.

Theorem 32.3 *If D is a* UFD, *then so is $D[x]$.*

Proof. Let $f(x) \in D[x]$, where $f(x)$ is neither zero nor a unit. If $f(x)$ is of degree 0, we are done, since D is a UFD. Suppose that $(\text{degree } f(x)) > 0$, and let us view $f(x)$ as an element in $F[x]$, where F is a field of quotients of D. By Theorem 31.8, $f(x) = p_1(x) \cdots p_r(x)$ in $F[x]$, where $p_i(x)$ is irreducible in $F[x]$. Since F is a field of quotients of D, each coefficient in each $p_i(x)$ is of the form a/b for some $a, b \in D$. Clearing all denominators in the usual fashion, we arrive at

$$(d)f(x) = q_1(x) \cdots q_r(x)$$

for $d, q_i(x) \in D[x]$. Since each $p_i(x)$ was irreducible in $F[x]$, we see that $q_i(x)$, which is $p_i(x)$ multiplied by a *unit in F*, is also irreducible in $F[x]$. By Lemma 32.4, $f(x) = (c)g(x)$ and $q_i(x) = (c_i)q_i'(x)$ in $D[x]$ for $g(x)$ and $q_i'(x)$ primitive. Then

$$(dc)g(x) = (c_1 \cdots c_r)q_1'(x) \cdots q_r'(x),$$

where, by Lemma 32.5, the product $q_1'(x) \cdots q_r'(x)$ is primitive. By the uniqueness part of Lemma 32.4, we see that

$$c_1 \cdots c_r = dcu$$

for some unit u in D. Then

$$(dc)g(x) = (dcu)q_1'(x) \cdots q_r'(x),$$

so

$$f(x) = (c)g(x) = (cu)q_1'(x) \cdots q_r'(x).$$

Now cu can be factored into irreducibles in D. Also $q_1'(x), \ldots, q_r'(x)$ are irreducible in $D[x]$, since they are primitive and irreducible in $F[x]$ by construction. Thus we have shown that we can factor $f(x)$ into a product of irreducibles in $D[x]$.

The uniqueness of a factorization of $f(x) \in D[x]$ is clear for $f(x) \in D$ which is neither zero nor a unit. If $(\text{degree } f(x)) > 0$, we can view any factorization of $f(x)$ into irreducibles in $D[x]$ as a factorization in $F[x]$ into units (i.e., the factors in D) and irreducible polynomials in $F[x]$ by Lemma 32.6. By Theorem 31.8, these polynomials are unique, except for possible constant factors in F. But as an irreducible in $D[x]$, each polynomial of degree > 0 appearing in the factorization of $f(x)$ in $D[x]$ is primitive. By the uniqueness part of Lemma 32.4, this shows that these polynomials are unique in $D[x]$ up to unit factors, i.e., associates. The product of the irreducibles in D in the factorization of $f(x)$ is the content of $f(x)$, which is again unique up to a unit factor by Lemma 32.4. Thus all irreducibles in $D[x]$ appearing in the factorization are unique up to order and associates. ∎

Corollary. *If* F *is a field and* $x_1, \ldots, x_n$ *are indeterminates, then* $F[x_1, \ldots, x_n]$ *is a* UFD.

Proof. By Theorem 38.1, $F[x_1]$ is a UFD. By Theorem 32.3, so is $(F[x_1])[x_2] = F[x_1, x_2]$. Continuing in this procedure, we see (by induction) that $F[x_1, \ldots, x_n]$ is a UFD. ∎

We have seen that a PID is a UFD. The corollary of Theorem 32.3 makes it easy for us to give an example which shows that *not every* UFD *is a* PID.

Example 32.5 Let F be a field and let x and y be indeterminates. Then $F[x, y]$ is a UFD by the corollary of Theorem 32.3. Consider the set N of all polynomials in x and y in $F[x, y]$ having constant term 0. Clearly, N is an ideal, but not a principal ideal. Thus $F[x, y]$ is not a PID. ‖

Another example of a UFD which is not a PID is $Z[x]$, as shown in Exercise 33.5.

EXERCISES

*32.1 State which of the following elements are irreducibles of the indicated integral domains.

a) 5 in Z
b) -17 in Z
c) 14 in Z
d) $2x - 3$ in $Z[x]$
e) $2x - 10$ in $Z[x]$
f) $2x - 3$ in $Q[x]$
g) $2x - 10$ in $Q[x]$
h) $2x - 10$ in $Z_{11}[x]$

*32.2 If possible, give 4 different associates of $2x - 7$ viewed as an element of $Z[x]$; of $Q[x]$; of $Z_{11}[x]$.

*32.3 Factor the polynomial $4x^2 - 4x + 8$ into a product of irreducibles viewing it as an element of the integral domain $Z[x]$; of the integral domain $Q[x]$; of the integral domain $Z_{11}[x]$.

*32.4 Express each of the given polynomials as the product of its content with a primitive polynomial in the indicated UFD.

a) $18x^2 - 12x + 48$ in $Z[x]$
b) $18x^2 - 12x + 48$ in $Q[x]$
c) $2x^2 - 3x + 6$ in $Z[x]$
d) $2x^2 - 3x + 6$ in $Z_7[x]$

†*32.5 Prove the following.

a) If p is a prime in an integral domain D, then p is an irreducible.
b) If q is an irreducible in a UFD, then q is a prime.

*32.6 Mark each of the following true or false.

— a) Every field is a UFD.
— b) Every field is a PID.
— c) Every PID is a UFD.
— d) Every UFD is a PID.
— e) $Z[x]$ is a UFD.
— f) Any two irreducibles in any UFD are associates.
— g) If D is a PID, then $D[x]$ is a PID.
— h) If D is a UFD, then $D[x]$ is a UFD.
— i) In any UFD, if $p \mid a$ for an irreducible p, then p itself appears in every factorization of a.
— j) A UFD has no divisors of zero.

---

*32.7 For an integral domain D, show that the relation $a \sim b$ if a is an associate of b, i.e., if $a = bu$ for u a unit in D, is an equivalence relation on D.

*32.8 Let D be a UFD. Describe the irreducibles in $D[x]$ in terms of the irreducibles in D and the irreducibles in $F[x]$, where F is a field of quotients of D.

*32.9 Let D be an integral domain. Exercise 23.4 showed that $\langle U, \cdot \rangle$ is a group where U is the set of units of D. Show that the set $D^* - U$ of nonunits of D excluding 0 is closed under multiplication. Is this set a group under the multiplication of D?

*32.10 Lemma 32.6 states that if D is a UFD with a field of quotients F, then an irreducible $f(x)$ of $D[x]$ is also an irreducible of $F[x]$. Show by an example that a $g(x) \in D[x]$ which is an irreducible of $F[x]$ need not be an irreducible of $D[x]$.

*32.11 All our work in this section was restricted to integral domains. Taking the same definitions in Section 32.1 but for a commutative ring with unity, consider factorizations into irreducibles in $Z + Z$. What can happen? Consider in particular $(1, 0)$.

*32.12 Let D be a UFD. Show that a nonconstant divisor of a primitive polynomial in $D[x]$ is again a primitive polynomial.

*32.13 Show that in a PID, every ideal is contained in a maximal ideal. [*Hint:* Use Lemma 32.1.]

*32.14 Factor $x^3 - y^3$ into irreducibles in $Q[x, y]$ and prove that each of the factors is irreducible.

There are several other concepts often considered which are similar in character to the ascending chain condition on ideals in a ring. The following three exercises concern some of these concepts.

*32.15 Let R be any ring. The **ascending chain condition** (ACC) **for ideals** holds in R if every strictly increasing sequence $N_1 \subset N_2 \subset N_3 \subset \cdots$ of ideals in R is of finite length. The **maximum condition** (MC) **for ideals** holds in R if every nonempty set S of ideals in R contains an ideal not properly contained in any other ideal of the set S. The **finite basis condition** (FBC) **for ideals** holds in R if for each ideal N in R, there is a finite set $B_N = \{b_1, \ldots, b_n\} \subseteq N$ such that N is the intersection of all ideals of R containing B_N. The set B_N is a **finite basis for** N.

Show that for every ring R, the conditions ACC, MC, and FBC are equivalent.

*32.16 Let R be any ring. The **descending chain condition** (DCC) **for ideals** holds in R if every strictly decreasing sequence $N_1 \supset N_2 \supset N_3 \supset \cdots$ of ideals in R is of finite length. The **minimum condition** (mC) **for ideals** holds in R if given any set S of ideals of R, there is an ideal of S which does not properly contain any other ideal in the set S.

Show that for every ring, the conditions DCC and mC are equivalent.

*32.17 Give an example of a ring in which ACC holds but DCC does not hold. (See Exercises 32.15 and 32.16.)

33 | Euclidean Domains

INTRODUCTION AND DEFINITION

We have remarked several times on the importance of division algorithms. Our first contact with them was the *division algorithm for* **Z** (Lemma 6.1). This algorithm was immediately used to prove the important theorem that a subgroup of a cyclic group is cyclic, i.e., has a single generator. The *division algorithm for* $F[x]$ appeared in Theorem 31.1 and was used in a completely analogous way to show that $F[x]$ is a PID, i.e., that every ideal in $F[x]$ has a single generator. Now a modern technique of mathematics is to take some clearly related situations and to try to bring them under one roof by abstracting the important ideas common to them. It should be evident to the student how the following definition is an illustration of this technique. Let us see what we can develop by starting with the existence of a fairly general division algorithm in an integral domain.

> **Definition.** A *Euclidean valuation on an integral domain* D is a function ν mapping the nonzero elements of D into the nonnegative integers such that the following conditions are satisfied:
>
> 1) For all $a, b \in D$ with $b \neq 0$, there exist q and r in D such that $a = bq + r$, where either $r = 0$ or $\nu(r) < \nu(b)$.
>
> 2) For all $a, b \in D$, where neither a nor b is 0, $\nu(a) \leq \nu(ab)$.
>
> An integral domain D is a *Euclidean domain* if there exists a Euclidean valuation on D.

The importance of condition (1) is clear from our discussion. The importance of condition (2) lies in the fact that it will enable us to characterize the units of a Euclidean domain D.

Example 33.1 The integral domain **Z** is a Euclidean domain, for the valuation ν defined by $\nu(n) = |n|$ for $n \neq 0$ in **Z** is a Euclidean valuation on **Z**. Condition (1) holds by Lemma 6.1, and condition (2) is obvious. ∥

Example 33.2 If F is a field, then $F[x]$ is a Euclidean domain, for the valuation ν defined by $\nu(f(x)) = (\text{degree } f(x))$ for $f(x) \in F[x]$ and $f(x) \neq 0$ is a Euclidean valuation. Condition (1) holds by Theorem 31.1, and condition (2) is obvious. ∥

Of course, we should give some examples of Euclidean domains other than these familiar ones which motivated the definition. We shall do this in the next section. In view of our opening remarks, the student must surely be waiting for the following theorem.

Theorem 33.1 *Every Euclidean domain is a PID.*

Proof. Let D be a Euclidean domain with a Euclidean valuation ν, and let N be an ideal in D. If $N = \{0\}$, then $N = \langle 0 \rangle$ and N is principal. Suppose that $N \neq \{0\}$. Then there exists $b \neq 0$ in N. Let us choose b such that $\nu(b)$ is minimal among all $\nu(n)$ for $n \in N$. We claim that $N = \langle b \rangle$. Let $a \in N$. Then by condition (1) for a Euclidean domain, there exist q and r in D such that

$$a = bq + r,$$

where either $r = 0$ or $\nu(r) < b$. Now $r = a - bq$ and $a, b \in N$, so that $r \in N$, since N is an ideal. Thus $\nu(r) < b$ is impossible by our choice of b. Hence $r = 0$, so $a = bq$. Since a was any element of N, we see that $N = \langle b \rangle$. ∎

Corollary. *A Euclidean domain is a UFD.*

Proof. By Theorem 33.1, a Euclidean domain is a PID and by Theorem 32.2, a PID is a UFD. ∎

Finally, we should mention that while a Euclidean domain is a PID by Theorem 33.1, not every PID is a Euclidean domain. Examples of PID's which are not Euclidean are not easily found, however.

*33.2   ARITHMETIC IN EUCLIDEAN DOMAINS

We shall now investigate some properties of Euclidean domains related to their multiplicative structure. It should be made clear that the arithmetic structure of a Euclidean domain is *intrinsic to the domain* and is not affected in any way by a Euclidean valuation ν on the domain. A Euclidean valuation is merely a useful tool for possibly throwing some light on this arithmetic structure of the domain. The arithmetic structure of a domain D is completely determined by the set D and the two binary operations $+$ and $\cdot$ on D.

Let D be a Euclidean domain with a Euclidean valuation ν. We can use property (2) of a Euclidean valuation to characterize the units of D.

Theorem 33.2 *For a Euclidean domain with a Euclidean valuation ν, $\nu(1)$ is minimal among all $\nu(a)$ for nonzero $a \in D$, and $u \in D$ is a unit if and only if $\nu(u) = \nu(1)$.*

Proof. Condition (2) for ν tells us at once that for $a \neq 0$,

$$\nu(1) \leq \nu(1a) = \nu(a).$$

On the other hand, if u is a unit in D, then

$$\nu(u) \leq \nu(uu^{-1}) = \nu(1).$$

Thus

$$\nu(u) = \nu(1)$$

for a unit u in D.

Conversely, suppose that a nonzero $u \in D$ is such that $\nu(u) = \nu(1)$. Then by the division algorithm, there exist q and r in D such that

$$1 = uq + r,$$

where either $r = 0$ or $\nu(r) < \nu(u)$. But since $\nu(u) = \nu(1)$ is minimal over all $\nu(d)$ for nonzero $d \in D$, $\nu(r) < \nu(u)$ is impossible. Hence $r = 0$ and $1 = uq$, so u is a unit. ∎

Example 33.3 For **Z** with $\nu(n) = |n|$, the minimum of $\nu(n)$ for nonzero $n \in \mathbf{Z}$ is 1. Clearly, 1 and -1 are the only elements of **Z** with $\nu(n) = 1$. Of course, 1 and -1 are exactly the units of **Z**. ∥

Example 33.4 For $F[x]$ with $\nu(f(x)) = (\text{degree } f(x))$ for $f(x) \neq 0$, the minimum value of $\nu(f(x))$ for all nonzero $f(x) \in F[x]$ is 0. The nonzero polynomials of degree 0 are exactly the nonzero elements of F, and these are precisely the units of $F[x]$. ∥

The student should realize that everything we prove here holds in *every* Euclidean domain, in particular, in **Z** and $F[x]$. We are going to prove some nice, classical results about greatest common divisors in a Euclidean domain. You surely have a naive idea of what a greatest common divisor (gcd) of two elements a and b in a UFD should be. You simply take a and b and factor them both, adjusting the factorizations by units, so that if any irreducible divides both a and b, it either appears in both factorizations, and no other of its associates appear, or it does not appear in either factorization, i.e., one of its associates appears instead. We then obtain a gcd of a and b by multiplying together all irreducibles appearing in both factorizations, taking each irreducible to the highest power for which it divides both a and b. That is sloppily stated, but you probably get the idea. We have used the concept freely for **Z** in group theory already. Since an irreducible appearing in a factorization is only defined up to a unit factor, we see that a gcd must also be defined only up to a unit factor in a UFD. It is for this reason that we say "a" gcd, rather than "the" gcd, of a and b. It is customary to give the following more elegant definition of a gcd in a UFD. It should be obvious to you that for a UFD, this definition describes the same concept which we have just discussed.

Definition. Let D be a UFD. An element $d \in D$ is a **greatest common divisor** (abbreviated gcd) **of elements a and b in D** if $d \mid a$, $d \mid b$, and also $c \mid d$ for all c dividing both a and b.

Example 33.5 In **Z**, a gcd of 18 and 48 is 6. Another one is -6. In **Q**[x], a gcd of $x^2 - 2x + 1$ and $x^2 + x - 2$ is $x - 1$. Another one is $2(x - 1)$, since 2 is a unit in **Q**[x]. Still another one is $(15/13)(x - 1)$. However, in **Z**[x], the only gcd's of $x^2 - 2x + 1$ and $x^2 + x - 2$ are $x - 1$ and $-(x - 1)$, for 1 and -1 are the only units in **Z**[x]. ‖

The next theorem proves (from our definition) the existence of a gcd for two nonzero elements a and b in a Euclidean domain, and gives a useful property of a gcd.

Theorem 33.3 *If D is a Euclidean domain with Euclidean valuation v and a and b are nonzero elements of D, then there exists a gcd of a and b. Furthermore, each gcd of a and b can be expressed in the form $\lambda a + \mu b$ for some $\lambda, \mu \in D$.*

Proof. Consider the set

$$N = \{ra + sb \mid r, s \in D\}.$$

Since

$$(r_1 a + s_1 b) \pm (r_2 a + s_2 b) = (r_1 \pm r_2)a + (s_1 \pm s_2)b$$

and

$$t(ra + sb) = (tr)a + (ts)b$$

for $t \in D$, it is immediate that N is an ideal of D. By Theorem 33.1, $N = \langle d \rangle$ for some $d \in D$. Then $d \mid (ra + sb)$ for all $r, s \in D$, and taking first $s = 0$ with $r = 1$ and then $r = 0$ with $s = 1$, we see that $d \mid a$ and $d \mid b$. Also if $c \mid a$ and $c \mid b$, then $c \mid (ra + sb)$ for all $ra + sb$, that is, $c \mid n$ for all $n \in N$. Hence $c \mid d$. Thus d is a gcd of a and b.

For d as just constructed, $d \in N$ implies that there exist $\lambda, \mu \in D$ such that $d = \lambda a + \mu b$. But the definition of a gcd shows that if d_1 is also a gcd of a and b, then $d \mid d_1$ and $d_1 \mid d$. Thus

$$d_1 = vd = (v\lambda)a + (v\mu)b = \lambda_1 a + \mu_1 b. \ \blacksquare$$

The preceding proof is extremely elegant, but not at all constructive. Let us give the *Euclidean algorithm* for a Euclidean valuation on a Euclidean domain. This will not only give another proof of the existence of a gcd for $a, b \neq 0$ in D, but will also provide us with a convenient method for computing a gcd.

Theorem 33.4 (Euclidean algorithm). *Let D be a Euclidean domain with a Euclidean valuation v, and let a and b be nonzero elements of D. Let r_1 be as in condition (1) for a Euclidean valuation, that is,*

$$a = bq_1 + r_1,$$

where either $r_1 = 0$ or $v(r_1) < v(b)$. If $r_1 \neq 0$, let r_2 be such that

$$b = r_1 q_2 + r_2,$$

where either $r_2 = 0$ *or* $v(r_2) < v(r_1)$. *In general, let* r_{i+1} *be such that*

$$r_{i-1} = r_i q_{i+1} + r_{i+1},$$

where either $r_{i+1} = 0$ *or* $v(r_{i+1}) < v(r_i)$. *Then the sequence* $r_1, r_2, \ldots$ *must terminate with some* $r_s = 0$. *If* $r_1 = 0$, *then b is a gcd of a and b. If* $r_1 \neq 0$ *and* r_s *is the first* $r_i = 0$, *then a gcd of a and b is* r_{s-1}.

Proof. Since $v(r_i) < v(r_{i-1})$ and $v(r_i)$ is a nonnegative integer, it is clear that after some finite number of steps we must arrive at some $r_s = 0$.

If $r_1 = 0$, then $a = bq_1$, and obviously b is a gcd of a and b. Suppose $r_1 \neq 0$. Then if $d \mid a$ and $d \mid b$, we have

$$d \mid (a - bq_1),$$

so $d \mid r_1$. However, if $d_1 \mid r_1$ and $d_1 \mid b$, then

$$d_1 \mid (bq_1 + r_1),$$

so $d_1 \mid a$. Thus the set of common divisors of a and b is the same set as the set of common divisors of b and r_1. By a similar argument, if $r_2 \neq 0$, the set of common divisors of b and r_1 is the same set as the set of common divisors of r_1 and r_2. Continuing this process, we see finally that the set of common divisors of a and b is the same set as the set of common divisors of r_{s-2} and r_{s-1}, where r_s is the first r_i equal to 0. Thus a gcd of r_{s-2} and r_{s-1} is also a gcd of a and b. But the equation

$$r_{s-2} = q_s r_{s-1} + r_s = q_s r_{s-1}$$

shows that a gcd of r_{s-2} and r_{s-1} is r_{s-1}. ∎

Example 33.6 Let us illustrate the Euclidean algorithm for the Euclidean valuation $| \; |$ on $\mathbf{Z}$ by computing a gcd of 22,471 and 3,266. We just apply the division algorithm over and over again, and the last nonzero remainder is a gcd. We label the numbers obtained as in Theorem 33.4 to further illustrate the statement and proof of the theorem. The computations are easily checked.

$$
\begin{aligned}
a &= 22{,}471 \\
b &= 3{,}266 \\
22{,}471 = (3{,}266)6 + 2{,}875 \qquad r_1 &= 2{,}875 \\
3{,}266 = (2{,}875)1 + 391 \qquad r_2 &= 391 \\
2{,}875 = (391)7 + 138 \qquad r_3 &= 138 \\
391 = (138)2 + 115 \qquad r_4 &= 115 \\
138 = (115)1 + 23 \qquad r_5 &= 23 \\
115 = (23)5 + 0 \qquad r_6 &= 0
\end{aligned}
$$

Thus $r_5 = 23$ is a gcd of 22,471 and 3,266. We found a gcd without factoring! This is important, for sometimes it is very difficult to find a factorization of an integer into primes. ∥

Example 33.7 Note that the division algorithm (1) of the definition of a Euclidean valuation says nothing about r being "positive." In computing a gcd in $\mathbf{Z}$ by the Euclidean algorithm for $|\ |$, as in Example 33.6, it is surely to our interest to make $|r_i|$ as small as possible in each division. Thus, repeating Example 33.6, it would be more efficient to write

$$
\begin{aligned}
a &= 22{,}471\\
b &= 3{,}266
\end{aligned}
$$

$$
\begin{aligned}
22{,}471 &= (3{,}266)7 - 391 & r_1 &= -391\\
3{,}266 &= (391)8 + 138 & r_2 &= 138\\
391 &= (138)3 - 23 & r_3 &= -23\\
138 &= (23)6 + 0 & r_4 &= 0
\end{aligned}
$$

The fact that we can change the sign of r_i from negative to positive when we wish is due to the fact that the divisors of r_i and $-r_i$ are the same. ∥

EXERCISES

*33.1 State which of the given functions ν are Euclidean valuations for the given integral domains.

a) The function ν for $\mathbf{Z}$ given by $\nu(n) = n^2$ for nonzero $n \in \mathbf{Z}$
b) The function ν for $\mathbf{Z}[x]$ given by $\nu(f(x)) = $ (degree of $f(x)$) for nonzero $f(x) \in \mathbf{Z}[x]$
c) The function ν for $\mathbf{Z}[x]$ given by $\nu(f(x)) = $ (the absolute value of the coefficient of the highest degree nonzero term of $f(x)$) for nonzero $f(x) \in \mathbf{Z}[x]$
d) The function ν for $\mathbf{Q}$ given by $\nu(a) = a^2$ for nonzero $a \in \mathbf{Q}$
e) The function ν for $\mathbf{Q}$ given by $\nu(a) = 50$ for nonzero $a \in \mathbf{Q}$

*33.2 Find a gcd of 49,349 and 15,555 in $\mathbf{Z}$.

*33.3 Find a gcd of

$$x^{10} - 3x^9 + 3x^8 - 11x^7 + 11x^6 - 11x^5 + 19x^4 - 13x^3 + 8x^2 - 9x + 3$$

and

$$x^6 - 3x^5 + 3x^4 - 9x^3 + 5x^2 - 5x + 2$$

in $\mathbf{Q}[x]$.

*33.4 By referring to Example 33.7 of the text, actually express 23 in the form $\lambda(22{,}471) + \mu(3{,}266)$ for $\lambda, \mu \in \mathbf{Z}$. [*Hint:* From the next to the last line of the computation in Example 33.7, $23 = (138)3 - 391$. From the line before that, $138 = 3{,}266 - (391)8$, so substituting, you get $23 = [3{,}266 - (391)8]3 - 391$, etc. That is, work your way back up to actually find values for λ and μ.]

*33.5 Let us consider $\mathbf{Z}[x]$.

a) Is $\mathbf{Z}[x]$ a UFD? Why?
b) Show that $\{a + xf(x) \mid a \in 2\mathbf{Z},\ f(x) \in \mathbf{Z}[x]\}$ is an ideal in $\mathbf{Z}[x]$.
c) Is $\mathbf{Z}[x]$ a PID? (Consider part (b).)
d) Is $\mathbf{Z}[x]$ a Euclidean domain? Why?

†*33.6 Let D be a Euclidean domain and let ν be a Euclidean valuation on D. Show that if a and b are associates in D, then $\nu(a) = \nu(b)$.

*33.7 Mark each of the following true or false.

— a) Every Euclidean domain is a PID.

— b) Every PID is a Euclidean domain.

— c) Every Euclidean domain is a UFD.

— d) Every UFD is a Euclidean domain.

— e) A gcd of 2 and 3 in $\mathbf{Q}$ is $\frac{1}{2}$.

— f) The Euclidean algorithm gives a constructive method for finding a gcd of two integers.

— g) If ν is a Euclidean valuation on a Euclidean domain D, then $\nu(1) \leq \nu(a)$ for all nonzero $a \in D$.

— h) If ν is a Euclidean valuation on a Euclidean domain D, then $\nu(1) < \nu(a)$ for all nonzero $a \in D$, $a \neq 1$.

— i) If ν is a Euclidean valuation on a Euclidean domain D, then $\nu(1) < \nu(a)$ for all nonzero nonunits $a \in D$.

— j) For any field F, $F[x]$ is a Euclidean domain.

---

*33.8 Does the choice of a particular Euclidean valuation ν on a Euclidean domain D influence the arithmetic structure of D in any way? Explain.

*33.9 Following the idea of Exercise 33.4 and referring to Exercise 33.2, express the positive gcd of 49,349 and 15,555 in $\mathbf{Z}$ in the form $\lambda(49,349) + \mu(15,555)$ for $\lambda, \mu \in \mathbf{Z}$.

*33.10 Let D be a Euclidean domain and let ν be a Euclidean valuation on D. Show that for nonzero $a, b \in D$, one has $\nu(a) < \nu(ab)$ if and only if b is not a unit of D. [*Hint:* Argue from Exercise 33.6 that $\nu(a) < \nu(ab)$ implies b is not a unit of D. Using the Euclidean algorithm, show that $\nu(a) = \nu(ab)$ implies $\langle a \rangle = \langle ab \rangle$. Conclude that if b is not a unit, then $\nu(a) < \nu(ab)$.]

*33.11 Prove or disprove the following statement: If ν is a Euclidean valuation on a Euclidean domain D, then $\{a \in D \mid \nu(a) > \nu(1)\}$ is an ideal of D.

*33.12 Show that every field is a Euclidean domain.

*33.13 Let ν be a Euclidean valuation on a Euclidean domain D.

a) Show that if $s \in \mathbf{Z}$ such that $s + \nu(1) > 0$, then $\eta: D^* \to \mathbf{Z}$ defined by $\eta(a) = \nu(a) + s$ for nonzero $a \in D$ is a Euclidean valuation on D. As usual, D^* is the set of nonzero elements of D.

b) Show that for $r \in \mathbf{Z}^+$, $\lambda: D^* \to \mathbf{Z}$ given by $\lambda(a) = r(\nu(a))$ for nonzero $a \in D$ is a Euclidean valuation on D.

c) Show that there exists a Euclidean valuation μ on D such that $\mu(1) = 1$ and $\mu(a) > 100$ for all nonzero nonunits $a \in D$.

*33.14 Let D be a UFD. An element c in D is a **least common multiple** (abbreviated lcm) **of two elements** a **and** b **in** D if $a \mid c$, $b \mid c$, and if c divides every element of D which is divisible by both a and b. Show that every two nonzero elements a and b of a Euclidean domain D have a lcm in D. [*Hint:* Show that all common multiples, in the obvious sense, of both a and b form an ideal of D.]

*33.15 Use the last statement in Theorem 33.3 to show that two nonzero elements $r, s \in \mathbf{Z}$ generate the group $\langle \mathbf{Z}, + \rangle$ if and only if r and s, viewed as integers in the domain $\mathbf{Z}$, are **relatively prime**, that is, have a gcd of 1.

*33.16 Using the last statement in Theorem 33.3, show that for nonzero $a, b, n \in \mathbf{Z}$, the congruence $ax \equiv b \pmod{n}$ has a solution in $\mathbf{Z}$ if a and n are relatively prime.

*33.17 Generalize Exercise 33.16 by showing that for nonzero $a, b, n \in \mathbf{Z}$, the congruence $ax \equiv b \pmod{n}$ has a solution in $\mathbf{Z}$ if and only if the positive gcd of a and n in $\mathbf{Z}$ divides b. Interpret this result in the ring $\mathbf{Z}_n$.

*33.18 Following the idea of Exercises 33.17 and 33.4, outline a constructive method for finding a solution in $\mathbf{Z}$ of the congruence $ax \equiv b \pmod{n}$ for nonzero $a, b, n \in \mathbf{Z}$, if the congruence does have a solution. Use this method to find a solution of the congruence $22x \equiv 18 \pmod{42}$.

34 | Gaussian Integers and Norms

*34.1 GAUSSIAN INTEGERS

We should give an example of a Euclidean domain different from $\mathbf{Z}$ and $F[x]$.

Definition. A *Gaussian integer* is a complex number $a + bi$, where $a, b \in \mathbf{Z}$. For a Gaussian integer $\alpha = a + bi$, the *norm* $N(\alpha)$ *of* α is $a^2 + b^2$.

We shall let $\mathbf{Z}[i]$ be the set of all Gaussian integers. The following lemma gives some basic properties of the norm function N on $\mathbf{Z}[i]$, and leads to a demonstration that the function ν defined by $\nu(\alpha) = N(\alpha)$ for nonzero $\alpha \in \mathbf{Z}[i]$ is a Euclidean valuation on $\mathbf{Z}[i]$. Note that the Gaussian integers include all the **rational integers**, i.e., all the elements of $\mathbf{Z}$.

Lemma 34.1 *In* $\mathbf{Z}[i]$, *the following properties of the norm function N hold for all $\alpha, \beta \in \mathbf{Z}[i]$:*

1) $N(\alpha) \geq 0$.
2) $N(\alpha) = 0$ *if and only if* $\alpha = 0$.
3) $N(\alpha\beta) = N(\alpha)N(\beta)$.

Proof. If we let $\alpha = a_1 + a_2 i$ and $\beta = b_1 + b_2 i$, these results are all either obvious or straightforward computations. We leave the proof of these properties as an exercise (see Exercise 34.8). ∎

Lemma 34.2 $\mathbf{Z}[i]$ *is an integral domain.*

Proof. It is obvious that $\mathbf{Z}[i]$ is a commutative ring with unity. We show that there are no divisors of zero. Let $\alpha, \beta \in \mathbf{Z}[i]$. Using Lemma 34.1, if $\alpha\beta = 0$, then

$$N(\alpha)N(\beta) = N(\alpha\beta) = N(0) = 0.$$

Thus $\alpha\beta = 0$ implies that $N(\alpha) = 0$ or $N(\beta) = 0$. By Lemma 34.1 again, this implies that either $\alpha = 0$ or $\beta = 0$. Thus $\mathbf{Z}[i]$ has no divisors of zero, so $\mathbf{Z}[i]$ is an integral domain. ∎

Of course, since $\mathbf{Z}[i]$ is a subring of $\mathbf{C}$, where $\mathbf{C}$ is the field of complex numbers, it is really obvious that $\mathbf{Z}[i]$ has no zero divisors. We gave the argument of Lemma 34.2 to illustrate the use of the multiplicative property (3) of the norm function N and to avoid going outside of $\mathbf{Z}[i]$ in our argument.

286

Theorem 34.1 *The function v given by $v(\alpha) = N(\alpha)$ for nonzero $\alpha \in \mathbf{Z}[i]$ is a Euclidean valuation on $\mathbf{Z}[i]$. Thus $\mathbf{Z}[i]$ is a Euclidean domain.*

Proof. Note that for $\beta = b_1 + b_2 i \neq 0$, $N(b_1 + b_2 i) = b_1{}^2 + b_2{}^2$, so $N(\beta) \geq 1$. Then for all $\alpha, \beta \neq 0$ in $\mathbf{Z}[i]$, $N(\alpha) \leq N(\alpha)N(\beta) = N(\alpha\beta)$. This proves condition (2) for a Euclidean valuation.

It remains to prove the division algorithm, condition (1), for N. Let $\alpha, \beta \in \mathbf{Z}[i]$, with $\alpha = a_1 + a_2 i$ and $\beta = b_1 + b_2 i$, where $\beta \neq 0$. We must find σ and ρ in $\mathbf{Z}[i]$ such that $\alpha = \beta\sigma + \rho$, where either $\rho = 0$ or $N(\rho) < N(\beta) = b_1{}^2 + b_2{}^2$. Let us put $\sigma = q_1 + q_2 i$, where q_1 and q_2 are *rational integers* in $\mathbf{Z}$ to be determined. Then ρ will have to have the form

$$\rho = (a_1 + a_2 i) - (b_1 + b_2 i)(q_1 + q_2 i)$$
$$= (a_1 - b_1 q_1 + b_2 q_2) + (a_2 - b_1 q_2 - b_2 q_1)i.$$

We have to try to find *rational integers* q_1 and q_2 such that

$$N(\rho) = (a_1 - b_1 q_1 + b_2 q_2)^2 + (a_2 - b_1 q_2 - b_2 q_1)^2 < b_1{}^2 + b_2{}^2,$$

that is, such that

$$\frac{(a_1 - b_1 q_1 + b_2 q_2)^2}{b_1{}^2 + b_2{}^2} + \frac{(a_2 - b_1 q_2 - b_2 q_1)^2}{b_1{}^2 + b_2{}^2} < 1.$$

Now the student will recall that

$$\frac{(a_1 - b_1 q_1 + b_2 q_2)^2}{b_1{}^2 + b_2{}^2}$$

is exactly the square of the distance d in the Euclidean plane from a point (q_1, q_2) to the line l with equation $a_1 - b_1 X + b_2 Y = 0$. Similarly,

$$\frac{(a_2 - b_1 q_2 - b_2 q_1)^2}{b_1{}^2 + b_2{}^2}$$

is the square of the distance d' from (q_1, q_2) to the line l' with equation $a_2 - b_2 X - b_1 Y = 0$. Note that l is perpendicular to l'. Let P be the point of intersection of these two lines as shown in Fig. 34.1. From this figure, we see that $d^2 + (d')^2$ is the square of the distance from (q_1, q_2) to P.

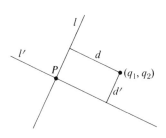

Fig. 34.1

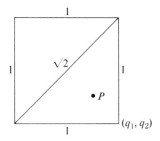

Fig. 34.2

Thus we must show that there is a point (q_1, q_2) *with integral coordinates* and with a distance from P whose square is less than 1. Since P is contained within or on the boundary of some square of unit side such that both coordinates of each vertex are integers, it is clear that if (q_1, q_2) is chosen to be the point with integer coordinates *closest* to P, its distance from P can be at most half the length of a diagonal of the square, i.e., at most $\sqrt{2}/2$ (see Fig. 34.2). Thus the square of this distance from P is at most $\frac{1}{2}$, which is less than 1. ∎

We could have proved the division algorithm for the function N purely algebraically; the algebraic proof is easier, shorter to describe, and more useful in applying the Euclidean algorithm for N (see Exercises 34.4 and 34.12). But we confess a certain fondness for our geometric argument. Besides, it is nice to have variety in the proofs in a text. We leave an algebraic proof to the exercises (see Exercise 34.11).

Example 34.1 We can now apply all our results of Section 33 to $\mathbf{Z}[i]$. In particular, since $N(1) = 1$, the units of $\mathbf{Z}[i]$ are exactly the $\alpha = a_1 + a_2 i$ with $N(\alpha) = a_1{}^2 + a_2{}^2 = 1$. From the fact that a_1 and a_2 are integers, it follows that the only possibilities are $a_1 = \pm 1$ with $a_2 = 0$, or $a_1 = 0$ with $a_2 = \pm 1$. Thus the units of $\mathbf{Z}[i]$ are ± 1 and $\pm i$. One can also use the Euclidean algorithm to compute a gcd of two nonzero elements. We leave such computations to the exercises. Finally, note that while 5 is an irreducible in $\mathbf{Z}$, 5 is no longer an irreducible in $\mathbf{Z}[i]$, for $5 = (1 + 2i)(1 - 2i)$, and neither $1 + 2i$ nor $1 - 2i$ is a unit. ‖

*34.2 MULTIPLICATIVE NORMS

Let us point out again that for an integral domain D, *the arithmetic concepts of irreducibles and units are intrinsic to the integral domain itself*, and are not affected in any way by a valuation or norm which may be defined on the domain. However, as the preceding section and our work thus far in this section shows, a suitably defined valuation or norm may be of help in determining the arithmetic structure of D. This is strikingly illustrated in *algebraic number theory*, where for a domain of *algebraic integers*, one considers many different valuations of the domain, each doing its part in helping to determine the arithmetic structure of the domain. In a domain of algebraic integers, one has essentially one valuation for each irreducible (up to associates), and each such valuation gives information concerning the behavior in the integral domain of the irreducible to which it corresponds. This is an example of the importance of studying properties of elements in an algebraic structure by means of mappings associated with them. We shall be doing this for zeros of polynomials in the following sections.

Let us study integral domains which have a multiplicative norm satisfying the properties of N on $\mathbf{Z}[i]$ given in Lemma 34.1.

Definition. Let D be an integral domain. A ***multiplicative norm N on D*** is a function mapping D into the integers $\mathbf{Z}$ such that the following conditions are satisfied:

1) $N(\alpha) \geq 0$ for all $\alpha \in D$.
2) $N(\alpha) = 0$ if and only if $\alpha = 0$.
3) $N(\alpha\beta) = N(\alpha)N(\beta)$ for all $\alpha, \beta \in D$.

Theorem 34.2 *If D is an integral domain with a multiplicative norm N, then $N(1) = 1$ and $N(u) = 1$ for every unit u in D. If, furthermore, every α such that $N(\alpha) = 1$ is a unit in D, then an element π in D, with $N(\pi) = p$ for a prime $p \in \mathbf{Z}$, is an irreducible of D.*

Proof. Let D be an integral domain with a multiplicative norm N. Then

$$N(1) = N((1)(1)) = N(1)N(1)$$

shows that $N(1) = 1$. Also, if u is a unit in D, then

$$1 = N(1) = N(uu^{-1}) = N(u)N(u^{-1}).$$

Since $N(u)$ is a nonnegative integer, this implies that $N(u) = 1$.

Now suppose that the units of D are *exactly* the elements of norm 1. Let $\pi \in D$ be such that $N(\pi) = p$, where p is a prime in $\mathbf{Z}$. Then if $\pi = \alpha\beta$, we have

$$p = N(\pi) = N(\alpha)N(\beta),$$

so either $N(\alpha) = 1$ or $N(\beta) = 1$. By assumption, this means that either α or β is a unit of D. Thus π is an irreducible of D. $\blacksquare$

Example 34.2 On $\mathbf{Z}[i]$, the function N defined by $N(a + bi) = a^2 + b^2$ gives a multiplicative norm in the sense of our definition. We saw that the function ν given by $\nu(\alpha) = N(\alpha)$ for nonzero $\alpha \in \mathbf{Z}[i]$ is a Euclidean valuation on $\mathbf{Z}[i]$, so the units are precisely the elements α of $\mathbf{Z}[i]$ with $N(\alpha) = N(1) = 1$. Thus the second part of Theorem 34.2 applies in $\mathbf{Z}[i]$. We saw in Example 34.1 that 5 is not an irreducible in $\mathbf{Z}[i]$, for $5 = (1 + 2i)(1 - 2i)$. Since $N(1 + 2i) = N(1 - 2i) = 1^2 + 2^2 = 5$ and 5 is a prime in $\mathbf{Z}$, we see from Theorem 34.2 that $1 + 2i$ and $1 - 2i$ are both irreducibles in $\mathbf{Z}[i]$. $\parallel$

As an application of multiplicative norms, we shall now give an example of an integral domain which is *not* a UFD. We have as yet seen no such example. The following is the standard illustration.

Example 34.3 Let $\mathbf{Z}[\sqrt{-5}] = \{a + ib\sqrt{5} \mid a, b \in \mathbf{Z}\}$. As a subset of the complex numbers closed under addition, subtraction, and multiplication, and containing 0 and 1, $\mathbf{Z}[\sqrt{-5}]$ is an integral domain. Define N on $\mathbf{Z}[\sqrt{-5}]$ by

$$N(a + b\sqrt{-5}) = a^2 + 5b^2.$$

(Here $\sqrt{-5} = i\sqrt{5}$.) Clearly, $N(\alpha) \geq 0$ and $N(\alpha) = 0$ if and only if

$\alpha = a + b\sqrt{-5} = 0$. That $N(\alpha\beta) = N(\alpha)N(\beta)$ is a straightforward computation which we leave to the exercises (see Exercise 34.9). Let us find all candidates for units in $\mathbf{Z}[\sqrt{-5}]$ by finding all elements α in $\mathbf{Z}[\sqrt{-5}]$ with $N(\alpha) = 1$. If $\alpha = a + b\sqrt{-5}$, and $N(\alpha) = 1$, we must have $a^2 + 5b^2 = 1$ for *integers* a and b. This is only possible if $b = 0$ and $a = \pm1$. Hence ±1 are the only candidates for units. Since ±1 are units, they are then precisely the units in $\mathbf{Z}[\sqrt{-5}]$.

Now in $\mathbf{Z}[\sqrt{-5}]$, we have $21 = (3)(7)$ and also

$$21 = (1 + 2\sqrt{-5})(1 - 2\sqrt{-5}).$$

If we can show that $3, 7, 1 + 2\sqrt{-5}$, and $1 - 2\sqrt{-5}$ are all irreducibles in $\mathbf{Z}[\sqrt{-5}]$, we will then know that $\mathbf{Z}[\sqrt{-5}]$ cannot be a UFD, since neither 3 nor 7 is $\pm(1 + 2\sqrt{-5})$.

Suppose that $3 = \alpha\beta$. Then

$$9 = N(3) = N(\alpha)N(\beta)$$

shows that we must have $N(\alpha) = 1, 3$, or 9. If $N(\alpha) = 1$, then α is a unit. If $\alpha = a + b\sqrt{-5}$, then $N(\alpha) = a^2 + 5b^2$, and for no choice of integers a and b is $N(\alpha) = 3$. If $N(\alpha) = 9$, then $N(\beta) = 1$, so β is a unit. Thus from $3 = \alpha\beta$, we can conclude that either α or β is a unit. Therefore, 3 is an irreducible in $\mathbf{Z}[\sqrt{-5}]$. A similar argument shows that 7 is also an irreducible in $\mathbf{Z}[\sqrt{-5}]$.

If $1 + 2\sqrt{-5} = \gamma\delta$, we have

$$21 = N(1 + 2\sqrt{-5}) = N(\gamma)N(\delta),$$

so $N(\gamma) = 1, 3, 7$, or 21. We have seen that there is no element of $\mathbf{Z}[\sqrt{-5}]$ of norm 3 or 7. Thus either $N(\gamma) = 1$, and γ is a unit, or $N(\gamma) = 21$, so $N(\delta) = 1$, and δ is a unit. Therefore, $1 + 2\sqrt{-5}$ is an irreducible in $\mathbf{Z}[\sqrt{-5}]$. A parallel argument shows that $1 - 2\sqrt{-5}$ is also an irreducible in $\mathbf{Z}[\sqrt{-5}]$.

In summary, we have shown that

$$\mathbf{Z}[\sqrt{-5}] = \{a + ib\sqrt{5} \mid a, b \in \mathbf{Z}\}$$

is an integral domain but not a UFD. In particular, there are two *different* factorizations

$$21 = 3 \cdot 7 = (1 + 2\sqrt{-5})(1 - 2\sqrt{-5})$$

of 21 into irreducibles. These irreducibles cannot be primes, for the property of a prime enables one to prove uniqueness of factorization (see the proof of Theorem 32.2). ‖

EXERCISES

*34.1 Factor each of the following Gaussian integers into a product of irreducibles in $\mathbf{Z}[i]$. [*Hint:* Since an irreducible factor of $\alpha \in \mathbf{Z}[i]$ must have norm >1 and dividing $N(\alpha)$, there are only a finite number of Gaussian integers $a + bi$ to consider as possible irreducible factors of a given α. Divide α by each of them in **C**, and see for which ones the quotient is again in $\mathbf{Z}[i]$.]

a) 5   b) 7   c) $4 + 3i$ d) $6 - 7i$

*34.2 Show that 6 does not factor uniquely (up to associates) into irreducibles in $\mathbf{Z}[\sqrt{-5}]$. Exhibit two different factorizations.

*34.3 Consider $\alpha = 7 + 2i$ and $\beta = 3 - 4i$ in $\mathbf{Z}[i]$. Find σ and ρ in $\mathbf{Z}[i]$ such that

$$\alpha = \beta\sigma + \rho \qquad \text{with} \qquad N(\rho) < N(\beta).$$

[*Hint:* Use the construction in the hint of Exercise 34.11.]

*34.4 Use a Euclidean algorithm in $\mathbf{Z}[i]$ to find a gcd of $8 + 6i$ and $5 - 15i$ in $\mathbf{Z}[i]$. [*Hint:* Use the construction in the hint of Exercise 34.11.]

†*34.5 Let D be an integral domain with a multiplicative norm N such that $N(\alpha) = 1$ for $\alpha \in D$ if and only if α is a unit of D. Let π be such that $N(\pi)$ is minimal among all $N(\beta) > 1$ for $\beta \in D$. Show that π is an irreducible of D.

*34.6 Mark each of the following true or false.

___ a) $\mathbf{Z}[i]$ is a PID.

___ b) $\mathbf{Z}[i]$ is a Euclidean domain.

___ c) Every integer in **Z** is a Gaussian integer.

___ d) Every complex number is a Gaussian integer.

___ e) A Euclidean algorithm holds in $\mathbf{Z}[i]$.

___ f) A multiplicative norm on an integral domain is sometimes an aid in finding irreducibles of the domain.

___ g) If N is a multiplicative norm on an integral domain D, then $N(u) = 1$ for every unit u of D.

___ h) If F is a field, then the function N defined by $N(f(x)) = $ (degree of $f(x)$) is a multiplicative norm on $F[x]$.

___ i) If F is a field, then the function defined by $N(f(x)) = 2^{(\text{degree of } f(x))}$ for $f(x) \neq 0$ and $N(0) = 0$ is a multiplicative norm on $F[x]$ according to our definition.

___ j) $\mathbf{Z}[\sqrt{-5}]$ is an integral domain but not a UFD.

---

*34.7 Show that $1 + i$ is an irreducible of $\mathbf{Z}[i]$. [*Hint:* Apply Theorem 34.2.] (For a description of all Gaussian primes, see Pollard [33].)

*34.8 Prove Lemma 34.1.

*34.9 Prove that N of Example 34.3 is multiplicative, i.e., that $N(\alpha\beta) = N(\alpha)N(\beta)$ for $\alpha, \beta \in \mathbf{Z}[\sqrt{-5}]$.

*34.10 Let D be an integral domain with a multiplicative norm N such that $N(\alpha) = 1$ for $\alpha \in D$ if and only if α is a unit of D. Show that every nonzero non-unit of D has a factorization into irreducibles in D.

*34.11 Prove algebraically that the division algorithm holds in $\mathbf{Z}[i]$ for ν given by $\nu(\alpha) = N(\alpha)$ for nonzero $\alpha \in \mathbf{Z}[i]$. [*Hint:* For α and β in $\mathbf{Z}[i]$ with $\beta \neq 0$, $\alpha/\beta = r + si$ in **C** for $r, s \in \mathbf{Q}$. Let q_1 and q_2 be rational integers in **Z** as close as possible to the rational numbers r and s respectively. Show that for $\sigma = q_1 + q_2 i$ and

$\rho = \alpha - \beta\sigma$, one has $N(\rho) < N(\beta)$, by showing that

$$N(\rho)/N(\beta) = |(\alpha/\beta) - \sigma|^2 < 1.$$

Here $|\ |$ is the usual absolute value for elements of **C**.]

*34.12 Use a Euclidean algorithm in **Z**[i] to find a gcd of $16 + 7i$ and $10 - 5i$ in **Z**[i]. [*Hint:* Use the construction in the hint of Exercise 34.11.]

*34.13 Let $\langle\alpha\rangle$ be a principal ideal in **Z**[i].

a) Show that **Z**[i]/$\langle\alpha\rangle$ is a finite ring. [*Hint:* Use the division algorithm.]
b) Show that if π is an irreducible of **Z**[i], then **Z**[i]/$\langle\pi\rangle$ is a field.
c) Referring to (b), find the order and characteristic of each of the following fields.

  i) **Z**[i]/$\langle 3\rangle$ ii) **Z**[i]/$\langle 1 + i\rangle$ iii) **Z**[i]/$\langle 1 + 2i\rangle$

*34.14 Let $n \in$ **Z**$^+$ be square free, i.e., not divisible by the square of any prime integer. Let **Z**[$\sqrt{-n}$] $= \{a + ib\sqrt{n} \mid a, b \in$ **Z**$\}$.

a) Show that the norm N, defined by $N(\alpha) = a^2 + nb^2$ for $\alpha = a + ib\sqrt{n}$, is a multiplicative norm on **Z**[$\sqrt{-n}$].
b) Show that $N(\alpha) = 1$ for $\alpha \in$ **Z**[$\sqrt{-n}$] if and only if α is a unit of **Z**[$\sqrt{-n}$].
c) Show that every nonzero $\alpha \in$ **Z**[$\sqrt{-n}$] which is not a unit has a factorization into irreducibles in **Z**[$\sqrt{-n}$]. [*Hint:* Use (b).]

*34.15 Repeat Exercise 34.14 for **Z**[$\sqrt{n}$] $= \{a + b\sqrt{n} \mid a, b \in$ **Z**$\}$, with N defined by $N(\alpha) = |a^2 - nb^2|$ for $\alpha = a + b\sqrt{n}$ in **Z**[$\sqrt{n}$].

*34.16 Show by a construction analogous to that given in the hint of Exercise 34.11 that the division algorithm holds in the integral domains **Z**[$\sqrt{-2}$], **Z**[$\sqrt{2}$], and **Z**[$\sqrt{3}$] for $\nu(\alpha) = N(\alpha)$ for nonzero α in one of these domains (see Exercises 34.14 and 34.15). (Thus these domains are Euclidean. See Hardy and Wright [27] for a discussion of which domains **Z**[$\sqrt{n}$] and **Z**[$\sqrt{-n}$] are Euclidean.)

35 | Introduction to Extension Fields

35.1 OUR BASIC GOAL ACHIEVED

We are now in a position to achieve our *Basic Goal* which, loosely stated, is to show that every nonconstant polynomial has a zero. This will be stated more precisely and proved in Theorem 35.1. We first introduce some new terminology for some old ideas.

Definition. A field E is an ***extension field of a field*** F if $F \leq E$.

Thus $\mathbf{R}$ is an extension field of $\mathbf{Q}$, and $\mathbf{C}$ is an extension field of both $\mathbf{R}$ and $\mathbf{Q}$. As in the study of groups, it will often be convenient to use lattice diagrams to picture extension fields, the larger field being on top. We illustrate this in Fig. 35.1. A configuration where there is just one single column of fields, as at the left in Fig. 35.1, is often referred to, without any precise definition, as a "**tower of fields**". We shall use this term freely.

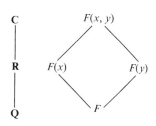

Fig. 35.1

Now for our *Basic Goal!* This great and important result follows easily and elegantly from the techniques we now have at our disposal. Find a nice quiet place where you can have a quarter hour to read it, digest it, and marvel to your heart's content without interruption. We write out the proof for you in our usual hideous detail. (A graduate text would give it about three lines at this point.)

Theorem 35.1 (Kronecker) (Basic Goal). *Let F be a field and let $f(x)$ be a nonconstant polynomial in $F[x]$. Then there exists an extension field E of F and an $\alpha \in E$ such that $f(\alpha) = 0$.*

Proof. By Theorem 31.8, $f(x)$ has a factorization in $F[x]$ into polynomials which are irreducible over F. Let $p(x)$ be an irreducible polynomial in such a factorization. It is clearly sufficient to find an extension field E of F containing an element α such that $p(\alpha) = 0$.

By Theorem 31.6, $\langle p(x) \rangle$ is a maximal ideal in $F[x]$, so $F[x]/\langle p(x) \rangle$ is a field. We claim that F can be identified with a subfield of $F[x]/\langle p(x) \rangle$ in a natural way by use of the map $\psi: F \to F[x]/\langle p(x) \rangle$ given by

$$a\psi = a + \langle p(x) \rangle$$

293

for $a \in F$. This map is one to one, for if $a\psi = b\psi$, that is, if $a + \langle p(x) \rangle = b + \langle p(x) \rangle$ for some $a, b \in F$, then $(a - b) \in \langle p(x) \rangle$, so $a - b$ must be a multiple of the polynomial $p(x)$, which is of degree ≥ 1. Now $a, b \in F$ implies that $a - b$ is in F. Thus we must have $a - b = 0$, so $a = b$. We defined addition and multiplication in $F[x]/\langle p(x) \rangle$ by choosing any representatives, so we may choose $a \in (a + \langle p(x) \rangle)$. It is then clear that this map ψ is an isomorphism of F into $F[x]/\langle p(x) \rangle$. We identify F with $\{a + \langle p(x) \rangle \mid a \in F\}$ by means of this isomorphism. Thus we shall view $E = F[x]/\langle p(x) \rangle$ as an extension field of F. We have now manufactured our desired extension field E of F. It remains for us to show that E contains a zero of $p(x)$.

Let us set

$$\alpha = x + \langle p(x) \rangle,$$

so $\alpha \in E$. Consider the evaluation homomorphism $\phi_\alpha : F[x] \to E$, given by Theorem 30.2. If $p(x) = a_0 + a_1 x + \cdots + a_n x^n$, where $a_i \in F$, then we have

$$p(x)\phi_\alpha = a_0 + a_1(x + \langle p(x) \rangle) + \cdots + a_n(x + \langle p(x) \rangle)^n$$

in $E = F[x]/\langle p(x) \rangle$. *But we can compute in $F[x]/\langle p(x) \rangle$ by choosing representatives, and x is a representative of the coset* $\alpha = x + \langle p(x) \rangle$. Therefore,

$$p(\alpha) = (a_0 + a_1 x + \cdots + a_n x^n) + \langle p(x) \rangle$$
$$= p(x) + \langle p(x) \rangle = \langle p(x) \rangle = 0$$

in $F[x]/\langle p(x) \rangle$. We have found an element α in $E = F[x]/\langle p(x) \rangle$ such that $p(\alpha) = 0$, and therefore $f(\alpha) = 0$. ∎

We illustrate the construction involved in the proof of Theorem 35.1 by two examples.

Example 35.1 Let $F = \mathbf{R}$, and let $f(x) = x^2 + 1$, which is well known to have no zeros in $\mathbf{R}$, and thus is irreducible over $\mathbf{R}$ by Theorem 31.2. Then $\langle x^2 + 1 \rangle$ is a maximal ideal in $\mathbf{R}[x]$, so $\mathbf{R}[x]/\langle x^2 + 1 \rangle$ is a field. Identifying $r \in \mathbf{R}$ with $r + \langle x^2 + 1 \rangle$ in $\mathbf{R}[x]/\langle x^2 + 1 \rangle$, we can view $\mathbf{R}$ as a subfield of $E = \mathbf{R}[x]/\langle x^2 + 1 \rangle$. Let

$$\alpha = x + \langle x^2 + 1 \rangle.$$

Computing in $\mathbf{R}[x]/\langle x^2 + 1 \rangle$, we find

$$\alpha^2 + 1 = (x + \langle x^2 + 1 \rangle)^2 + (1 + \langle x^2 + 1 \rangle)$$
$$= (x^2 + 1) + \langle x^2 + 1 \rangle = 0.$$

Thus α is a zero of $x^2 + 1$. We shall identify $\mathbf{R}[x]/\langle x^2 + 1 \rangle$ with $\mathbf{C}$ at the close of this section. ‖

Example 35.2 Let $F = \mathbf{Q}$, and let $f(x) = x^4 - 5x^2 + 6$. This time $f(x)$ factors in $\mathbf{Q}[x]$ into $(x^2 - 2)(x^2 - 3)$, both factors being irreducible over $\mathbf{Q}$, as we have seen. We can start with $x^2 - 2$ and construct an extension field

E of $\mathbf{Q}$ containing α such that $\alpha^2 - 2 = 0$, or we can construct an extension field K of $\mathbf{Q}$ containing an element β such that $\beta^2 - 3 = 0$. The construction in either case is just as in Example 35.1. ‖

35.2 ALGEBRAIC AND TRANSCENDENTAL ELEMENTS

As we said before, the rest of the unstarred portion of this text is devoted to the study of zeros of polynomials. We commence this study by putting an element of an extension field E of a field F into one of two categories.

Definition. An element α of an extension field E of a field F is *algebraic over F* if $f(\alpha) = 0$ for some nonzero $f(x) \in F[x]$. If α is not algebraic over F, then α is *transcendental over F*.

Example 35.3 $\mathbf{C}$ is an extension field of $\mathbf{Q}$. Since $\sqrt{2}$ is a zero of $x^2 - 2$, $\sqrt{2}$ is an algebraic element over $\mathbf{Q}$. Also, i is an algebraic element over $\mathbf{Q}$, being a zero of $x^2 + 1$. ‖

Example 35.4 It is well known (but not easy to prove) that the real numbers π and e are transcendental over $\mathbf{Q}$. Here e is the base for the natural logarithms. ‖

Just as we do not speak simply of an *irreducible polynomial*, but rather of an *irreducible polynomial over F*, similarly we don't speak simply of an *algebraic element*, but rather of an *element algebraic over F*. The following illustration shows the reason for this.

Example 35.5 The real number π is transcendental over $\mathbf{Q}$, as we stated in Example 35.4. However, π is algebraic over $\mathbf{R}$, for it is a zero of $(x - \pi) \in \mathbf{R}[x]$. ‖

Example 35.6 It is easy to see that the real number $\sqrt{1 + \sqrt{3}}$ is algebraic over $\mathbf{Q}$. For if $\alpha = \sqrt{1 + \sqrt{3}}$, then $\alpha^2 = 1 + \sqrt{3}$, so $\alpha^2 - 1 = \sqrt{3}$ and $(\alpha^2 - 1)^2 = 3$. Therefore $\alpha^4 - 2\alpha^2 - 2 = 0$, so α is a zero of $x^4 - 2x^2 - 2$, which is in $\mathbf{Q}[x]$. ‖

To connect these ideas with those of number theory, we give the following definition.

Definition. An element of $\mathbf{C}$ which is algebraic over $\mathbf{Q}$ is an *algebraic number*. A *transcendental number* is an element of $\mathbf{C}$ which is transcendental over $\mathbf{Q}$.

There is an extensive and elegant theory of algebraic numbers.

The next theorem gives a useful characterization of algebraic and transcendental elements over F in an extension field E of F. It also illustrates the importance of our evaluation homomorphisms ϕ_α. *Note that once more we are phrasing our concepts in terms of mappings.*

Theorem 35.2 *Let E be an extension field of a field F and let $\alpha \in E$. Let $\phi_\alpha : F[x] \rightarrow E$ be the evaluation homomorphism of $F[x]$ into E such that $a\phi_\alpha = a$ for $a \in F$ and $x\phi_\alpha = \alpha$. Then α is transcendental over F if and only if ϕ_α is an isomorphism mapping $F[x]$ into E, that is, if and only if ϕ_α is a one-to-one map.*

Proof. Now α is transcendental over F if and only if $f(\alpha) \neq 0$ for all nonconstant $f(x) \in F[x]$, which is true (by definition) if and only if $f(x)\phi_\alpha \neq 0$ for all nonconstant $f(x) \in F[x]$, which is true if and only if the kernel of ϕ_α is $\{0\}$, that is, if and only if ϕ_α is an isomorphism mapping $F[x]$ into E. ∎

35.3 THE IRREDUCIBLE POLYNOMIAL FOR α OVER F

Consider the extension field **R** of **Q**. We know that $\sqrt{2}$ is algebraic over **Q**, being a zero of $x^2 - 2$. Of course, $\sqrt{2}$ is also a zero of $x^3 - 2x$ and of $x^4 - 3x^2 + 2 = (x^2 - 2)(x^2 - 1)$. All these other polynomials having $\sqrt{2}$ as a zero were multiples of $x^2 - 2$. The next theorem shows that this is an illustration of a general situation. This theorem plays a central role in our later work.

Theorem 35.3 *Let E be an extension field of F, and let $\alpha \in E$, where $\alpha \neq 0$ and α is algebraic over F. Then there is an irreducible polynomial $p(x) \in F[x]$ such that $p(\alpha) = 0$. This irreducible polynomial $p(x)$ is uniquely determined up to a constant factor in F and is a polynomial of minimal degree ≥ 1 in $F[x]$ having α as a zero. If $f(\alpha) = 0$ for $f(x) \in F[x]$, with $f(x) \neq 0$, then $p(x)$ divides $f(x)$.*

Proof. Let ϕ_α be the evaluation homomorphism of $F[x]$ into E, given by Theorem 30.2. The kernel of ϕ_α is an ideal and by Theorem 31.5, it must be a principal ideal generated by some element $p(x) \in F[x]$. Clearly, $\langle p(x) \rangle$ consists precisely of those elements of $F[x]$ having α as a zero. Thus, if $f(\alpha) = 0$ for $f(x) \neq 0$, then $f(x) \in \langle p(x) \rangle$, so $p(x)$ divides $f(x)$. Clearly, $p(x)$ is a polynomial of minimal degree ≥ 1 having α as a zero, and any other such polynomial of the same degree as $p(x)$ must be of the form $(a)p(x)$ for some $a \in F$.

It only remains for us to show that $p(x)$ is irreducible. If $p(x) = r(x)s(x)$ were a factorization of $p(x)$ into polynomials of lower degree, then $p(\alpha) = 0$ would imply that $r(\alpha)s(\alpha) = 0$, so either $r(\alpha) = 0$ or $s(\alpha) = 0$, since E is a field. This would contradict the fact that $p(x)$ is of minimal degree ≥ 1 such that $p(\alpha) = 0$. Thus $p(x)$ is irreducible. ∎

By multiplying by a suitable constant in F, we can assume that the coefficient of the highest power of x appearing in $p(x)$ of Theorem 35.3 is 1. Such a polynomial having the coefficient of the highest power of x appearing equal to 1 is a **monic polynomial**.

Definition. Let E be an extension field of a field F, and let $\alpha \in E$ be algebraic over F. The unique monic polynomial $p(x)$ of Theorem 35.3 is the *irreducible polynomial for α over F* and will be denoted by "irr(α, F)". The degree of irr(α, F) is the *degree of α over F*, denoted by "deg(α, F)".

Example 35.7 Clearly, irr$(\sqrt{2}, \mathbf{Q}) = x^2 - 2$. Referring to Example 35.6, we see that for $\alpha = \sqrt{1 + \sqrt{3}}$ in $\mathbf{R}$, α is a zero of $x^4 - 2x^2 - 2$ which is in $\mathbf{Q}[x]$. Since $x^4 - 2x^2 - 2$ is irreducible over $\mathbf{Q}$ (by Eisenstein with $p = 2$, or by application of the technique of Example 31.6), we see that

$$\mathrm{irr}\left(\sqrt{1 + \sqrt{3}}, \mathbf{Q}\right) = x^4 - 2x^2 - 2.$$

Thus $\sqrt{1 + \sqrt{3}}$ is algebraic of degree 4 over $\mathbf{Q}$. ‖

Just as we must speak of an element α as *algebraic over F* rather than simply as *algebraic*, we must speak of the *degree of α over F* rather than the *degree of α*. To take a trivial illustration, $\sqrt{2} \in \mathbf{R}$ is algebraic of degree 2 over $\mathbf{Q}$ but algebraic of degree 1 over $\mathbf{R}$, for irr$(\sqrt{2}, \mathbf{R}) = x - \sqrt{2}$.

We hope that the student is impressed with the beauty and elegance of this theory. He should appreciate that its easy development here is due to the machinery of homomorphisms and ideal theory which we now have at our disposal. Note especially our constant use of the evaluation homomorphisms ϕ_α.

35.4 SIMPLE EXTENSIONS

Let E be an extension field of a field F, and let $\alpha \in E$. Let ϕ_α be the evaluation homomorphism of $F[x]$ into E with $a\phi_\alpha = a$ for $a \in F$ and $x\phi_\alpha = \alpha$, as in Theorem 30.2. We consider two cases.

CASE I. *Suppose α is algebraic over F.* Then as in Theorem 35.3, the kernel of ϕ_α is $\langle \mathrm{irr}(\alpha, F) \rangle$ and by Theorem 31.6, $\langle \mathrm{irr}(\alpha, F) \rangle$ is a maximal ideal of $F[x]$. Therefore, $F[x]/\langle \mathrm{irr}(\alpha, F) \rangle$ is a field and is isomorphic to the image $(F[x])\phi_\alpha$ in E. This subfield $(F[x])\phi_\alpha$ of E is clearly the smallest subfield of E containing F and α. We shall denote this field by "$F(\alpha)$".

CASE II. *Suppose α is transcendental over F.* Then by Theorem 35.2, ϕ_α is an isomorphism mapping $F[x]$ into E. Thus in this case $(F[x])\phi_\alpha$ is *not* a field, but an integral domain which we shall denote by "$F[\alpha]$". By Corollary 1 of Theorem 26.2, E contains a field of quotients of $F[\alpha]$, which is clearly the smallest subfield of E containing F and α. As in Case I, we denote this field by "$F(\alpha)$".

Example 35.8 Since π is transcendental over $\mathbf{Q}$, the field $\mathbf{Q}(\pi)$ is isomorphic to the field $\mathbf{Q}(x)$ of rational functions over $\mathbf{Q}$ in the indeterminate x. Thus from a structural viewpoint, an element which is transcendental over a field F behaves as though it were an indeterminate over F. ‖

Definition. An extension field E of a field F is a *simple extension of F* if $E = F(\alpha)$ for some $\alpha \in E$.

Many important results appear throughout this section. You see, we have now developed so much machinery that results are starting to pour out of our efficient plant at an alarming rate. The next theorem gives us insight into the nature of the field $F(\alpha)$ in the case where α is algebraic over F.

Theorem 35.4 *Let E be a simple extension $F(\alpha)$ of a field F, and let α be algebraic over F. Let the degree of $\text{irr}(\alpha, F)$ be $n \geq 1$. Then every element β of $E = F(\alpha)$ can be uniquely expressed in the form*

$$\beta = b_0 + b_1\alpha + \cdots + b_{n-1}\alpha^{n-1},$$

where the b_i are in F.

Proof. For the usual evaluation homomorphism ϕ_α, every element of

$$F(\alpha) = (F[x])\phi_\alpha$$

is of the form $(f(x))\phi_\alpha = f(\alpha)$, a formal polynomial in α with coefficients in F. Let

$$\text{irr}(\alpha, F) = p(x) = x^n + a_{n-1}x^{n-1} + \cdots + a_0.$$

Then $p(\alpha) = 0$, so

$$\alpha^n = -a_{n-1}\alpha^{n-1} - \cdots - a_0.$$

This equation in $F(\alpha)$ can be used to express every monomial α^m for $m \geq n$ in terms of powers of α which are less than n. For example,

$$\alpha^{n+1} = \alpha\alpha^n = -a_{n-1}\alpha^n - a_{n-2}\alpha^{n-1} - \cdots - a_0\alpha$$
$$= -a_{n-1}(-a_{n-1}\alpha^{n-1} - \cdots - a_0) - a_{n-2}\alpha^{n-1} - \cdots - a_0\alpha.$$

Thus, if $\beta \in F(\alpha)$, β can be expressed in the required form

$$\beta = b_0 + b_1\alpha + \cdots + b_{n-1}\alpha^{n-1}.$$

For uniqueness, if

$$b_0 + b_1\alpha + \cdots + b_{n-1}\alpha^{n-1} = b_0' + b_1'\alpha + \cdots + b_{n-1}'\alpha^{n-1}$$

for $b_i' \in F$, then

$$(b_0 - b_0') + (b_1 - b_1')x + \cdots + (b_{n-1} - b_{n-1}')x^{n-1} = g(x)$$

is in $F[x]$ and $g(\alpha) = 0$. Also, the degree of $g(x)$ is less than the degree of $\text{irr}(\alpha, F)$. Since $\text{irr}(\alpha, F)$ is a nonzero polyncmial of minimal degree in $F[x]$ having α as a zero, we must have $g(x) = 0$. Therefore, $b_i - b_i' = 0$, so

$$b_i = b_i',$$

and the uniqueness of the b_i is established. ∎

+	0	1	α	$1 + \alpha$
0	0	1	α	$1 + \alpha$
1	1	0	$1 + \alpha$	α
α	α	$1 + \alpha$	0	1
$1 + \alpha$	$1 + \alpha$	α	1	0

$\cdot$	0	1	α	$1 + \alpha$
0	0	0	0	0
1	0	1	α	$1 + \alpha$
α	0	α	$1 + \alpha$	1
$1 + \alpha$	0	$1 + \alpha$	1	α

Fig. 35.2

We give an impressive example illustrating Theorem 35.4.

Example 35.9 The polynomial $p(x) = x^2 + x + 1$ in $\mathbf{Z}_2[x]$ is irreducible over $\mathbf{Z}_2$ by Theorem 31.2, since neither element 0 nor element 1 of $\mathbf{Z}_2$ is a zero of $p(x)$. By Theorem 35.1, we know that there is an extension field E of $\mathbf{Z}_2[x]$ containing a zero α of $x^2 + x + 1$. By Theorem 35.4, $\mathbf{Z}_2(\alpha)$ has as elements $0 + 0\alpha$, $1 + 0\alpha$, $0 + 1\alpha$, and $1 + 1\alpha$, that is, 0, 1, α, and $1 + \alpha$. *This gives us a new finite field of four elements!* The addition and multiplication tables for this field are shown in Fig. 35.2. For example, to compute $(1 + \alpha)(1 + \alpha)$ in $\mathbf{Z}_2(\alpha)$, one observes that since $p(\alpha) = \alpha^2 + \alpha + 1 = 0$, then,

$$\alpha^2 = -\alpha - 1 = \alpha + 1.$$

Therefore,

$$(1 + \alpha)(1 + \alpha) = 1 + \alpha + \alpha + \alpha^2 = 1 + \alpha^2 = 1 + \alpha + 1 = \alpha. \; \|$$

Finally, we can use Theorem 35.4 to fulfill our promise of Example 35.1 and show that $\mathbf{R}[x]/\langle x^2 + 1 \rangle$ is isomorphic to the field $\mathbf{C}$ of complex numbers. We saw in Example 35.1 that we can view $\mathbf{R}[x]/\langle x^2 + 1 \rangle$ as an extension field of $\mathbf{R}$. Let

$$\alpha = x + \langle x^2 + 1 \rangle.$$

Then $\mathbf{R}(\alpha) = \mathbf{R}[x]/\langle x^2 + 1 \rangle$ and consists of all elements of the form $a + b\alpha$ for $a, b \in \mathbf{R}$, by Theorem 35.4. But since $\alpha^2 + 1 = 0$, we see that α plays the role of $i \in \mathbf{C}$, and $a + b\alpha$ plays the role of $(a + bi) \in \mathbf{C}$. Thus $\mathbf{R}(\alpha) \simeq \mathbf{C}$. *This is the elegant algebraic way to construct $\mathbf{C}$ from $\mathbf{R}$.*

EXERCISES

35.1 For each of the given numbers $\alpha \in \mathbf{C}$, show that α is algebraic over $\mathbf{Q}$ by finding $f(x) \in \mathbf{Q}[x]$ such that $f(\alpha) = 0$.

a) $1 + \sqrt{2}$ 　　　　　　　　　　b) $\sqrt{2} + \sqrt{3}$

c) $1 + i$ 　　　　　　　　　　　　d) $\sqrt{1 + \sqrt[3]{2}}$

e) $\sqrt{\sqrt[3]{2} - i}$

35.2 For each of the given algebraic numbers $\alpha \in \mathbf{C}$, find irr$(\alpha, \mathbf{Q})$ and deg$(\alpha, \mathbf{Q})$. Be prepared to prove that your polynomials are irreducible over $\mathbf{Q}$ if challenged to do so.

a) $\sqrt{3 - \sqrt{6}}$ 　　　　b) $\sqrt{(\frac{1}{3}) + \sqrt{7}}$ 　　　　c) $\sqrt{2} + i$

35.3 Classify each of the given $\alpha \in \mathbf{C}$ as algebraic or transcendental over the given field F. If α is algebraic over F, find deg(α, F).

a) $\alpha = i,\ F = \mathbf{Q}$ 　　　　　　　　b) $\alpha = 1 + i,\ F = \mathbf{R}$

c) $\alpha = \sqrt{\pi},\ F = \mathbf{Q}$ 　　　　　　　d) $\alpha = \sqrt{\pi},\ F = \mathbf{R}$

e) $\alpha = \sqrt{\pi},\ F = \mathbf{Q}(\pi)$ 　　　　　f) $\alpha = \sqrt{\pi} + 1,\ F = \mathbf{Q}(\pi^2)$

g) $\alpha = \pi^2,\ F = \mathbf{Q}$ 　　　　　　　h) $\alpha = \pi^2,\ F = \mathbf{Q}(\pi)$

i) $\alpha = \pi^2,\ F = \mathbf{Q}(\pi^3)$ 　　　　　j) $\alpha = \sqrt{2} + \sqrt[3]{\pi},\ F = \mathbf{Q}(\pi)$

35.4 Refer to Example 35.9 of the text. The polynomial $x^2 + x + 1$ has a zero α in $\mathbf{Z}_2(\alpha)$, and thus must factor into a product of linear factors in $(\mathbf{Z}_2(\alpha))[x]$. Find this factorization. [*Hint:* Divide $x^2 + x + 1$ by $x - \alpha$ by long division, using the fact that $\alpha^2 = \alpha + 1$.]

35.5 Let E be an extension field of a field F and let $\alpha \in E$ be algebraic over F. The polynomial irr(α, F) is sometimes referred to as the "**minimal polynomial for α over** F." Why is this designation appropriate?

†**35.6** Let E be an extension field of a finite field F, where F has q elements. Let $\alpha \in E$ be algebraic over F of degree n. Prove that $F(\alpha)$ has q^n elements.

35.7 a) Show that the polynomial $x^2 + 1$ is irreducible in $\mathbf{Z}_3[x]$.

b) Let α be a zero of $x^2 + 1$ in an extension field of $\mathbf{Z}_3$. As in Example 35.9, give the multiplication and addition tables for the nine elements of $\mathbf{Z}_3(\alpha)$, written in the order $0, 1, 2, \alpha, 2\alpha, 1 + \alpha, 1 + 2\alpha, 2 + \alpha$, and $2 + 2\alpha$.

35.8 Mark each of the following true or false.

___ a) The number π is transcendental over $\mathbf{Q}$.

___ b) $\mathbf{C}$ is a simple extension of $\mathbf{R}$.

___ c) Every element of a field F is algebraic over F.

___ d) $\mathbf{R}$ is an extension field of $\mathbf{Q}$.

___ e) $\mathbf{Q}$ is an extension field of $\mathbf{Z}_2$.

___ f) Let $\alpha \in \mathbf{C}$ be algebraic over $\mathbf{Q}$ of degree n. If $f(\alpha) = 0$ for nonzero $f(x) \in \mathbf{Q}[x]$, then (degree $f(x)$) $\geq n$.

___ g) Let $\alpha \in \mathbf{C}$ be algebraic over $\mathbf{Q}$ of degree n. If $f(\alpha) = 0$ for nonzero $f(x) \in \mathbf{R}[x]$, then (degree $f(x)$) $\geq n$.

___ h) Every nonconstant polynomial in $F[x]$ has a zero in some extension field of F.

___ i) Every nonconstant polynomial in $F[x]$ has a zero in every extension field of F.

___ j) If x is an indeterminate, $\mathbf{Q}[\pi] \simeq \mathbf{Q}[x]$.

35.9 a) Show that there exists an irreducible polynomial of degree 3 in $\mathbf{Z}_3[x]$.

b) Show from (a) that there exists a finite field of 27 elements. [*Hint:* Use Exercise 35.6.]

35.10 Consider the prime field $\mathbf{Z}_p$ of characteristic $p \neq 0$.

a) Show that, for $p \neq 2$, not every element in $\mathbf{Z}_p$ is a square of an element of $\mathbf{Z}_p$. [*Hint:* $1^2 = (p - 1)^2 = 1$ in $\mathbf{Z}_p$. Deduce the desired conclusion *by counting*.]

b) Using part (a), show that there exist finite fields of p^2 elements for every prime p in $\mathbf{Z}^+$.

35.11 We have stated without proof that π and e are transcendental over $\mathbf{Q}$.

a) Find a subfield F of $\mathbf{R}$ such that π is algebraic of degree 3 over F.

b) Find a subfield E of $\mathbf{R}$ such that $\alpha = e + \pi$ is algebraic of degree 5 over E.

35.12 Let E be an extension field of a field F and let $\alpha \in E$ be transcendental over F. Show that every element of $F(\alpha)$ which is not in F is also transcendental over F.

35.13 a) Show that $x^3 + x^2 + 1$ is irreducible over $\mathbf{Z}_2$.

b) Let α be a zero of $x^3 + x^2 + 1$ in an extension field of $\mathbf{Z}_2$. Show that $x^3 + x^2 + 1$ factors into three linear factors in $(\mathbf{Z}_2(\alpha))[x]$ by actually finding this factorization. [*Hint:* Every element of $\mathbf{Z}_2(\alpha)$ is of the form

$$a_0 + a_1\alpha + a_2\alpha^2 \qquad \text{for} \quad a_i = 0, 1.$$

Divide $x^3 + x^2 + 1$ by $x - \alpha$ by long division. Show that the quotient also has a zero in $\mathbf{Z}_2(\alpha)$ by simply trying the 8 possible elements. Then complete the factorization.]

35.14 Show that $\{a + b\sqrt[3]{2} + c(\sqrt[3]{2})^2 \mid a, b, c \in \mathbf{Q}\}$ is a subfield of $\mathbf{R}$ by using the ideas of this section, rather than by a formal verification of the field axioms. [*Hint:* Use Theorem 35.4.]

35.15 Let E be an extension field of $\mathbf{Z}_2$ and let $\alpha \in E$ be algebraic of degree 3 over $\mathbf{Z}_2$. Classify the groups $\langle \mathbf{Z}_2(\alpha), + \rangle$ and $\langle (\mathbf{Z}_2(\alpha))^*, \cdot \rangle$ according to the Fundamental Theorem of Finitely Generated Abelian Groups. As usual, $(\mathbf{Z}_2(\alpha))^*$ is the set of nonzero elements of $\mathbf{Z}_2(\alpha)$.

35.16 Following the idea of Exercise 35.9, show that there exists a field of 8 elements; of 16 elements; of 25 elements.

35.17 Let F be a finite field of characteristic p. Show that every element of F is algebraic over the prime field $\mathbf{Z}_p \leq F$. [*Hint:* Let F^* be the set of nonzero elements of F. Apply group theory to the group $\langle F^*, \cdot \rangle$ to show that every $\alpha \in F^*$ is a zero of some polynomial in $\mathbf{Z}_p[x]$ of the form $x^n - 1$.]

35.18 Use Exercises 35.6 and 35.17 to show that every finite field is of prime power order, i.e., has a prime power number of elements.

36 | Vector Spaces

36.1 DEFINITION AND ELEMENTARY PROPERTIES

The topic of vector spaces is the cornerstone of linear algebra. Since linear algebra is not the subject for study in this text, our treatment of vector spaces will be brief, designed to develop only the concepts of linear independence and dimension that we need for our field theory.

The student is probably familiar with the terms *vector* and *scalar* from his course in calculus. Here we allow scalars to be elements of any field, not just the real numbers, and develop the theory by axioms just as for the other algebraic structures we have studied. The properties appearing in these axioms should look familiar to the student.

Definition. A *vector space* consists of an abelian group V under addition and a field F, together with an operation of scalar multiplication of each element of V by each element of F on the left, such that for all $a, b \in F$ and $\alpha, \beta \in V$ the following conditions are satisfied:

$\mathcal{V}_1$. $a\alpha \in V$.
$\mathcal{V}_2$. $a(b\alpha) = (ab)\alpha$.
$\mathcal{V}_3$. $(a + b)\alpha = (a\alpha) + (b\alpha)$.
$\mathcal{V}_4$. $a(\alpha + \beta) = (a\alpha) + (a\beta)$.
$\mathcal{V}_5$. $1\alpha = \alpha$.

The elements of V are *vectors* and the elements of F are *scalars*. We shall somewhat incorrectly say that V is a *vector space over* F.

Note that multiplication for a vector space is not a binary operation on one set in the sense we defined it in Section 1. It is rather a rule which associates an element $a\alpha$ of V with each ordered pair (a, α), consisting of an element a of F and an element α of V. This can be viewed as a *function* mapping $F \times V$ into V. The *nice* way to define a binary operation on a set S is similarly to say that it is a function from $S \times S$ into S, but we wished to be more naive in Section 1. Both the additive identity for V, the 0-vector, and the additive identity for F, the 0-scalar, will be denoted by "0".

Example 36.1 Consider the abelian group $\langle \mathbf{R}^n, + \rangle = \mathbf{R} \times \mathbf{R} \times \cdots \times \mathbf{R}$ for n factors which consists of ordered n-tuples under addition by components. Define scalar multiplication for scalars in **R** by

$$r\alpha = (ra_1, \ldots, ra_n)$$

for $r \in \mathbf{R}$ and $\alpha = (a_1, \ldots, a_n) \in \mathbf{R}^n$. With these operations, $\mathbf{R}^n$ becomes a vector space over **R**. The axioms for a vector space are easily checked. In particular, for $n = 2$, the student will have no trouble convincing himself that $\mathbf{R}^2 = \mathbf{R} \times \mathbf{R}$ as a vector space over **R** can be viewed as all "vectors whose starting points are the origin of the Euclidean plane" in the sense often studied in calculus courses. ‖

Example 36.2 For any field F, $F[x]$ can be viewed as a vector space over F, where addition of vectors is ordinary addition of polynomials in $F[x]$ and scalar multiplication of an element of $F[x]$ by an element of F is ordinary multiplication in $F[x]$. The axioms $\mathcal{V}_1$ through $\mathcal{V}_5$ for a vector space then follow immediately from the axioms for the integral domain $F[x]$. ‖

Example 36.3 Let E be an extension field of a field F. Then E can be regarded as a vector space over F, where addition of vectors is the usual addition in E and scalar multiplication is the usual field multiplication in E with $a \in F$ and $\alpha \in E$. The axioms follow at once from the field axioms for E. Here our field of scalars is actually a subset of our space of vectors. *It is this example that is the important one for us.* ‖

We are assuming nothing about vector spaces from previous work, and shall prove everything we need from the definition, even though the student may be familiar with the results from his calculus course.

Theorem 36.1 *If V is a vector space over F, then $0\alpha = 0$, $a0 = 0$ and $(-a)\alpha = a(-\alpha) = -(a\alpha)$ for all $a \in F$ and $\alpha \in V$.*

Proof. The equation $0\alpha = 0$ is to be read "(0-scalar)α = 0-vector", and likewise, $a0 = 0$ is to be read "a(0-vector) = 0-vector". The proofs here are very similar to those in Theorem 23.1 for a ring, and again depend heavily on the distributive laws $\mathcal{V}_3$ and $\mathcal{V}_4$. Now

$$(0\alpha) = (0 + 0)\alpha = (0\alpha) + (0\alpha)$$

is an equation in the abelian group $\langle V, + \rangle$, so by the group cancellation law, $0 = 0\alpha$. Likewise, from

$$a0 = a(0 + 0) = a0 + a0,$$

we conclude that $a0 = 0$. Then

$$0 = 0\alpha = (a + (-a))\alpha = a\alpha + (-a)\alpha,$$

so $(-a)\alpha = -(a\alpha)$. Likewise, from

$$0 = a0 = a(\alpha + (-\alpha)) = a\alpha + a(-\alpha),$$

we conclude that $a(-\alpha) = -(a\alpha)$ also. ∎

36.2 LINEAR INDEPENDENCE AND BASES

Definition. Let V be a vector space over F. The vectors in a subset $S = \{\alpha_i \mid i \in I\}$ of V *span* (or *generate*) V if for every $\beta \in V$, we have

$$\beta = a_1 \alpha_{i_1} + a_2 \alpha_{i_2} + \cdots + a_n \alpha_{i_n}$$

for some $a_j \in F$ and $\alpha_{i_j} \in S$, $j = 1, \ldots, n$. A vector $\sum_{j=1}^{n} a_j \alpha_{i_j}$ is a *linear combination of the* α_{i_j}.

Example 36.4 In the vector space $\mathbf{R}^n$ over $\mathbf{R}$ of Example 36.1, the vectors

$$(1, 0, \ldots, 0), (0, 1, \ldots, 0), \ldots, (0, 0, \ldots, 1)$$

clearly span $\mathbf{R}^n$, for

$(a_1, a_2, \ldots, a_n)$
$$= a_1(1, 0, \ldots, 0) + a_2(0, 1, \ldots, 0) + \cdots + a_n(0, 0, \ldots, 1).$$

Also, the monomials x^m for $m \geq 0$ span $F[x]$ over F, the vector space of Example 36.2. ‖

Example 36.5 Let F be a field and E an extension field of F. Let $\alpha \in E$ be algebraic over F. Then $F(\alpha)$ is a vector space over F and by Theorem 35.4, it is spanned by the vectors in $\{1, \alpha, \ldots, \alpha^{n-1}\}$, where $n = \deg(\alpha, F)$. *This is the important example for us.* ‖

Definition. A vector space V over a field F is *finite dimensional* if there is a finite subset of V whose vectors span V.

Example 36.6 Example 36.4 shows that $\mathbf{R}^n$ is finite dimensional. The vector space $F[x]$ over F is *not* finite dimensional, since polynomials of arbitrarily large degree could obviously not be linear combinations of elements of any *finite* set of polynomials. ‖

Example 36.7 If $F \leq E$ and $\alpha \in E$ is algebraic over the field F, Example 36.5 shows that $F(\alpha)$ is a finite-dimensional vector space over F. *This is the most important example for us.* ‖

The next definition contains the most important idea in this section.

Definition. The vectors in a subset $S = \{\alpha_i \mid i \in I\}$ of a vector space V over a field F are *linearly independent over* F if $\sum_{j=1}^{n} a_j \alpha_{i_j} = 0$ implies that $a_j = 0$ for $j = 1, \ldots, n$. If the vectors are not linearly independent over F, they are *linearly dependent over F.*

Thus the vectors in $\{\alpha_i \mid i \in I\}$ are linearly independent over F if the only way the 0-vector can be expressed as a linear combination of the vectors α_i is to have all scalar coefficients equal to 0. If the vectors are linearly dependent over F, then there exist $a_j \in F$ for $j = 1, \ldots, n$ such that $\sum_{j=1}^{n} a_j \alpha_{i_j} = 0$, where not all $a_j = 0$.

Example 36.8 It is clear that the vectors of the set of vectors spanning the space $\mathbf{R}^n$ which are given in Example 36.4 are linearly independent over $\mathbf{R}$. Likewise, the vectors in $\{x^m \mid m \geq 0\}$ are linearly independent vectors of $F[x]$ over F. Note that $(1, -1)$, $(2, 1)$, and $(-3, 2)$ are linearly dependent in $\mathbf{R}^2$ over $\mathbf{R}$, since

$$7(1, -1) + (2, 1) + 3(-3, 2) = (0, 0) = 0. \; \|$$

Example 36.9 Let E be an extension field of a field F, and let $\alpha \in E$ be algebraic over F. If $\deg(\alpha, F) = n$, then by Theorem 35.4, every element of $F(\alpha)$ can be *uniquely* expressed in the form

$$b_0 + b_1\alpha + \cdots + b_{n-1}\alpha^{n-1}$$

for $b_i \in F$. In particular, $0 = 0 + 0\alpha + \cdots + 0\alpha^{n-1}$ must be a *unique* such expression for 0. Thus the elements $1, \alpha, \ldots, \alpha^{n-1}$ are linearly independent vectors in $F(\alpha)$ over the field F. They also span $F(\alpha)$, so by the next definition, $1, \alpha, \ldots, \alpha^{n-1}$ form a *basis* for $F(\alpha)$ over F. *This is the important example for us.* In fact, this is the reason we are doing this material on vector spaces. $\|$

Definition. If V is a vector space over a field F, the vectors in a subset $B = \{\beta_i \mid i \in I\}$ of V form a **basis for V over F** if they span V and are linearly independent.

36.3 DIMENSION

The only other results we wish to prove about vector spaces are that every finite-dimensional vector space has a basis, and that any two bases of a finite-dimensional vector space have the same number of elements. Both these facts are true without the assumption that the vector space is finite dimensional, but the proofs require more knowledge of set theory than we are assuming, and the finite-dimensional case is all we need. First we give an easy lemma.

Lemma 36.1 *Let V be a vector space over a field F, and let $\alpha \in V$. If α is a linear combination of vectors β_i for $i = 1, \ldots, m$ and each β_i is a linear combination of vectors γ_j for $j = 1, \ldots, n$, then α is a linear combination of the γ_j.*

Proof. Let $\alpha = \sum_{i=1}^{m} a_i\beta_i$, and let $\beta_i = \sum_{i=1}^{n} b_{ij}\gamma_j$, where a_i and b_{ij} are in F. Then

$$\alpha = \sum_{i=1}^{m} a_i \left(\sum_{j=1}^{n} b_{ij}\gamma_j \right) = \sum_{j=1}^{n} \left(\sum_{i=1}^{m} a_i b_{ij} \right) \gamma_j,$$

and $(\sum_{i=1}^{m} a_i b_{ij}) \in F$. ∎

Theorem 36.2 *In a finite-dimensional vector space, every finite set of vectors spanning the space contains a subset which is a basis.*

Proof. Let V be finite dimensional over F, and let vectors $\alpha_1, \ldots, \alpha_n$ in V span V. Let us list the α_i in a row. Examine each α_i in succession, starting at the left with $i = 1$, and discard the first α_j which is some linear combination of the preceding α_i for $i < j$. Then continue, starting with the following α_{j+1}, and discard the next α_k which is some linear combination of its remaining predecessors, etc. When we reach α_n after a finite number of steps, those α_i remaining in our list are such that none is a linear combination of the preceding α_i in this reduced list. Lemma 36.1 shows that any vector which is a linear combination of the original collection of α_i is still a linear combination of our reduced, and possibly smaller, set in which no α_i is a linear combination of its predecessors. Thus the vectors in the reduced set of α_i again span V.

For the reduced set, suppose that

$$a_1\alpha_{i_1} + \cdots + a_r\alpha_{i_r} = 0$$

for $i_1 < i_2 < \cdots < i_r$ and that some $a_j \neq 0$. We may assume from Theorem 36.1 that $a_r \neq 0$, or we could drop $a_r\alpha_{i_r}$ from the left side of the equation. Then, using Theorem 36.1 again, we obtain

$$\alpha_{i_r} = \left(-\frac{a_1}{a_r}\right)\alpha_{i_1} + \cdots + \left(-\frac{a_{r-1}}{a_r}\right)\alpha_{i_{r-1}},$$

which shows that α_{i_r} is a linear combination of its predecessors, contradicting our construction. Thus the vectors α_i in the reduced set both span V and are linearly independent, so they form a basis for V over F. ∎

Corollary. *A finite-dimensional vector space has a finite basis.*

Proof. By definition, a finite-dimensional vector space has a finite set of vectors which span the space. Theorem 36.2 completes the proof. ∎

The next theorem is the culmination of our work on vector spaces.

Theorem 36.3 *Let* $S = \{\alpha_1, \ldots, \alpha_r\}$ *be a finite set of linearly independent vectors of a finite-dimensional vector space V over a field F. Then S can be enlarged to a basis for V over F. Furthermore, if $B = \{\beta_1, \ldots, \beta_n\}$ is any basis for V over F, then $r \leq n$.*

Proof. By the corollary of Theorem 36.2, there is a basis $B = \{\beta_1, \ldots, \beta_n\}$ for V over F. Consider the finite sequence of vectors

$$\alpha_1, \ldots, \alpha_r, \beta_1, \ldots, \beta_n.$$

These vectors span V, since B is a basis. Following the technique, used in Theorem 36.2, of discarding in turn each vector which is a linear combination of its remaining predecessors, working from left to right, we arrive at a basis for V. Clearly, no α_i is cast out, since the α_i are linearly independent. Thus S can be enlarged to a basis for V over F.

For the second part of the conclusion, consider the sequence

$$\alpha_1, \beta_1, \ldots, \beta_n.$$

These vectors are not linearly independent over F, because α_1 is a linear combination

$$\alpha_1 = b_1\beta_1 + \cdots + b_n\beta_n,$$

since the β_i form a basis. Thus

$$\alpha_1 + (-b_1)\beta_1 + \cdots + (-b_n)\beta_n = 0.$$

The vectors in the sequence do span V, and if we form a basis by the technique of working from left to right and casting out in turn each vector which is a linear combination of its remaining predecessors, at least one β_i must be cast out, giving a basis

$$\{\alpha_1, \beta_1^{(1)}, \ldots, \beta_m^{(1)}\},$$

where $m \leq n - 1$. Applying the same technique to the sequence of vectors

$$\alpha_1, \alpha_2, \beta_1^{(1)}, \ldots, \beta_m^{(1)},$$

we arrive at a new basis

$$\{\alpha_1, \alpha_2, \beta_1^{(2)}, \ldots, \beta_s^{(2)}\},$$

with $s \leq n - 2$. Continuing, we arrive finally at a basis

$$\{\alpha_1, \ldots, \alpha_r, \beta_1^{(r)}, \ldots, \beta_t^{(r)}\},$$

where $0 \leq t \leq n - r$. Thus $r \leq n$. ∎

Corollary. *Any two bases of a finite-dimensional vector space V over F have the same number of elements.*

Proof. Let $B = \{\beta_1, \ldots, \beta_n\}$ and $B' = \{\beta_1', \ldots, \beta_m'\}$ be two bases. Then by Theorem 36.3, regarding B as an independent set of vectors and B' as a basis, we see that $n \leq m$. A symmetric argument gives $m \leq n$, so $m = n$. ∎

Definition. If V is a finite-dimensional vector space over a field F, the number of elements in a basis (independent of the choice of basis, by Theorem 36.3) is the **dimension of V over F.**

Example 36.10 Let E be an extension field of a field F, and let $\alpha \in E$. Example 36.9 shows that if α is algebraic over F and $\deg(\alpha, F) = n$, then the dimension of $F(\alpha)$ as a vector space over F is n. *This is the important example for us.* ‖

36.4　AN APPLICATION TO FIELD THEORY

We collect the results of field theory contained in Examples 36.3, 36.5, 36.7, 36.9, and 36.10, and incorporate them into one theorem. The last sentence of this theorem gives an additional elegant application of these vector space ideas to field theory.

Theorem 36.4 *Let E be an extension field of F, and let $\alpha \in E$ be algebraic over F. If $\deg(\alpha, F) = n$, then $F(\alpha)$ is an n-dimensional vector space over F with basis $\{1, \alpha, \ldots, \alpha^{n-1}\}$. Furthermore, every element β of $F(\alpha)$ is algebraic over F, and $\deg(\beta, F) \leq \deg(\alpha, F)$.*

Proof. We have shown everything in the preceding examples except the *very important result* stated in the last sentence of the above theorem. Let $\beta \in F(\alpha)$, where α is algebraic over F of degree n. Consider the elements

$$1, \beta, \beta^2, \ldots, \beta^n.$$

These can't be $n + 1$ distinct elements of $F(\alpha)$ which are linearly independent over F, for by Theorem 36.3, any basis of $F(\alpha)$ over F would have to contain at least as many elements as are in any set of linearly independent vectors over F. However, the basis $\{1, \alpha, \ldots, \alpha^{n-1}\}$ has just n elements. If $\beta^i = \beta^j$, then $\beta^i - \beta^j = 0$, so in any case there exist $b_i \in F$ such that

$$b_0 + b_1\beta + b_2\beta^2 + \cdots + b_n\beta^n = 0,$$

where not all $b_i = 0$. Then $f(x) = b_nx^n + \cdots + b_1x + b_0$ is a nonzero element of $F[x]$ such that $f(\beta) = 0$. Therefore, β is algebraic over F and $\deg(\beta, F)$ is at most n. ∎

EXERCISES

36.1 Find three bases for $\mathbf{R}^2$ over $\mathbf{R}$, no two of which have a vector in common.

36.2 Determine whether or not the given set of vectors is a basis for $\mathbf{R}^3$ over $\mathbf{R}$.

a) $\{(1, 1, 0), (1, 0, 1), (0, 1, 1)\}$
b) $\{(-1, 1, 2), (2, -3, 1), (10, -14, 0)\}$

36.3 According to Theorem 36.4, the element $1 + \alpha$ of $\mathbf{Z}_2(\alpha)$ of Example 35.9 is algebraic over $\mathbf{Z}_2$. Find the irreducible polynomial for $1 + \alpha$ in $\mathbf{Z}_2[x]$.

36.4 Give a basis for each of the following vector spaces over the indicated fields.

a) $\mathbf{Q}(\sqrt{2})$ over $\mathbf{Q}$ b) $\mathbf{R}(\sqrt{2})$ over $\mathbf{R}$
c) $\mathbf{Q}(\sqrt[3]{2})$ over $\mathbf{Q}$ d) $\mathbf{C}$ over $\mathbf{R}$
e) $\mathbf{Q}(i)$ over $\mathbf{Q}$ f) $\mathbf{Q}(\sqrt[4]{2})$ over $\mathbf{Q}$

†**36.5** Prove that if V is a finite-dimensional vector space over a field F, then a subset $\{\beta_i\}$ of V is a basis for V over F if and only if every vector in V can be expressed *uniquely* as a linear combination of the β_i.

36.6 Mark each of the following true or false.

— a) The sum of two vectors is a vector.
— b) The sum of two scalars is a vector.
— c) The product of two scalars is a scalar.
— d) The product of a scalar and a vector is a vector.
— e) Every vector space has a finite basis.

— f) The vectors in a basis are linearly dependent.
— g) The 0-vector may be part of a basis.
— h) If $F \leq E$ and $\alpha \in E$ is algebraic over the field F, then α^2 is algebraic over F.
— i) If $F \leq E$ and $\alpha \in E$ is algebraic over the field F, then $\alpha + \alpha^2$ is algebraic over F.
— j) Every vector space has a basis.

The exercises that follow deal with the further study of vector spaces. In many cases, we ask the student to *define* for vector spaces some concept which is analogous to one which we have studied for other algebraic structures. These exercises should improve the student's ability to recognize parallel and related situations in algebra. Any of these exercises may assume that the student has knowledge of concepts defined in the preceding exercises.

36.7 Let V be a vector space over a field F.

a) Define a *subspace of the vector space V over F.*
b) Prove that an intersection of subspaces of V is again a subspace of V over F.

36.8 Let V be a vector space over a field F, and let $S = \{\alpha_i \mid i \in I\}$ be a collection of vectors in V.

a) Using Exercise 36.7(b), define the *subspace of V generated by S.*
b) Prove that the vectors in the subspace of V generated by S are precisely the (finite) linear combinations of vectors in S. (Compare with Theorem 9.1.)

36.9 Let $V_1, \ldots, V_n$ be vector spaces over the same field F. Define the *direct sum* $V_1 + \cdots + V_n$ *of the vector spaces V_i* for $i = 1, \ldots, n$, and show that the direct sum is again a vector space over F.

36.10 Generalize Example 36.1 to obtain the vector space F^n of ordered n-tuples of elements of F over the field F, for any field F. What is a basis for F^n?

36.11 Let F be any field. Consider the "system of m simultaneous linear equations in n unknowns"

$$a_{11}X_1 + a_{12}X_2 + \cdots + a_{1n}X_n = b_1,$$
$$a_{21}X_1 + a_{22}X_2 + \cdots + a_{2n}X_n = b_2,$$
$$\vdots$$
$$a_{m1}X_1 + a_{m2}X_2 + \cdots + a_{mn}X_n = b_m,$$

where $a_{ij}, b_i \in F$.

a) Show that the "system has a solution" if and only if the vector $\beta = (b_1, \ldots, b_m)$ of F^m lies in the subspace of F^m generated by the vectors $\alpha_j = (a_{1j}, \ldots, a_{mj})$. (This result is trivial to prove, being practically the definition of a solution, but should really be regarded as the *fundamental existence theorem for a simultaneous solution of a system of linear equations.*)
b) From (a), show that if $n = m$ and $\{\alpha_j \mid j = 1, \ldots, n\}$ is a basis for F^n, then the system always has a unique solution.

36.12 Define an *isomorphism of a vector space V over a field F with a vector space V' over the same field F.*

36.13 Prove that every finite-dimensional vector space V of dimension n over a field F is isomorphic to the vector space F^n of Exercise 36.10.

36.14 Let V and V' be vector spaces over the same field F. A function $\phi: V \to V'$ is a **linear transformation of V into** V' if the following conditions are satisfied for all $\alpha, \beta \in V$ and $a \in F$:

1) $(\alpha + \beta)\phi = \alpha\phi + \beta\phi$.
2) $(a\alpha)\phi = a(\alpha\phi)$.

a) If $\{\beta_i \mid i \in I\}$ is a basis for V over F, show that a linear transformation $\phi: V \to V'$ is completely determined by the vectors $\beta_i\phi \in V'$.
b) Let $\{\beta_i \mid i \in I\}$ be a basis for V, and let $\{\beta_i' \mid i \in I\}$ be any set of vectors, not necessarily distinct, of V'. Show that there exists exactly one linear transformation $\phi: V \to V'$ such that $\beta_i\phi = \beta_i'$.

36.15 Let V and V' be vector spaces over the same field F, and let $\phi: V \to V'$ be a linear transformation.

a) To what concept which we have studied for the algebraic structures of groups and rings does the concept of a *linear transformation* correspond?
b) Define the *kernel* (or *null space*) *of* ϕ, and show that it is a subspace of V.
c) Describe when ϕ is an isomorphism of V with V'.

36.16 Let V be a vector space over a field F, and let S be a subspace of V. Define the *quotient space* V/S, and show that it is a vector space over F.

36.17 Let V and V' be vector spaces over the same field F, and let V be finite dimensional over F. Let $\dim(V)$ be the dimension of the vector space V over F. Let $\phi: V \to V'$ be a linear transformation.

a) Show that $V\phi$ is a subspace of V'.
b) Show that $\dim(V\phi) = \dim(V) - \dim(\text{kernel } \phi)$. [*Hint:* Choose a convenient basis for V, using Theorem 36.3, e.g., enlarge a basis for (kernel ϕ) to a basis for V.]

*36.18 Let S and T be subspaces of a vector space V over a field F.

a) Define the *join* $S \vee T$ *of S and T*, and show that $S \vee T$ is also a subspace of V over F. (Compare with Section 8.2. $S \vee T$ is usually denoted by "$S + T$" in the literature.)
b) Describe the elements in $S \vee T$ in terms of those in S and those in T.

*36.19 Let V be a finite-dimensional vector space over a field F, and let $\dim(V)$ be the dimension of V over F. Show that if S and T are subspaces of V over F, then

$$\dim(S \vee T) = \dim(S) + \dim(T) - \dim(S \cap T).$$

[*Hint:* The dimension of a space is the number of elements in a basis. Use Theorem 36.3 to choose convenient bases, so that the proof becomes easy. First choose any basis $\{\alpha_i\}$ for $S \cap T$, and then augment it with vectors β_j so that $\{\alpha_i, \beta_j\}$ is a basis for S. Repeat the process so that $\{\alpha_i, \gamma_k\}$ is a basis for T. Show that $\{\alpha_i, \beta_j, \gamma_k\}$ is a basis for $S \vee T$.]

37 | Further Algebraic Structures

This is an expository section designed to give the reader an idea of some additional important algebraic structures and their relation to the structures we have studied.

*37.1 GROUPS WITH OPERATORS

Definition. A *group with operators* consists of a group G and a set $\mathcal{O}$, the *set of operators*, together with an operation of external multiplication of each element of G by each element of $\mathcal{O}$ on the right such that for all $\alpha, \beta \in G$ and $a \in \mathcal{O}$, the following conditions are satisfied:

1) $(\alpha a) \in G$.
2) $(\alpha\beta)a = (\alpha a)(\beta a)$.

We shall somewhat incorrectly speak of the "$\mathcal{O}$-*group G.*"

Note how we have followed closely the form of our definition of a vector space from Section 36. The operation of external multiplication is really a function $\phi: G \times \mathcal{O} \to G$, where $(\alpha, a)\phi$ is denoted by "αa" for $\alpha \in G$ and $a \in \mathcal{O}$. Of course, whether you write the operator a on the right of α, as we did here, or on the left is a matter of preference and convenience.

While we did not require any structure on the set $\mathcal{O}$ in our definition, it frequently happens that $\mathcal{O}$ has some natural algebraic structure. Let us give a few examples.

Example 37.1 Every abelian group G gives rise to a natural group with operators. Let us write the group operation of G multiplicatively. Letting $\mathcal{O} = \mathbf{Z}$, for $\alpha \in G$ and $n \in \mathbf{Z}$, define $\alpha n = \alpha^n$. Since G is abelian, we have

$$(\alpha\beta)n = (\alpha\beta)^n = \alpha^n\beta^n = (\alpha n)(\beta n).$$

Thus every abelian group G can be regarded as a **Z**-*group.* Here **Z** has a natural ring structure. ‖

Example 37.2 If V is a vector space over a field F, V can be regarded in a natural way as a (left) F-group. Here F has a field structure. ‖

Consider an $\mathcal{O}$-group G. For a *fixed* $a \in \mathcal{O}$, the mapping $\rho_a: G \to G$ defined by $\alpha\rho_a = \alpha a$ for $\alpha \in G$ is a homomorphism of G into G, since $(\alpha\beta)a = (\alpha a)(\beta a)$. This suggests our next example.

311

Example 37.3 Let G be any group, and let $\mathcal{O}$ be any set of homomorphisms of G into itself (such homomorphisms are *endomorphisms of G*). For $\alpha, \beta \in G$ and $\phi \in \mathcal{O}$, the property $(\alpha\beta)\phi = (\alpha\phi)(\beta\phi)$ for the endomorphism ϕ shows that G can be naturally viewed as an $\mathcal{O}$-group. ‖

A substantial chunk of our abelian group theory could have been done for $\mathcal{O}$-groups. Starting back with subgroups, an **admissible subgroup**, or **$\mathcal{O}$-subgroup** of an $\mathcal{O}$-group G, is a subgroup H of $\langle G, \cdot \rangle$ such that $\alpha a \in H$ for all $\alpha \in H$ and $a \in \mathcal{O}$, that is, such that H is closed under external multiplication by elements of $\mathcal{O}$. If one is constantly working with $\mathcal{O}$-groups, one simply drops the term *admissible* and speaks of a subgroup of the $\mathcal{O}$-group G, always meaning an $\mathcal{O}$-subgroup. A couple more examples will show how elegantly these ideas relate to our past work.

Example 37.4 Let G be any group, and let $\mathcal{I}$ be the set of all inner automorphisms of G. As in Example 37.3, G becomes an $\mathcal{I}$-group in a natural way. An (admissible) subgroup H of G then must have the property that $ai_g = gag^{-1}$ is in H for all $a \in H$ and all $g \in G$. Thus the $\mathcal{I}$-subgroups H of the $\mathcal{I}$-group G are essentially the normal subgroups of G. ‖

Example 37.5 Let R be a ring. The additive group $\langle R, + \rangle$ of R can be regarded as an R-group, where for a group element a in $\langle R, + \rangle$ and $r \in R$, one defines ar by the ring multiplication. The right distributive law in R gives $(a + b)r = (ar) + (br)$, which is exactly the condition that $\langle R, + \rangle$ be an R-group. An R-subgroup N of the R-group $\langle R, + \rangle$ then must be a subgroup of $\langle R, + \rangle$ satisfying $ar \in N$ for all $a \in N$ and $r \in R$. Thus the R-subgroups are essentially the right ideals of R. If R is a commutative ring, then the R-subgroups are essentially the ideals of R. ‖

One can form the factor group of an $\mathcal{O}$-group G modulo a normal $\mathcal{O}$-subgroup; the factor group also becomes an $\mathcal{O}$-group in a natural way when one defines external multiplication on cosets by using representatives. One has the concept of an $\mathcal{O}$-homomorphism from one $\mathcal{O}$-group into another. We shall ask the student to try to come up with the proper definitions for these ideas in the exercises.

Finally, we state without proof the Jordan-Hölder theorem for an $\mathcal{O}$-group. The definitions of subnormal series and composition series are analogous to our definitions of Part I; you need only require that every subgroup be an $\mathcal{O}$-subgroup.

Theorem 37.1 (Jordan-Hölder). *Any two composition series of an $\mathcal{O}$-group G are isomorphic.*

Taking $\mathcal{O} = \{\iota\}$, where ι is the identity map of a group G onto itself, we recover the Jordan-Hölder theorem of Section 14 for composition series. Taking $\mathcal{O} = \mathcal{I}$, the set of inner automorphisms of G, we recover the Jordan-

Hölder theorem of Section 14 for a principal series. And finally, but most impressively, taking the Jordan-Hölder theorem for the F-group V, where V is a finite-dimensional vector space over F, we recover the invariance of dimension of V, for a simple F-factor group of the F-group V will be a vector space (an F-group) of dimension 1 over F.

*37.2 MODULES

Definition. A (*left*) R-*module* consists of an abelian group M and a ring R, together with an operation of external multiplication of each element of M by each element of R on the left such that for all $\alpha, \beta \in M$ and $r, s \in R$, the following conditions are satisfied:

1) $(r\alpha) \in M$.
2) $r(\alpha + \beta) = r\alpha + r\beta$.
3) $(r + s)\alpha = r\alpha + s\alpha$.
4) $(rs)\alpha = r(s\alpha)$.

We shall somewhat incorrectly speak of the "R-*module* M."

An R-module is very much like a vector space except that the "scalars" need only form a ring. If R is a ring with unity and $1\alpha = \alpha$ for all $\alpha \in M$, then M is a **unitary R-module**.

Example 37.6 Every abelian group G can be regarded as a **Z**-module if we define $n\alpha = \alpha^n$ for $\alpha \in G$ and $n \in \mathbf{Z}$. We have used multiplicative notation for the operation in G. The axioms for a module are easily verified. ‖

Example 37.7 For an ideal N in R, $\langle N, + \rangle$ can be viewed as an R-module, where for $\alpha \in N$ and $r \in R$, $r\alpha$ is the ordinary ring multiplication of r and α, both viewed as elements of the ring R. ‖

One can speak of submodules, quotient modules, and R-homomorphisms from one R-module into another, all by the natural definitions. One can also take a direct sum of R-modules, arriving again at an R-module.

Definition. An R-module M is *cyclic* if there exists $\alpha \in M$ such that $M = \{r\alpha \mid r \in R\}$.

Thus a cyclic R-module is generated by a single element. The idea of a *set of generators for an R-module* is the natural generalization of the idea of a spanning set of vectors of a vector space. A nice result for us to state without proof and then to illustrate for some special cases, in light of our past work, is the following theorem.

Theorem 37.2 *If R is a* **PID**, *then every finitely generated R-module is isomorphic to a direct sum of cyclic R-modules.*

Taking $R = \mathbf{Z}$ and referring to Example 37.6, we see from the theorem that every finitely generated abelian group is isomorphic to a direct sum of cyclic groups. This is a large part of the Fundamental Theorem of Finitely Generated Abelian Groups. For $R = F$, where F is a field, applying Theorem 37.2 to a finite-dimensional vector space V over F, we see that V is isomorphic to a direct sum of vector spaces of dimension 1 over F.

*37.3   ALGEBRAS

Definition. An *algebra* consists of a vector space V over a field F, together with a binary operation of multiplication on the set V of vectors, such that for all $a \in F$ and $\alpha, \beta, \gamma \in V$, the following conditions are satisfied:

1) $(a\alpha)\beta = a(\alpha\beta) = \alpha(a\beta)$.
2) $(\alpha + \beta)\gamma = \alpha\gamma + \beta\gamma$.
3) $\alpha(\beta + \gamma) = \alpha\beta + \alpha\gamma$.

We shall somewhat incorrectly speak of an "*algebra V over F.*" Also, V is an *associative algebra over F*, if in addition

4) $(\alpha\beta)\gamma = \alpha(\beta\gamma)$ for all $\alpha, \beta, \gamma \in V$.

Example 37.8 If E is an extension field of a field F, then $V = E$ can be viewed as an associative algebra over F, where addition and multiplication of elements of V are field addition and multiplication in E, and scalar multiplication by elements of F is again field multiplication in E. ‖

Example 37.9 For any group G and field F, the *group algebra* $F(G)$ defined in Section 25.3 is an associative algebra over F. ‖

Definition. An algebra V over a field F is a *division algebra over F* if V has a unity for multiplication and contains a multiplicative inverse of each nonzero element.   (Note that associativity of multiplication is *not* assumed.)

Example 37.10 An extension field E of a field F can be viewed as an associative division algebra over F. Also the quaternions $\mathbf{Q}$ of Section 25.4 form an associative division algebra over the real numbers. ‖

We conclude with a statement, for the reader's information, of some famous results regarding division algebras over the real numbers.

Theorem 37.3 *The real numbers, the complex numbers, and the quaternions are the only (up to isomorphism) associative division algebras over the real numbers (Frobenius, 1878). The only additional division algebra over the real numbers is the Cayley algebra, which is a vector space of dimension 8 over* $\mathbf{R}$ *(Bott and Milnor, 1957).*

EXERCISES

*37.1 Our definition of an $\mathcal{O}$-group did *not* start

$$\text{``An } \mathcal{O}\text{-group } \langle \quad , \quad , \dots , \quad \rangle \text{ is } \dots \text{''}$$

in a fashion similar to our definitions of a group and a ring. If an $\mathcal{O}$-group were defined in this way, what would $\langle \quad , \quad , \dots , \quad \rangle$ be? Be sure that you get *all* the sets and *all* the operations involved.

*37.2 Repeat Exercise 37.1 for an R-module.

*37.3 Repeat Exercise 37.1 for an algebra.

*37.4 Show that an intersection of admissible subgroups of an $\mathcal{O}$-group G is again an admissible subgroup of G.

*37.5 Let G be any group, and let $\mathcal{A}$ be the set of *all* automorphisms of G. By Example 37.3, G can be regarded as an $\mathcal{A}$-group. An $\mathcal{A}$-subgroup of G is a **characteristic subgroup** of G.

Every subgroup of every abelian group is a normal subgroup, i.e., an $\mathcal{J}$-subgroup, but give an example showing that not every subgroup of every abelian group is a characteristic subgroup.

*37.6 Show that the factor group of an $\mathcal{O}$-group G modulo an admissible normal subgroup can again be considered an $\mathcal{O}$-group in a natural way. That is, define the external multiplication on the cosets, show it is well defined, and check the axioms for an $\mathcal{O}$-group.

*37.7 Define the concept of an *$\mathcal{O}$-homomorphism of an $\mathcal{O}$-group G into an $\mathcal{O}$-group G'*. Show that the kernel is an $\mathcal{O}$-subgroup of G.

*37.8 Prove that for a left R-module M,

$$0\alpha = 0, \qquad r0 = 0,$$

and

$$(-r)\alpha = -(r\alpha) = r(-\alpha)$$

for every $\alpha \in M$ and $r \in R$.

*37.9 Let M be a left R-module, and let $\alpha \in M$. Show that $L_\alpha = \{a \in R \mid a\alpha = 0\}$ is a left ideal of R.

*37.10 Define a *submodule of a (left) R-module M* and a *quotient module of a (left) R-module M modulo a submodule N*.

*37.11 Define an *R-homomorphism of a (left) R-module M into a (left) R-module M'*.

*37.12 Show that the ring $\langle M_n(F), +, \cdot \rangle$ of Section 25.1 becomes an algebra of dimension n^2 over F, if one defines scalar multiplication by $b(a_{ij}) = (ba_{ij})$ for $(a_{ij}) \in M_n(F)$ and $b \in F$.

*37.13 Let V be an algebra of finite dimension and with a basis

$$B = \{\beta_i \mid i = 1, \dots, n\}$$

over a field F. Show that the multiplication of vectors in V is completely determined by the n^2 products $\beta_r\beta_s$ for each ordered pair (β_r, β_s) of basis vectors from B.

*37.14 Let V be an algebra of finite dimension and with a basis

$$B = \{\beta_i \mid i = 1, \ldots, n\}$$

over a field F. Show that V is an associative algebra over F if and only if

$$\beta_r(\beta_s\beta_t) = (\beta_r\beta_s)\beta_t$$

for each of the n^3 ordered triples $(\beta_r, \beta_s, \beta_t)$ of basis vectors from B.

*37.15 Let V be a finite-dimensional vector space over a field F with a basis $B = \{\beta_i \mid i = 1, \ldots, n\}$ over F. Let $\{c_{rst} \mid r, s, t = 1, \ldots, n\}$ be any collection of n^3 scalars in F. Show that there exists exactly one binary operation of multiplication on V such that V is an algebra over F under this multiplication and such that

$$\beta_r\beta_s = \sum_t c_{rst}\beta_t$$

for every ordered pair (β_r, β_s) of basis vectors from B. The scalars c_{rst} are the **structure constants of the algebra**.

38 | Algebraic Extensions

38.1 FINITE EXTENSIONS

In Theorem 36.4 we saw that if E is an extension field of a field F and $\alpha \in E$ is algebraic over F, then every element of $F(\alpha)$ is algebraic over F. In studying zeros of polynomials in $F[x]$, we shall be interested almost exclusively in extensions of F containing only elements algebraic over F.

Definition. An extension field E of a field F is an *algebraic extension of F* if every element in E is algebraic over F.

Definition. If an extension field E of a field F is of finite dimension n as a vector space over F, then E is a *finite extension of degree n over F*. We shall let $[E : F]$ be the degree n of E over F.

We shall often use the fact that if E is a finite extension of F, then $[E : F] = 1$ if and only if $E = F$. We need only observe that by Theorem 36.3, $\{1\}$ can always be enlarged to a basis for E over F. Then $[E : F] = 1$ implies that $E = F(1) = F$. The converse is obvious.

Let us repeat the argument of Theorem 36.4 to show that a finite extension E of a field F must be an algebraic extension of F.

Theorem 38.1 *A finite extension field E of a field F is an algebraic extension of F.*

Proof. We must show that for $\alpha \in E$, α is algebraic over F. By Theorem 36.3, if $[E : F] = n$, then

$$1, \alpha, \ldots, \alpha^n$$

cannot be linearly independent elements, so there exist $a_i \in F$ such that

$$a_n \alpha^n + \cdots + a_1 \alpha + a_0 = 0,$$

and not all $a_i = 0$. Then $f(x) = a_n x^n + \cdots + a_1 x + a_0$ is a nonzero polynomial in $F[x]$, and $f(\alpha) = 0$. Therefore, α is algebraic over F. ∎

We cannot overemphasize the importance of our next theorem. It plays a role in field theory analogous to the role of the theorem of Lagrange in group theory. While its proof follows easily from our brief work with vector spaces, it is a tool of incredible power. We shall later be using the theorem constantly in our Galois theory arguments. Also, an elegant application of it in the starred section which follows shows the impossibility of performing certain geometric constructions with a straightedge and a compass. *Never underestimate a theorem which counts something.*

   Theorem 38.2 *If E is a finite extension field of a field F and K is a finite extension field of E, then K is a finite extension of F, and*

$$[K : F] = [K : E][E : F].$$

Proof. Let $\{\alpha_i \mid i = 1, \ldots, n\}$ be a basis for E as a vector space over F, and let $\{\beta_j \mid j = 1, \ldots m\}$ be a basis for K as a vector space over E. The theorem will be proved if we can show that the mn elements $\alpha_i\beta_j$ form a basis for K, viewed as a vector space over F.

   Let γ be any element of K. Since the β_j form a basis for K over E, we have

$$\gamma = \sum_{j=1}^{m} b_j\beta_j$$

for $b_j \in E$. Since the α_i form a basis for E over F, we have

$$b_j = \sum_{i=1}^{n} a_{ij}\alpha_i$$

for $a_{ij} \in F$. Then

$$\gamma = \sum_{j=1}^{m}\left(\sum_{i=1}^{n} a_{ij}\alpha_i\right)\beta_j = \sum_{i,j} a_{ij}(\alpha_i\beta_j),$$

so the mn vectors $\alpha_i\beta_j$ span K over F.

   It remains for us to show that the mn elements $\alpha_i\beta_j$ are independent over F. Suppose that $\sum_{i,j} c_{ij}(\alpha_i\beta_j) = 0$, with $c_{ij} \in F$. Then

$$\sum_{j=1}^{m}\left(\sum_{i=1}^{n} c_{ij}\alpha_i\right)\beta_j = 0,$$

and $(\sum_{i=1}^{n} c_{ij}\alpha_i) \in E$. Since the elements β_j are independent over E, we must have

$$\sum_{i=1}^{n} c_{ij}\alpha_i = 0$$

for all j. But now the α_i are independent over F, so $\sum_{i=1}^{n} c_{ij}\alpha_i = 0$ implies that $c_{ij} = 0$ for all i and j. Thus the $\alpha_i\beta_j$ not only span K over F but also are independent over F. Thus they form a basis for K over F. ∎

   Note that we proved this theorem by actually exhibiting a basis. It is worth remembering that if $\{\alpha_i \mid i = 1, \ldots, n\}$ is a basis for E over F and $\{\beta_j \mid j = 1, \ldots, m\}$ is a basis for K over E, for fields $F \leq E \leq K$, then the set $\{\alpha_i\beta_j\}$ of mn products is a basis for K over F. Figure 38.1 gives a diagram for this situation. We shall illustrate this further in a moment.

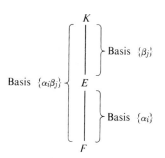

Basis $\{\alpha_i \beta_j\}$ Basis $\{\beta_j\}$ Basis $\{\alpha_i\}$

Fig. 38.1

Corollary 1 *If F_i is a field for $i = 1, \ldots, r$ and F_{i+1} is a finite extension of F_i, then F_r is a finite extension of F_1, and*

$$[F_r : F_1] = [F_r : F_{r-1}][F_{r-1} : F_{r-2}] \cdots [F_2 : F_1].$$

Proof. This is an obvious extension of Theorem 38.2, by induction. ∎

Corollary 2 *If E is an extension field of F, $\alpha \in E$ is algebraic over F, and $\beta \in F(\alpha)$, then $\deg(\beta, F)$ divides $\deg(\alpha, F)$.*

Proof. By Theorem 36.4, $\deg(\alpha, F) = [F(\alpha) : F]$ and $\deg(\beta, F) = [F(\beta) : F]$. We have $F \leq F(\beta) \leq F(\alpha)$, so by Theorem 38.2, $[F(\beta) : F]$ divides $[F(\alpha) : F]$. ∎

The following example illustrates a type of argument one often makes using Theorem 38.2 or its corollaries.

Example 38.1 By Corollary 2 of Theorem 38.2, there is no element of $Q(\sqrt{2})$ which is a zero of $x^3 - 2$. For $\deg(\sqrt{2}, Q) = 2$, while a zero of $x^3 - 2$ is of degree 3 over Q, but 3 does not divide 2. ∥

Let E be an extension field of a field F, and let α_1, α_2 be elements of E, not necessarily algebraic over F. By definition, $F(\alpha_1)$ is the smallest extension field of F in E which contains α_1. Similarly, $\big(F(\alpha_1)\big)(\alpha_2)$ can be characterized as the smallest extension field of F in E containing both α_1 and α_2. We could equally well have started with α_2, so clearly $\big(F(\alpha_1)\big)(\alpha_2) = \big(F(\alpha_2)\big)(\alpha_1)$. We denote this field by "$F(\alpha_1, \alpha_2)$". Similarly, for $\alpha_i \in E$, $F(\alpha_1, \ldots, \alpha_n)$ is the smallest extension field of F containing all the α_i for $i = 1, \ldots, n$. We obtain the field $F(\alpha_1, \ldots, \alpha_n)$ from the field F by **adjoining to F the elements** α_i **in E.** The student will have no difficulty in checking that, analogous to an intersection of subgroups of a group, an intersection of subfields of a field E is again a subfield of E. It is obvious that $F(\alpha_1, \ldots, \alpha_n)$ is the intersection of all subfields of E containing F and all the α_i for $i = 1, \ldots, n$.

Example 38.2 Consider $Q(\sqrt{2})$. Theorem 36.4 shows that $\{1, \sqrt{2}\}$ is a basis for $Q(\sqrt{2})$ over Q. Similarly, $\{1, \sqrt{3}\}$ is a basis for $Q(\sqrt{2}, \sqrt{3}) = \big(Q(\sqrt{2})\big)(\sqrt{3})$ over $Q(\sqrt{2})$. The proof of Theorem 38.2 (see the comment following the theorem) then shows that $\{1, \sqrt{2}, \sqrt{3}, \sqrt{6}\}$ is a basis for $Q(\sqrt{2}, \sqrt{3})$ over Q. ∥

Example 38.3 Let $2^{1/3}$ be the real cube root of 2 and $2^{1/2}$ be the positive square root of 2. Then, as we saw in Example 38.1, $2^{1/3} \notin \mathbf{Q}(2^{1/2})$. Thus $[\mathbf{Q}(2^{1/2}, 2^{1/3}) : \mathbf{Q}(2^{1/2})] = 3$. Then $\{1, 2^{1/2}\}$ is a basis for $\mathbf{Q}(2^{1/2})$ over $\mathbf{Q}$, and $\{1, 2^{1/3}, 2^{2/3}\}$ is a basis for $\mathbf{Q}(2^{1/2}, 2^{1/3})$ over $\mathbf{Q}(2^{1/2})$. Furthermore, by Theorem 38.2 (see the comment following the theorem),

$$\{1, 2^{1/2}, 2^{1/3}, 2^{5/6}, 2^{2/3}, 2^{7/6}\}$$

is a basis for $\mathbf{Q}(2^{1/2}, 2^{1/3})$ over $\mathbf{Q}$. Clearly, since $2^{7/6} = (2)2^{1/6}$, we have $2^{1/6} \in \mathbf{Q}(2^{1/2}, 2^{1/3})$. Now $2^{1/6}$ is a zero of $x^6 - 2$, which is irreducible over $\mathbf{Q}$, by Eisenstein's criterion, with $p = 2$. Thus

$$\mathbf{Q} \le \mathbf{Q}(2^{1/6}) \le \mathbf{Q}(2^{1/2}, 2^{1/3})$$

and by Theorem 38.2,

$$6 = [\mathbf{Q}(2^{1/2}, 2^{1/3}) : \mathbf{Q}] = [\mathbf{Q}(2^{1/2}, 2^{1/3}) : \mathbf{Q}(2^{1/6})][\mathbf{Q}(2^{1/6}) : \mathbf{Q}]$$
$$= [\mathbf{Q}(2^{1/2}, 2^{1/3}) : \mathbf{Q}(2^{1/6})](6).$$

Therefore, we must have

$$[\mathbf{Q}(2^{1/2}, 2^{1/3}) : \mathbf{Q}(2^{1/6})] = 1,$$

so $\mathbf{Q}(2^{1/2}, 2^{1/3}) = \mathbf{Q}(2^{1/6})$, by the comment preceding Theorem 38.1. ‖

Example 38.3 shows that it is possible for an extension $F(\alpha_1, \ldots, \alpha_n)$ of a field F to be actually a simple extension, even though $n > 1$.

Let us characterize extensions of F of the form $F(\alpha_1, \ldots, \alpha_n)$ in the case that all the α_i are algebraic over F.

> **Theorem 38.3** *Let E be an algebraic extension of a field F. Then there exist a finite number of elements $\alpha_1, \ldots, \alpha_n$ in E such that $E = F(\alpha_1, \ldots, \alpha_n)$ if and only if E is a finite-dimensional vector space over F, i.e., if and only if E is a finite extension of F.*

Proof. Suppose that $E = F(\alpha_1, \ldots, \alpha_n)$. Since E is an algebraic extension of F, each α_i is algebraic over F, so clearly each α_i is algebraic over every extension field of F in E. Thus $F(\alpha_1)$ is algebraic over F, and in general, $F(\alpha_1, \ldots, \alpha_j)$ is algebraic over $F(\alpha_1, \ldots, \alpha_{j-1})$ for $j = 2, \ldots, n$. Corollary 1 of Theorem 38.2 applied to the sequence of finite extensions

$$F, F(\alpha_1), F(\alpha_1, \alpha_2), \ldots, F(\alpha_1, \ldots, \alpha_n) = E$$

then shows that E is a finite extension of F.

Conversely, suppose that E is a finite algebraic extension of F. If $[E : F] = 1$, then $E = F(1) = F$, and we are done. If $E \ne F$, let $\alpha_1 \in E$, where $\alpha_1 \notin F$. Then $[F(\alpha_1) : F] > 1$. If $F(\alpha_1) = E$, we are done; if not, let $\alpha_2 \in E$, where $\alpha_2 \notin F(\alpha_1)$. Continuing this process, we see from Theorem 38.2 that since $[E : F]$ is finite, we must arrive at α_n such that

$$F(\alpha_1, \ldots, \alpha_n) = E. \blacksquare$$

38.2   ALGEBRAICALLY CLOSED FIELDS AND ALGEBRAIC CLOSURES

We have not yet observed that if E is an extension of a field F and $\alpha, \beta \in E$ are algebraic over F, then so are $\alpha + \beta$, $\alpha\beta$, $\alpha - \beta$, and α/β, if $\beta \neq 0$. This follows easily from Theorem 38.3 and is also included in the following theorem.

Theorem 38.4 *Let E be an extension field of F. Then*

$$\overline{F}_E = \{\alpha \in E \mid \alpha \text{ is algebraic over } F\}$$

is a subfield of E, the **algebraic closure** *of F in E.*

Proof. Let $\alpha, \beta \in \overline{F}_E$. Then Theorem 38.3 shows that $F(\alpha, \beta)$ is a finite extension of F, and by Theorem 38.1, every element of $F(\alpha, \beta)$ is algebraic over F, that is, $F(\alpha, \beta) \subseteq \overline{F}_E$. Thus $\overline{F}_E$ contains $\alpha + \beta$, $\alpha\beta$, $\alpha - \beta$, and also α/β for $\beta \neq 0$, so $\overline{F}_E$ is a subfield of E. ∎

Corollary. *The set of all algebraic numbers forms a field.*

Proof. This is immediate from Theorem 38.4, for the set of all algebraic numbers is the algebraic closure of **Q** in **C**. ∎

It is well known that the complex numbers have the property that every nonconstant polynomial in $\mathbf{C}[x]$ has a zero in **C**. This is known as the *"Fundamental Theorem of Algebra."* An analytic proof of this theorem is given later in this section under a starred heading. We now give a definition generalizing this important concept to other fields.

Definition. A field F is **algebraically closed** if every nonconstant polynomial in $F[x]$ has a zero in F.

The next theorem shows that the concept of a field being algebraically closed can also be defined in terms of factorization of polynomials over the field.

Theorem 38.5 *A field F is algebraically closed if and only if every nonconstant polynomial in $F[x]$ factors in $F[x]$ into linear factors.*

Proof. Let F be algebraically closed, and let $f(x)$ be a nonconstant polynomial in $F[x]$. Then $f(x)$ has a zero $a \in F$. By Corollary 1 of Theorem 31.1, $x - a$ is a factor of $f(x)$, so $f(x) = (x - a)g(x)$. Then if $g(x)$ is nonconstant, it has a zero $b \in F$, and $f(x) = (x - a)(x - b)h(x)$. Continuing, we get a factorization of $f(x)$ in $F[x]$ into linear factors.

Conversely, suppose that every nonconstant polynomial of $F[x]$ has a factorization into linear factors. If $ax - b$ is a linear factor of $f(x)$, then b/a is a zero of $f(x)$. Thus F is algebraically closed. ∎

Corollary. *An algebraically closed field F has no proper algebraic extensions, i.e., no algebraic extensions E with $F < E$.*

Proof. Let E be an algebraic extension of F, so $F \leq E$. Then if $\alpha \in E$, we have $\text{irr}(\alpha, F) = x - \alpha$, by Theorem 38.5, since F is algebraically closed. Thus $\alpha \in F$, and we must have $F = E$. ∎

In starred Section 38.3 we shall show that just as there exists an algebraically closed extension **C** of the real numbers **R**, for any field F, there similarly exists an algebraic extension $\overline{F}$ of F, with the property that $\overline{F}$ is algebraically closed. Section 41 will show that such an extension $\overline{F}$ is unique, up to isomorphism, of course. Naively, to find $\overline{F}$ you proceed as follows. If not every polynomial $f(x)$ in $F[x]$ has a zero, then adjoin a zero α of such an $f(x)$ to F, thus obtaining the field $F(\alpha)$. *Theorem 35.1, Kronecker's theorem, is strongly used here, of course.* If $F(\alpha)$ is still not algebraically closed, then continue the process further. The trouble is that, contrary to the situation for the algebraic closure **C** of **R**, you may have to do this a (possibly large) infinite number of times. It can easily be shown (see Exercises 38.13 and 38.16) that $\overline{\mathbf{Q}}$ is isomorphic to the field of all algebraic numbers, and that one cannot obtain $\overline{\mathbf{Q}}$ from **Q** by adjoining a finite number of algebraic numbers. We shall have to first discuss some set-theoretic machinery, *Zorn's lemma*, in order to be able to handle such a situation. This machinery is a bit complex, so we are putting it under a starred heading. However, the existence theorem for $\overline{F}$ is very important, and we state it here so that the student will realize that he should know this fact. There is nothing wrong with his simply assuming its validity.

Theorem 38.6 *Every field F has an* **algebraic closure**, *i.e., an algebraic extension $\overline{F}$ which is algebraically closed.*

*38.3 THE EXISTENCE OF AN ALGEBRAIC CLOSURE

We shall prove that every field has an algebraic extension which is algebraically closed. We feel that every student should have the opportunity to see some proof involving the *Axiom of Choice* by the time he finishes a course in algebra. This is a natural place for such a proof. We shall use an equivalent form, *Zorn's lemma*, of the Axiom of Choice. To state Zorn's lemma, we have to give a set-theoretic definition.

Definition. A *partial ordering of a set* S is given by a relation $\leq$ defined for certain ordered pairs of elements of S such that the following conditions are satisfied:

1) $a \leq a$ for all $a \in S$ (*reflexive law*).
2) If $a \leq b$ and $b \leq a$, then $a = b$ (*antisymmetric law*).
3) If $a \leq b$ and $b \leq c$, then $a \leq c$ (*transitive law*).

In a *partially* ordered set, not every two elements need be **comparable**, i.e., for $a, b \in S$, you need not have either $a \leq b$ or $b \leq a$. As usual, "$a < b$" denotes $a \leq b$ but $a \neq b$.

A subset T of a partially ordered set S is a **chain** if every two elements a and b in T are comparable, i.e., either $a \leq b$ or $b \leq a$ (or both). An element $u \in S$ is an **upper bound for a subset** A of a partially ordered set S if $a \leq u$ for all $a \in A$. Finally, an element m of a partially ordered set S is **maximal** if there is no $s \in S$ such that $m < s$.

Example 38.4 The collection of all subsets of a set forms a partially ordered set under the relation $\leq$ given by $\subseteq$. For example, if the whole set is $\mathbf{R}$, we have $\mathbf{Z} \subseteq \mathbf{Q}$. Note, however, that for $\mathbf{Z}$ and $\mathbf{Q}^{+}$, neither $\mathbf{Z} \subseteq \mathbf{Q}^{+}$ nor $\mathbf{Q}^{+} \subseteq \mathbf{Z}$. ‖

Zorn's Lemma. *If S is a partially ordered set such that every chain in S has an upper bound in S, then S has at least one maximal element.*

There is no question of *proving* Zorn's lemma. The lemma is equivalent to the *Axiom of Choice*. Thus we are really taking Zorn's lemma here as an *axiom* for our set theory. We refer the student to the literature for a statement of the Axiom of Choice and a proof of its equivalence to Zorn's lemma.

Zorn's lemma is often useful when you want to show the existence of a largest or maximal structure of some kind. If a field F has an algebraic extension $\overline{F}$ which is algebraically closed, then $\overline{F}$ will certainly be a maximal algebraic extension of F, for since $\overline{F}$ is algebraically closed, it can have no proper algebraic extensions.

The idea of our proof of Theorem 38.6 is very simple. Given a field F, we shall first describe a class of algebraic extensions of F which is so large that it must contain (up to isomorphism) any conceivable algebraic extension of F. We then define a partial ordering, the ordinary subfield ordering, on this class, and show that the hypotheses of Zorn's lemma are satisfied. By Zorn's lemma, there will exist a maximal algebraic extension $\overline{F}$ of F in this class. We shall then argue that this $\overline{F}$ can have no proper algebraic extensions, so it must be algebraically closed.

Our proof differs a bit from the one found in many texts. We like it because it uses no algebra other than that of Theorems 35.1 and 38.2. Thus it throws into sharp relief the tremendous strength of both Kronecker's theorem and Zorn's lemma. The proof looks long, but only because we are writing out every little step in hideous detail. To the professional mathematician, the construction of the proof from the information in the preceding paragraph is a routine matter. This proof was suggested to the author during his graduate student days by a fellow graduate student, Norman Shapiro, who also had a strong preference for it.

We are now ready to carry out our proof of Theorem 38.6 which we restate here.

Theorem 38.6 *Every field F has an algebraic closure $\overline{F}$.*

Proof. It can be shown in set theory that given any set, there exists a set with *strictly more* elements. Suppose we form a set

$$A = \{\omega_{f_i} \mid f \in F[x]; \, i = 0, \ldots, (\text{degree } f)\}$$

which has an element for every possible zero of any $f(x) \in F[x]$. Let Ω be a set with strictly more elements than A. By forming $\Omega \cup F$ if necessary, we can assume $F \subset \Omega$. Consider all possible fields which are algebraic extensions of F and which, as sets, consist of elements of Ω. One such algebraic extension is F itself. If E is any extension field of F, and if $\gamma \in E$ is a zero of $f(x) \in F[x]$ for $\gamma \notin F$ and $\deg(\gamma, F) = n$, then renaming γ by "ω" for $\omega \in \Omega$ and $\omega \notin F$, and renaming elements $a_0 + a_1\gamma + \cdots + a_{n-1}\gamma^{n-1}$ of $F(\gamma)$ by distinct elements of Ω as the a_i range over F, we can consider our "renamed $F(\gamma)$" to be an algebraic extension field $F(\omega)$ of F, with $F(\omega) \subset \Omega$ and $f(\omega) = 0$. The set Ω has enough elements to form $F(\omega)$, since Ω has more than enough elements to provide n different zeros for each element of each degree n in any subset of $F[x]$.

All algebraic extension fields E_j of F, with $E_j \subseteq \Omega$, form a set

$$S = \{E_j \mid j \in J\}$$

which is partially ordered under our usual subfield inclusion $\leq$. One element of S is F itself. The preceding paragraph shows that if F is far away from being algebraically closed, there will be many fields E_j in S.

Let $T = \{E_{j_k}\}$ be a chain in S, and let $W = \bigcup_k E_{j_k}$. We now make W into a field. Let $\alpha, \beta \in W$. Then there exist $E_{j_1}, E_{j_2} \in S$, with $\alpha \in E_{j_1}$ and $\beta \in E_{j_2}$. Since T is a chain, one of the fields E_{j_1} and E_{j_2} is a subfield of the other, say $E_{j_1} \leq E_{j_2}$. Then $\alpha, \beta \in E_{j_2}$, and we use the field operations of E_{j_2} to *define* the sum of α and β in W as $(\alpha + \beta) \in E_{j_2}$ and, likewise, the product as $(\alpha\beta) \in E_{j_2}$. These operations are well defined in W; they are independent of our choice of E_{j_2}, since if $\alpha, \beta \in E_{j_3}$ also, for E_{j_3} in T, then one of the fields E_{j_2} and E_{j_3} is a subfield of the other, since T is a chain. Thus we have operations of addition and multiplication defined on W.

All the field axioms for W under these operations now follow from the fact that these operations were defined in terms of addition and multiplication in fields. Thus, for example, $1 \in F$ serves as multiplicative identity in W, since for $\alpha \in W$, if $1, \alpha \in E_{j_1}$, then we have $1\alpha = \alpha$ in E_{j_1}, so $1\alpha = \alpha$ in W, by definition of multiplication in W. Also, as further illustration, to check the distributive laws let $\alpha, \beta, \gamma \in W$. Since T is a chain, we can find one field in T containing all three elements α, β, and γ, and in this field the distributive laws for α, β, and γ hold. Thus they hold in W. Therefore, we can view W as a field, and by construction, $E_{j_k} \leq W$ for every $E_{j_k} \in T$.

If we can show that W is algebraic over F, then $W \in S$ will be an upper bound for T. But if $\alpha \in W$, then $\alpha \in E_{j_1}$ for some E_{j_1} in T, so α is algebraic over F. Hence W is an algebraic extension of F and is an upper bound for T.

The hypotheses of Zorn's lemma are thus fulfilled, so there is a maximal element $\overline{F}$ of S. We claim that $\overline{F}$ is algebraically closed. Let $f(x) \in \overline{F}[x]$, where $f(x) \notin \overline{F}$. Suppose that $f(x)$ has no zero in $\overline{F}$. Since Ω has many more elements than $\overline{F}$ has, we can take $\omega \in \Omega$, where $\omega \notin \overline{F}$, and form a field $\overline{F}(\omega) \subseteq \Omega$, with ω a zero of $f(x)$, as we saw in the first paragraph of this proof. Let β be in $\overline{F}(\omega)$. Then by Theorem 36.4, β is a zero of a polynomial

$$g(x) = \alpha_0 + \alpha_1 x + \cdots + \alpha_n x^n$$

in $\overline{F}[x]$, with $\alpha_i \in \overline{F}$, and hence α_i algebraic over F. Then by Theorem 38.3, $F(\alpha_0, \ldots, \alpha_n)$ is a finite extension of F, and since β is algebraic over $F(\alpha_0, \ldots, \alpha_n)$, we also see that $F(\alpha_0, \ldots, \alpha_n, \beta)$ is a finite extension over $F(\alpha_0, \ldots, \alpha_n)$. Theorem 38.2 then shows that $F(\alpha_0, \ldots, \alpha_n, \beta)$ is a finite extension of F, so by Theorem 38.1, β is algebraic over F. Hence $\overline{F}(\omega) \in S$ and $\overline{F} < \overline{F}(\omega)$, which contradicts the choice of $\overline{F}$ as maximal in S. Thus $f(x)$ must have had a zero in $\overline{F}$, so $\overline{F}$ is algebraically closed. ∎

The mechanics of the preceding proof are largely routine to the professional mathematician. Since it may be the first proof that the student has ever seen using Zorn's lemma, we wrote the proof out in simply hideous detail. However, the construction and arguments involved should properly be regarded as easy and routine.

It is well known that $\mathbf{C}$ is an algebraically closed field. Although there are more algebraic proofs of this fact, we give an analytic proof as the one most accessible to the student who has had a course in functions of a complex variable.

Theorem 38.7 (Fundamental Theorem of Algebra). *The field $\mathbf{C}$ of complex numbers is an algebraically closed field.*

Proof. Let the polynomial $f(z) \in \mathbf{C}[z]$ have no zero in $\mathbf{C}$. Then $1/f(z)$ gives an entire function, i.e., $1/f$ is analytic everywhere. Also if $f \notin \mathbf{C}$, $\lim_{|c| \to \infty} |f(c)| = \infty$, so $\lim_{|c| \to \infty} |1/f(c)| = 0$. Thus $1/f$ must be bounded in the plane. Hence by Liouville's theorem of complex function theory, $1/f$ is constant, and thus f is constant. Therefore, a nonconstant polynomial in $\mathbf{C}[z]$ must have a zero in $\mathbf{C}$, so $\mathbf{C}$ is algebraically closed. ∎

EXERCISES

38.1 Find the degree and a basis for each of the given field extensions.

a) $\mathbf{Q}(\sqrt{2})$ over $\mathbf{Q}$

b) $\mathbf{Q}(\sqrt{2}, \sqrt{3})$ over $\mathbf{Q}$

c) $\mathbf{Q}(\sqrt{2}, \sqrt{3}, \sqrt{5})$ over $\mathbf{Q}$

d) $\mathbf{Q}(\sqrt[3]{2}, \sqrt{3})$ over $\mathbf{Q}$

e) $\mathbf{Q}(\sqrt{2}, \sqrt[3]{2})$ over $\mathbf{Q}$

38.2 Find the degree of each of the following field extensions. Be prepared to justify your answers.

a) $\mathbf{Q}(\sqrt{2} + \sqrt{3})$ over $\mathbf{Q}$

b) $\mathbf{Q}(\sqrt{2}\sqrt{3})$ over $\mathbf{Q}$

c) $\mathbf{Q}(\sqrt{2}, \sqrt[3]{5})$ over $\mathbf{Q}$

d) $\mathbf{Q}(\sqrt[3]{2}, \sqrt[3]{6}, \sqrt[3]{24})$ over $\mathbf{Q}$

e) $Q(\sqrt{2}, \sqrt{6})$ over $Q(\sqrt{3})$ f) $Q(\sqrt{2} + \sqrt{3})$ over $Q(\sqrt{3})$

g) $Q(\sqrt{2}, \sqrt{3})$ over $Q(\sqrt{2} + \sqrt{3})$

h) $Q(\sqrt{2}, \sqrt{6} + \sqrt{10})$ over $Q(\sqrt{3} + \sqrt{5})$

38.3 Find a basis for each of the field extensions given in (e) through (h) of Exercise 38.2.

38.4 Show by an example that for a proper extension field E of a field F, the algebraic closure of F in E need not be algebraically closed.

†**38.5** Show that if E is a finite extension of a field F and $[E : F]$ is a prime number, then E is a simple extension of F and indeed, $E = F(\alpha)$ for every $\alpha \in E$ not in F.

38.6 Mark each of the following true or false.

  —   a) Every finite extension of a field is an algebraic extension.

  —   b) Every algebraic extension of a field is a finite extension.

  —   c) The top field of a finite tower of finite extensions of fields is a finite extension of the bottom field.

  —   d) R is algebraically closed.

  —   e) Q is its own algebraic closure in R, that is, Q is **algebraically closed in R**.

  —   f) C is algebraically closed in $C(x)$, where x is an indeterminate.

  —   g) $C(x)$ is algebraically closed, where x is an indeterminate.

  —   h) The field $C(x)$ has no algebraic closure, since C already contains all algebraic numbers.

  —   i) An algebraically closed field must be of characteristic 0.

  —   j) If E is an algebraically closed extension field of F, then E is an algebraic extension of F.

38.7 Prove that $x^2 - 3$ is irreducible over $Q(\sqrt[3]{2})$.

38.8 What degree field extensions can we obtain by successively adjoining to a field F a square root of an element of F not a square in F, then a square root of some nonsquare in this new field, etc.? Argue from this that a zero of $x^{14} - 3x^2 + 12$ over Q can never be expressed as a rational function of square roots, square roots of rational functions of square roots, etc. of elements of Q.

38.9 Prove in detail that $Q(\sqrt{3} + \sqrt{7}) = Q(\sqrt{3}, \sqrt{7})$.

38.10 Generalizing Exercise 38.9, show that $Q(\sqrt{a} + \sqrt{b}) = Q(\sqrt{a}, \sqrt{b})$ for all a and b in Q.

38.11 Let E be a finite extension of a field F, and let $p(x) \in F[x]$ be irreducible over F and have degree relatively prime to $[E : F]$. Show that $p(x)$ has no zeros in E.

38.12 Show that if F, E, and K are fields with $F \leq E \leq K$, then K is algebraic over F if and only if E is algebraic over F and K is algebraic over E. (You must *not* assume the extensions are finite.)

38.13 Let E be an extension field of a field F. Prove that every $\alpha \in E$ which is not in the algebraic closure $\bar{F}_E$ of F in E is transcendental over $\bar{F}_E$.

38.14 Let E be an algebraically closed extension field of a field F. Show that the algebraic closure $\bar{F}_E$ of F in E is algebraically closed. (Applying this exercise to

C and **Q**, we see that the field of all algebraic numbers is an algebraically closed field.)

38.15 Show that if E is an algebraic extension of a field F and contains all zeros in $\bar{F}$ of every $f(x) \in F[x]$, then E is an algebraically closed field.

38.16 Show that no finite field of odd characteristic is algebraically closed. (Actually, no finite field of characteristic 2 is algebraically closed either.) [*Hint:* By counting, show that for such a finite field F, some polynomial $x^2 - a$, for some $a \in F$, has no zero in F. See Exercise 35.10.]

38.17 Prove that, as was asserted in the text, the algebraic closure of **Q** in **C** is not a finite extension of **Q**.

*38.18 Argue that every finite extension field of **R** either is **R** itself or is isomorphic to **C**.

*38.19 Use Zorn's lemma to show that every proper ideal of a ring R with unity is contained in some maximal ideal.

39 | Geometric Constructions

In this section we digress briefly to give an application demonstrating the power of Theorem 38.2. For a more detailed study of geometric constructions, the student is referred to Courant and Robbins [43, Chapter III].

We are interested in what types of figures can be constructed with a compass and a straightedge in the sense of classical Euclidean plane geometry. The student no doubt remembers being told that "it is impossible to trisect the angle." We shall discuss this and other classical questions.

*39.1 CONSTRUCTIBLE NUMBERS

Let us imagine that we are given only a single line segment which we shall define to be *one unit* in length. A real number α is **constructible** if one can construct a line segment of length $|\alpha|$ in a finite number of steps from this given segment of unit length by using a straightedge and a compass. Recall that using a straightedge and a compass, it is possible, among other things, to erect a perpendicular to a given line at a known point on the line and to find a line passing through a given point and parallel to a given line. Our first result is the following theorem.

Theorem 39.1 *If α and β are constructible real numbers, then so are $\alpha + \beta$, $\alpha - \beta$, $\alpha\beta$, and α/β, if $\beta \neq 0$.*

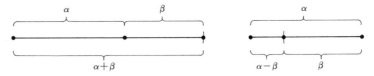

Fig. 39.1

Proof. We are given that α and β are constructible, so there are line segments of lengths $|\alpha|$ and $|\beta|$ available to us. For $\alpha, \beta > 0$, extend a line segment of length α with the straightedge. Start at one end of the original segment of length α, and lay off on the extension the length β with the compass. This constructs a line segment of length $\alpha + \beta$; $\alpha - \beta$ is similarly constructible (see Fig. 39.1). If α and β are not both positive, an obvious breakdown into cases according to their signs shows that $\alpha + \beta$ and $\alpha - \beta$ are still constructible.

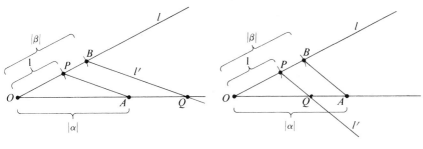

Fig. 39.2   **Fig. 39.3**

The construction of $\alpha\beta$ is indicated in Fig. 39.2. We shall let $\overline{OA}$ be the line segment from the point O to the point A, and shall let $|\overline{OA}|$ be the length of this line segment. If $\overline{OA}$ is of length $|\alpha|$, find a line l through O not containing $\overline{OA}$. Then find the points P and B on l such that $\overline{OP}$ is of length 1 and $\overline{OB}$ is of length $|\beta|$. Draw $\overline{PA}$ and construct l' through B, parallel to $\overline{PA}$, and intersecting $\overline{OA}$ extended at Q. By similar triangles, we have

$$\frac{1}{|\alpha|} = \frac{|\beta|}{|\overline{OQ}|},$$

so $\overline{OQ}$ is of length $|\alpha\beta|$.

Finally, Fig. 39.3 shows that α/β is constructible, if $\beta \neq 0$. Let $\overline{OA}$ be of length $|\alpha|$, and find l through O not containing $\overline{OA}$. Then find B and P on l such that $\overline{OB}$ is of length $|\beta|$ and $\overline{OP}$ is of length 1. Draw $\overline{BA}$ and construct l' through P, parallel to $\overline{BA}$, and intersecting $\overline{OA}$ at Q. Again by similar triangles, we have

$$\frac{|\overline{OQ}|}{1} = \frac{|\alpha|}{|\beta|},$$

so $\overline{OQ}$ is of length $|\alpha/\beta|$. ∎

Corollary. *The set of all constructible real numbers forms a subfield F of the field of real numbers.*

Proof. This is immediate from Theorem 39.1. ∎

Clearly, the field F of all constructible real numbers contains **Q**, the field of rational numbers, since **Q** is the smallest subfield of **R**.

From now on, we proceed analytically. We can construct any rational number. Regarding our given segment

$$0 \underline{\quad\quad\quad} 1$$

of length 1 as the basic unit on an x-axis, we can locate any point (q_1, q_2) in the plane with both coordinates rational. Any further point in the plane which we can locate by using a compass and a straightedge can be found in one of the following three ways:

1) as an intersection of two lines, each of which passes through two known points having rational coordinates,

2) as an intersection of a line which passes through two points having rational coordinates and a circle whose center has rational coordinates and the square of whose radius is rational,

3) as an intersection of two circles whose centers have rational coordinates and the squares of whose radii are rational.

Equations of lines and circles of the type discussed in (1), (2), and (3) are of the form

$$ax + by + c = 0$$

and

$$x^2 + y^2 + dx + ey + f = 0,$$

where $a, b, c, d, e,$ and f are all in $\mathbf{Q}$. Since in case (3), the intersection of two circles with equations

$$x^2 + y^2 + d_1 x + e_1 y + f_1 = 0$$

and

$$x^2 + y^2 + d_2 x + e_2 y + f_2 = 0$$

is the same as the intersection of the first circle, having equation

$$x^2 + y^2 + d_1 x + e_1 y + f_1 = 0,$$

and the line (the common chord), having equation

$$(d_1 - d_2)x + (e_1 - e_2)y + f_1 - f_2 = 0,$$

it is clear that case (3) can be reduced to case (2). For case (1), a simultaneous solution of two linear equations with rational coefficients can only lead to rational values of x and y, giving us no new points. However, finding a simultaneous solution of a linear equation with rational coefficients and a quadratic equation with rational coefficients, as in case (2), leads, upon substitution, to a quadratic equation. Such an equation, when solved by the quadratic formula, may have solutions involving square roots of numbers which are not squares in $\mathbf{Q}$.

In the preceding argument, nothing was really used involving $\mathbf{Q}$ except field axioms. If H is the smallest field containing those real numbers constructed so far, the argument shows that the "next new number" constructed lies in a field $H(\sqrt{\alpha})$ for some $\alpha \in H$, where $\alpha > 0$. We have proved half of our next theorem.

Theorem 39.2 *The field F of constructible real numbers consists precisely of all real numbers which we can obtain from* $\mathbf{Q}$ *by taking square roots of positive numbers a finite number of times and applying a finite number of field operations.*

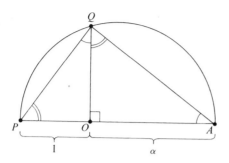

Fig. 39.4

Proof. We have shown that F can contain no numbers except those we obtain from $\mathbf{Q}$ by taking a finite number of square roots of positive numbers and applying a finite number of field operations. However, if $\alpha > 0$ is constructible, then Fig. 39.4 shows that $\sqrt{\alpha}$ is constructible. Let $\overline{OA}$ have length α, and find P on $\overline{OA}$ extended so that $\overline{OP}$ has length 1. Find the midpoint of $\overline{PA}$, and draw a semicircle with $\overline{PA}$ as diameter. Erect a perpendicular to $\overline{PA}$ at O, intersecting the semicircle at Q. Then the triangles OPQ and OQA are similar, so

$$\frac{|\overline{OQ}|}{|\overline{OA}|} = \frac{|\overline{OP}|}{|\overline{OQ}|},$$

and $|\overline{OQ}|^2 = 1\alpha = \alpha$. Thus $\overline{OQ}$ is of length $\sqrt{\alpha}$. Therefore square roots of constructible numbers are constructible.

Theorem 39.1 showed that field operations are possible by construction. ∎

Corollary. *If γ is constructible and $\gamma \notin \mathbf{Q}$, then there is a finite sequence of real numbers $\alpha_1, \ldots, \alpha_n = \gamma$ such that $\mathbf{Q}(\alpha_1, \ldots, \alpha_i)$ is an extension of $\mathbf{Q}(\alpha_1, \ldots, \alpha_{i-1})$ of degree 2. In particular, $[\mathbf{Q}(\gamma) : \mathbf{Q}] = 2^r$ for some integer $r \geq 0$.*

Proof. The existence of the α_i is immediate from Theorem 39.2. Then

$$2^n = [\mathbf{Q}(\alpha_1, \ldots, \alpha_n) : \mathbf{Q}]$$
$$= [\mathbf{Q}(\alpha_1, \ldots, \alpha_n) : \mathbf{Q}(\gamma)][\mathbf{Q}(\gamma) : \mathbf{Q}],$$

by Theorem 38.2, which completes the proof. ∎

Hopefully, the student realizes that the preceding ideas, although rather messy to write down in detail, are very simple.

*39.2 THE IMPOSSIBILITY OF CERTAIN CONSTRUCTIONS

We now can show the impossibility of certain geometric constructions.

Theorem 39.3 *"Doubling the cube is impossible," i.e., given a side of a cube, it is not always possible to construct with a straightedge and a compass the side of a cube which has double the volume of the original cube.*

Proof. Let the given cube have a side of length 1, and hence a volume of 1. The cube being sought would have to have a volume of 2, and hence a side of length $\sqrt[3]{2}$. But $\sqrt[3]{2}$ is a zero of irreducible $x^3 - 2$ over $\mathbf{Q}$, so

$$[\mathbf{Q}(\sqrt[3]{2}) : \mathbf{Q}] = 3.$$

The corollary of Theorem 39.2 shows that to double this cube of volume 1, we would need to have $3 = 2^r$ for some integer r. Clearly, no such r exists. ∎

Theorem 39.4 *"Squaring the circle is impossible," i.e., given a circle, it is not always possible to construct with a straightedge and a compass a square having area equal to the area of the given circle.*

Proof. Let the given circle have a radius of 1, and hence an area of π. We would need to construct a square of side $\sqrt{\pi}$. But π is transcendental over $\mathbf{Q}$, so $\sqrt{\pi}$ is transcendental over $\mathbf{Q}$ also. ∎

Theorem 39.5 *"Trisecting the angle is impossible," i.e., there exists an angle that cannot be trisected with a straightedge and a compass.*

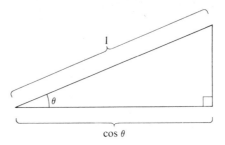

cos θ   **Fig. 39.5**

Proof. Figure 39.5 indicates that the angle θ can be constructed if and only if a segment of length $|\cos \theta|$ can be constructed. Now 60° is a constructible angle, and we shall show that it cannot be trisected. Note that

$$
\begin{aligned}
\cos 3\theta &= \cos (2\theta + \theta) \\
&= \cos 2\theta \cos \theta - \sin 2\theta \sin \theta \\
&= (2 \cos^2 \theta - 1) \cos \theta - 2 \sin \theta \cos \theta \sin \theta \\
&= (2 \cos^2 \theta - 1) \cos \theta - 2 \cos \theta (1 - \cos^2 \theta) \\
&= 4 \cos^3 \theta - 3 \cos \theta.
\end{aligned}
$$

Let $\theta = 20°$, so that $\cos 3\theta = \frac{1}{2}$, and let $\alpha = \cos 20°$. From our identity $4 \cos^3 \theta - 3 \cos \theta = \cos 3\theta$, we see that

$$4\alpha^3 - 3\alpha = \tfrac{1}{2}.$$

Thus α is a zero of $8x^3 - 6x - 1$. This polynomial is irreducible in $\mathbf{Q}[x]$, since, by Theorem 31.3, it is enough to show that it does not factor in $\mathbf{Z}[x]$. But a factorization in $\mathbf{Z}[x]$ would entail a linear factor of the form $(8x \pm 1)$, $(4x \pm 1)$, $(2x \pm 1)$, or $(x \pm 1)$. One can easily check that none of the numbers $\pm \frac{1}{8}, \pm \frac{1}{4}, \pm \frac{1}{2}$, and ± 1 is a zero of $8x^3 - 6x - 1$. Thus

$$[Q(\alpha) : Q] = 3,$$

so by the corollary of Theorem 39.2, α is not constructible. Hence $60°$ cannot be trisected. ∎

Note that the regular n-gon is constructible for $n \geq 3$ if and only if the angle $2\pi/n$ is constructible, which is the case if and only if a line segment of length $\cos (2\pi/n)$ is constructible. We shall return to the study of the constructibility of certain regular n-gons in Section 48.

EXERCISES

*39.1 Using Theorem 39.5, show that the regular 9-gon is not constructible.

*39.2 Show *algebraically* that it is possible to construct an angle of $30°$.

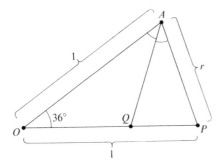

Fig. 39.6

*39.3 Referring to Fig. 39.6, where $\overline{AQ}$ bisects angle OAP, show that the regular 10-gon is constructible (and therefore that the regular pentagon is also). [*Hint:* Triangle OAP is similar to triangle APQ. Show algebraically that r is constructible.]

*39.4 Using the results of Exercise 39.3 where needed, show that the following are true.

a) The regular 20-gon is constructible.
b) The regular 30-gon is constructible.
c) The angle $72°$ can be trisected.
d) The regular 15-gon can be constructed.

*39.5 Mark each of the following true or false.

— a) It is impossible to double any cube of constructible edge by compass and straightedge constructions.
— b) It is impossible to double every cube of constructible edge by compass and straightedge constructions.
— c) It is impossible to square any circle of constructible radius by straightedge and compass constructions.
— d) No constructible angle can be trisected by straightedge and compass constructions.

  ___ e) Every constructible number is of degree 2^r over $\mathbf{Q}$ for some integer $r \geq 0$.

  ___ f) We have shown that every real number of degree 2^r over $\mathbf{Q}$ for some integer $r \geq 0$ is constructible.

  ___ g) The fact that $\mathbf{Z}$ is a UFD was used strongly at the conclusion of Theorems 39.3 and 39.5.

  ___ h) Counting arguments are exceedingly powerful mathematical tools.

  ___ i) One can find any constructible number in a finite number of steps by starting with a given segment of unit length and using a straightedge and a compass.

  ___ j) One can find the totality of all constructible numbers in a finite number of steps by starting with a given segment of unit length and using a straightedge and a compass.

40 | Automorphisms of Fields

40.1 THE BASIC ISOMORPHISMS OF ALGEBRAIC FIELD THEORY

Let F be a field, and let $\bar{F}$ be an algebraic closure of F, that is, an algebraic extension of F which is algebraically closed. Such a field $\bar{F}$ exists, by Theorem 38.6. Our selection of a particular $\bar{F}$ is not critical, since, as we shall show in Section 41, any two algebraic closures of F are isomorphic under a map leaving F fixed. *From now on in our work, we shall assume that all algebraic extensions and all elements algebraic over a field F under consideration are contained in one fixed algebraic closure $\bar{F}$ of F.*

Remember that we are engaged in the study of zeros of polynomials. In the terminology of Section 38, studying zeros of polynomials in $F[x]$ amounts to studying the structure of algebraic extensions of F and of elements algebraic over F. We shall show that if E is an algebraic extension of F with $\alpha, \beta \in E$, then α and β have the same algebraic properties if and only if $\mathrm{irr}(\alpha, F) = \mathrm{irr}(\beta, F)$. We shall phrase this fact in terms of mappings, as we have been doing all along in field theory. We achieve this by showing the existence of an isomorphism $\psi_{\alpha,\beta}$ of $F(\alpha)$ onto $F(\beta)$ which maps each element of F onto itself and maps α onto β, in the case that $\mathrm{irr}(\alpha, F) = \mathrm{irr}(\beta, F)$. The next theorem exhibits this isomorphism $\psi_{\alpha,\beta}$. These isomorphisms will become our fundamental tools for the study of algebraic extensions; they supplant the *evaluation homomorphisms* ϕ_α of Section 30, which make their last contribution in defining these isomorphisms. For this reason, we shall refer to the isomorphism $\psi_{\alpha,\beta}$ as a *"basic isomorphism of algebraic field theory."* Before stating and proving this theorem, let us introduce some more terminology.

Definition. Let E be an algebraic extension of a field F. Two elements $\alpha, \beta \in E$ are **conjugate over** F if $\mathrm{irr}(\alpha, F) = \mathrm{irr}(\beta, F)$, that is, if α and β are zeros of the same irreducible polynomial over F.

Example 40.1 The concept of conjugate elements just defined conforms with the classical idea of *conjugate complex numbers* if we understand that by conjugate complex numbers we mean numbers which are *conjugate over* $\mathbf{R}$. If $a, b \in \mathbf{R}$ and $b \neq 0$, the conjugate complex numbers $a + bi$ and $a - bi$ are both zeros of $x^2 - 2ax + a^2 + b^2$, which is irreducible in $\mathbf{R}[x]$. ‖

Theorem 40.1 (The Basic Isomorphisms of Algebraic Field Theory). *Let F be a field, and let α and β be algebraic over F with $\deg(\alpha, F) = n$. The*

335

map $\psi_{\alpha,\beta}: F(\alpha) \to F(\beta)$ *defined by*

$$(c_0 + c_1\alpha + \cdots + c_{n-1}\alpha^{n-1})\psi_{\alpha,\beta} = c_0 + c_1\beta + \cdots + c_{n-1}\beta^{n-1}$$

for $c_i \in F$ *is an isomorphism of* $F(\alpha)$ *onto* $F(\beta)$ *if and only if* α *and* β *are conjugate over* F.

Proof. Suppose that $\psi_{\alpha,\beta}: F(\alpha) \to F(\beta)$ as defined in the statement of the theorem is an isomorphism. Let $\mathrm{irr}(\alpha, F) = a_0 + a_1x + \cdots + a_nx^n$. Then $a_0 + a_1\alpha + \cdots + a_n\alpha^n = 0$, so

$$(a_0 + a_1\alpha + \cdots + a_n\alpha^n)\psi_{\alpha,\beta} = a_0 + a_1\beta + \cdots + a_n\beta^n = 0.$$

By the last assertion in the statement of Theorem 35.3, this implies that $\mathrm{irr}(\beta, F)$ divides $\mathrm{irr}(\alpha, F)$. A similar argument using the isomorphism $(\psi_{\alpha,\beta})^{-1} = \psi_{\beta,\alpha}$ shows that $\mathrm{irr}(\alpha, F)$ divides $\mathrm{irr}(\beta, F)$. Therefore, since both polynomials are monic, $\mathrm{irr}(\alpha, F) = \mathrm{irr}(\beta, F)$, so α and β are conjugate over F.

Conversely, suppose $\mathrm{irr}(\alpha, F) = \mathrm{irr}(\beta, F) = p(x)$. Then the evaluation homomorphisms $\phi_\alpha: F[x] \to F(\alpha)$ and $\phi_\beta: F[x] \to F(\beta)$ both have the same kernel $\langle p(x)\rangle$. By Theorem 29.3, corresponding to $\phi_\alpha: F[x] \to F(\alpha)$, there is a natural isomorphism ψ_α mapping $F[x]/\langle p(x)\rangle$ onto $(F[x])\phi_\alpha = F(\alpha)$. Similarly, ϕ_β gives rise to an isomorphism ψ_β mapping $F[x]/\langle p(x)\rangle$ onto $F(\beta)$. Let $\psi_{\alpha,\beta} = (\psi_\alpha)^{-1}\psi_\beta$. These mappings are diagrammed in Fig. 40.1, where the dashed lines indicate corresponding elements under the mappings. As the composition of two isomorphisms, $\psi_{\alpha,\beta}$ is again an isomorphism and maps $F(\alpha)$ onto $F(\beta)$. Also, for $(c_0 + c_1\alpha + \cdots + c_{n-1}\alpha^{n-1}) \in F(\alpha)$, we have

$$(c_0 + c_1\alpha + \cdots + c_{n-1}\alpha^{n-1})\psi_{\alpha,\beta}$$
$$= (c_0 + c_1\alpha + \cdots + c_{n-1}\alpha^{n-1})(\psi_\alpha^{-1}\psi_\beta)$$
$$= [(c_0 + c_1x + \cdots + c_{n-1}x^{n-1}) + \langle p(x)\rangle]\psi_\beta$$
$$= c_0 + c_1\beta + \cdots + c_{n-1}\beta^{n-1}.$$

Thus $\psi_{\alpha,\beta}$ is the map defined in the statement of the theorem. ∎

The following corollary of Theorem 40.1 is the cornerstone of our proof of the important *Isomorphism Extension Theorem* of Section 41 and of most of the rest of our work.

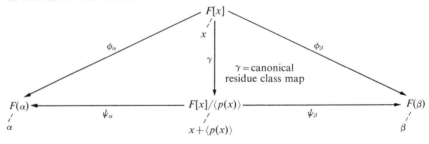

Fig. 40.1

Corollary 1 *Let α be algebraic over a field F. Every isomorphism ψ mapping $F(\alpha)$ into $\bar{F}$ such that $a\psi = a$ for $a \in F$ maps α onto a conjugate β of α over F. Conversely, for each conjugate β of α over F, there exists exactly one isomorphism $\psi_{\alpha,\beta}$ of $F(\alpha)$ into $\bar{F}$ mapping α onto β and mapping each $a \in F$ onto itself.*

Proof. Let ψ be an isomorphism mapping $F(\alpha)$ into $\bar{F}$ such that $a\psi = a$ for $a \in F$. Let $\mathrm{irr}(\alpha, F) = a_0 + a_1 x + \cdots + a_n x^n$. Then

$$a_0 + a_1\alpha + \cdots + a_n\alpha^n = 0,$$

so

$$0 = (a_0 + a_1\alpha + \cdots + a_n\alpha^n)\psi = a_0 + a_1(\alpha\psi) + \cdots + a_n(\alpha\psi)^n,$$

and $\beta = \alpha\psi$ is a conjugate of α.

Conversely, for each conjugate β of α over F, the isomorphism $\psi_{\alpha,\beta}$ of Theorem 40.1 is an isomorphism with the desired properties. That $\psi_{\alpha,\beta}$ is the only such isomorphism follows from the fact that an isomorphism of $F(\alpha)$ is completely determined by its values on elements of F and its value on α. ∎

As a second corollary of Theorem 40.1, we can prove a result with which the student is probably already familiar.

Corollary 2 *Let $f(x) \in \mathbf{R}[x]$. If $f(a + bi) = 0$ for $(a + bi) \in \mathbf{C}$, where $a, b \in \mathbf{R}$, then $f(a - bi) = 0$ also. Loosely, complex zeros of polynomials with real coefficients occur in conjugate pairs.*

Proof. We have seen that $\mathbf{C} = \mathbf{R}(i)$, and, of course, $\mathbf{C} = \mathbf{R}(-i)$ also. Now

$$\mathrm{irr}(i, \mathbf{R}) = \mathrm{irr}(-i, \mathbf{R}) = x^2 + 1,$$

so i and $-i$ are conjugate over $\mathbf{R}$. By Theorem 40.1, the map $\psi_{i,-i}: \mathbf{C} \to \mathbf{C}$ given by $(a + bi)\psi_{i,-i} = a - bi$ is an isomorphism. Thus, if for $a_i \in \mathbf{R}$,

$$f(a + bi) = a_0 + a_1(a + bi) + \cdots + a_n(a + bi)^n = 0,$$

then

$$0 = (f(a + bi))\psi_{i,-i} = a_0 + a_1(a - bi) + \cdots + a_n(a - bi)^n$$
$$= f(a - bi),$$

that is, $f(a - bi) = 0$ also. ∎

Example 40.2 Consider $\mathbf{Q}(\sqrt{2})$ over $\mathbf{Q}$. The zeros of $\mathrm{irr}(\sqrt{2}, \mathbf{Q}) = x^2 - 2$ are $\sqrt{2}$ and $-\sqrt{2}$, so $\sqrt{2}$ and $-\sqrt{2}$ are conjugate over $\mathbf{Q}$. According to Theorem 40.1, the map $\psi_{\sqrt{2},-\sqrt{2}}: \mathbf{Q}(\sqrt{2}) \to \mathbf{Q}(\sqrt{2})$ defined by

$$(a + b\sqrt{2})\psi_{\sqrt{2},-\sqrt{2}} = a - b\sqrt{2}$$

is an isomorphism of $\mathbf{Q}(\sqrt{2})$ onto itself. ‖

40.2 AUTOMORPHISMS AND FIXED FIELDS

As illustrated in the preceding corollary and example, a field may have a nontrivial isomorphism onto itself. *Such maps will be of utmost importance in the work which follows.*

> **Definition.** An isomorphism of a field onto itself is an ***automorphism of the field.***

> **Definition.** If σ is an isomorphism of a field E into some field, then an element a of E is ***left fixed by*** σ, if $a\sigma = a$. A collection S of isomorphisms of E ***leaves a subfield F of E fixed*** if each $a \in F$ is left fixed by every $\sigma \in S$. If $\{\sigma\}$ leaves F fixed, then σ ***leaves F fixed.***

It is our purpose to study the structure of an algebraic extension E of a field F by studying the automorphisms of E which leave fixed each element of F. We shall presently show that these automorphisms form a group in a natural way. We can then apply the results of Part I concerning group structure to get information about the structure of our field extension. Thus much of our preceding work is now being brought together. The next three theorems are very easy to prove, but the ideas contained in them form the foundation for everything that follows. These theorems are therefore of great importance to us. They really amount to observations, rather than theorems; it is the *ideas* contained in them that are important. A big step in mathematics does not always consist of proving a *hard* theorem, but may consist of noticing how certain known mathematics may relate to new situations. Here we are bringing group theory into our study of zeros of polynomials. Be sure that you understand the concepts involved. Unlikely as it may seem, they are the key to the solution of our *Final Goal* in this text.

> **Final Goal** (To be more precisely stated later): *To show that not all zeros of every quintic (degree 5) polynomial $f(x)$ can be expressed in terms of radicals starting with elements in the field of coefficients of $f(x)$.*

If $\{\sigma_i \mid i \in I\}$ is a collection of automorphisms of a field E, the elements of E about which $\{\sigma_i \mid i \in I\}$ gives the least information are those $a \in E$ left fixed by every σ_i for $i \in I$. This first of our three theorems contains almost all that can be said about these fixed elements of E.

> ***Theorem 40.2*** *Let $\{\sigma_i \mid i \in I\}$ be a collection of automorphisms of a field E. Then the set $E_{\{\sigma_i\}}$ of all $a \in E$ left fixed by every σ_i for $i \in I$ forms a subfield of E.*

Proof. If $a\sigma_i = a$ and $b\sigma_i = b$ for all $i \in I$, then

$$(a \pm b)\sigma_i = a\sigma_i \pm b\sigma_i = a \pm b$$

and

$$(ab)\sigma_i = (a\sigma_i)(b\sigma_i) = ab$$

for all $i \in I$. Also, if $b \neq 0$, then

$$(a/b)\sigma_i = (a\sigma_i)/(b\sigma_i) = a/b$$

for all $i \in I$. Since the σ_i are automorphisms, we have

$$0\sigma_i = 0 \quad \text{and} \quad 1\sigma_i = 1$$

for all $i \in I$. Hence $0, 1 \in E_{\{\sigma_i\}}$. Thus $E_{\{\sigma_i\}}$ is a subfield of E. ∎

Definition. The field $E_{\{\sigma_i\}}$ of Theorem 40.2 is the *fixed field of* $\{\sigma_i \mid i \in I\}$. For a single automorphism σ, we shall refer to $E_{\{\sigma\}}$ as "the *fixed field of σ*".

Example 40.3 Consider the automorphism $\psi_{\sqrt{2}, -\sqrt{2}}$ of $\mathbf{Q}(\sqrt{2})$ given in Example 40.2. For $a, b \in \mathbf{Q}$, we have

$$(a + b\sqrt{2})\psi_{\sqrt{2}, -\sqrt{2}} = a - b\sqrt{2}$$

and $a - b\sqrt{2} = a + b\sqrt{2}$ if and only if $b = 0$. Thus the fixed field of $\psi_{\sqrt{2}, -\sqrt{2}}$ is $\mathbf{Q}$. ‖

Note that an automorphism of a field E is in particular a one-to-one mapping of E onto E, that is, a *permutation of E*. If σ and τ are automorphisms of E, then the permutation $\sigma\tau$ is again an automorphism of E, since, in general, a compositum of homomorphisms is again a homomorphism. This is how group theory makes its entrance.

Theorem 40.3 *The set of all automorphisms of a field E is a group under function composition.*

Proof. Multiplication of automorphisms of E is defined by function composition, and is thus associative (it is *permutation multiplication*). The identity permutation $\iota: E \rightarrow E$ given by $\alpha\iota = \alpha$ for $\alpha \in E$ is obviously an automorphism of E. If σ is an automorphism, then the permutation σ^{-1} is also obviously an automorphism. Thus all automorphisms of E form a subgroup of S_E, the group of all permutations of E given by Theorem 4.1. ∎

Theorem 40.4 *Let E be a field, and let F be a subfield of E. Then the set $G(E/F)$ of all automorphisms of E leaving F fixed forms a subgroup of the group of all automorphisms of E. Furthermore, $F \leq E_{G(E/F)}$.*

Proof. For $\sigma, \tau \in G(E/F)$ and $a \in F$, we have

$$a(\sigma\tau) = (a\sigma)\tau = a\tau = a,$$

so $\sigma\tau \in G(E/F)$. Clearly, the identity automorphism ι is in $G(E/F)$. Also, if $a\sigma = a$ for $a \in F$, then $a = a\sigma^{-1}$, so $\sigma \in G(E/F)$ implies that $\sigma^{-1} \in G(E/F)$. Thus $G(E/F)$ is a subgroup of the group of all automorphisms of E.

Since every element of F is left fixed by every element of $G(E/F)$, it follows immediately that the field $E_{G(E/F)}$ of *all* elements of E left fixed by $G(E/F)$ contains F. ∎

Definition. The group $G(E/F)$ of the preceding theorem is the **group of automorphisms of E leaving F fixed**, or more briefly, the **group of E over F**.

Do not think of "E/F" as denoting a quotient space of some sort, but rather as meaning that E is an extension field of the field F.

The ideas contained in the preceding three theorems are illustrated in the following example. We urge the student to study carefully this example.

Example 40.4 Consider the field $Q(\sqrt{2}, \sqrt{3})$. If we view $Q(\sqrt{2}, \sqrt{3})$ as $(Q(\sqrt{3}))(\sqrt{2})$, the basic isomorphism $\psi_{\sqrt{2},-\sqrt{2}}$ of Theorem 40.1 defined by

$$(a + b\sqrt{2})\psi_{\sqrt{2},-\sqrt{2}} = a - b\sqrt{2}$$

for $a, b \in Q(\sqrt{3})$ is an automorphism of $Q(\sqrt{2}, \sqrt{3})$, having $Q(\sqrt{3})$ as fixed field. Similarly, we have the automorphism $\psi_{\sqrt{3},-\sqrt{3}}$ of $Q(\sqrt{2}, \sqrt{3})$, having $Q(\sqrt{2})$ as fixed field. Since the product of two automorphisms is an automorphism, we can consider $\psi_{\sqrt{2},-\sqrt{2}} \psi_{\sqrt{3},-\sqrt{3}}$, which **moves** both $\sqrt{2}$ and $\sqrt{3}$, that is, leaves neither number fixed. Let

$$\iota = \text{the identity automorphism,}$$
$$\sigma_1 = \psi_{\sqrt{2},-\sqrt{2}},$$
$$\sigma_2 = \psi_{\sqrt{3},-\sqrt{3}}, \quad \text{and}$$
$$\sigma_3 = \psi_{\sqrt{2},-\sqrt{2}} \psi_{\sqrt{3},-\sqrt{3}}.$$

The group of all automorphisms of $Q(\sqrt{2}, \sqrt{3})$ has a fixed field, by Theorem 40.2. This fixed field must contain Q, since every automorphism of a field leaves 1 and hence the prime subfield fixed. A basis for $Q(\sqrt{2}, \sqrt{3})$ over Q is $\{1, \sqrt{2}, \sqrt{3}, \sqrt{6}\}$. Since $\sqrt{2}\sigma_1 = -\sqrt{2}$, $\sqrt{6}\sigma_1 = -\sqrt{6}$ and $\sqrt{3}\sigma_2 = -\sqrt{3}$, we see that Q is exactly the fixed field of $\{\iota, \sigma_1, \sigma_2, \sigma_3\}$. It is easily seen that $G = \{\iota, \sigma_1, \sigma_2, \sigma_3\}$ is a group under automorphism multiplication (function composition). The group table for G is given in Fig. 40.2. For example,

$$\sigma_1\sigma_3 = \psi_{\sqrt{2},-\sqrt{2}}(\psi_{\sqrt{2},-\sqrt{2}} \psi_{\sqrt{3},-\sqrt{3}}) = \psi_{\sqrt{3},-\sqrt{3}} = \sigma_2.$$

The group G is isomorphic to the Klein 4-group. We can show that G is the full group $G(Q(\sqrt{2}, \sqrt{3})/Q)$, for every automorphism τ of $Q(\sqrt{2}, \sqrt{3})$ maps $\sqrt{2}$ onto either $\pm\sqrt{2}$, by Corollary 1 of Theorem 40.1. Similarly, τ maps $\sqrt{3}$ onto either $\pm\sqrt{3}$. But since $\{1, \sqrt{2}, \sqrt{3}, \sqrt{2}\sqrt{3}\}$ is a basis for $Q(\sqrt{2}, \sqrt{3})$ over Q, an automorphism of $Q(\sqrt{2}, \sqrt{3})$ leaving Q fixed is determined by its values on $\sqrt{2}$ and $\sqrt{3}$. Clearly, $\iota, \sigma_1, \sigma_2$, and σ_3 then give all possible combinations of values on $\sqrt{2}$ and $\sqrt{3}$, and hence are all possible automorphisms of $Q(\sqrt{2}, \sqrt{3})$.

	ι	σ_1	σ_2	σ_3
ι	ι	σ_1	σ_2	σ_3
σ_1	σ_1	ι	σ_3	σ_2
σ_2	σ_2	σ_3	ι	σ_1
σ_3	σ_3	σ_2	σ_1	ι

Fig. 40.2

Note that $G(\mathbf{Q}(\sqrt{2}, \sqrt{3})/\mathbf{Q})$ has order 4, and $[\mathbf{Q}(\sqrt{2}, \sqrt{3}) : \mathbf{Q}] = 4$. *This is no accident*, but rather an instance of a quite general situation, as we shall see later. ∥

40.3 THE FROBENIUS AUTOMORPHISM

Let F be a finite field. We shall show later that the group of all automorphisms of F is cyclic. Now a cyclic group has by definition a generating element, and may have several generating elements. For an abstract cyclic group, there is no way of distinguishing any one generator as being more important than any other. However, for the cyclic group of all automorphisms of a finite field, there is a canonical (natural) generator, the *Frobenius automorphism* (classically, the *Frobenius substitution*). This fact is of considerable importance in some advanced work in algebra. The next theorem exhibits this Frobenius automorphism.

Theorem 40.5 *Let F be a finite field of characteristic p. Then the map $\sigma_p: F \rightarrow F$ defined by $a\sigma_p = a^p$ for $a \in F$ is an automorphism, the* **Frobenius automorphism,** *of F. Also, $F_{\{\sigma_p\}} \simeq \mathbf{Z}_p$.*

Proof. Let $a, b \in F$. Applying the binomial theorem to $(a + b)^p$, we have

$$(a + b)^p = a^p + (p \cdot 1)a^{p-1}b + \left(\frac{p(p - 1)}{2} \cdot 1\right) a^{p-2}b^2$$
$$+ \cdots + (p \cdot 1)ab^{p-1} + b^p$$
$$= a^p + 0a^{p-1}b + 0a^{p-2}b^2 + \cdots + 0ab^{p-1} + b^p$$
$$= a^p + b^p.$$

Thus we have

$$(a + b)\sigma_p = (a + b)^p = a^p + b^p = a\sigma_p + b\sigma_p.$$

Of course,

$$(ab)\sigma_p = (ab)^p = a^p b^p = (a\sigma_p)(b\sigma_p),$$

so σ_p is at least a homomorphism. If $a\sigma_p = 0$, then $a^p = 0$, and $a = 0$, so the kernel of σ_p is $\{0\}$, and σ_p is an isomorphic mapping. Finally, since F is finite, σ_p is onto, by counting. Thus σ_p is an automorphism of F.

The prime field $\mathbf{Z}_p$ must be contained (up to isomorphism) in F, since F is of characteristic p. For $c \in \mathbf{Z}_p$, we have $c\sigma_p = c^p = c$, by Fermat's theorem (see the corollary of Theorem 24.6). Thus the polynomial $x^p - x$ has p zeros in F, namely the elements of $\mathbf{Z}_p$. By Corollary 2 of Theorem 31.1, a polynomial of degree n over a field can have at most n zeros in the field. Since the elements fixed under σ_p are precisely the zeros in F of $x^p - x$, we see that

$$\mathbf{Z}_p = F_{\{\sigma_p\}}. \quad \blacksquare$$

Freshmen in college still sometimes make the error of saying that $(a + b)^n = a^n + b^n$. Here we see that this *freshman exponentiation* $(a + b)^p = a^p + b^p$ with exponent p is actually valid in a field F of characteristic p.

EXERCISES

40.1 Find all conjugates of each of the given numbers over the given fields.

a) $\sqrt{2}$ over $\mathbf{Q}$

b) $\sqrt{2}$ over $\mathbf{R}$

c) $3 + \sqrt{2}$ over $\mathbf{Q}$

d) $\sqrt{2} - \sqrt{3}$ over $\mathbf{Q}$

e) $\sqrt{2} + i$ over $\mathbf{Q}$

f) $\sqrt{2} + i$ over $\mathbf{R}$

g) $\sqrt{1 + \sqrt{2}}$ over $\mathbf{Q}$

h) $\sqrt{1 + \sqrt{2}}$ over $\mathbf{Q}(\sqrt{2})$

40.2 Consider the field $E = \mathbf{Q}(\sqrt{2}, \sqrt{3}, \sqrt{5})$. In the notation of Theorem 40.1, we have the following basic isomorphisms (which are here automorphisms of E):

$$\psi_{\sqrt{2},-\sqrt{2}}: (\mathbf{Q}(\sqrt{3}, \sqrt{5}))(\sqrt{2}) \to (\mathbf{Q}(\sqrt{3}, \sqrt{5}))(-\sqrt{2}),$$

$$\psi_{\sqrt{3},-\sqrt{3}}: (\mathbf{Q}(\sqrt{2}, \sqrt{5}))(\sqrt{3}) \to (\mathbf{Q}(\sqrt{2}, \sqrt{5}))(-\sqrt{3}),$$

$$\psi_{\sqrt{5},-\sqrt{5}}: (\mathbf{Q}(\sqrt{2}, \sqrt{3}))(\sqrt{5}) \to (\mathbf{Q}(\sqrt{2}, \sqrt{3}))(-\sqrt{5}).$$

For shorter notation, let $\tau_2 = \psi_{\sqrt{2},-\sqrt{2}}$, $\tau_3 = \psi_{\sqrt{3},-\sqrt{3}}$, and $\tau_5 = \psi_{\sqrt{5},-\sqrt{5}}$. Compute each of the following.

a) $\sqrt{3}\tau_2$

b) $(\sqrt{2} + \sqrt{5})\tau_2$

c) $(\sqrt{2} + 3\sqrt{5})(\tau_2\tau_3)$

d) $\dfrac{\sqrt{2} - 3\sqrt{5}}{2\sqrt{3} - \sqrt{2}} (\tau_3\tau_5)$

e) $(\sqrt{2} + \sqrt{45})(\tau_2\tau_3\tau_5{}^2)$

f) $[(\sqrt{2} - \sqrt{3})\tau_5 + \sqrt{30}(\tau_5\tau_2)]\tau_3$

40.3 The fields $\mathbf{Q}(\sqrt{2})$ and $\mathbf{Q}(3 + \sqrt{2})$ are the same, of course. Let $\alpha = 3 + \sqrt{2}$.

a) Find a conjugate $\beta \neq \alpha$ of α over $\mathbf{Q}$.

b) Referring to (a), compare the basic automorphism $\psi_{\sqrt{2},-\sqrt{2}}$ of $\mathbf{Q}(\sqrt{2})$ with the basic automorphism $\psi_{\alpha,\beta}$.

40.4 Referring to Example 40.4, find the following fixed fields in $E = \mathbf{Q}(\sqrt{2}, \sqrt{3})$.

a) $E_{\{\sigma_1,\sigma_3\}}$

b) $E_{\{\sigma_3\}}$

c) $E_{\{\sigma_2,\sigma_3\}}$

40.5 Referring to Exercise 40.2, find the fixed field of each of the following automorphisms or sets of automorphisms of E.

a) τ_3

b) $\tau_3{}^2$

c) $\{\tau_2, \tau_3\}$

d) $\tau_2\tau_5$

e) $\tau_2\tau_3\tau_5$

f) $\{\tau_2, \tau_3, \tau_5\}$

†**40.6** Let α be algebraic of degree n over F. Show from Corollary 1 of Theorem 40.1 that there are at most n different isomorphisms of $F(\alpha)$ into $\overline{F}$.

40.7 Mark each of the following true or false.

— a) For all $\alpha, \beta \in E$, there is always an automorphism of E mapping α onto β.

— b) For α, β algebraic over a field F, there is always an isomorphism of $F(\alpha)$ onto $F(\beta)$.

— c) For α, β algebraic and conjugate over a field F, there is always an isomorphism of $F(\alpha)$ onto $F(\beta)$.

— d) Every automorphism of every field E leaves fixed every element of the prime subfield of E.

— e) Every automorphism of every field E leaves fixed an infinite number of elements of E.

— f) Every automorphism of every field E leaves fixed at least two elements of E.

— g) Every automorphism of every field E of characteristic 0 leaves fixed an infinite number of elements of E.

— h) All automorphisms of a field E form a group under function composition.

— i) The set of all elements of a field E left fixed by a single automorphism of E forms a subfield of E.

— j) For fields $F \leq E \leq K$, $G(K/E) \leq G(K/F)$.

40.8 Refer to Exercise 40.2 for the following:

a) Show that each of the automorphisms τ_2, τ_3, and τ_5 is of order 2 in $G(E/\mathbf{Q})$. (Remember what is meant by the *order* of an element of a group.)

b) Find the subgroup H of $G(E/\mathbf{Q})$ generated by the elements τ_2, τ_3, and τ_5, and give the group table. [*Hint:* There are 8 elements.]

c) Just as was done in Example 40.4, argue that the group H of (b) is the full group $G(E/\mathbf{Q})$.

40.9 Describe the value of the Frobenius automorphism σ_2 on each element of the finite field of 4 elements given in Example 35.9. Find the fixed field of σ_2.

40.10 Describe the value of the Frobenius automorphism σ_3 on each element of the finite field of 9 elements given in Exercise 35.7. Find the fixed field of σ_3.

40.11 Let F be a field of characteristic $p \neq 0$. Give an example to show that the map $\sigma_p: F \to F$ given by $a\sigma_p = a^p$ for $a \in F$ need not be an automorphism in the case that F is infinite. What may go wrong?

40.12 Let $F(\alpha_1, \ldots, \alpha_n)$ be an extension field of F. Show that any automorphism σ of $F(\alpha_1, \ldots, \alpha_n)$ leaving F fixed is completely determined by the n values $\alpha_i\sigma$.

40.13 Let E be an algebraic extension of a field F, and let σ be an automorphism of E leaving F fixed. Let $\alpha \in E$. Show that σ induces a permutation of the set of all zeros of irr(α, F) which are in E.

40.14 Let E be an algebraic extension of a field F. Let $S = \{\sigma_i \mid i \in I\}$ be a collection of automorphisms of E such that every σ_i leaves each element of F fixed. Show that if S generates the subgroup H of $G(E/F)$, then $E_S = E_H$.

40.15 We saw in the corollary of Theorem 31.4 that the cyclotomic polynomial

$$\Phi_p(x) = \frac{x^p - 1}{x - 1} = x^{p-1} + x^{p-2} + \cdots + x + 1$$

is irreducible over $\mathbf{Q}$ for every prime p. Let ζ be a zero of $\Phi_p(x)$, and consider the field $\mathbf{Q}(\zeta)$.

a) Show that ζ, $\zeta^2, \ldots, \zeta^{p-1}$ are distinct zeros of $\Phi_p(x)$, and conclude that they are all the zeros of $\Phi_p(x)$.

b) Deduce from Corollary 1 of Theorem 40.1 and (a) that $G(\mathbf{Q}(\zeta)/\mathbf{Q})$ is abelian of order $p - 1$.

c) Show that the fixed field of $G(\mathbf{Q}(\zeta)/\mathbf{Q})$ is $\mathbf{Q}$. [*Hint:* Show that

$$\{\zeta, \zeta^2, \ldots, \zeta^{p-1}\}$$

is a basis for $\mathbf{Q}(\zeta)$ over $\mathbf{Q}$, and consider which linear combinations of $\zeta, \zeta^2, \ldots,$ ζ^{p-1} are left fixed by all elements of $G(\mathbf{Q}(\zeta)/\mathbf{Q})$.]

40.16 Theorem 40.1 described basic isomorphisms for the case where α and β were conjugate algebraic elements over F. Is there a similar isomorphism of $F(\alpha)$ with $F(\beta)$ in the case that α and β are both transcendental over F?

40.17 Let F be a field, and let x be an indeterminate over F. Determine all automorphisms of $F(x)$ leaving F fixed, by describing their values on x.

40.18 Prove the following sequence of theorems.

a) An automorphism of a field E carries elements which are squares of elements in E onto elements which are squares of elements of E.
b) An automorphism of the field $\mathbf{R}$ of real numbers carries positive numbers onto positive numbers.
c) If σ is an automorphism of $\mathbf{R}$ and $a < b$, where $a, b \in \mathbf{R}$, then $a\sigma < b\sigma$.
d) An automorphism of $\mathbf{R}$ is completely determined by its values on elements of $\mathbf{Q}$.
e) The only automorphism of $\mathbf{R}$ is the identity automorphism.

41 | The Isomorphism Extension Theorem

41.1 THE EXTENSION THEOREM

Let us continue studying automorphisms of fields. In this section and the next, we shall be concerned with both the existence and the number of automorphisms of a field E.

Suppose that E is an algebraic extension of F and that we want to find some automorphisms of E. We know from Theorem 40.1 that if $\alpha, \beta \in E$ are conjugate over F, then there is an isomorphism $\psi_{\alpha,\beta}$ of $F(\alpha)$ onto $F(\beta)$. Of course, $\alpha, \beta \in E$ implies both $F(\alpha) \leq E$ and $F(\beta) \leq E$. It is natural to wonder whether the domain of definition of $\psi_{\alpha,\beta}$ can be enlarged from $F(\alpha)$ to a bigger field, perhaps all of E, and whether this might perhaps lead to an automorphism of E. A mapping diagram of this situation is shown in Fig. 41.1. Rather than speak of "enlarging the domain of definition of $\psi_{\alpha,\beta}$," it is customary to speak of "**extending the map $\psi_{\alpha,\beta}$ to a map τ**," which is a mapping of all of E. (Compare with the situation in Theorem 26.2.)

Remember that we are always assuming that all algebraic extensions of F under consideration are contained in a fixed algebraic closure $\bar{F}$ of F. The *Isomorphism Extension Theorem* shows that the mapping $\psi_{\alpha,\beta}$ can indeed always be extended to an *isomorphism* of E into $\bar{F}$. Whether this extension gives an *automorphism* of E, that is, maps E onto itself, is a question we shall study in Section 42. Thus this extension theorem, used in conjunction with our basic isomorphisms $\psi_{\alpha,\beta}$, will guarantee the existence of lots of *isomorphic mappings*, at least, for many fields. Except for Theorem 26.2, this theorem may well be the only extension theorem the student has encountered. Such theorems are very important in mathematics, particularly in algebraic and topological situations.

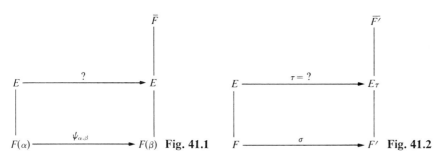

Fig. 41.1   Fig. 41.2

Let us take a more general look at this situation. Suppose that E is an algebraic extension of a field F and that we have an isomorphism σ of F onto a field F'. Let $\overline{F'}$ be an algebraic closure of F'. We would like to extend σ to an isomorphism τ of E into $\overline{F'}$. This situation is shown in Fig. 41.2. Naively, we pick $\alpha \in E$ but not in F, and try to extend σ to $F(\alpha)$. If

$$p(x) = \mathrm{irr}(\alpha, F) = a_0 + a_1 x + \cdots + a_n x^n,$$

let β be a zero in $\overline{F'}$ of

$$q(x) = a_0\sigma + (a_1\sigma)x + \cdots + (a_n\sigma)x^n.$$

Here $q(x) \in F'[x]$. Since σ is an isomorphism, it is clear that $q(x)$ is irreducible in $F'[x]$. It is also pretty clear that $F(\alpha)$ can be mapped isomorphically onto $F'(\beta)$ by a map extending σ and mapping α onto β. (This isn't quite Theorem 40.1, but it is close to it; a few elements have been renamed by the isomorphism σ.) If $F(\alpha) = E$, we are done. If $F(\alpha) \neq E$, we have to find another element in E not in $F(\alpha)$ and continue the process. It is a situation very much like that in the construction of an algebraic closure $\overline{F}$ of a field F. Again, the trouble is that, in general, where E is not a finite extension, the process may have to be repeated a (possibly large) infinite number of times, so we need Zorn's lemma to handle it. For this reason, we give the general proof of Theorem 41.1 under a starred heading at the end of this section.

Theorem 41.1 (Isomorphism Extension Theorem). *Let E be an algebraic extension of a field F. Let σ be an isomorphism of F onto a field F'. Let $\overline{F'}$ be an algebraic closure of F'. Then σ can be extended to an isomorphism τ of E into $\overline{F'}$ such that $a\tau = a\sigma$ for all $a \in F$.*

We give as a corollary the existence of an extension of one of our basic isomorphisms $\psi_{\alpha,\beta}$, as discussed at the start of this section.

Corollary 1 *If $E \le \overline{F}$ is an algebraic extension of F and $\alpha, \beta \in E$ are conjugate over F, then the basic isomorphism $\psi_{\alpha,\beta}: F(\alpha) \to F(\beta)$, given by Theorem 40.1, can be extended to an isomorphism of E into $\overline{F}$.*

Proof. This is immediate from Theorem 41.1, if in the statement of the theorem we replace F by $F(\alpha)$, F' by $F(\beta)$, and $\overline{F'}$ by $\overline{F}$. ∎

As another corollary, we can show, as we promised earlier, that an algebraic closure of F is unique, up to an isomorphism leaving F fixed.

Corollary 2 *Let $\overline{F}$ and $\overline{F'}$ be two algebraic closures of F. Then $\overline{F}$ is isomorphic to $\overline{F'}$ under an isomorphism leaving each element of F fixed.*

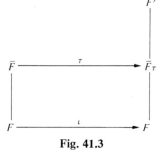

Fig. 41.3

Proof. By Theorem 41.1, the identity isomorphism of F onto F can be extended to an isomorphism τ mapping $\overline{F}$ into $\overline{F}'$ which leaves F fixed (see Fig. 41.3). We need only show τ is onto $\overline{F}'$. But by Theorem 41.1, the map $\tau^{-1}\colon \overline{F}\tau \to \overline{F}$ can be extended to an isomorphism of $\overline{F}'$ into $\overline{F}$. Since τ^{-1} is already onto $\overline{F}$, we must have $\overline{F}\tau = \overline{F}'$. ∎

41.2　THE INDEX OF A FIELD EXTENSION

Having discussed the question of *existence*, we turn now to the question of *how many*. For a *finite* extension E of a field F, we would like to count how many isomorphisms there are of E into $\overline{F}$ which leave F fixed. We shall show that there are only a finite number of such isomorphisms. Since every automorphism in $G(E/F)$ is such an isomorphism, a count of these isomorphisms will include all these automorphisms. Example 40.4 showed that $G(Q(\sqrt{2}, \sqrt{3})/Q)$ has 4 elements, and that $4 = [Q(\sqrt{2}, \sqrt{3}) : Q]$. While such an equality is not always true, it is true in a very important case. The next theorem takes the first big step in proving this. We state the theorem in more general terms than we shall need, but it does not make the proof any harder.

Theorem 41.2 *Let E be a finite extension of a field F. Let σ be an isomorphism of F onto a field F', and let $\overline{F}'$ be an algebraic closure of F'. Then the number of extensions of σ to an isomorphism τ of E into $\overline{F}'$ is finite, and independent of $F', \overline{F}'$, and σ. That is, this number of extensions is completely determined by the two fields E and F; it is intrinsic to them.*

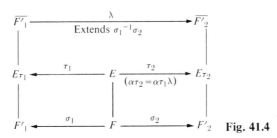

Fig. 41.4

Proof. The diagram in Fig. 41.4 may help you to follow the construction that we are about to make. This diagram is constructed in the following way. Consider two isomorphisms

$$\sigma_1\colon F \xrightarrow{\text{onto}} F'_1, \qquad \sigma_2\colon F \xrightarrow{\text{onto}} F'_2,$$

where $\overline{F'_1}$ and $\overline{F'_2}$ are algebraic closures of F'_1 and F'_2, respectively. Now $\sigma_1^{-1}\sigma_2$ is an isomorphism of F'_1 onto F'_2. Then by Theorem 41.1 and its second corollary, there is an isomorphism

$$\lambda\colon \overline{F'_1} \xrightarrow{\text{onto}} \overline{F'_2}$$

extending this isomorphism $\sigma_1{}^{-1}\sigma_2 \colon F'_1 \xrightarrow{\text{onto}} F'_2$. Referring to Fig. 41.4, corresponding to each $\tau_1 \colon E \to \overline{F'_1}$ which extends σ_1, one obtains an isomorphism $\tau_2 \colon E \to \overline{F'_2}$, by starting at E and going first to the left, then up, and then to the right. Written algebraically,

$$\alpha\tau_2 = \alpha\tau_1\lambda$$

for $\alpha \in E$. Clearly τ_2 extends σ_2. The fact that we could have *started* with τ_2 and recovered τ_1 by defining

$$\alpha\tau_1 = \alpha\tau_2\lambda^{-1},$$

that is, by chasing the other way around the diagram, shows that the correspondence between $\tau_1 \colon E \to \overline{F'_1}$ and $\tau_2 \colon E \to \overline{F'_2}$ is one to one. In view of this one-to-one correspondence, the number of τ extending σ is independent of F', $\overline{F'}$, and σ.

That the number of mappings extending σ is finite follows from the fact that since E is a finite extension of F, $E = F(\alpha_1, \ldots, \alpha_n)$ for some $\alpha_1, \ldots, \alpha_n$ in E, by Theorem 38.3. There are only a finite number of possible candidates for the images $\alpha_i\tau$ in $\overline{F'}$, for if

$$\mathrm{irr}(\alpha_i, F) = a_{i0} + a_{i1}x + \cdots + a_{im_i}x^{m_i},$$

where $a_{ik} \in F$, then $\alpha_i\tau$ must be one of the zeros in $\overline{F'}$ of

$$[a_{i0}\sigma + (a_{i1}\sigma)x + \cdots + (a_{im_i}\sigma)x^{m_i}] \in F'[x]. \quad \blacksquare$$

Definition. Let E be a finite extension of a field F. The number of isomorphisms of E into $\overline{F}$ leaving F fixed is the **index** $\{E : F\}$ **of E over F**.

Corollary. *If $F \leq E \leq K$, where K is a finite extension field of the field F, then $\{K : F\} = \{K : E\}\{E : F\}$.*

Proof. It follows from Theorem 41.2 that each of the $\{E : F\}$ isomorphisms $\tau_i \colon E \to \overline{F}$ leaving F fixed has $\{K : E\}$ extensions to an isomorphism of K into $\overline{F}$. $\quad \blacksquare$

The preceding corollary was really the main thing we were after. Note that it counts something. *Never underestimate a result which counts something,* even if it is only called a "corollary."

We shall show in Section 43 that unless F is an infinite field of characteristic $p \neq 0$, we always have $[E : F] = \{E : F\}$, for every finite extension field E of F. For the case $E = F(\alpha)$, the $\{F(\alpha) : F\}$ extensions of the identity map $\iota \colon F \to F$ to maps of $F(\alpha)$ into $\overline{F}$ are given by the basic isomorphisms $\psi_{\alpha,\beta}$, for each conjugate β in $\overline{F}$ of α over F. Thus if $\mathrm{irr}(\alpha, F)$ has n *distinct* zeros in $\overline{F}$, we have $\{E : F\} = n$. We shall show later that unless F is infinite and of characteristic $p \neq 0$, the number of distinct zeros of $\mathrm{irr}(\alpha, F)$ is $\deg(\alpha, F) = [F(\alpha) : F]$.

Example 41.1 Consider $E = \mathbf{Q}(\sqrt{2}, \sqrt{3})$ over $\mathbf{Q}$, as in Example 40.4. Our work in Example 40.4 shows that $\{E : \mathbf{Q}\} = [E : \mathbf{Q}] = 4$. Also, $\{E : \mathbf{Q}(\sqrt{2})\} = 2$, and $\{\mathbf{Q}(\sqrt{2}) : \mathbf{Q}\} = 2$, so

$$4 = \{E : \mathbf{Q}\} = \{E : \mathbf{Q}(\sqrt{2})\} \{\mathbf{Q}(\sqrt{2}) : \mathbf{Q}\} = (2)(2);$$

this illustrates the corollary of Theorem 41.2. ‖

*41.3 PROOF OF THE EXTENSION THEOREM

We restate the Extension Theorem.

> **Theorem 41.1** (Isomorphism Extension Theorem). *Let E be an algebraic extension of a field F. Let σ be an isomorphism of F onto a field F'. Let $\overline{F'}$ be an algebraic closure of F'. Then σ can be extended to an isomorphism τ of E into $\overline{F'}$ such that $a\tau = a\sigma$ for $a \in F$.*

Proof. Consider all pairs (L, λ), where L is a field such that $F \leq L \leq E$ and λ is an isomorphism of L into $\overline{F'}$ such that $a\lambda = a\sigma$ for $a \in F$. The set S of such pairs (L, λ) is nonempty, since (F, σ) is such a pair. Define a partial ordering on S by $(L_1, \lambda_1) \leq (L_2, \lambda_2)$, if $L_1 \leq L_2$ and $a\lambda_2 = a\lambda_1$ for $a \in L_1$. It is easily checked that this relation $\leq$ does give a partial ordering of S.

Let $T = \{(H_i, \lambda_i) \mid i \in I\}$ be a chain of S. We claim that $H = \bigcup_{i \in I} H_i$ is a subfield of E. Let $a, b \in H$, where $a \in H_1$ and $b \in H_2$; then either $H_1 \leq H_2$ or $H_2 \leq H_1$, since T is a chain. If say $H_1 \leq H_2$, then $a, b \in H_2$, so $a \pm b$, ab, and a/b for $b \neq 0$ are all in H_2 and hence in H. Since for each $i \in I$, $F \subseteq H_i \subseteq E$, we have $F \subseteq H \subseteq E$. Thus H is a subfield of E.

Define $\lambda : H \to \overline{F'}$ as follows. Let $c \in H$. Then $c \in H_i$ for some $i \in I$, and let

$$c\lambda = c\lambda_i.$$

The map λ is well defined, because if $c \in H_1$ and $c \in H_2$, then either $(H_1, \lambda_1) \leq (H_2, \lambda_2)$ or $(H_2, \lambda_2) \leq (H_1, \lambda_1)$, since T is a chain. In either case, $c\lambda_1 = c\lambda_2$. We claim that λ is an isomorphism of H into $\overline{F'}$. If $a, b \in H$, then there is an H_i such that $a, b \in H_i$, and

$$(a + b)\lambda = (a + b)\lambda_i = a\lambda_i + b\lambda_i = a\lambda + b\lambda.$$

Similarly,

$$(ab)\lambda = (ab)\lambda_i = (a\lambda_i)(b\lambda_i) = (a\lambda)(b\lambda).$$

If $a\lambda = 0$, then $a \in H_i$ for some i implies that $a\lambda_i = 0$, so $a = 0$. Therefore, λ is an isomorphism. Thus $(H, \lambda) \in S$, and it is clear from our definitions of H and λ that (H, λ) is an upper bound for T.

We have shown that every chain of S has an upper bound in S, so the hypotheses of Zorn's lemma are satisfied. Hence there exists a maximal element (K, τ) of S. Let $K\tau = K'$, where $K' \leq \overline{F'}$. Now if $K \neq E$, let $\alpha \in E$ but $\alpha \notin K$. Now α is algebraic over F, so α is algebraic over K. Also, let $p(x) = \text{irr}(\alpha, K)$. Let ψ_α be the canonical isomorphism

$$\psi_\alpha: K[x]/\langle p(x) \rangle \to K(\alpha),$$

corresponding to the basic homomorphism $\phi_\alpha: K[x] \to K(\alpha)$. If

$$p(x) = a_0 + a_1 x + \cdots + a_n x^n,$$

consider

$$q(x) = a_0 \tau + (a_1 \tau)x + \cdots + (a_n \tau)x^n$$

in $K'[x]$. Obviously, since τ is an isomorphism, $q(x)$ is irreducible in $K'[x]$. Since $K' \leq \overline{F'}$, there is a zero α' of $q(x)$ in $\overline{F'}$. Let

$$\psi_{\alpha'}: K'[x]/\langle q(x) \rangle \to K'(\alpha')$$

be the isomorphism analogous to ψ_α. Finally, let

$$\overline{\tau}: K[x]/\langle p(x) \rangle \to K'[x]/\langle q(x) \rangle$$

be the obvious isomorphism extending τ on K and mapping $x + \langle p(x) \rangle$ onto $x + \langle q(x) \rangle$. (See Fig. 41.5.) Then the composition of maps

$$(\psi_\alpha)^{-1}\overline{\tau}\psi_{\alpha'}: K(\alpha) \to K'(\alpha')$$

is an isomorphism of $K(\alpha)$ into $\overline{F'}$. Clearly, $(K, \tau) < \left(K(\alpha), (\psi_\alpha)^{-1}\overline{\tau}\psi_{\alpha'}\right)$, which contradicts that (K, τ) is maximal. Therefore we must have had $K = E$. ∎

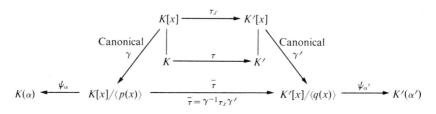

Fig. 41.5

Unfortunately, the student probably considers this last proof to be hopelessly complicated. To the professional algebraist, this construction is so standard that he usually just writes down something like "the proof follows by Zorn's lemma" as soon as he has defined the partial ordering $\leq$ on the set S. We wrote out the proof in hideous detail, since it may be only the second proof the student has seen involving Zorn's lemma.

EXERCISES

41.1 Let $E = \mathbf{Q}(\sqrt{2}, \sqrt{3}, \sqrt{5})$. For each isomorphic mapping of a subfield of E given below, give all extensions of the mapping to an isomorphic mapping of E into $\overline{\mathbf{Q}}$. Describe the extensions by giving values on the generating set $\{\sqrt{2}, \sqrt{3}, \sqrt{5}\}$ for E over $\mathbf{Q}$.

a) $\iota: \mathbf{Q}(\sqrt{2}, \sqrt{15}) \to \mathbf{Q}(\sqrt{2}, \sqrt{15})$, where ι is the identity map
b) $\sigma: \mathbf{Q}(\sqrt{2}, \sqrt{15}) \to \mathbf{Q}(\sqrt{2}, \sqrt{15})$, where $\sqrt{2}\sigma = \sqrt{2}$ and $\sqrt{15}\sigma = -\sqrt{15}$
c) $\psi_{\sqrt{30}, -\sqrt{30}}: \mathbf{Q}(\sqrt{30}) \to \mathbf{Q}(\sqrt{30})$

41.2 It is a fact, which you can verify by cubing, that the zeros of $x^3 - 2$ in $\mathbf{Q}$ are

$$\alpha_1 = \sqrt[3]{2}, \qquad \alpha_2 = \sqrt[3]{2}\,\frac{-1 + i\sqrt{3}}{2}, \qquad \text{and} \qquad \alpha_3 = \sqrt[3]{2}\,\frac{-1 - i\sqrt{3}}{2},$$

where $\sqrt[3]{2}$, as usual, is the real cube root of 2.

a) Describe all extensions of the identity map of $\mathbf{Q}$ to an isomorphism mapping $\mathbf{Q}(\sqrt[3]{2})$ into $\overline{\mathbf{Q}}$.
b) Describe all extensions of the identity map of $\mathbf{Q}$ to an isomorphism mapping $\mathbf{Q}(\sqrt[3]{2}, \sqrt{3})$ into $\overline{\mathbf{Q}}$.
c) Describe all extensions of the automorphism $\psi_{\sqrt{3}, -\sqrt{3}}$ of $\mathbf{Q}(\sqrt{3})$ to an isomorphism mapping $\mathbf{Q}(i, \sqrt{3}, \sqrt[3]{2})$ into $\overline{\mathbf{Q}}$.

41.3 Let σ be the automorphism of $\mathbf{Q}(\pi)$ which maps π onto $-\pi$.

a) Describe the fixed field of σ.
b) Describe all extensions of σ to an isomorphism mapping the field $\mathbf{Q}(\sqrt{\pi})$ into $\overline{\mathbf{Q}(\pi)}$.

†**41.4** Let K be an algebraically closed field. Show that every isomorphism σ of K into itself such that K is algebraic over $K\sigma$ is an automorphism of K, i.e., is an onto map. [*Hint:* Apply Theorem 41.1 to σ^{-1}.]

41.5 Mark each of the following true or false.

— a) Let $F(\alpha)$ be any simple extension of a field F. Then every isomorphism of F into $\overline{F}$ has an extension to an isomorphism of $F(\alpha)$ into $\overline{F}$.
— b) Let $F(\alpha)$ be any simple algebraic extension of a field F. Then every isomorphism of F into $\overline{F}$ has an extension to an isomorphism of $F(\alpha)$ into $\overline{F}$.
— c) An isomorphism of F into $\overline{F}$ has the same number of extensions to each simple algebraic extension of F.
— d) Algebraic closures of isomorphic fields are always isomorphic.
— e) Algebraic closures of fields which are not isomorphic are never isomorphic.
— f) Any algebraic closure of $\mathbf{Q}(\sqrt{2})$ is isomorphic to any algebraic closure of $\mathbf{Q}(\sqrt{17})$.
— g) The index of a finite extension E over a field F is finite.
— h) The index behaves multiplicatively with respect to finite towers of finite extensions of fields.
— i) Our remarks prior to the statement of Theorem 41.1 in Section 41.1 essentially constitute a proof of this theorem for a finite extension E over F.
— j) Corollary 2 of Theorem 41.1 shows that $\mathbf{C}$ is isomorphic to $\overline{\mathbf{Q}}$.

41.6 Let E be an algebraic extension of a field F. Show that every isomorphism of E into $\overline{F}$ leaving F fixed can be extended to an automorphism of $\overline{F}$.

41.7 Prove that if E is an algebraic extension of a field F, then two algebraic closures $\overline{F}$ and $\overline{E}$ of F and E, respectively, are isomorphic.

41.8 Prove that the algebraic closure of $\mathbf{Q}(\sqrt{\pi})$ in $\mathbf{C}$ is isomorphic to any algebraic closure of $\overline{\mathbf{Q}}(x)$, where $\overline{\mathbf{Q}}$ is the field of algebraic numbers and x is an indeterminate.

41.9 Prove that if E is a finite extension of a field F, then $\{E : F\} \leq [E : F]$. [*Hint:* The remarks preceding Example 41.1 essentially showed this for a simple algebraic extension $F(\alpha)$ of F. Use the fact that a finite extension is a tower of simple extensions, together with the multiplicative properties of the index and degree.]

42 | Splitting Fields

We are going to be interested chiefly in *automorphisms* of a field E, rather than mere isomorphic mappings of E. You see, it is the *automorphisms* of a field which form a group. We wonder whether for some extension field E of a field F, *every* isomorphic mapping of E into $\bar{F}$ leaving F fixed is actually an automorphism of E.

Suppose E is an algebraic extension of a field F. If $\alpha \in E$ and $\beta \in \bar{F}$ is a conjugate of α over F, then there is a basic isomorphism

$$\psi_{\alpha,\beta}: F(\alpha) \rightarrow F(\beta).$$

By Corollary 1 of Theorem 41.1, $\psi_{\alpha,\beta}$ can be extended to an isomorphic mapping of E into $\bar{F}$. Now if $\beta \notin E$, such an isomorphic mapping of E can't be an automorphism of E. *Thus, if an algebraic extension E of a field F is such that all its isomorphic mappings into $\bar{F}$ leaving F fixed are actually automorphisms of E, then for every $\alpha \in E$, all conjugates of α over F must be in E also.* This observation seemed to come very easily. We point out that we used a lot of power, namely the existence of the basic isomorphisms and the Isomorphism Extension Theorem.

These ideas suggest the formulation of the following definition.

Definition. Let F be a field with algebraic closure $\bar{F}$. Let $\{f_i(x) \mid i \in I\}$ be a collection of polynomials in $F[x]$. A field $E \le \bar{F}$ is the *splitting field* *of* $\{f_i(x) \mid i \in I\}$ *over* F if E is the smallest subfield of $\bar{F}$ containing F and all the zeros in $\bar{F}$ of each of the $f_i(x)$ for $i \in I$. A field $K \le \bar{F}$ is a *splitting field over* F if it is the splitting field of some set of polynomials in $F[x]$.

For one polynomial $f(x) \in F[x]$, we shall often refer to the splitting field of $\{f(x)\}$ over F as the "**splitting field of** $f(x)$ **over** F." It is clear that the splitting field of $\{f_i(x) \mid i \in I\}$ over F in $\bar{F}$ is the intersection of all subfields of $\bar{F}$ containing F and all zeros in $\bar{F}$ of each $f_i(x)$ for $i \in I$. Thus such a splitting field surely does exist.

We now show that splitting fields over F are precisely those fields $E \le \bar{F}$ with the property that all isomorphic mappings of E into $\bar{F}$ leaving F fixed are automorphisms of E. This will be a corollary of the next theorem. *Once more, we are characterizing a concept in terms of mappings.* Remember, we are always assuming that all algebraic extensions of a field F under consideration are in one fixed algebraic closure $\bar{F}$ of F.

Theorem 42.1 *A field E, where $F \leq E \leq \overline{F}$, is a splitting field over F if and only if every automorphism of $\overline{F}$ leaving F fixed maps E onto itself and thus induces an automorphism of E leaving F fixed.*

Proof. Let E be a splitting field over F in $\overline{F}$ of $\{f_i(x) \mid i \in I\}$, and let σ be an automorphism of $\overline{F}$ leaving F fixed. Let $\{\alpha_j \mid j \in J\}$ be the collection of all zeros in $\overline{F}$ of all the $f_i(x)$ for $i \in I$. Now our previous work shows that for a fixed α_j, $F(\alpha_j)$ has as elements all expressions of the form

$$g(\alpha_j) = a_0 + a_1\alpha_j + \cdots + a_{n_j - 1}\alpha_j^{n_j - 1},$$

where n_j is the degree of $\mathrm{irr}(\alpha_j, F)$ and $a_k \in F$. Consider the set S of all *finite* sums of *finite* products of elements of the form $g(\alpha_j)$ for all $j \in J$. The set S is a subset of E obviously closed under addition and multiplication and containing 0, 1 and the additive inverse of each element. Since each element of S is in some $F(\alpha_{j_1}, \ldots, \alpha_{j_r}) \subseteq S$, we see that S also contains the multiplicative inverse of each nonzero element. Thus S is a subfield of E containing all α_j for $j \in J$. By definition of the splitting field E of $\{f_i(x) \mid i \in I\}$, we see that we must have $S = E$. All this work was just to show that $\{\alpha_j \mid j \in J\}$ *generates* E over F, in the sense of taking *finite* sums and *finite* products. Knowing this, we see immediately that the value of σ on any element of E is completely determined by the values $\alpha_j\sigma$. But by Corollary 1 of Theorem 40.1, $\alpha_j\sigma$ must also be a zero of $\mathrm{irr}(\alpha_j, F)$. By Theorem 35.3, $\mathrm{irr}(\alpha_j, F)$ divides the $f_i(x)$ for which $f_i(\alpha_j) = 0$, so $\alpha_j\sigma \in E$ also. Thus σ maps E into E isomorphically. However, the same is true of the automorphism σ^{-1} of $\overline{F}$. Since for $\beta \in E$,

$$\beta = (\beta\sigma^{-1})\sigma,$$

we see that σ maps E onto E, and thus induces an automorphism of E.

Suppose, conversely, that every automorphism of $\overline{F}$ leaving F fixed induces an automorphism of E. Let $g(x)$ be an *irreducible* polynomial in $F[x]$ having a zero α in E. If β is any zero of $g(x)$ in $\overline{F}$, then by Theorem 40.1, there is a basic isomorphism $\psi_{\alpha,\beta}$ of $F(\alpha)$ onto $F(\beta)$ leaving F fixed. By Theorem 41.1, $\psi_{\alpha,\beta}$ can be extended to an isomorphism τ of $\overline{F}$ into $\overline{F}$. But then

$$\tau^{-1} : \overline{F}\tau \to \overline{F}$$

can be extended to an isomorphism mapping $\overline{F}$ into $\overline{F}$. Since the image of τ^{-1} is already all of $\overline{F}$, we see that τ must have been onto $\overline{F}$, so τ is an automorphism of $\overline{F}$ leaving F fixed. Then by assumption, τ induces an automorphism of E, so $\alpha\tau = \beta$ is in E. We have shown that if $g(x)$ is an irreducible polynomial in $F[x]$ having one zero in E, then all zeros of $g(x)$ in $\overline{F}$ are in E. Hence if $\{g_k(x)\}$ is the set of *all* irreducible polynomials in $F[x]$ having a zero in E, then E is the splitting field of $\{g_k(x)\}$. ∎

Definition. Let E be an extension field of a field F. A polynomial $f(x) \in F[x]$ **splits in** E if it factors into a product of linear factors in $E[x]$.

Corollary 1 *If $E \leq \overline{F}$ is a splitting field over F, then every irreducible polynomial in F[x] having a zero in E splits in E.*

Proof. If E is a splitting field over F in $\overline{F}$, then every automorphism of $\overline{F}$ induces an automorphism of E. The second half of the proof of Theorem 42.1 then showed precisely that E is also the splitting field over F of the set $\{g_k(x)\}$ of *all* irreducible polynomials in $F[x]$ having a zero in E. Thus an irreducible polynomial $f(x)$ of $F[x]$ having a zero in E has all its zeros in $\overline{F}$ in E. Therefore, its factorization into linear factors in $\overline{F}[x]$, given by Theorem 38.5, actually takes place in $E[x]$, so $f(x)$ splits in E. ∎

Corollary 2 *If $E \leq \overline{F}$ is a splitting field over F, then every isomorphic mapping of E into $\overline{F}$ leaving F fixed is actually an automorphism of E. In particular, if E is a splitting field of finite degree over F, then*

$$\{E : F\} = |G(E/F)|.$$

Proof. Every isomorphism σ mapping E into $\overline{F}$ leaving F fixed can be extended to an automorphism τ of $\overline{F}$, by Theorem 41.1, together with the *onto* argument of the second half of the proof of Theorem 42.1. If E is a splitting field over F, then by Theorem 42.1, τ restricted to E, that is σ, is an automorphism of E. Thus for a splitting field E over F, every isomorphic mapping of E into $\overline{F}$ leaving F fixed is an automorphism of E.

The equation $\{E : F\} = |G(E/F)|$ then follows immediately for a splitting field E of finite degree over F, since $\{E : F\}$ was defined as the number of different isomorphic mappings of E into $\overline{F}$ leaving F fixed. ∎

Example 42.1 Obviously, $\mathbf{Q}(\sqrt{2}, \sqrt{3})$ is the splitting field of

$$\{x^2 - 2, x^2 - 3\}$$

over $\mathbf{Q}$. Example 40.4 showed that the mappings $\iota, \sigma_1, \sigma_2,$ and σ_3 are all the automorphisms of $\mathbf{Q}(\sqrt{2}, \sqrt{3})$ leaving $\mathbf{Q}$ fixed. (Actually, since every automorphism of a field must leave the prime subfield fixed, we see that these are the only automorphisms of $\mathbf{Q}(\sqrt{2}, \sqrt{3})$.) Then

$$\{\mathbf{Q}(\sqrt{2}, \sqrt{3}) : \mathbf{Q}\} = |G(\mathbf{Q}(\sqrt{2}, \sqrt{3})/\mathbf{Q})| = 4,$$

illustrating Corollary 2. ‖

We wish to determine conditions under which

$$|G(E/F)| = \{E : F\} = [E : F]$$

for finite extensions E of F. This is our next topic. We shall show in the following section that this equation always holds when E is a splitting field over a field F of characteristic 0 or when F is a finite field. This equation need not be true when F is an infinite field of characteristic $p \neq 0$.

Example 42.2 Let $\sqrt[3]{2}$ be the real cube root of 2, as usual. Now $x^3 - 2$ does not split in $Q(\sqrt[3]{2})$, for $Q(\sqrt[3]{2}) < R$, and only one zero of $x^3 - 2$ is real. Thus $x^3 - 2$ factors in $(Q(\sqrt[3]{2}))[x]$ into a linear factor $x - \sqrt[3]{2}$ and an irreducible quadratic factor. The splitting field E of $x^3 - 2$ over Q is therefore of degree 2 over $Q(\sqrt[3]{2})$. Then

$$[E : Q] = [E : Q(\sqrt[3]{2})][Q(\sqrt[3]{2}) : Q] = (2)(3) = 6.$$

We have shown that the splitting field over Q of $x^3 - 2$ is of degree 6 over Q. The student can verify by cubing that

$$\sqrt[3]{2}\,\frac{-1 + i\sqrt{3}}{2} \qquad \text{and} \qquad \sqrt[3]{2}\,\frac{-1 - i\sqrt{3}}{2}$$

are the other zeros of $x^3 - 2$ in C. Thus the splitting field E of $x^3 - 2$ over Q is $Q(\sqrt[3]{2}, i\sqrt{3})$. (This is *not* the same field as $Q(\sqrt[3]{2}, i, \sqrt{3})$, which is of degree 12 over Q.) Further study of this interesting example is left to the exercises (see Exercises 42.3, 42.8, 42.12, and 42.14). ‖

EXERCISES

42.1 For each of the given polynomials in $Q[x]$, find the degree over Q of the splitting field over Q of the polynomial.

a) $x^2 + 3$ b) $x^4 - 1$ c) $(x^2 - 2)(x^2 - 3)$
d) $x^3 - 3$ e) $x^3 - 1$ f) $(x^2 - 2)(x^3 - 2)$

42.2 Let $f(x)$ be a polynomial in $F[x]$ of degree n. Let $E \leq \bar{F}$ be the splitting field of $f(x)$ over F in $\bar{F}$. What bounds can be put on $[E : F]$?

42.3 Refer to Example 42.2 to answer the following questions.

a) What is the order of $G(Q(\sqrt[3]{2})/Q)$?
b) What is the order of $G(Q(\sqrt[3]{2}, i\sqrt{3})/Q)$?
c) What is the order of $G(Q(\sqrt[3]{2}, i\sqrt{3})/Q(\sqrt[3]{2}))$?

42.4 Let α be a zero of $x^3 + x^2 + 1$ over Z_2. Show that $x^3 + x^2 + 1$ splits in $Z_2(\alpha)$. [*Hint:* There are 8 elements in $Z_2(\alpha)$. Exhibit two more zeros of $x^3 + x^2 + 1$, in addition to α, among these 8 elements.]

†**42.5** Show that if a finite extension E of a field F is a splitting field over F, then E is a splitting field of one polynomial in $F[x]$.

42.6 Mark each of the following true or false.

—   a) Let $\alpha, \beta \in E$, where $E \leq \bar{F}$ is a splitting field over F. Then there exists an automorphism of E leaving F fixed and mapping α onto β if and only if $\mathrm{irr}(\alpha, F) = \mathrm{irr}(\beta, F)$.
—   b) R is a splitting field over Q.
—   c) R is a splitting field over R.
—   d) C is a splitting field over R.
—   e) $Q(i)$ is a splitting field over Q.
—   f) $Q(\pi)$ is a splitting field over $Q(\pi^2)$.
—   g) For every splitting field E over F, where $E \leq \bar{F}$, every isomorphic mapping of E is an automorphism of E.

— h) For every splitting field E over F, where $E \leq \overline{F}$, every isomorphism mapping E into $\overline{F}$ is an automorphism of E.

— i) For every splitting field E over F, where $E \leq \overline{F}$, every isomorphism mapping E into $\overline{F}$ and leaving F fixed is an automorphism of E.

— j) Every algebraic closure $\overline{F}$ of a field F is a splitting field over F.

42.7 Show by an example that Corollary 1 of Theorem 42.1 is no longer true if the word *irreducible* is deleted.

42.8 a) Is $|G(E/F)|$ multiplicative for finite towers of finite extensions, i.e., is

$$|G(K/F)| = |G(K/E)| \, |G(E/F)| \qquad \text{for} \quad F \leq E \leq K \leq \overline{F}?$$

Why? [*Hint:* Use Exercise 42.3.]

b) Is $|G(E/F)|$ multiplicative for finite towers of finite extensions, each of which is a splitting field over the bottom field? Why?

42.9 Show that if $[E : F] = 2$, then E is a splitting field over $\overline{F}$.

42.10 Show that for $F \leq E \leq \overline{F}$, E is a splitting field over F if and only if E contains all conjugates over F in $\overline{F}$ of each of its elements.

42.11 Show that $\mathbf{Q}(\sqrt[3]{2})$ has only the identity automorphism.

42.12 Referring to Example 42.2, show that

$$G(\mathbf{Q}(\sqrt[3]{2}, i\sqrt{3})/\mathbf{Q}(i\sqrt{3})) \simeq \langle \mathbf{Z}_3, + \rangle.$$

42.13 a) Show that an automorphism of a splitting field E over F of a polynomial $f(x) \in F[x]$ permutes the zeros of $f(x)$ in E.

b) Show that an automorphism of a splitting field E over F of a polynomial $f(x) \in F[x]$ is completely determined by the permutation of the zeros of $f(x)$ in E given in (a).

c) Show that if E is a splitting field over F of a polynomial $f(x) \in F[x]$, then $G(E/F)$ can be viewed in a natural way as a certain group of permutations.

42.14 Let E be the splitting field of $x^3 - 2$ over $\mathbf{Q}$, as in Example 42.2.

a) What is the order of $G(E/\mathbf{Q})$? [*Hint:* Use Corollary 2 of Theorem 42.2 and the corollary of Theorem 41.1 applied to the tower $\mathbf{Q} \leq \mathbf{Q}(i\sqrt{3}) \leq E$.]

b) Show that $G(E/\mathbf{Q}) \simeq S_3$, the symmetric group on 3 letters. [*Hint:* Use Exercise 42.13, together with (a).]

42.15 Show that for a prime p, the splitting field over $\mathbf{Q}$ of $x^p - 1$ is of degree $p - 1$ over $\mathbf{Q}$. [*Hint:* Refer to the corollary of Theorem 31.4.]

42.16 Let $\overline{F}$ and $\overline{F}'$ be two algebraic closures of a field F, and let $f(x) \in F[x]$. Show that the splitting field E over F of $f(x)$ in $\overline{F}$ is isomorphic to the splitting field E' over F of $f(x)$ in $\overline{F}'$. [*Hint:* Use Theorem 41.1.]

43 | Separable Extensions

43.1 MULTIPLICITY OF ZEROS OF A POLYNOMIAL

Remember that we are now always assuming that all algebraic extensions of a field F under consideration are contained in one fixed algebraic closure $\overline{F}$ of F.

Our next aim is to determine, for a finite extension E of F, under what conditions $\{E : F\} = [E : F]$. The key to answering this question is to consider the multiplicity of zeros of polynomials.

Definition. Let $f(x) \in F[x]$. An element α of $\overline{F}$ such that $f(\alpha) = 0$ is a ***zero of*** $f(x)$ ***of multiplicity*** ν if ν is the greatest integer such that $(x - \alpha)^\nu$ is a factor of $f(x)$ in $\overline{F}[x]$.

The next theorem shows that the multiplicities of the zeros of one given *irreducible* polynomial over a field are all the same. The ease with which we can prove this theorem is a further indication of the power of our basic isomorphisms and of our whole approach to the study of zeros of polynomials by means of mappings.

Theorem 43.1 *Let $f(x)$ be irreducible in $F[x]$. Then all zeros of $f(x)$ in $\overline{F}$ have the same multiplicity.*

Proof. Let α and β be zeros of $f(x)$ in $\overline{F}$. Then by Theorem 40.1, there is a basic isomorphism $\psi_{\alpha,\beta} : F(\alpha) \xrightarrow{\text{onto}} F(\beta)$. By Corollary 1 of Theorem 41.1, $\psi_{\alpha,\beta}$ can be extended to an isomorphism $\tau : \overline{F} \to \overline{F}$. Then τ induces a natural isomorphism $\tau_x : \overline{F}[x] \to \overline{F}[x]$, with $x\tau_x = x$. Now τ_x leaves $f(x)$ fixed, since $f(x) \in F[x]$ and $\psi_{\alpha,\beta}$ leaves F fixed. However,

$$((x - \alpha)^\nu)\tau_x = (x - \beta)^\nu,$$

which shows that the multiplicity of β in $f(x)$ is greater than or equal to the multiplicity of α. A symmetric argument gives the reverse inequality, so the multiplicity of α equals that of β. ∎

Corollary. *If $f(x)$ is irreducible in $F[x]$, then $f(x)$ has a factorization in $\overline{F}[x]$ of the form*

$$a \prod_i (x - \alpha_i)^\nu,$$

where the α_i are the distinct zeros of $f(x)$ in $\overline{F}$ and $a \in F$.

Proof. This is immediate from Theorem 43.1. ∎

At this point, we should probably show by an example that the phenomenon of a zero of multiplicity greater than 1 of an irreducible polynomial can occur. We shall show later in this section that it can only occur for a polynomial over an infinite field of characteristic $p \neq 0$.

Example 43.1 Let $E = \mathbf{Z}_p(y)$, where y is an indeterminate. Let $t = y^p$, and let F be the subfield $\mathbf{Z}_p(t)$ of E. (See Fig. 43.1.) Now $E = F(y)$ is algebraic over F, for y is a zero of $(x^p - t) \in F[x]$. By Theorem 35.3, $\mathrm{irr}(y, F)$ must divide $x^p - t$ in $F[x]$. [Actually, $\mathrm{irr}(y, F) = x^p - t$. We leave a proof of this to the exercises (see Exercise 43.4).] Since $F(y)$ is clearly not equal to F, we must have the degree of $\mathrm{irr}(y, F) \geq 2$. But note that

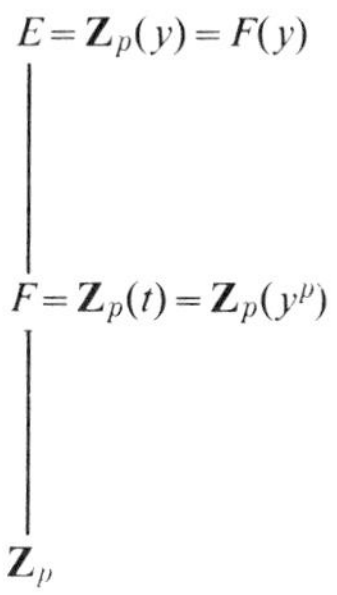

$$E = \mathbf{Z}_p(y) = F(y)$$
$$F = \mathbf{Z}_p(t) = \mathbf{Z}_p(y^p)$$
$$\mathbf{Z}_p$$

$$x^p - t = x^p - y^p = (x - y)^p,$$

Fig. 43.1

since E has characteristic p (see Theorem 40.5 and the following comment). Thus y is a zero of $\mathrm{irr}(y, F)$ of multiplicity > 1. Actually, $x^p - t = \mathrm{irr}(y, F)$, so the multiplicity of y is p. ∥

From here on we rely heavily on Theorem 41.2 and its corollary. Theorem 40.1 and its corollary show that for a simple algebraic extension $F(\alpha)$ of F, there is one extension of the identity isomorphism ι mapping F into F for every distinct zero of $\mathrm{irr}(\alpha, F)$, and that these are the only extensions of ι. *Thus $\{F(\alpha) : F\}$ is the number of distinct zeros of* $\mathrm{irr}(\alpha, F)$.

In view of our work with the Theorem of Lagrange and Theorem 38.2, the student should recognize the potential of a theorem like this next one.

Theorem 43.2 *If E is a finite extension of F, then $\{E : F\}$ divides $[E : F]$.*

Proof. By Theorem 38.3, if E is finite over F, then $E = F(\alpha_1, \ldots, \alpha_n)$, where $\alpha_i \in \overline{F}$. Let $\mathrm{irr}(\alpha_i, F(\alpha_1, \ldots, \alpha_{i-1}))$ have α_i as one of n_i distinct zeros which are all of a common multiplicity ν_i, by Theorem 43.1. Then

$$[F(\alpha_1, \ldots, \alpha_i) : F(\alpha_1, \ldots, \alpha_{i-1})] = n_i \nu_i$$
$$= \{F(\alpha_1, \ldots, \alpha_i) : F(\alpha_1, \ldots, \alpha_{i-1})\} \nu_i.$$

By Theorem 38.2 and the corollary of Theorem 41.2,

$$[E : F] = \prod_i n_i \nu_i,$$

and

$$\{E : F\} = \prod_i n_i.$$

Therefore, $\{E : F\}$ divides $[E : F]$. ∎

43.2 SEPARABLE EXTENSIONS

Definition. A finite extension E of F is a **separable extension of** F if $\{E : F\} = [E : F]$. An element α of $\overline{F}$ is **separable over** F if $F(\alpha)$ is a separable extension of F. An irreducible polynomial $f(x) \in F[x]$ is **separable over** F if every zero of $f(x)$ in $\overline{F}$ is separable over F.

To make things a little easier for the student, we have restricted our definition of a separable extension of a field F to *finite* extensions E of F. For the corresponding definition for infinite extensions, see Exercise 43.6.

We know that $\{F(\alpha) : F\}$ is the number of distinct zeros of $\mathrm{irr}(\alpha, F)$. Also, the multiplicity of α in $\mathrm{irr}(\alpha, F)$ is the same as the multiplicity of each conjugate of α over F, by Theorem 43.1. *Thus α is separable over F if and only if* $\mathrm{irr}(\alpha, F)$ *has all zeros of multiplicity 1.* This tells us at once that *an irreducible polynomial $f(x) \in F[x]$ is separable over F if and only if $f(x)$ has all zeros of multiplicity 1.*

Theorem 43.3 *If K is a finite extension of E and E is a finite extension of F, that is, $F \le E \le K$, then K is separable over F if and only if K is separable over E and E is separable over F.*

Proof. Now
$$[K : F] = [K : E][E : F],$$
and
$$\{K : F\} = \{K : E\}\{E : F\}.$$

Then if K is separable over F, so that $[K : F] = \{K : F\}$, we must have $[K : E] = \{K : E\}$ and $[E : F] = \{E : F\}$, since in each case the index divides the degree, by Theorem 43.2. Thus, if K is separable over F, then K is separable over E and E is separable over F.

The converse is equally easy, for $[K : E] = \{K : E\}$ and $[E : F] = \{E : F\}$ implies that

$$[K : F] = [K : E][E : F] = \{K : E\}\{E : F\} = \{K : F\}. \quad \blacksquare$$

Theorem 43.3 can be extended in the obvious way, by induction, to any finite tower of finite extensions. The top field is a separable extension of the bottom one if and only if each field is a separable extension of the one immediately under it.

Corollary. *If E is a finite extension of F, then E is separable over F if and only if each α in E is separable over F.*

Proof. Suppose that E is separable over F, and let $\alpha \in E$. Then

$$F \le F(\alpha) \le E,$$

and Theorem 43.3 shows that $F(\alpha)$ is separable over F.

Suppose, conversely, that every $\alpha \in E$ is separable over F. Since E is finite over F, there exist $\alpha_1, \ldots, \alpha_n$ such that

$$F < F(\alpha_1) < F(\alpha_1, \alpha_2) < \cdots < E = F(\alpha_1, \ldots, \alpha_n).$$

Now, clearly, since α_i is separable over F, α_i is separable over $F(\alpha_1, \ldots, \alpha_{i-1})$, since

$$q(x) = \mathrm{irr}\big(\alpha_i, F(\alpha_1, \ldots, \alpha_{i-1})\big)$$

divides $\mathrm{irr}(\alpha_i, F)$, so that α_i is a zero of $q(x)$ of multiplicity 1. Thus $F(\alpha_1, \ldots, \alpha_i)$ is separable over $F(\alpha_1, \ldots, \alpha_{i-1})$, so E is separable over F, by Theorem 43.3, extended by induction. ∎

43.3 PERFECT FIELDS

We now turn to the task of proving that α can fail to be separable over F only if F is an infinite field of characteristic $p \neq 0$. One method is to introduce formal derivatives of polynomials. While this is an elegant technique, and also a useful one, we shall, for the sake of brevity, use the lemma which follows instead. Formal derivatives are developed in Exercises 43.8 through 43.15.

Lemma 43.1 *Let $\overline{F}$ be an algebraic closure of F, and let*

$$f(x) = x^n + a_{n-1}x^{n-1} + \cdots + a_1x + a_0$$

be any monic polynomial in $\overline{F}[x]$. If $(f(x))^m \in F[x]$ and $m \cdot 1 \neq 0$ in F, then $f(x) \in F[x]$, that is, all $a_i \in F$.

Proof. We must show that $a_i \in F$, and we proceed, by induction on r, to show that $a_{n-r} \in F$. For $r = 1$,

$$(f(x))^m = x^{mn} + (m \cdot 1)a_{n-1}x^{mn-1} + \cdots + a_0{}^m.$$

Since $(f(x))^m \in F[x]$, we have, in particular,

$$(m \cdot 1)a_{n-1} \in F.$$

Thus $a_{n-1} \in F$, since $m \cdot 1 \neq 0$ in F.

As induction hypothesis, suppose that $a_{n-r} \in F$ for $r = 1, 2, \ldots, k$. Then the coefficient of $x^{mn-(k+1)}$ in $(f(x))^m$ is of the form

$$(m \cdot 1)a_{n-(k+1)} + g_{k+1}(a_{n-1}, a_{n-2}, \ldots, a_{n-k}),$$

where $g_{k+1}(a_{n-1}, a_{n-2}, \ldots, a_{n-k})$ is a formal polynomial in $a_{n-1}, a_{n-2}, \ldots, a_{n-k}$. By the induction assumption, $g_{k+1}(a_{n-1}, a_{n-2}, \ldots, a_{n-k}) \in F$, so again $a_{n-(k+1)} \in F$, since $m \cdot 1 \neq 0$ in F. ∎

We are now in a position to handle fields F of characteristic zero and to show that for a finite extension E of F, we have $\{E : F\} = [E : F]$. By definition, this amounts to proving that every finite extension of a field of characteristic zero is a separable extension. First, we give a definition.

Definition. A field is **perfect** if every finite extension is a separable extension.

Theorem 43.4 *Every field of characteristic zero is perfect.*

Proof. Let E be a finite extension of a field F of characteristic zero, and let $\alpha \in E$. Then $f(x) = \text{irr}(\alpha, F)$ factors in $\bar{F}[x]$ into $\prod_i (x - \alpha_i)^\nu$, where the α_i are the distinct zeros of $\text{irr}(\alpha, F)$, and say $\alpha = \alpha_1$. Thus

$$f(x) = \left(\prod_i (x - \alpha_i) \right)^\nu,$$

and since $\nu \cdot 1 \neq 0$ for a field F of characteristic 0, we must have

$$\left(\prod_i (x - \alpha_i) \right) \in F[x],$$

by Lemma 43.1. Since $f(x)$ is irreducible and of minimal degree in $F[x]$ having α as a zero, we then see that $\nu = 1$. Therefore, α is separable over F for all $\alpha \in E$. By the corollary of Theorem 43.3, this means that E is a separable extension of F. ∎

Lemma 43.1 will also get us through for the case of a finite field, although the proof is a bit harder. Don't feel that you can't understand the meaning of a theorem just because you get lost in the proof.

Theorem 43.5 *Every finite field is perfect.*

Proof. Let F be a finite field of characteristic p, and let E be a finite extension of F. Let $\alpha \in E$. We need to show that α is separable over F. Now $f(x) = \text{irr}(\alpha, F)$ factors in $\bar{F}$ into $\prod_i (x - \alpha_i)^\nu$, where the α_i are the distinct zeros of $f(x)$, and say $\alpha = \alpha_1$. Let $\nu = p^t e$, where p does not divide e. Then

$$f(x) = \prod_i (x - \alpha_i)^\nu = \left(\prod_i (x - \alpha_i)^{p^t} \right)^e$$

is in $F[x]$, and by Lemma 43.1, $\prod_i (x - \alpha_i)^{p^t}$ is in $F[x]$, since $e \cdot 1 \neq 0$ in F. Since $f(x) = \text{irr}(\alpha, F)$ is of minimal degree over F having α as a zero, we must have $e = 1$.

Theorem 40.5 and the remark following it show then that

$$f(x) = \prod_i (x - \alpha_i)^{p^t} = \prod_i (x^{p^t} - \alpha_i^{p^t}).$$

Thus, if we regard $f(x)$ as $g(x^{p^t})$, we must have $g(x) \in F[x]$. Now $g(x)$ is separable over F with distinct zeros $\alpha_i^{p^t}$. Consider $F(\alpha_1^{p^t}) = F(\alpha^{p^t})$. Then $F(\alpha^{p^t})$ is separable over F. Since $x^{p^t} - \alpha^{p^t} = (x - \alpha)^{p^t}$, we see that α is the only zero of $x^{p^t} - \alpha^{p^t}$ in $\bar{F}$. As a finite-dimensional vector space over a finite field F, $F(\alpha^{p^t})$ must be again a finite field. Hence the map

$$\sigma_p \colon F(\alpha^{p^t}) \to F(\alpha^{p^t})$$

given by $a\sigma_p = a^p$ for $a \in F(\alpha^{p^t})$ is an automorphism of $F(\alpha^{p^t})$, by Theorem 40.5. Consequently, $(\sigma_p)^t$ is also an automorphism of $F(\alpha^{p^t})$, and

$$a((\sigma_p)^t) = a^{p^t}.$$

Since an automorphism of $F(\alpha^{p^t})$ is an onto map, there is $\beta \in F(\alpha^{p^t})$ such that $\beta((\sigma_p)^t) = \alpha^{p^t}$. But then $\beta^{p^t} = \alpha^{p^t}$, and we saw that α was the only zero of $x^{p^t} - \alpha^{p^t}$, so we must have $\beta = \alpha$. Since $\beta \in F(\alpha^{p^t})$, we have $F(\alpha) = F(\alpha^{p^t})$. Since $F(\alpha^{p^t})$ was separable over F, we now see that $F(\alpha)$ is separable over F. Therefore, α is separable over F and $t = 0$.

We have shown that for $\alpha \in E$, α is separable over F. Then by the corollary of Theorem 43.3, E is a separable extension of F. ∎

We have completed our aim, which was to show that fields of characteristic 0 and finite fields have only separable finite extensions, i.e., these fields are perfect. *For finite extensions E of such perfect fields F, we have then $[E : F] = \{E : F\}$.*

*43.4 THE PRIMITIVE ELEMENT THEOREM

The following interesting theorem will be of use to us later under another starred heading.

Theorem 43.6 (Primitive Element Theorem). *Let E be a finite separable extension of an infinite field F. Then there exists $\alpha \in E$ such that $E = F(\alpha)$. (Such an element α is a **primitive element**.) That is, a finite separable extension of an infinite field is a simple extension.*

Proof. We prove this in the case that $E = F(\beta, \gamma)$. The induction argument is obvious. Let $irr(\beta, F)$ have distinct zeros $\beta = \beta_1, \ldots, \beta_n$, and let $irr(\gamma, F)$ have distinct zeros $\gamma = \gamma_1, \ldots, \gamma_m$ in $\bar{F}$, where all zeros are of multiplicity 1, since E is a separable extension of F. Since F is infinite, we can find $a \in F$ such that

$$a \neq (\beta_i - \beta)/(\gamma - \gamma_j)$$

for all i and j, with $j \neq 1$. That is, $a(\gamma - \gamma_j) \neq \beta_i - \beta$. Letting $\alpha = \beta + a\gamma$, we have $\alpha = \beta + a\gamma \neq \beta_i + a\gamma_j$, so

$$\alpha - a\gamma_j \neq \beta_i$$

for all i and all $j \neq 1$. Let $f(x) = irr(\beta, F)$, and consider

$$h(x) = f(\alpha - ax) \in (F(\alpha))[x].$$

Now $h(\gamma) = f(\beta) = 0$. However, $h(\gamma_j) \neq 0$ for $j \neq 1$ by construction, since the β_i were the only zeros of $f(x)$. Hence $h(x)$ and $g(x) = irr(\gamma, F)$ have a common factor in $(F(\alpha))[x]$, namely $irr(\gamma, F(\alpha))$, which must be linear, since γ is the only common zero of $g(x)$ and $h(x)$. Thus $\gamma \in F(\alpha)$, and therefore $\beta = \alpha - a\gamma$ is in $F(\alpha)$. Hence $F(\beta, \gamma) = F(\alpha)$. ∎

Corollary. *A finite extension of a field of characteristic zero is a simple extension.*

Proof. This follows at once from Theorem 43.6 and the fact that every field of characteristic zero is infinite and perfect. ∎

Actually, a finite extension of a finite field F is also a simple extension of F. This will be shown in Section 45. Thus, once again, the only possible "bad case" is an extension of an infinite field of characteristic $p \neq 0$. In summary, *a finite separable extension of a field F is a simple extension of F.*

EXERCISES

43.1 Give an example of an $f(x) \in \mathbf{Q}[x]$ which has no zeros in $\mathbf{Q}$ but whose zeros in $\mathbf{C}$ are all of multiplicity 2. Explain how this is consistent with Theorem 43.4, which shows that $\mathbf{Q}$ is perfect.

†**43.2** Show that if $\alpha, \beta \in \overline{F}$ are both separable over F, then $\alpha \pm \beta$, $\alpha\beta$, and α/β, if $\beta \neq 0$, are all separable over F. [*Hint:* Use Theorem 43.3 and its corollary.]

43.3 Mark each of the following true or false.

___ a) Every finite extension of every field F is separable over F.
___ b) Every finite extension of every finite field F is separable over F.
___ c) Every field of characteristic 0 is perfect.
___ d) Every polynomial of degree n over every field F always has n distinct zeros in $\overline{F}$.
___ e) Every polynomial of degree n over every perfect field F always has n distinct zeros in $\overline{F}$.
___ f) Every irreducible polynomial of degree n over every perfect field F always has n distinct zeros in $\overline{F}$.
___ g) Every algebraically closed field is perfect.
___ h) Every field F has an algebraic extension E which is perfect.
___ i) If E is a finite separable splitting field extension of F, then $|G(E/F)| = [E : F]$.
___ j) If E is a finite splitting field extension of F, then $|G(E/F)|$ divides $[E : F]$.

43.4 Show that $\{1, y, \ldots, y^{p-1}\}$ is a basis for $\mathbf{Z}_p(y)$ over $\mathbf{Z}_p(y^p)$, where y is an indeterminate. Referring to Example 43.1, conclude by a degree argument that $x^p - t$ is irreducible over $\mathbf{Z}_p(t)$, where $t = y^p$.

43.5 Prove that if E is an algebraic extension of a perfect field F, then E is perfect.

43.6 A (possibly infinite) algebraic extension E of a field F is a **separable extension** of F if for every $\alpha \in E$, $F(\alpha)$ is a separable extension of F, in the sense defined in the text. Show that if E is a (possibly infinite) separable extension of F and K is a (possibly infinite) separable extension of E, then K is a separable extension of F.

43.7 Let E be an algebraic extension of a field F. Show that the set of all elements in E which are separable over F forms a subfield of E, the **separable closure of F in E**. [*Hint:* Use Exercise 43.2.]

Exercises 43.8 through 43.15 introduce formal derivatives in $F[x]$.

43.8 Let F be any field, and let $f(x) = a_0 + a_1x + \cdots + a_ix^i + \cdots + a_nx^n$ be in $F[x]$. The **derivative** $f'(x)$ **of** $f(x)$ is the polynomial

$$f'(x) = a_1 + \cdots + (i \cdot 1)a_ix^{i-1} + \cdots + (n \cdot 1)a_nx^{n-1},$$

where $i \cdot 1$ has its usual meaning for $i \in \mathbf{Z}^+$ and $1 \in F$. *These are formal derivatives; no "limits" are involved here.*

a) Prove that the map $D: F[x] \to F[x]$ given by $D(f(x)) = f'(x)$ is a homomorphism of $\langle F[x], + \rangle$.
b) Find the kernel of D in the case that F is of characteristic 0.
c) Find the kernel of D in the case that F is of characteristic $p \neq 0$.

43.9 Continuing the ideas of Exercise 43.8, show that:

a) $D(af(x)) = aD(f(x))$ for all $f(x) \in F[x]$ and $a \in F$.
b) $D(f(x)g(x)) = f(x)g'(x) + f'(x)g(x)$ for all $f(x), g(x) \in F[x]$. [*Hint:* Use (a) of this and the preceding exercise, and proceed by induction on the degree of $f(x)g(x)$.]
c) $D((f(x))^m) = (m \cdot 1)f(x)^{m-1}f'(x)$ for all $f(x) \in F[x]$. [*Hint:* Use (b).]

43.10 Let $f(x) \in F[x]$, and let $\alpha \in \bar{F}$ be a zero of $f(x)$ of multiplicity ν. Show that $\nu > 1$ if and only if α is also a zero of $f'(x)$. [*Hint:* Apply (b) and (c) of Exercise 43.9 to the factorization $f(x) = (x - \alpha)^\nu g(x)$ of $f(x)$ in $\bar{F}[x]$.]

43.11 Show from Exercise 43.10 that every irreducible polynomial over a field F of characteristic 0 is separable. [*Hint:* Use the fact that $\mathrm{irr}(\alpha, F)$ is the *minimal* polynomial for α over F.]

43.12 Show from Exercise 43.10 that an irreducible polynomial $q(x)$ over a field F of characteristic $p \neq 0$ is not separable if and only if each exponent of each term of $q(x)$ is divisible by p.

43.13 Generalize Exercise 43.10, showing that $f(x) \in F[x]$ has no zero of multiplicity > 1 if and only if $f(x)$ and $f'(x)$ have no common nonconstant factor in $\bar{F}[x]$.

*43.14 Working a bit harder than in Exercise 43.13, show that $f(x) \in F[x]$ has no zero of multiplicity >1 if and only if $f(x)$ and $f'(x)$ have no common nonconstant factor in $F[x]$. [*Hint:* Use Theorem 33.3 to show that if 1 is a gcd of $f(x)$ and $f'(x)$ in $F[x]$, it is a gcd of these polynomials in $\bar{F}[x]$ also.]

*43.15 Describe a feasible computational procedure for determining whether $f(x) \in F[x]$ has a zero of multiplicity >1, without actually finding the zeros of $f(x)$. [*Hint:* Use Exercise 43.14.]

*43.16 Find $\alpha \in \mathbf{Q}(\sqrt{2}, \sqrt{3})$ such that $\mathbf{Q}(\sqrt{2}, \sqrt{3}) = \mathbf{Q}(\alpha)$. Verify by direct computation that $\sqrt{2}$ and $\sqrt{3}$ can indeed be expressed as formal polynomials in your α, with coefficients in $\mathbf{Q}$.

*43.17 Observe where the hypothesis that F was infinite was used in Theorem 43.6. Show that if F is finite with s elements, and if β and γ are algebraic over F and of degrees n and m, respectively, then there exists α such that $F(\beta, \gamma) = F(\alpha)$, provided $s > mn$. (This result will be superseded by the work in Section 45.)

44 | Totally Inseparable Extensions

*44.1 TOTALLY INSEPARABLE EXTENSIONS

We develop our theory of totally inseparable extensions in a fashion parallel to our development of separable extensions.

Definition. A finite extension E of a field F is a ***totally inseparable extension of*** F if $\{E : F\} = 1 < [E : F]$. An element α of $\bar{F}$ is ***totally inseparable over*** F if $F(\alpha)$ is totally inseparable over F.

We know that $\{F(\alpha) : F\}$ is the number of distinct zeros of $\text{irr}(\alpha, F)$. *Thus α is totally inseparable over F if and only if $\text{irr}(\alpha, F)$ has only one zero which is of multiplicity > 1.*

Example 44.1 With reference to Example 43.1, it is clear that $\mathbf{Z}_p(y)$ is totally inseparable over $\mathbf{Z}_p(y^p)$, where y is an indeterminate. ∥

Theorem 44.1 (Counterpart of Theorem 43.3). *If K is a finite extension of E, E is a finite extension of F, and $F < E < K$, then K is totally inseparable over F if and only if K is totally inseparable over E and E is totally inseparable over F.*

Proof. Since $F < E < K$, we have $[K : E] > 1$ and $[E : F] > 1$. Suppose K is totally inseparable over F. Then $\{K : F\} = 1$, and

$$\{K : F\} = \{K : E\}\{E : F\},$$

so we must have

$$\{K : E\} = 1 < [K : E] \quad \text{and} \quad \{E : F\} = 1 < [E : F].$$

Thus K is totally inseparable over E, and E is totally inseparable over F.

Conversely, if K is totally inseparable over E and E is totally inseparable over F, then

$$\{K : F\} = \{K : E\}\{E : F\} = (1)(1) = 1,$$

and $[K : F] < 1$. Thus K is totally inseparable over F. ∎

Theorem 44.1 can be extended in the obvious way, by induction, to any finite proper tower of finite extensions. The top field is a totally inseparable extension of the bottom one if and only if each field is a totally inseparable extension of the one immediately under it.

Corollary (Counterpart of the corollary of Theorem 43.3). *If E is a finite extension of F, then E is totally inseparable over F if and only if each α in E, $\alpha \notin F$, is totally inseparable over F.*

Proof. Suppose that E is totally inseparable over F, and let $\alpha \in E$, with $\alpha \notin F$. Then

$$F < F(\alpha) \leq E.$$

If $F(\alpha) = E$, we are done, by the definition of α totally inseparable over F. If $F < F(\alpha) < E$, then Theorem 44.1 shows that since E is totally inseparable over F, $F(\alpha)$ is totally inseparable over F.

Conversely, suppose that for every $\alpha \in E$, with $\alpha \notin F$, α is totally inseparable over F. Since E is finite over F, there exist $\alpha_1, \ldots, \alpha_n$ such that

$$F < F(\alpha_1) < F(\alpha_1, \alpha_2) < \cdots < E = F(\alpha_1, \ldots, \alpha_n).$$

Now since α_i is totally inseparable over F, α_i is totally inseparable over $F(\alpha_1, \ldots, \alpha_{i-1})$, for $q(x) = \mathrm{irr}\big(\alpha_i, F(\alpha_1, \ldots, \alpha_{i-1})\big)$ divides $\mathrm{irr}(\alpha_i, F)$ so that α_i is the only zero of $q(x)$ and is of multiplicity > 1. Thus $F(\alpha_1, \ldots, \alpha_i)$ is totally inseparable over $F(\alpha_1, \ldots, \alpha_{i-1})$, and E is totally inseparable over F, by Theorem 44.1, extended by induction. ∎

Obviously, thus far we have so closely paralleled our work in Section 43 that we could easily have handled these ideas together.

*44.2   SEPARABLE CLOSURES

We now come to our main reason for including this material.

Theorem 44.2 *Let F have characteristic $p \neq 0$, and let E be a finite extension of F. Then $\alpha \in E$, $\alpha \notin F$, is totally inseparable over F if and only if there is some integer $t \geq 1$ such that $\alpha^{p^t} \in F$. Furthermore, there is a unique extension K of F, with $F \leq K \leq E$, such that K is separable over F, and either $E = K$ or E is totally inseparable over K.*

Proof. Let $\alpha \in E$, $\alpha \notin F$, be totally inseparable over F. Then $\mathrm{irr}(\alpha, F)$ has just one zero α of multiplicity > 1, and, as shown in the proof of Theorem 43.5, $\mathrm{irr}(\alpha, F)$ must be of the form

$$x^{p^t} - \alpha^{p^t}.$$

Hence $\alpha^{p^t} \in F$ for some $t \geq 1$.

Conversely, if $\alpha^{p^t} \in F$ for some $t \geq 1$, where $\alpha \in E$ and $\alpha \notin F$, then

$$x^{p^t} - \alpha^{p^t} = (x - \alpha)^{p^t},$$

and $(x^{p^t} - \alpha^{p^t}) \in F[x]$, showing that $\mathrm{irr}(\alpha, F)$ divides $(x - \alpha)^{p^t}$. Thus $\mathrm{irr}(\alpha, F)$ has α as its only zero and this zero is of multiplicity > 1, so α is totally inseparable over F.

For the second part of the theorem, let $E = F(\alpha_1, \ldots, \alpha_n)$. Then if

$$\text{irr}(\alpha_i, F) = \prod_j (x^{p^{t_i}} - \alpha_{ij}^{p^{t_i}}),$$

with $\alpha_{i1} = \alpha_i$, let $\beta_{ij} = \alpha_{ij}^{p^{t_i}}$. We have $F(\beta_{11}, \beta_{21}, \ldots, \beta_{n1}) \le E$, and β_{i1} is a zero of

$$f_i(x) = \prod_j (x - \beta_{ij}),$$

where $f_i(x) \in F[x]$. Now since raising to the power p is an isomorphism σ_p of E into E, raising to the power p^t is the isomorphic mapping $(\sigma_p)^t$ of E into E. Thus since the α_{ij} are all distinct for a fixed i, so are the β_{ij} for a fixed i. Therefore, β_{ij} is separable over F, because it is a zero of a polynomial $f_i(x)$ in $F[x]$ with zeros of multiplicity 1. Then

$$K = F(\beta_{11}, \beta_{21}, \ldots, \beta_{n1})$$

is separable over F, by the proof of the corollary of Theorem 43.3. If all $p^{t_i} = 1$, then $K = E$. If some $p^{t_i} \ne 1$, then $K \ne E$, and $\alpha_i^{p^{t_i}} = \beta_{i1}$ is in K, showing that each $\alpha_i \notin K$ is totally inseparable over K, by the first part of this theorem. Hence $E = K(\alpha_1, \ldots, \alpha_n)$ is totally inseparable over K, by the proof of the corollary of Theorem 44.1.

It follows from the corollaries of Theorems 43.3 and 44.1 that the field K of Theorem 44.2 consists of all elements α in E which are separable over F. Thus K is unique. ∎

Definition. The unique field K of Theorem 44.2 is the *separable closure of F in E*.

The preceding theorem shows the precise structure of totally inseparable extensions of a field of characteristic p. Such an extension can be obtained by repeatedly adjoining pth roots of elements (which are not already pth powers) to obtain larger and larger fields.

We remark that Theorem 44.2 is true for infinite algebraic extensions E of F. The proof of the first assertion of the theorem is valid for the case of infinite extensions also. For the second part, since $\alpha \pm \beta$, $\alpha\beta$, and α/β, for $\beta \ne 0$, are all contained in the field $F(\alpha, \beta)$, all elements of E separable over F form a subfield K of E, the **separable closure of F in E**. It follows that an $\alpha \in E$, $\alpha \notin K$, is totally inseparable over K, since α and all coefficients of $\text{irr}(\alpha, K)$ are in a finite extension of F, and then Theorem 44.2 can be applied.

EXERCISES

*44.1 Let y and z be indeterminates, and let $u = y^{12}$ and $v = z^{18}$. Describe the separable closure of $\mathbf{Z}_3(u, v)$ in $\mathbf{Z}_3(y, z)$.

*44.2 Let y and z be indeterminates, and let $u = y^{12}$ and $v = y^2 z^{18}$. Describe the separable closure of $\mathbf{Z}_3(u, v)$ in $\mathbf{Z}_3(y, z)$.

*44.3 Show that if E is an algebraic extension of a field F, then the set of all elements of E totally inseparable over F forms a subfield of E, the **totally inseparable closure of F in E**.

*44.4 Referring to Exercise 44.1, describe the totally inseparable closure (see Exercise 44.3) of $Z_3(u, v)$ in $Z_3(y, z)$.

*44.5 Referring to Exercise 44.2, describe the totally inseparable closure of $Z_3(u, v)$ in $Z_3(y, z)$.

*44.6 Mark each of the following true or false.

— a) No proper algebraic extension of an infinite field of characteristic $p \neq 0$ is ever a separable extension.

— b) If $F(\alpha)$ is totally inseparable over F of characteristic $p \neq 0$, then $\alpha^{p^t} \in F$ for some $t > 0$.

— c) For an indeterminate y, $Z_5(y)$ is separable over $Z_5(y^5)$.

— d) For an indeterminate y, $Z_5(y)$ is separable over $Z_5(y^{10})$.

— e) For an indeterminate y, $Z_5(y)$ is totally inseparable over $Z_5(y^{10})$.

— f) If F is a field and α is algebraic over F, then α is either separable or totally inseparable over F.

— g) If E is an algebraic extension of a field F, then F has a separable closure in E.

— h) If E is an algebraic extension of a field F, then E is totally inseparable over the separable closure of F in E.

— i) If E is an algebraic extension of a field F and E is not a separable extension of F, then E is totally inseparable over the separable closure of F in E.

— j) If α is totally inseparable over F, then α is the only zero of $\mathrm{irr}(\alpha, F)$.

---

*44.7 Show that a field F of characteristic $p \neq 0$ is perfect if and only if $F^p = F$, that is, every element of F is a pth power of some element of F.

*44.8 Let E be a finite extension of a field F of characteristic p. In the notation of Exercise 44.7, show that $E^p = E$ if and only if $F^p = F$. [*Hint:* The map $\sigma_p \colon E \to E$ defined by $\alpha\sigma_p = \alpha^p$ for $\alpha \in E$ is an isomorphism. Consider the diagram in Fig. 44.1, and make degree arguments.]

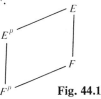

Fig. 44.1

45 | Finite Fields

It is the purpose of this section to determine the structure of all finite fields. We shall show that for every prime p and positive integer n, there is exactly one finite field (up to isomorphism) of order p^n. This field $GF(p^n)$ is usually referred to as the "**Galois field of order** p^n." We shall be using quite a bit of our material on cyclic groups. The proofs are simple and elegant.

45.1 THE STRUCTURE OF A FINITE FIELD

It is easy to see that all finite fields must have prime power order.

Theorem 45.1 *Let E be a finite extension of degree n over a finite field F. If F has q elements, then E has q^n elements.*

Proof. Let $\{\alpha_1, \ldots, \alpha_n\}$ be a basis for E as a vector space over F. Then every $\beta \in E$ can be *uniquely* written in the form

$$\beta = b_1\alpha_1 + \cdots + b_n\alpha_n$$

for $b_i \in F$. Since each b_i may be any of the q elements of F, the total number of such distinct linear combinations of the α_i is q^n. ∎

Corollary. *If E is a finite field of characteristic p, then E contains exactly p^n elements for some positive integer n.*

Proof. Every finite field E is a finite extension of a prime field isomorphic to the field $\mathbf{Z}_p$, where p is the characteristic of E. The corollary follows at once from Theorem 45.1. ∎

We now turn to the study of the multiplicative structure of a finite field. The following theorem will show us how any finite field can be formed from the prime subfield.

Theorem 45.2 *A finite field E of p^n elements is the splitting field of $x^{p^n} - x$ over its prime subfield $\mathbf{Z}_p$ (up to isomorphism).*

Proof. Let E be a finite field with p^n elements, where p is the characteristic of E. The set E^* of nonzero elements of E forms a multiplicative group of order $p^n - 1$ under the field multiplication. For $\alpha \in E^*$, the order of α in this group divides the order $p^n - 1$ of the group. Thus for $\alpha \in E^*$, we have

$\alpha^{p^n-1} = 1$, so $\alpha^{p^n} = \alpha$. Therefore, every element in E is a zero of $x^{p^n} - x$. Since $x^{p^n} - x$ can have at most p^n zeros, we see that E is the splitting field of $x^{p^n} - x$ over $\mathbf{Z}_p$. ∎

Definition. An element α of a field is an **nth root of unity** if $\alpha^n = 1$. It is a **primitive nth root of unity** if $\alpha^n = 1$ and $\alpha^m \neq 1$ for $0 < m < n$.

Thus the nonzero elements of a finite field of p^n elements are all $(p^n - 1)$th roots of unity.

Let F be *any field*, and let U_n be the set of all nth roots of unity in F. It is easy for us to see that U_n is a group under field multiplication. If $a^n = 1$ and $b^n = 1$, then

$$(ab)^n = a^n b^n = 1,$$

so multiplication is closed on U_n. Checking the group axioms is equally trivial. We claim that U_n is a cyclic group. In fact, the following, more general, result is true.

Theorem 45.3 *If G is a finite multiplicative subgroup of the multiplicative group $\langle F^*, \cdot \rangle$ of nonzero elements of a field F, then G is cyclic.*

Proof. By Theorem 9.3, as a finite abelian group, G is isomorphic to a direct product $\mathbf{Z}_{m_1} \times \mathbf{Z}_{m_2} \times \cdots \times \mathbf{Z}_{m_r}$ of cyclic groups, where m_i divides m_{i+1}. Let us think of each of the $\mathbf{Z}_{m_i}$ as a cyclic group of order m_i in *multiplicative* notation. Then for $a_i \in \mathbf{Z}_{m_i}$, $a_i^{m_i} = 1$, so $a_i^{m_r} = 1$, since m_i divides m_r. Thus for all $\alpha \in G$, we have $\alpha^{m_r} = 1$, so every element of G is a zero of $x^{m_r} - 1$. But G has $\prod_{i=1}^{r} m_i$ elements, while $x^{m_r} - 1$ can have at most m_r zeros in a field. Therefore, we must have $r = 1$, so G is cyclic. ∎

Corollary 1 *The multiplicative group of all nonzero elements of a finite field under field multiplication is cyclic.*

Proof. This is immediate from Theorem 45.3. ∎

Corollary 2 *A finite extension E of a finite field F is a simple extension of F.*

Proof. Let α be a generator for the cyclic group E^* of nonzero elements of E. Then, obviously, $E = F(\alpha)$. ∎

Example 45.1 Consider the finite field $\mathbf{Z}_{11}$. By Corollary 1 of Theorem 45.3, $\langle \mathbf{Z}_{11}^*, \cdot \rangle$ is cyclic. Let us try to find a generator of $\mathbf{Z}_{11}^*$ by brute force and ignorance. We start by trying 2. Since $|\mathbf{Z}_{11}^*| = 10$, 2 must be an element of $\mathbf{Z}_{11}^*$ of order dividing 10, that is, either 2, 5, or 10. Now

$$2^2 = 4, \qquad 2^4 = 4^2 = 5, \qquad \text{and} \qquad 2^5 = (2)(5) = 10 = -1.$$

Thus neither 2^2 nor 2^5 is 1, but, of course, $2^{10} = 1$, so 2 is a generator of $\mathbf{Z}_{11}^*$, that is, 2 is a primitive 10th root of unity in $\mathbf{Z}_{11}$. We were lucky.

By the theory of cyclic groups, all the generators of $\mathbf{Z}_{11}^*$, i.e., all the primitive 10th roots of unity in $\mathbf{Z}_{11}$, are then of the form 2^n, where n is relatively prime to 10. These elements are

$$2^1 = 2, \qquad 2^3 = 8,$$
$$2^7 = 7, \qquad 2^9 = 6.$$

The primitive 5th roots of unity in $\mathbf{Z}_{11}$ are of the form 2^m, where the gcd of m and 10 is 2, that is,

$$2^2 = 4, \qquad 2^4 = 5,$$
$$2^6 = 9, \qquad 2^8 = 3.$$

The primitive square root of unity in $\mathbf{Z}_{11}$ is $2^5 = 10 = -1$. $\parallel$

45.2 THE EXISTENCE OF GF(p^n)

We turn now to the question of the existence of a finite field of order p^r for every prime power p^r, $r > 0$. We need the following lemma.

Lemma 45.1 *If F is a finite field of characteristic p, then $x^{p^n} - x$ has p^n distinct zeros in the splitting field $K \leq \bar{F}$ of $x^{p^n} - x$ over F.*

Proof. Let F be a finite field of characteristic p, and let K be the splitting field in $\bar{F}$ of the polynomial $x^{p^n} - x$ over F. We are to show that $x^{p^n} - x$ has p^n distinct zeros in K. Since we have not taken time to introduce an algebraic theory of derivatives, this elegant technique is not available to us, so we proceed by brute force. Obviously, 0 is a zero of $x^{p^n} - x$ of multiplicity 1. Suppose $\alpha \neq 0$ is a zero of $x^{p^n} - x$, and hence is a zero of $f(x) = x^{p^n-1} - 1$. Then $x - \alpha$ is a factor of $f(x)$ in $K[x]$, and by long division, we find that

$$\frac{f(x)}{(x - \alpha)} = g(x)$$
$$= x^{p^n-2} + \alpha x^{p^n-3} + \alpha^2 x^{p^n-4} + \cdots + \alpha^{p^n-3} x + \alpha^{p^n-2}.$$

Now $g(x)$ has $p^n - 1$ summands, and in $g(\alpha)$, each summand is

$$\alpha^{p^n-2} = \frac{\alpha^{p^n-1}}{\alpha} = \frac{1}{\alpha}.$$

Thus

$$g(\alpha) = [(p^n - 1) \cdot 1]\frac{1}{\alpha} = -\frac{1}{\alpha},$$

since we are in a field of characteristic p. Therefore, $g(\alpha) \neq 0$, so α is a zero of $f(x)$ of multiplicity 1. $\blacksquare$

Theorem 45.4 *A finite field GF(p^n) of p^n elements exists for every prime power p^n.*

Proof. Let $K \leq \bar{\mathbf{Z}}_p$ be the splitting field of $x^{p^n} - x$ over $\mathbf{Z}_p$, and let F be the subset of K consisting of all zeros of $x^{p^n} - x$ in K. Then for $\alpha, \beta \in F$, the equations

$$(\alpha \pm \beta)^{p^n} = \alpha^{p^n} \pm \beta^{p^n} = \alpha \pm \beta$$

and

$$(\alpha\beta)^{p^n} = \alpha^{p^n}\beta^{p^n} = \alpha\beta$$

show that F is closed under addition, subtraction, and multiplication. Clearly, 0 and 1 are zeros of $x^{p^n} - x$. For $\alpha \ne 0$, $\alpha^{p^n} = \alpha$ implies that $(1/\alpha)^{p^n} = 1/\alpha$. Thus F is a subfield of K containing $\mathbf{Z}_p$. Since K is the smallest extension of $\mathbf{Z}_p$ containing the zeros of $x^{p^n} - x$, we see that we must have $K = F$. Therefore, K is the desired field of p^n elements, since Lemma 45.1 showed that $x^{p^n} - x$ has p^n distinct zeros in $\overline{\mathbf{Z}_p}$. ∎

Corollary. *If F is any finite field, then for every positive integer n, there is an irreducible polynomial in $F[x]$ of degree n.*

Proof. Let F have $q = p^r$ elements, where p is the characteristic of F. By Theorem 45.4, there is a field $K \le \overline{F}$ containing $\mathbf{Z}_p$ (up to isomorphism) and consisting precisely of the zeros of $x^{p^{rn}} - x$. Every element of F is a zero of $x^{p^r} - x$, by Theorem 45.2. Now $p^{rs} = p^r p^{r(s-1)}$. Applying this equation repeatedly to the exponents and using the fact that for $\alpha \in F$ we have $\alpha^{p^r} = \alpha$, we see that for $\alpha \in F$,

$$\alpha^{p^{rn}} = \alpha^{p^{r(n-1)}} = \alpha^{p^{r(n-2)}} = \cdots = \alpha^{p^r} = \alpha.$$

Thus $F \le K$. Then Theorem 45.1 shows that we must have $[K : F] = n$. We have seen that K is simple over F in Corollary 2 of Theorem 45.3, so $K = F(\beta)$ for some $\beta \in K$. Therefore, $\mathrm{irr}(\beta, F)$ must be of degree n. ∎

EXERCISES

45.1 Find all generators of each of the following cyclic groups. (Recall the theory of cyclic groups.)

a) $\langle \mathbf{Z}_7^*, \cdot \rangle$ b) $\langle \mathbf{Z}_{17}^*, \cdot \rangle$ c) $\langle \mathbf{Z}_{23}^*, \cdot \rangle$

45.2 (Recall the theory of cyclic groups.)

a) Find the number of primitive 8th roots of unity in $\mathrm{GF}(9)$.
b) Find the number of primitive 18th roots of unity in $\mathrm{GF}(19)$.
c) Find the number of primitive 15th roots of unity in $\mathrm{GF}(31)$.
d) Find the number of primitive 10th roots of unity in $\mathrm{GF}(23)$.

45.3 Let $\overline{\mathbf{Z}_2}$ be an algebraic closure of $\mathbf{Z}_2$, and let $\alpha, \beta \in \overline{\mathbf{Z}_2}$ be zeros of $x^3 + x^2 + 1$ and $x^3 + x + 1$, respectively. Using the results of this section, show that $\mathbf{Z}_2(\alpha) = \mathbf{Z}_2(\beta)$.

†**45.4** Show that every irreducible polynomial in $\mathbf{Z}_p[x]$ is a divisor of $x^{p^n} - x$ for some n.

45.5 Mark each of the following true or false.

— a) The nonzero elements of every finite field form a cyclic group under multiplication.

— b) The elements of every finite field form a cyclic group under addition.
— c) The zeros in $\mathbf{C}$ of $(x^{28} - 1) \in \mathbf{Q}[x]$ form a cyclic group under multiplication.
— d) There exists a finite field of 60 elements.
— e) There exists a finite field of 125 elements.
— f) There exists a finite field of 36 elements.
— g) The complex number i is a primitive 4th root of unity.
— h) There exists an irreducible polynomial of degree 58 in $\mathbf{Z}_2[x]$.
— i) The nonzero elements of $\mathbf{Q}$ form a cyclic group $\mathbf{Q}^*$ under field multiplication.
— j) If F is a finite field, then every isomorphism mapping F into an algebraic closure $\overline{F}$ of F is an automorphism of F.

45.6 Let F be a finite field of p^n elements containing the prime subfield $\mathbf{Z}_p$. Show that if $\alpha \in F$ is a generator of the cyclic group $\langle F^*, \cdot \rangle$ of nonzero elements of F, then $\deg(\alpha, \mathbf{Z}_p) = n$.

45.7 Show that a finite field of p^n elements has exactly one subfield of p^m elements for each divisor m of n.

45.8 Show that $x^{p^n} - x$ is the product of all monic irreducible polynomials in $\mathbf{Z}_p[x]$ of a degree d dividing n.

45.9 Let p be an odd prime.

a) Show that for $a \in \mathbf{Z}$, where $a \not\equiv 0 \pmod{p}$, the congruence $x^2 \equiv a \pmod{p}$ has a solution in $\mathbf{Z}$ if and only if $a^{(p-1)/2} \equiv 1 \pmod{p}$. [*Hint:* Formulate an equivalent statement in the finite field $\mathbf{Z}_p$, and use the theory of cyclic groups.]
b) Using part (a), determine whether or not the polynomial $x^2 - 6$ is irreducible in $\mathbf{Z}_{17}[x]$.

45.10 Show that two finite fields of the same order are isomorphic.

45.11 Use Exercise 43.10 to show that $x^{p^n} - x$ has no zeros of multiplicity > 1 in $\overline{\mathbf{Z}}_p$. (See the proof of Lemma 45.1.)

45.12 Let E be a finite field of order p^n.

a) Show that the Frobenius automorphism σ_p has order n.
b) Deduce from (a) that $G(E/\mathbf{Z}_p)$ is cyclic of order n with generator σ_p. [*Hint:* Remember that

$$|G(E/F)| = \{E : F\} = [E : F]$$

for a finite separable splitting field extension E over F.]

46 | GaloisTheory

46.1 RÉSUMÉ

This section is perhaps the climax in elegance of the subject matter of the entire text. The Galois theory gives a beautiful interplay of group and field theory. Starting with Section 40, our work has been aimed at this goal. We shall start by recalling for you the main ideas we have developed which you should have well in mind.

1) Let $F \leq E \leq \overline{F}$, $\alpha \in E$, and let β be a conjugate of α over F, i.e., irr(α, F) has β as a zero also. Then there is an isomorphism $\psi_{\alpha,\beta}$ mapping $F(\alpha)$ onto $F(\beta)$ which leaves F fixed and maps α onto β.

2) If $F \leq E \leq \overline{F}$ and $\alpha \in E$, then an automorphism σ of $\overline{F}$ which leaves F fixed *must* map α onto some conjugate of α over F.

3) If $F \leq E$, the collection of all automorphisms of E leaving F fixed forms a group $G(E/F)$. For any subset S of $G(E/F)$, the set of all elements of E left fixed by all elements of S is a field E_S. Also, $F \leq E_{G(E/F)}$.

4) A field E, $F \leq E \leq \overline{F}$, is a splitting field over F if and only if every isomorphism of E into $\overline{F}$ leaving F fixed is an automorphism of E. If E is a finite extension and a splitting field over F, then $|G(E/F)| = \{E : F\}$.

5) If E is a finite extension of F, then $\{E : F\}$ divides $[E : F]$. If E is also separable over F, then $\{E : F\} = [E : F]$. Also, E is separable over F if and only if irr(α, F) has all zeros of multiplicity 1 for every $\alpha \in E$.

6) If E is a finite extension of F and is a separable splitting field over F, then $|G(E/F)| = \{E : F\} = [E : F]$.

46.2 NORMAL EXTENSIONS

We are going to be interested in finite extensions K of F such that every isomorphism of K leaving F fixed is an automorphism of K and such that

$$[K : F] = \{K : F\}.$$

In view of (4) and (5), these are the finite extensions of F which are separable splitting fields over F.

Definition. A finite extension K of F is a *finite normal extension of F* if K is a separable splitting field over F.

375

Suppose that K is a finite normal extension of F, where $K \leq \bar{F}$, as usual. Then by (4), every automorphism of $\bar{F}$ leaving F fixed induces an automorphism of K. As before, we let $G(K/F)$ be the group of all automorphisms of K leaving F fixed. After one more result, we shall be ready to illustrate the main theorem.

Theorem 46.1 *Let K be a finite normal extension of F, and let E be an extension of F, where $F \leq E \leq K \leq \bar{F}$. Then K is a finite normal extension of E, and $G(K/E)$ is precisely the subgroup of $G(K/F)$ consisting of all those automorphisms which leave E fixed. Moreover, two automorphisms σ and τ in $G(K/F)$ induce the same isomorphism of E into $\bar{F}$ if and only if they are in the same right coset of $G(K/E)$ in $G(K/F)$.*

Proof. If K is the splitting field of a set $\{f_i(x) \mid i \in I\}$ of polynomials in $F[x]$, then clearly K is the splitting field over E of this same set of polynomials viewed as elements of $E[x]$. Theorem 43.3 shows that K is separable over E, since K is separable over F. Thus K is a normal extension of E. This establishes our first contention.

Clearly, every element of $G(K/E)$ is an automorphism of K leaving F fixed, since it even leaves the possibly larger field E fixed. Thus $G(K/E)$ can be viewed as a subset of $G(K/F)$. Since $G(K/E)$ is a group under function composition also, we see that $G(K/E) \leq G(K/F)$.

Finally, for σ and τ in $G(K/F)$, σ and τ are in the same right coset of $G(K/E)$ if and only if $\sigma\tau^{-1} \in G(K/E)$ or if and only if $\sigma = \mu\tau$ for $\mu \in G(K/E)$. But if $\sigma = \mu\tau$ for $\mu \in G(K/E)$, then for $\alpha \in E$, we have

$$\alpha\sigma = \alpha(\mu\tau) = (\alpha\mu)\tau = \alpha\tau,$$

since $\alpha\mu = \alpha$ for $\alpha \in E$. Conversely, if $\alpha\sigma = \alpha\tau$ for all $\alpha \in E$, then

$$\alpha(\sigma\tau^{-1}) = \alpha$$

for all $\alpha \in E$, so $\sigma\tau^{-1}$ leaves E fixed, and $\mu = \sigma\tau^{-1}$ is thus in $G(K/E)$. ∎

The preceding theorem shows that there is a one-to-one correspondence between right cosets of $G(K/E)$ in $G(K/F)$ and isomorphisms of E leaving F fixed. Note that we can't say that these right cosets correspond to *automorphisms* of E over F, since E may not be a splitting field over F. Of course, if E is a *normal* extension of F, then these isomorphisms would be automorphisms of E over F. The student might guess that this will happen if and only if $G(K/E)$ is a *normal* subgroup of $G(K/F)$, and this is indeed the case. That is, the two different uses of the word *normal* are really closely related. Thus if E is a normal extension of F, then the right cosets of $G(K/E)$ in $G(K/F)$ can be viewed as elements of the *factor group* $G(K/F)/G(K/E)$, which is then a group of automorphisms acting on E and leaving F fixed. We shall show that this factor group is isomorphic to $G(E/F)$.

46.3 THE MAIN THEOREM

The *Main Theorem of Galois Theory* states that for a finite normal extension K of a field F, there is a one-to-one correspondence between the subgroups of $G(K/F)$ and the intermediate fields E, where $F \leq E \leq K$. *This correspondence associates with each intermediate field E the subgroup $G(K/E)$.* Of course, we can also go the other way and start with a subgroup H of $G(K/F)$ and associate with H its fixed field K_H. We shall illustrate this with an easy example, then state the theorem, and discuss its proof.

Example 46.1 Let $K = \mathbf{Q}(\sqrt{2}, \sqrt{3})$. Now K is a normal extension of $\mathbf{Q}$, and Example 40.4 showed that there are four automorphisms of K leaving $\mathbf{Q}$ fixed. We recall them by giving their values on the basis $\{1, \sqrt{2}, \sqrt{3}, \sqrt{6}\}$ for K over $\mathbf{Q}$.

  ι: The identity map.
  σ_1: Maps $\sqrt{2}$ onto $-\sqrt{2}$, $\sqrt{6}$ onto $-\sqrt{6}$, and leaves the others fixed.
  σ_2: Maps $\sqrt{3}$ onto $-\sqrt{3}$, $\sqrt{6}$ onto $-\sqrt{6}$, and leaves the others fixed.
  σ_3: Maps $\sqrt{2}$ onto $-\sqrt{2}$, $\sqrt{3}$ onto $-\sqrt{3}$, and leaves the others fixed.

We saw that $\{\iota, \sigma_1, \sigma_2, \sigma_3\}$ is isomorphic to the Klein 4-group. The complete list of subgroups, with each subgroup paired off with the corresponding intermediate field which it leaves fixed, is as follows:

$$\{\iota, \sigma_1, \sigma_2, \sigma_3\} \leftrightarrow \mathbf{Q},$$
$$\{\iota, \sigma_1\} \leftrightarrow \mathbf{Q}(\sqrt{3}),$$
$$\{\iota, \sigma_2\} \leftrightarrow \mathbf{Q}(\sqrt{2}),$$
$$\{\iota, \sigma_3\} \leftrightarrow \mathbf{Q}(\sqrt{6}),$$
$$\{\iota\} \leftrightarrow \mathbf{Q}(\sqrt{2}, \sqrt{3}).$$

All subgroups of the abelian group $\{\iota, \sigma_1, \sigma_2, \sigma_3\}$ are normal subgroups, and clearly all the intermediate fields are normal extensions of $\mathbf{Q}$. Isn't that elegant?

Note that if one subgroup is contained in another, then the larger of the two subgroups corresponds to the smaller of the two corresponding fixed fields. The reason for this is clear. The larger the subgroup, i.e., the more automorphisms, the smaller the fixed field, i.e., the fewer elements left fixed. In Fig. 46.1, we give the corresponding lattice diagrams for the subgroups and intermediate fields. *Note again that the groups near the top correspond to the fields near the bottom.* That is, one lattice looks like the other *inverted* or turned upside down. Since here each lattice actually looks like itself turned upside down, this is not a good example for us to use to illustrate this *lattice inversion principle.* If you will turn ahead to Fig. 47.3, you will see diagrams in which the lattices do not look like their own inversions. ‖

Definition. If K is a finite normal extension of a field F, then $G(K/F)$ is the *Galois group of K over F.*

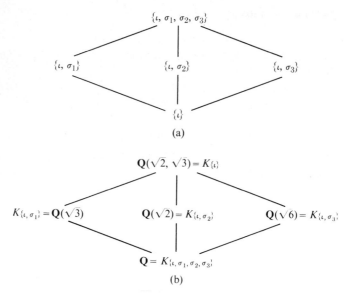

Fig. 46.1. (a) Group lattice diagram. (b) Field lattice diagram.

We shall now state the main theorem, then give another example, and finally, under a starred heading, complete the proof of the main theorem.

Theorem 46.2 (Main Theorem of Galois Theory). *Let K be a finite normal extension of a field F, with Galois group $G(K/F)$. For a field E, where $F \leq E \leq K$, let $E\lambda$ be the subgroup of $G(K/F)$ leaving E fixed. Then λ is a one-to-one map of the set of all such intermediate fields E onto the set of all subgroups of $G(K/F)$. The following properties hold for λ:*

1) *$E\lambda = G(K/E)$.*

2) *$E = K_{G(K/E)} = K_{E\lambda}$.*

3) *For $H \leq G(K/F)$, $K_H\lambda = H$.*

4) *$[K : E] = |E\lambda|$; $[E : F] = \{G(K/F) : E\lambda\}$, the number of cosets of $E\lambda$ in $G(K/F)$.*

5) *E is a normal extension of F if and only if $E\lambda$ is a normal subgroup of $G(K/F)$. When $E\lambda$ is a normal subgroup of $G(K/F)$, then*

$$G(E/F) \simeq G(K/F)/G(K/E).$$

6) *The lattice of subgroups of $G(K/F)$ is the inverted lattice of intermediate fields of K over F.*

Observations on the Proof. We have really already proved a substantial part of this theorem. Let us see just how much we have left to prove.

Property (1) is just the definition of λ found in the statement of the theorem.

For property (2), Theorem 40.4 shows that

$$E \leq K_{G(K/E)}.$$

Let $\alpha \in K$, where $\alpha \notin E$. Since K is a normal extension of E, by using a basic isomorphism and the Isomorphism Extension Theorem, we can find an automorphism of K leaving E fixed and mapping α onto a different zero of $\mathrm{irr}(\alpha, F)$. This implies that

$$K_{G(K/E)} \leq E,$$

so $E = K_{G(K/E)}$. This disposes of (2) and also tells us that λ is one to one, for if $E_1\lambda = E_2\lambda$, then by (2), we have

$$E_1 = K_{E_1\lambda} = K_{E_2\lambda} = E_2.$$

Now (3) is going to be our main job. This amounts exactly to showing that λ is an onto map. Of course, for $H \leq G(K/F)$, we have $H \leq K_H\lambda$, for H surely is included in the set of all automorphisms leaving K_H fixed. Here we will be using strongly our property $[K : E] = \{K : E\}$.

Property (4) is clear from $[K : E] = \{K : E\}$, $[E : F] = \{E : F\}$, and the last statement in Theorem 46.1.

We still have to show that the two senses of the word *normal* correspond for property (5).

We have already disposed of property (6) in Example 46.1.

Thus only properties (3) and (5) remain to be proved.

The Main Theorem of Galois Theory is a strong tool in the study of zeros of polynomials. If $f(x) \in F[x]$ is such that every irreducible factor of $f(x)$ is separable over F, then the splitting field K of $f(x)$ over F is a normal extension of F. The Galois group $G(K/F)$ is the **group of the polynomial** $f(x)$ **over** F. The structure of this group may give considerable information regarding the zeros of $f(x)$. This will be strikingly illustrated in Section 49 when we achieve our *Final Goal.*

46.4   GALOIS GROUPS OVER FINITE FIELDS

Let K be a finite extension of a *finite field* F. We have seen that K is a separable extension of F (a finite field is perfect). Suppose that the order of F is p^r and $[K : F] = n$, so the order of K is p^{rn}. Then we have seen that K is the splitting field of $x^{p^{rn}} - x$ over F. Hence K is a normal extension of F.

Now one automorphism of K which leaves F fixed is σ_{p^r}, where for $\alpha \in K$, $\alpha\sigma_{p^r} = \alpha^{p^r}$. Note that $\alpha(\sigma_{p^r})^i = \alpha^{p^{ri}}$. Since a polynomial of degree p^{ri} can have at most p^{ri} zeros in a field, we see that the smallest power of σ_{p^r} which could possibly leave all p^{rn} elements of K fixed is the nth power. That is, the order of the element σ_{p^r} in $G(K/F)$ is at least n. Therefore, since $|G(K/F)| = [K : F] = n$, it must be that $G(K/F)$ is cyclic and generated by σ_{p^r}. We summarize these arguments in a theorem.

Theorem 46.3 *Let K be a finite extension of degree n of a finite field F of p^r elements. Then $G(K/F)$ is cyclic of order n, and is generated by σ_{p^r}, where for $\alpha \in K$, $\alpha\sigma_{p^r} = \alpha^{p^r}$.*

We use this theorem to give another illustration of the Main Theorem of Galois Theory.

Example 46.2 Let $F = \mathbf{Z}_p$, and let $K = \mathrm{GF}(p^{12})$, so $[K : F] = 12$. Then $G(K/F)$ is isomorphic to the cyclic group $\langle \mathbf{Z}_{12}, + \rangle$. The lattice diagram for the subgroups and for the intermediate fields is given in Fig. 46.2. Again, each lattice is not only the inversion of the other, but unfortunately, also looks like the inversion of itself. Examples where the lattices do not look like their own inversions are given in the next (starred) section. We describe the cyclic subgroups of $G(K/F) = \langle \sigma_p \rangle$ by giving generators, e.g.,

$$\langle {\sigma_p}^4 \rangle = \{\iota, {\sigma_p}^4, {\sigma_p}^8\}. \; \|$$

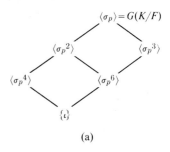

(a)

(b)

Fig. 46.2. (a) Group lattice diagram.
(b) Field lattice diagram.

*46.5 PROOF OF THE MAIN THEOREM COMPLETED

We saw that properties (3) and (5) are all that remain to be proved in the Main Theorem of Galois Theory.

Proof. Turning to property (3), we must show that for $H \leq G(K/F)$, $K_H\lambda = H$. We know that $H \leq K_H\lambda \leq G(K/F)$. Thus what we really must show is that it is impossible to have H a *proper* subgroup of $K_H\lambda$. We shall suppose that

$$H < K_H\lambda,$$

and shall derive a contradiction. If K_H is infinite, then as a finite separable extension of an infinite field, $K = K_H(\alpha)$ for some $\alpha \in K$, by Theorem 43.6. On the other hand, if K_H is finite, then we still have $K = K_H(\alpha)$ for some $\alpha \in K$, by Corollary 2 of Theorem 45.3. Let

$$n = [K : K_H] = \{K : K_H\} = |G(K/K_H)|.$$

Then $H < G(K/K_H)$ implies that $|H| < |G(K/K_H)| = n$. Thus we would have to have $|H| < [K : K_H] = n$. Let the elements of H be $\sigma_1, \ldots, \sigma_{|H|}$, and consider the polynomial

$$f(x) = \prod_{i=1}^{|H|} (x - \alpha\sigma_i).$$

Then $f(x)$ is of degree $|H| < n$. Now the coefficients of each power of x in $f(x)$ are *symmetric* expressions in the $\alpha\sigma_i$. For example, the coefficient of $x^{|H|-1}$ is $-\alpha\sigma_1 - \alpha\sigma_2 - \cdots - \alpha\sigma_{|H|}$. Thus these coefficients are invariant under each isomorphism $\sigma_i \in H$, since if $\sigma \in H$, then

$$\sigma_1\sigma, \ldots, \sigma_{|H|}\sigma$$

is again the sequence $\sigma_1, \ldots, \sigma_{|H|}$, except for order, H being a group. Hence $f(x)$ has coefficients in K_H, and since some σ_i is ι, we see that some $\alpha\sigma_i$ is α, so $f(\alpha) = 0$. Therefore, we would have

$$\deg(\alpha, K_H) \le |H| < n = [K : K_H] = [K_H(\alpha) : K_H].$$

This is impossible. Thus we have proved property (3).

We turn to property (5). Every extension E of F, $F \le E \le K$, is separable over F, by Theorem 43.3. Thus E is normal over F if and only if E is a splitting field over F. By the Isomorphism Extension Theorem, every isomorphism of E into $\bar{F}$ leaving F fixed can be extended to an *automorphism* of K, since K is *normal* over F. Thus the automorphisms of $G(K/F)$ induce all possible isomorphisms of E into $\bar{F}$ leaving F fixed. By Theorem 42.1, this shows that E is a splitting field over F, and hence normal over F, if and only if for all $\sigma \in G(K/F)$ and $\alpha \in E$,

$$(\alpha\sigma) \in E.$$

By property (2), E is the fixed field of $G(K/E)$, so $(\alpha\sigma) \in E$ if and only if for all $\tau \in G(K/E)$,

$$(\alpha\sigma)\tau = \alpha\sigma.$$

This in turn holds if and only if

$$\alpha(\sigma\tau\sigma^{-1}) = \alpha$$

for all $\alpha \in E$, $\sigma \in G(K/F)$, and $\tau \in G(K/E)$. But this means that for all $\sigma \in G(K/F)$ and $\tau \in G(K/E)$, $\sigma\tau\sigma^{-1}$ leaves every element of E fixed, i.e., that

$$(\sigma\tau\sigma^{-1}) \in G(K/E).$$

This is precisely the condition that $G(K/E)$ be a normal subgroup of $G(K/F)$.

It remains for us to show that when E is a normal extension of F, $G(E/F) \simeq G(K/F)/G(K/E)$. For $\sigma \in G(K/F)$, let σ_E be the *automorphism* of E induced by σ (we are assuming that E is a *normal* extension of F). Thus

$\sigma_E \in G(E/F)$. The map $\phi: G(K/F) \to G(E/F)$ given by

$$\sigma\phi = \sigma_E$$

for $\sigma \in G(K/F)$ is obviously a homomorphism. By the Isomorphism Extension Theorem, every automorphism of E leaving F fixed can be extended to some automorphism of K, i.e., it is τ_E for some $\tau \in G(K/F)$. Thus ϕ is onto $G(E/F)$. The kernel of ϕ is clearly $G(K/E)$. Therefore, by the Fundamental Isomorphism Theorem, $G(E/F) \simeq G(K/F)/G(K/E)$. Furthermore, this isomorphism is a natural one. ∎

EXERCISES

46.1 The field $K = \mathbf{Q}(\sqrt{2}, \sqrt{3}, \sqrt{5})$ is a finite normal extension of $\mathbf{Q}$. Fill in the blanks below. The notation is that of Theorem 46.2.

a) $[K : \mathbf{Q}] =$ _____.   b) $|G(K/\mathbf{Q})| =$ _____.

c) $|\mathbf{Q}\lambda| =$ _____.   d) $|(\mathbf{Q}(\sqrt{2}, \sqrt{3}))\lambda| =$ _____.

e) $|(\mathbf{Q}(\sqrt{6}))\lambda| =$ _____.   f) $|(\mathbf{Q}(\sqrt{30}))\lambda| =$ _____.

g) $|(\mathbf{Q}(\sqrt{2} + \sqrt{6}))\lambda| =$ _____.   h) $|K\lambda| =$ _____.

46.2 Describe the group of the polynomial $(x^4 - 1) \in \mathbf{Q}[x]$ over $\mathbf{Q}$.

46.3 Give the order and describe a generator of the group $G(\mathrm{GF}(729)/\mathrm{GF}(9))$.

46.4 Give an example of two finite normal extensions K_1 and K_2 of the same field F such that K_1 and K_2 are not isomorphic fields but $G(K_1/F) \simeq G(K_2/F)$.

46.5 Let K be the splitting field of $x^3 - 2$ over $\mathbf{Q}$. (Refer to Example 42.2.)

a) Describe the six elements of $G(K/\mathbf{Q})$ by giving their values on $\sqrt[3]{2}$ and $i\sqrt{3}$. (By Example 42.2, $K = \mathbf{Q}(\sqrt[3]{2}, i\sqrt{3})$.)

b) To what group we have seen before is $G(K/\mathbf{Q})$ isomorphic?

c) Using the notation given in the answer to (a) in the back of the text, give the lattice diagrams for the subfields of K and for the subgroups of $G(K/\mathbf{Q})$, indicating corresponding intermediate fields and subgroups, as we did in Fig. 46.1.

†**46.6** A finite normal extension K of a field F is **abelian over** F if $G(K/F)$ is an abelian group. Show that if K is abelian over F and E is a normal extension of F, where $F \le E \le K$, then K is abelian over E and E is abelian over F.

46.7 Mark each of the following true or false.

___ a) Two different subgroups of a Galois group may have the same fixed field.

___ b) In the notation of Theorem 46.2, if $F \le E < L \le K$, then $E\lambda < L\lambda$.

___ c) If K is a finite normal extension of F, then K is a normal extension of E, where $F \le E \le K$.

___ d) If two finite normal extensions E and L of a field F have isomorphic Galois groups, then $[E : F] = [L : F]$.

___ e) If E is a finite normal extension of F and H is a normal subgroup of $G(E/F)$, then E_H is a normal extension of F.

___ f) If E is any finite normal simple extension of a field F, then the Galois group $G(E/F)$ is a simple group.

___ g) No Galois group is simple.

___ h) The Galois group of a finite extension of a finite field is abelian.

— i) An extension E of degree 2 over a field F is always a normal extension of F.
— j) An extension E of degree 2 over a field F is always a normal extension of F if the characteristic of F is not 2.

46.8 Let K be a finite normal extension of a field F. Prove that for every $\alpha \in K$, the **norm of α over** F, given by

$$N_{K/F}(\alpha) = \prod_{\sigma \in G(K/F)} \alpha\sigma,$$

and the **trace of α over** F, given by

$$Tr_{K/F}(\alpha) = \sum_{\sigma \in G(K/F)} \alpha\sigma,$$

are elements of F.

46.9 Consider $K = \mathbf{Q}(\sqrt{2}, \sqrt{3})$. Referring to Exercise 46.8, compute each of the following (see Example 46.1).

a) $N_{K/\mathbf{Q}}(\sqrt{2})$ b) $N_{K/\mathbf{Q}}(\sqrt{2} + \sqrt{3})$
c) $N_{K/\mathbf{Q}}(\sqrt{6})$ d) $N_{K/\mathbf{Q}}(2)$
e) $Tr_{K/\mathbf{Q}}(\sqrt{2})$ f) $Tr_{K/\mathbf{Q}}(\sqrt{2} + \sqrt{3})$
g) $Tr_{K/\mathbf{Q}}(\sqrt{6})$ h) $Tr_{K/\mathbf{Q}}(2)$

46.10 Let K be a normal extension of F, and let $K = F(\alpha)$. Let

$$\mathrm{irr}(\alpha, F) = x^n + a_{n-1}x^{n-1} + \cdots + a_1x + a_0.$$

Referring to Exercise 46.8, show that

a) $N_{K/F}(\alpha) = (-1)^n a_0$,   b) $Tr_{K/F}(\alpha) = -a_{n-1}$.

46.11 Describe the group of the polynomial $(x^4 - 5x^2 + 6) \in \mathbf{Q}[x]$ over $\mathbf{Q}$.

46.12 Describe the group of the polynomial $(x^3 - 1) \in \mathbf{Q}[x]$ over $\mathbf{Q}$.

46.13 Let $f(x) \in F[x]$ be a polynomial of degree n such that each irreducible factor is separable over F. Show that the order of the group of $f(x)$ over F divides $n!$.

46.14 Let $f(x) \in F[x]$ be a polynomial such that every irreducible factor of $f(x)$ is a separable polynomial over F. Show that the group of $f(x)$ over F can be viewed in a natural way as a group of permutations of the zeros of $f(x)$ in $\bar{F}$.

46.15 Let F be a field and let ζ be a primitive nth root of unity in $\bar{F}$, where the characteristic of F is either 0 or does not divide n.

a) Show that $F(\zeta)$ is a normal extension of F.
b) Show that $G(F(\zeta)/F)$ is abelian. [*Hint:* Every $\sigma \in G(F(\zeta)/F)$ maps ζ onto some ζ^r and is completely determined by this value r.]

46.16 A finite normal extension K of a field F is **cyclic over** F if $G(K/F)$ is a cyclic group.

a) Show that if K is cyclic over F and E is a normal extension of F, where $F \leq E \leq K$, then E is cyclic over F and K is cyclic over E.
b) Show that if K is cyclic over F, then there exists exactly one field E, $F \leq E \leq K$, of degree d over F for each divisor d of $[K : F]$.

46.17 Let K be a finite normal extension of F.

a) For $\alpha \in K$, show that

$$f(x) = \prod_{\sigma \in G(K/F)} (x - \alpha\sigma)$$

is in $F[x]$.

b) Referring to (a), show that $f(x)$ is a power of irr(α, F), and $f(x) = $ irr(α, F) if and only if $E = F(\alpha)$.

46.18 Let K be a finite normal extension of a field F, and let E and L be extensions of F contained in K, as shown in Fig. 46.3. Describe $G\{K/(E \vee L)\}$ in terms of $G(K/E)$ and $G(K/L)$.

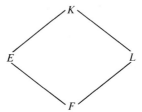

Fig. 46.3

*46.19 With reference to the situation in Exercise 46.18, describe $G\{K/(E \cap L)\}$ in terms of $G(K/E)$ and $G(K/L)$.

47 | Illustrations of Galois Theory

*47.1 SYMMETRIC FUNCTIONS

Let F be a field, and let $y_1, \ldots, y_n$ be n indeterminates. There are some obvious automorphisms of $F(y_1, \ldots, y_n)$ leaving F fixed, namely those defined by permutations of $\{y_1, \ldots, y_n\}$. To be less clear but more explicit, let σ be a permutation of $\{1, \ldots, n\}$, that is, $\sigma \in S_n$. Then σ gives rise to a natural map $\bar{\sigma}: F(y_1, \ldots, y_n) \to F(y_1, \ldots, y_n)$ given by

$$\frac{f(y_1, \ldots, y_n)}{g(y_1, \ldots, y_n)}\bar{\sigma} = \frac{f(y_{1\sigma}, \ldots, y_{n\sigma})}{g(y_{1\sigma}, \ldots, y_{n\sigma})}$$

for $f(y_1, \ldots, y_n), g(y_1, \ldots, y_n) \in F[y_1, \ldots, y_n]$, with $g(y_1, \ldots, y_n) \neq 0$. It is immediate that $\bar{\sigma}$ is an automorphism of $F(y_1, \ldots, y_n)$ leaving F fixed. The elements of $F(y_1, \ldots, y_n)$ left fixed by *all* $\bar{\sigma}$, for all $\sigma \in S_n$, are those rational functions which are *symmetric* in the indeterminates $y_1, \ldots, y_n$.

Definition. An element of $F(y_1, \ldots, y_n)$ is a **symmetric function in** $y_1, \ldots, y_n$ **over** F, if it is left fixed by all permutations of $y_1, \ldots, y_n$, in the sense just explained.

Let $\overline{S_n}$ be the group of all the automorphisms $\bar{\sigma}$ for $\sigma \in S_n$. Obviously, $\overline{S_n}$ is naturally isomorphic to S_n. Let K be the subfield of $F(y_1, \ldots, y_n)$, which is the fixed field of $\overline{S_n}$. Consider the polynomial

$$f(x) = \prod_{i=1}^{n} (x - y_i);$$

this polynomial $f(x) \in (F(y_1, \ldots, y_n))[x]$ is a **general polynomial of degree** n. Let $\bar{\sigma}_x$ be the extension of $\bar{\sigma}$, in the natural way, to $(F(y_1, \ldots, y_n))[x]$, where $x\bar{\sigma}_x = x$. Now clearly $f(x)$ is left fixed by each map $\bar{\sigma}_x$ for $\sigma \in S_n$, that is,

$$\prod_{i=1}^{n} (x - y_i) = \prod_{i=1}^{n} (x - y_{i\sigma}).$$

Thus the coefficients of $f(x)$ are in K; they are symmetric functions in the $y_1, \ldots, y_n$. As illustration, note that the constant term of $f(x)$ is

$$(-1)^n y_1 y_2 \cdots y_n,$$

385

the coefficient of x^{n-1} is $-(y_1 + y_2 + \cdots + y_n)$, etc. These are obviously symmetric functions in $y_1, \ldots, y_n$.

Definition. The *ith* **elementary symmetric function** in $y_1, \ldots, y_n$ is $s_i = (-1)^i a_i$, where a_i is the coefficient of x^{n-i} in the general polynomial $\prod_{i=1}^{n} (x - y_i)$.

Thus the first elementary symmetric function in $y_1, \ldots, y_n$ is

$$s_1 = y_1 + y_2 + \cdots + y_n,$$

the second is $s_2 = y_1 y_2 + y_1 y_3 + \cdots + y_{n-1} y_n$, etc., and the nth is $s_n = y_1 y_2 \cdots y_n$.

Consider the field $E = F(s_1, \ldots, s_n)$. Of course, $E \leq K$, where K is the field of all symmetric functions in $y_1, \ldots, y_n$ over F. But $F(y_1, \ldots, y_n)$ is a finite normal extension of E, namely the splitting field of

$$f(x) = \prod_{i=1}^{n} (x - y_i)$$

over E. Since the degree of $f(x)$ is n, we have at once

$$[F(y_1, \ldots, y_n) : E] \leq n!$$

(see Exercise 42.2). However, since K is the fixed field of $\overline{S_n}$ and

$$|\overline{S_n}| = |S_n| = n!,$$

we have also

$$n! \leq \{F(y_1, \ldots, y_n) : K\} \leq [F(y_1, \ldots, y_n) : K].$$

Therefore,

$$n! \leq [F(y_1, \ldots, y_n) : K] \leq [F(y_1, \ldots, y_n) : E] \leq n!,$$

so

$$K = E.$$

The full Galois group of $F(y_1, \ldots, y_n)$ over E is therefore $\overline{S_n}$. The fact that $K = E$ shows that every symmetric function can be expressed as a rational function of the elementary symmetric functions $s_1, \ldots, s_n$. We summarize these results in a theorem.

Theorem 47.1 *Let $s_1, \ldots, s_n$ be the elementary symmetric functions in the indeterminates $y_1, \ldots, y_n$. Then every symmetric function of $y_1, \ldots, y_n$ over F is a rational function of the elementary symmetric functions. Also, $F(y_1, \ldots, y_n)$ is a finite normal extension of degree $n!$ of $F(s_1, \ldots, s_n)$, and the Galois group of this extension is naturally isomorphic to S_n.*

In view of Cayley's theorem, it can be deduced from Theorem 47.1 that any finite group can occur as a Galois group (up to isomorphism). (See Exercise 47.13.)

Let us give our promised example of a finite normal extension having a Galois group whose lattice of subgroups does not look like its own inversion.

Example 47.1 Consider the splitting field in C of $x^4 - 2$ over Q. Now $x^4 - 2$ is irreducible over Q, by Eisenstein's criterion, with $p = 2$. Let $\alpha = \sqrt[4]{2}$ be the real positive zero of $x^4 - 2$. Then the four zeros of $x^4 - 2$ in C are obviously α, $-\alpha$, $i\alpha$, and $-i\alpha$, where i is the usual zero of $x^2 + 1$ in C. The splitting field K of $x^4 - 2$ over Q thus contains $(i\alpha)/\alpha = i$. Since α is a real number, $Q(\alpha) < R$, so $Q(\alpha) \neq K$. However, since $Q(\alpha, i)$ contains all zeros of $x^4 - 2$, we see that $Q(\alpha, i) = K$. Letting $E = Q(\alpha)$, we have the diagram in Fig. 47.1.

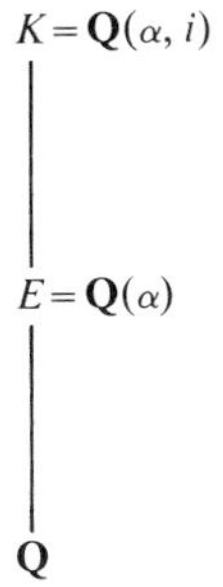

$K = Q(\alpha, i)$

$E = Q(\alpha)$

Q

Fig. 47.1

Now $\{1, \alpha, \alpha^2, \alpha^3\}$ is a basis for E over Q, and $\{1, i\}$ is a basis for K over E. Thus

$$\{1, \alpha, \alpha^2, \alpha^3, i, i\alpha, i\alpha^2, i\alpha^3\}$$

is a basis for K over Q. Since $[K : Q] = 8$, we must have $|G(K/Q)| = 8$, so we need to find eight automorphisms of K leaving Q fixed. We know that any such automorphism σ is completely determined by its values on elements of the basis $\{1, \alpha, \alpha^2, \alpha^3, i, i\alpha, i\alpha^2, i\alpha^3\}$, and these values are in turn determined by $\alpha\sigma$ and $i\sigma$. But $\alpha\sigma$ must always be a conjugate of α over Q, that is, one of the four zeros of $irr(\alpha, Q) = x^4 - 2$. Likewise, $i\sigma$ must be a zero of $irr(i, Q) = x^2 + 1$. Thus the four possibilities for $\alpha\sigma$, combined with the two possibilities for $i\sigma$, must give all eight automorphisms. We describe these by the table in Fig. 47.2.

	ρ_0	ρ_1	ρ_2	ρ_3	μ_1	δ_1	μ_2	δ_2
$\alpha \rightarrow$	α	$i\alpha$	$-\alpha$	$-i\alpha$	α	$i\alpha$	$-\alpha$	$-i\alpha$
$i \rightarrow$	i	i	i	i	$-i$	$-i$	$-i$	$-i$

Fig. 47.2

For example, $\alpha\rho_3 = -i\alpha$ and $i\rho_3 = i$, while ρ_0 is the identity automorphisn . Now

$$\alpha(\rho_1\mu_1) = (\alpha\rho_1)\mu_1 = (i\alpha)\mu_1 = (i\mu_1)(\alpha\mu_1) = -i\alpha,$$

and, similarly,

$$i(\rho_1\mu_1) = -i,$$

so $\rho_1\mu_1 = \delta_2$. A similar computation shows that

$$\alpha(\mu_1\rho_1) = i\alpha \qquad \text{and} \qquad i(\mu_1\rho_1) = -i.$$

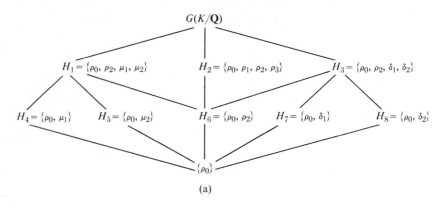

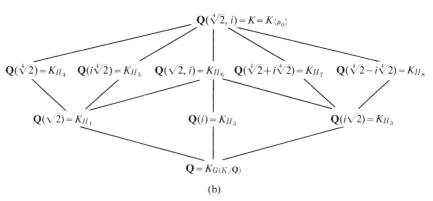

Fig. 47.3. (a) Group lattice diagram. (b) Field lattice diagram.

Thus $\mu_1\rho_1 = \delta_1$, so $\rho_1\mu_1 \neq \mu_1\rho_1$ and $G(K/\mathbf{Q})$ is not abelian. Therefore, $G(K/\mathbf{Q})$ must be isomorphic to one of the two nonabelian groups of order 8 described in Example 18.5. Computing from the table in Fig. 47.2, we see that ρ_1 is of order 4, μ_1 is of order 2, $\{\rho_1, \mu_1\}$ generates $G(K/\mathbf{Q})$, and $\mu_1\rho_1 = \rho_1{}^3\mu_1 = \delta_1$. Thus $G(K/\mathbf{Q})$ is isomorphic to the group G_1 of Example 18.5, the *octic group*. We chose our notation for the elements of $G(K/\mathbf{Q})$ so that its group table would coincide with the table for the octic group given in Fig. 4.8. The lattice of subgroups H_i of $G(K/\mathbf{Q})$ is that given in Fig. 4.9. We repeat it here in Fig. 47.3, and also give the corresponding lattice of intermediate fields between $\mathbf{Q}$ and K. This finally illustrates nicely that one lattice is the inversion of the other.

The determination of the fixed fields K_{H_i} sometimes requires a bit of ingenuity. Let's illustrate. Finding K_{H_2} is easy, for we merely have to find an extension of $\mathbf{Q}$ of degree 2 left fixed by $\{\rho_0, \rho_1, \rho_2, \rho_3\}$. Since all ρ_j leave i fixed, $\mathbf{Q}(i)$ is clearly the field we are after. To find K_{H_4}, we have to find an extension of $\mathbf{Q}$ of degree 4 left fixed by ρ_0 and μ_1. Since μ_1 leaves α

fixed and α is a zero of irr$(\alpha, \mathbf{Q}) = x^4 - 2$, we see that $\mathbf{Q}(\alpha)$ is of degree 4 over $\mathbf{Q}$ and is left fixed by $\{\rho_0, \mu_1\}$. *By Galois theory, it is the only such field.* Here we are using strongly the one-to-one correspondence given by the Galois theory. If we find one field that fits the bill, it is the one we are after. Finding K_{H_7} requires more ingenuity. Since $H_7 = \{\rho_0, \delta_1\}$ is a group, for any $\beta \in K$, we see that $\beta\rho_0 + \beta\delta_1$ is left fixed by ρ_0 and δ_1. Taking $\beta = \alpha$, we see that $\alpha\rho_0 + \alpha\delta_1 = \alpha + i\alpha$ is left fixed by H_7. We can check and see that ρ_0 and δ_1 are the only automorphisms leaving $\alpha + i\alpha$ fixed. Thus by the one-to-one correspondence, we must have

$$\mathbf{Q}(\alpha + i\alpha) = \mathbf{Q}(\sqrt[4]{2} + i\sqrt[4]{2}) = K_{H_7}.$$

Suppose one wished to find irr$(\alpha + i\alpha, \mathbf{Q})$. If $\gamma = \alpha + i\alpha$, then for every conjugate of γ over $\mathbf{Q}$, there exists an automorphism of K mapping γ into that conjugate. Thus we need only compute the various different values $\gamma\sigma$ for $\sigma \in G(K/\mathbf{Q})$ to find the other zeros of irr$(\gamma, \mathbf{Q})$. By Theorem 46.1, elements σ of $G(K/\mathbf{Q})$ giving these different values can be found by taking a set of representatives of the right cosets of $G(K/Q(\gamma)) = \{\rho_0, \delta_1\}$ in $G(K/\mathbf{Q})$. A set of representatives for these right cosets is

$$\{\rho_0, \rho_1, \rho_2, \rho_3\}.$$

The conjugates of $\gamma = \alpha + i\alpha$ are thus $\alpha + i\alpha$, $i\alpha - \alpha$, $-\alpha - i\alpha$, and $-i\alpha + \alpha$. Hence

$$\begin{aligned}
\text{irr}(\gamma, \mathbf{Q}) &= [(x - (\alpha + i\alpha))(x - (i\alpha - \alpha))] \\
&\quad \cdot [(x - (-\alpha - i\alpha))(x - (-i\alpha + \alpha))] \\
&= (x^2 - 2i\alpha x - 2\alpha^2)(x^2 + 2i\alpha x - 2\alpha^2) \\
&= x^4 + 4\alpha^4 = x^4 + 8. \ \|
\end{aligned}$$

We have seen examples in which the splitting field of a quartic (4th degree) polynomial over a field F is an extension of F of degree 8 (Example 47.1) and of degree 24 (Theorem 47.1, with $n = 4$). The degree of an extension of a field F which is a splitting field of a quartic over F must clearly always divide $4! = 24$. Of course, the splitting field of $(x - 2)^4$ over $\mathbf{Q}$ is $\mathbf{Q}$, an extension of degree 1, and the splitting field of $(x^2 - 2)^2$ over $\mathbf{Q}$ is $\mathbf{Q}(\sqrt{2})$, an extension of degree 2. Our last example will give an extension of degree 4 for the splitting field of a quartic.

	σ_1	σ_3	σ_5	σ_7
$\alpha \rightarrow$	α	α^3	α^5	α^7

Fig. 47.4

Example 47.2 Consider the splitting field of $x^4 + 1$ over $\mathbf{Q}$. By Theorem 31.3, one can show that $x^4 + 1$ is irreducible over $\mathbf{Q}$, by arguing that it does not factor in $\mathbf{Z}[x]$. This is easy to show (see Exercise 47.1). One can easily verify, by trial, that the zeros of $x^4 + 1$ are $(1 \pm i)/\sqrt{2}$ and $(-1 \pm i)/\sqrt{2}$. A computation shows that if

$$\alpha = \frac{1 + i}{\sqrt{2}},$$

then

$$\alpha^3 = \frac{-1 + i}{\sqrt{2}}, \qquad \alpha^5 = \frac{-1 - i}{\sqrt{2}}, \qquad \text{and} \qquad \alpha^7 = \frac{1 - i}{\sqrt{2}}.$$

(These facts which are pulled out of thin air are covered in the first few weeks of a course in functions of a complex variable, or may be derived from the work in the next section.) Thus the splitting field K of $x^4 + 1$ over $\mathbf{Q}$ is $\mathbf{Q}(\alpha)$, and $[K : \mathbf{Q}] = 4$. Let us compute $G(K/\mathbf{Q})$ and give the group and field lattice diagrams. Since there exist automorphisms of K mapping α onto each conjugate of α, and since an automorphism σ of $\mathbf{Q}(\alpha)$ is completely determined by $\alpha\sigma$, we see that the four elements of $G(K/\mathbf{Q})$ are defined by the table in Fig. 47.4. Since

$$\alpha(\sigma_j \sigma_k) = \alpha^j \sigma_k = (\alpha \sigma_k)^j = (\alpha^k)^j = \alpha^{jk}$$

and $\alpha^8 = 1$, we see that $G(K/\mathbf{Q})$ is isomorphic to the group $\{1, 3, 5, 7\}$ under multiplication modulo 8. This is the group G_8 of Theorem 24.7. Since $\sigma_j{}^2 = \sigma_1$, the identity, for all j, $G(K/\mathbf{Q})$ must be isomorphic to the Klein 4-group. The lattice diagrams are given in Fig. 47.5.

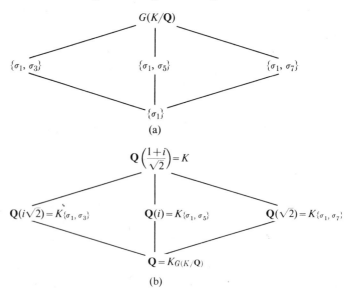

Fig. 47.5. (a) Group lattice diagram. (b) Field lattice diagram.

To find $K_{\{\sigma_1,\sigma_3\}}$, it is only necessary to find an element of K not in $\mathbf{Q}$ left fixed by $\{\sigma_1, \sigma_3\}$, since $[K_{\{\sigma_1,\sigma_3\}}: \mathbf{Q}] = 2$.   Clearly, $\alpha\sigma_1 + \alpha\sigma_3$ is left fixed by both σ_1 and σ_3, since $\{\sigma_1, \sigma_3\}$ is a group. We have

$$\alpha\sigma_1 + \alpha\sigma_3 = \alpha + \alpha^3 = i\sqrt{2}.$$

Similarly,

$$\alpha\sigma_1 + \alpha\sigma_7 = \alpha + \alpha^7 = \sqrt{2}$$

is left fixed by $\{\sigma_1, \sigma_7\}$. This technique is of no use in finding $E_{\{\sigma_1,\sigma_5\}}$, for

$$\alpha\sigma_1 + \alpha\sigma_5 = \alpha + \alpha^5 = 0,$$

and $0 \in \mathbf{Q}$. But by a similar argument, $(\alpha\sigma_1)(\alpha\sigma_5)$ is left fixed by both σ_1 and σ_5, and

$$(\alpha\sigma_1)(\alpha\sigma_5) = \alpha\alpha^5 = -i.$$

Thus $\mathbf{Q}(-i) = \mathbf{Q}(i)$ is the field we are after. $\|$

EXERCISES

*47.1 Show that $x^4 + 1$ is irreducible in $\mathbf{Q}[x]$, as we asserted in Example 47.2.

*47.2 Verify that the intermediate fields given in the field lattice diagram in Fig. 47.3 are correct. (Some are verified in the text. Verify the rest.)

*47.3 For each field in the field lattice diagram in Fig. 47.3, find a primitive element generating the field over $\mathbf{Q}$ (see Theorem 43.6), and give its irreducible polynomial over $\mathbf{Q}$.

*47.4 Let ζ be a primitive 5th root of unity in $\mathbf{C}$.

a) Show that $\mathbf{Q}(\zeta)$ is the splitting field of $x^5 - 1$ over $\mathbf{Q}$.
b) Show that every automorphism of $K = \mathbf{Q}(\zeta)$ maps ζ onto some power ζ^r of ζ.
c) Using (b), describe the elements of $G(K/\mathbf{Q})$.
d) Give the group and field lattice diagrams for $\mathbf{Q}(\zeta)$ over $\mathbf{Q}$, computing the intermediate fields as we did in Examples 47.1 and 47.2.

*47.5 Describe the group of the polynomial $(x^5 - 2) \in (\mathbf{Q}(\zeta))[x]$ over $\mathbf{Q}(\zeta)$, where ζ is a primitive 5th root of unity.

*47.6 Repeat Exercise 47.4 for ζ a primitive 7th root of unity in $\mathbf{C}$.

*47.7 In the easiest way possible, describe the group of the polynomial

$$(x^8 - 1) \in \mathbf{Q}[x]$$

over $\mathbf{Q}$.

*47.8 Find the splitting field K in $\mathbf{C}$ of the polynomial $(x^4 - 4x^2 - 1) \in \mathbf{Q}[x]$. Compute the group of the polynomial over $\mathbf{Q}$, and exhibit the correspondence between the subgroups of $G(K/\mathbf{Q})$ and the intermediate fields. In other words, do the complete job.

*47.9 Express each of the following symmetric functions in y_1, y_2, y_3 over $\mathbf{Q}$ as a rational function of the elementary symmetric functions s_1, s_2, s_3.

a) $y_1^2 + y_2^2 + y_3^2$

b) $(y_1 - y_2)^2(y_1 - y_3)^2(y_2 - y_3)^2$

c) $\dfrac{y_1}{y_2} + \dfrac{y_2}{y_1} + \dfrac{y_1}{y_3} + \dfrac{y_3}{y_1} + \dfrac{y_2}{y_3} + \dfrac{y_3}{y_2}$

*47.10 Let $\alpha_1, \alpha_2, \alpha_3$ be the zeros in $\mathbf{C}$ of the polynomial

$$(x^3 - 4x^2 + 6x - 2) \in \mathbf{Q}[x].$$

Find the polynomial having as zeros precisely the following.

a) $\alpha_1 + \alpha_2 + \alpha_3$
b) $\alpha_1{}^2, \alpha_2{}^2, \alpha_3{}^2$
c) $(\alpha_1 - \alpha_2)^2, (\alpha_1 - \alpha_3)^2, (\alpha_2 - \alpha_3)^2$

*47.11 Let $f(x) \in F[x]$ be a monic polynomial of degree n having all its irreducible factors separable over F. Let $K \le \overline{F}$ be the splitting field of $f(x)$ over F, and suppose that $f(x)$ factors in $K[x]$ into

$$\prod_{i=1}^{n} (x - \alpha_i).$$

Let

$$\Delta(f) = \prod_{i<j} (\alpha_i - \alpha_j);$$

the product $(\Delta(f))^2$ is the **discriminant of** $f(x)$.

a) Show that $\Delta(f) = 0$ if and only if $f(x)$ has as a factor the square of some irreducible polynomial in $F[x]$.
b) Show that $(\Delta(f))^2 \in F$.
c) $G(K/F)$ may be viewed as a subgroup of $\overline{S}_n$, where $\overline{S}_n$ is the group of all permutations of $\{\alpha_i \mid i = 1, \dots, n\}$. Show that $G(K/F)$, when viewed in this fashion, is a subgroup of $\overline{A}_n$, the group formed by all even permutations of $\{\alpha_i \mid i = 1, \dots, n\}$, if and only if $\Delta(f) \in F$.

*47.12 An element of $\mathbf{C}$ is an **algebraic integer** if it is a zero of some *monic* polynomial in $\mathbf{Z}[x]$. Show that the set of all algebraic integers forms a subring of $\mathbf{C}$.

*47.13 Show that every finite group is isomorphic to some Galois group $G(K/F)$ for some finite normal extension K of some field F.

48 | Cyclotomic Extensions

This section deals with extension fields of a field F obtained by adjoining to F some roots of unity. The case of a finite field F was covered in Section 45, so we shall be primarily concerned with the case where F is infinite.

Definition. The splitting field of $x^n - 1$ over F is the **nth cyclotomic extension of F.**

Suppose that F is any field, and consider $(x^n - 1) \in F[x]$. By long division, as in the proof of Lemma 45.1, one sees that if α is a zero of $x^n - 1$ and $g(x) = (x^n - 1)/(x - \alpha)$, then $g(\alpha) = (n \cdot 1)(1/\alpha) \neq 0$, provided that the characteristic of F does not divide n. Therefore, under this condition, the splitting field of $x^n - 1$ is a separable and thus a normal extension of F.

Assume from now on that this is the case, and let K be the splitting field of $x^n - 1$ over F. Then $x^n - 1$ has n distinct zeros in K, and by Theorem 45.3, these form a cyclic group of order n under the field multiplication. We saw in the corollary of Theorem 6.4 that a cyclic group of order n has $\varphi(n)$ generators, where φ is the Euler phi-function introduced prior to Theorem 24.8. For our situation here, these $\varphi(n)$ generators are exactly the primitive nth roots of unity.

Definition. The polynomial

$$\Phi_n(x) = \prod_{i=1}^{\varphi(n)} (x - \alpha_i),$$

where the α_i are the primitive nth roots of unity in $\overline{F}$, is the **nth cyclotomic polynomial over F.**

Since an automorphism of the Galois group $G(K/F)$ must permute the primitive nth roots of unity, we see that $\Phi_n(x)$ is left fixed under every element of $G(K/F)$ regarded as extended in the natural way to $K[x]$. Thus $\Phi_n(x) \in F[x]$. In particular, for $F = \mathbf{Q}$, $\Phi_n(x) \in \mathbf{Q}[x]$, and $\Phi_n(x)$ is a divisor of $x^n - 1$. Thus over $\mathbf{Q}$, we must actually have $\Phi_n(x) \in \mathbf{Z}[x]$, by Theorem 31.3. We have seen that $\Phi_p(x)$ is irreducible over $\mathbf{Q}$, in the corollary of Theorem 31.4. While $\Phi_n(x)$ need not be irreducible in the case of the fields $\mathbf{Z}_p$, it can be shown that over $\mathbf{Q}$, $\Phi_n(x)$ is irreducible.

Let us now limit our discussion to characteristic zero, in particular, to subfields of the complex numbers. Let i be the usual complex zero of $x^2 + 1$. Using trigonometric identities, the student can easily check formally that

$$(\cos \theta_1 + i \sin \theta_1)(\cos \theta_2 + i \sin \theta_2) = \cos (\theta_1 + \theta_2) + i \sin (\theta_1 + \theta_2).$$

It is then immediate by induction that

$$(\cos \theta + i \sin \theta)^n = \cos n\theta + i \sin n\theta.$$

In particular, if $\theta = 2\pi/n$, we have

$$\left(\cos \frac{2\pi}{n} + i \sin \frac{2\pi}{n}\right)^n = \cos 2\pi + i \sin 2\pi = 1,$$

so $\cos (2\pi/n) + i \sin (2\pi/n)$ is an nth root of unity. Figure 48.1 may help the student to visualize all this. It is quite obvious from this figure that the least integer m such that $(\cos (2\pi/n) + i \sin (2\pi/n))^m = 1$ is n. *Thus* $\cos (2\pi/n) + i \sin (2\pi/n)$ *is a primitive* nth *root of unity, a zero of*

$$\Phi_n(x) \in \mathbf{Q}[x].$$

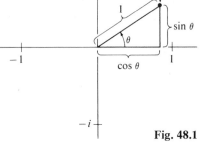

Fig. 48.1

Example 48.1 A primitive 8th root of unity in **C** is

$$\varsigma = \cos \frac{2\pi}{8} + i \sin \frac{2\pi}{8}$$

$$= \cos \frac{\pi}{4} + i \sin \frac{\pi}{4}$$

$$= \frac{1}{\sqrt{2}} + i \frac{1}{\sqrt{2}} = \frac{1 + i}{\sqrt{2}}.$$

By the theory of cyclic groups, in particular, by the corollary of Theorem 6.4, all the primitive 8th roots of unity in **Q** are ς, ς^3, ς^5, and ς^7, so

$$\Phi_8(x) = (x - \varsigma)(x - \varsigma^3)(x - \varsigma^5)(x - \varsigma^7).$$

The student may compute directly from this expression that $\Phi_8(x) = x^4 + 1$ (see Exercise 48.1). Compare this with Example 47.2. ‖

Let us still restrict our work to $F = \mathbf{Q}$, and let us assume, without proof, that $\Phi_n(x)$ is irreducible over **Q**. Let

$$\varsigma = \cos \frac{2\pi}{n} + i \sin \frac{2\pi}{n},$$

so that ς is a primitive nth root of unity. Note that ς is a generator of the cyclic multiplicative group of order n consisting of *all* nth roots of unity.

All the primitive nth roots of unity, i.e., all the generators of this group, are of the form ζ^m for $1 \leq m < n$ and m relatively prime to n. The field $\mathbf{Q}(\zeta)$ is the whole splitting field of $x^n - 1$ over $\mathbf{Q}$. Let $K = \mathbf{Q}(\zeta)$. If ζ^m is another primitive nth root of unity, then since ζ and ζ^m are conjugate over $\mathbf{Q}$, there is an automorphism τ_m in $G(K/\mathbf{Q})$ mapping ζ onto ζ^m. Let τ_r be the similar automorphism in $G(K/\mathbf{Q})$ corresponding to the primitive nth root of unity ζ^r. Then

$$\zeta(\tau_r\tau_m) = (\zeta^r)\tau_m = (\zeta\tau_m)^r = (\zeta^m)^r = \zeta^{rm}.$$

This shows that the Galois group $G(K/\mathbf{Q})$ is isomorphic to the group G_n of Theorem 24.7 consisting of elements of $\mathbf{Z}_n$ relatively prime to n under multiplication modulo n. This group has $\varphi(n)$ elements, and is, of course, abelian.

This material is easy. Special cases have appeared several times in the text and exercises. For example, α of Example 47.2 is a primitive 8th root of unity, and we made arguments in that example identical to those given here. We summarize these results in a theorem.

Theorem 48.1 *The Galois group of the nth cyclotomic extension of $\mathbf{Q}$ has $\varphi(n)$ elements and is isomorphic to the group consisting of the positive integers less than n and relatively prime to n under multiplication modulo n.*

Example 48.2 Example 47.2 illustrates this theorem, for it is easy to see that the splitting field of $x^4 + 1$ is the same as the splitting field of $x^8 - 1$ over $\mathbf{Q}$. This follows from the fact that $\Phi_8(x) = x^4 + 1$ (see Example 48.1 and Exercise 48.1). ‖

Corollary. *The Galois group of the pth cyclotomic extension of $\mathbf{Q}$ for a prime p is cyclic of order $p - 1$.*

Proof. By Theorem 48.1, the Galois group of the pth cyclotomic extension of $\mathbf{Q}$ has $\varphi(p) = p - 1$ elements, and is isomorphic to the group of positive integers less than p and relatively prime to p under multiplication modulo p. This is exactly the multiplicative group $\langle \mathbf{Z}_p{}^*, \cdot \rangle$ of nonzero elements of the field $\mathbf{Z}_p$ under field multiplication. By Corollary 1 of Theorem 45.3, this group is cyclic. ∎

*48.2 CONSTRUCTIBLE POLYGONS

We conclude with an application determining which regular n-gons are constructible with a compass and a straightedge. We saw in Section 39 that the regular n-gon is constructible if and only if $\cos(2\pi/n)$ is a constructible real number. Now let

$$\zeta = \cos\frac{2\pi}{n} + i\sin\frac{2\pi}{n}.$$

Then

$$\frac{1}{\zeta} = \cos\frac{2\pi}{n} - i\sin\frac{2\pi}{n},$$

for

$$\left(\cos\frac{2\pi}{n} + i\sin\frac{2\pi}{n}\right)\left(\cos\frac{2\pi}{n} - i\sin\frac{2\pi}{n}\right) = \cos^2\frac{2\pi}{n} + \sin^2\frac{2\pi}{n} = 1.$$

But then

$$\zeta + \frac{1}{\zeta} = 2\cos\frac{2\pi}{n}.$$

Thus the corollary of Theorem 39.2 shows that the regular n-gon is constructible only if $\zeta + 1/\zeta$ generates an extension of $\mathbf{Q}$ of degree a power of 2.

If K is the splitting field of $x^n - 1$ over $\mathbf{Q}$, then $[K : \mathbf{Q}] = \varphi(n)$, by Theorem 48.1. If $\sigma \in G(K/\mathbf{Q})$ and $\zeta\sigma = \zeta^r$, then

$$\left(\zeta + \frac{1}{\zeta}\right)\sigma = \zeta^r + \frac{1}{\zeta^r}$$

$$= \left(\cos\frac{2\pi r}{n} + i\sin\frac{2\pi r}{n}\right) + \left(\cos\frac{2\pi r}{n} - i\sin\frac{2\pi r}{n}\right)$$

$$= 2\cos\frac{2\pi r}{n}.$$

But for $1 < r < n$, we have $2\cos(2\pi r/n) = 2\cos(2\pi/n)$ only in the case that $r = n - 1$. Thus the only elements of $G(K/\mathbf{Q})$ carrying $\zeta + 1/\zeta$ onto itself are the identity automorphism and the automorphism τ, with $\zeta\tau = \zeta^{n-1} = 1/\zeta$. This shows that the subgroup of $G(K/\mathbf{Q})$ leaving $\mathbf{Q}(\zeta + 1/\zeta)$ fixed is of order 2, so by Galois theory,

$$\left[\mathbf{Q}\left(\zeta + \frac{1}{\zeta}\right) : \mathbf{Q}\right] = \frac{\varphi(n)}{2}.$$

Hence the regular n-gon is constructible only if $\varphi(n)/2$, and therefore also $\varphi(n)$, is a power of 2.

It can be shown by elementary arguments in number theory that if

$$n = 2^\nu p_1{}^{s_1}\cdots p_t{}^{s_t},$$

where the p_i are the distinct odd primes dividing n, then

$$\varphi(n) = 2^{\nu-1}p_1{}^{s_1-1}\cdots p_t{}^{s_t-1}(p_1 - 1)\cdots(p_t - 1). \qquad (1)$$

If $\varphi(n)$ is to be a power of 2, then every odd prime dividing n must appear only to the first power and must be one more than a power of 2. Thus we must have each

$$p_i = 2^m + 1$$

for some m. Since -1 is a zero of $x^q + 1$ for q an odd prime, $x + 1$ divides $x^q + 1$ for q an odd prime. Thus, if $m = qu$, where q is an odd prime, then $2^m + 1 = (2^u)^q + 1$ is divisible by $2^u + 1$. Therefore, for $p_i = 2^m + 1$ to be prime, it must be that m is divisible by 2 only, so p_i has to have the form

$$p_i = 2^{(2^k)} + 1,$$

a **Fermat prime.** Fermat conjectured that these numbers $2^{(2^k)} + 1$ were prime for all nonnegative integers k. Euler showed that while $k = 0, 1, 2, 3,$ and 4 give the primes 3, 5, 17, 257, and 65537, for $k = 5$, one finds that $2^{(2^5)} + 1$ is divisible by 641. It has been shown that for $5 \leq k \leq 16$, all the numbers $2^{(2^k)} + 1$ are composite. The case $k = 17$ is unsolved as this text goes to press. It is unknown whether the number of Fermat primes is finite or infinite.

We have thus shown that the only regular n-gons which might be constructible are those where the odd primes dividing n are Fermat primes whose squares do not divide n. In particular, the only regular p-gons which might be constructible for p a prime greater than 2 are those where p is a Fermat prime.

Example 48.3 The regular 7-gon is not constructible, since 7 is not a Fermat prime. Similarly, the regular 18-gon is not constructible, for while 3 is a Fermat prime, its square divides 18. ‖

It is a fact which we now demonstrate that all these regular n-gons which are candidates for being constructible are indeed actually constructible. Let ζ again be the primitive nth root of unity $\cos(2\pi/n) + i\sin(2\pi/n)$. We saw above that

$$2\cos\frac{2\pi}{n} = \zeta + \frac{1}{\zeta},$$

and that

$$\left[\mathbf{Q}\left(\zeta + \frac{1}{\zeta}\right) : \mathbf{Q}\right] = \frac{\varphi(n)}{2}.$$

Suppose now that $\varphi(n)$ is a power 2^s of 2. Let E be $\mathbf{Q}(\zeta + 1/\zeta)$. We saw above that $\mathbf{Q}(\zeta + 1/\zeta)$ is the subfield of $K = \mathbf{Q}(\zeta)$ left fixed by $H_1 = \{\iota, \tau\}$, where ι is the identity element of $G(K/\mathbf{Q})$ and $\zeta\tau = 1/\zeta$. By Sylow theory, there exist additional subgroups H_j of order 2^j of $G(\mathbf{Q}(\zeta)/\mathbf{Q})$ for $j = 0, 2, 3, \ldots, s$ such that

$$\{\iota\} = H_0 < H_1 < \cdots < H_s = G(\mathbf{Q}(\zeta)/\mathbf{Q}).$$

By Galois theory,

$$\mathbf{Q} = K_{H_s} < K_{H_{s-1}} < \cdots < K_{H_1} = \mathbf{Q}\left(\zeta + \frac{1}{\zeta}\right),$$

and $[K_{H_{j-1}} : K_{H_j}] = 2$. Note that $(\zeta + 1/\zeta) \in \mathbf{R}$, so $\mathbf{Q}(\zeta + 1/\zeta) < \mathbf{R}$.

If $K_{H_{j-1}} = K_{H_j}(\alpha_j)$, then α_j is a zero of some $(a_j x^2 + b_j x + c_j) \in K_{H_j}[x]$. By the familiar "quadratic formula," we have

$$K_{H_{j-1}} = K_{H_j}(\sqrt{b_j^2 - 4a_j c_j}).$$

Since we saw in Section 39 that construction of square roots of positive constructible numbers can be achieved by a straightedge and a compass, we see that every element in $Q(\zeta + 1/\zeta)$, in particular $\cos(2\pi/n)$, is constructible. Hence the regular n-gons where $\varphi(n)$ is a power of 2 are constructible.

We summarize our work under this heading in a theorem.

Theorem 48.2 *The regular n-gon is constructible with a compass and a straightedge if and only if all the odd primes dividing n are Fermat primes whose squares do not divide n.*

Example 48.4 The regular 60-gon is constructible, since $60 = (2^2)(3)(5)$ and 3 and 5 are both Fermat primes. ‖

EXERCISES

*48.1 Referring to Example 48.1, complete the indicated computation, showing that $\Phi_8(x) = x^4 + 1$. [*Suggestion:* Compute the product in terms of ζ, and then use the fact that $\zeta^8 = 1$ and $\zeta^4 = -1$ to simplify the coefficients.]

*48.2 Classify the group of the polynomial $(x^{20} - 1) \in Q[x]$ over Q according to the Fundamental Theorem of Finitely Generated Abelian Groups. [*Hint:* Use Theorem 48.1.]

*48.3 Using the formula for $\varphi(n)$ in terms of the factorization of n, as given in Eq. (1) of Section 48.2, compute the following.

a) $\varphi(60)$ b) $\varphi(1000)$ c) $\varphi(8100)$

*48.4 Give the first thirty values of $n \geq 3$ for which the regular n-gon is constructible with a straightedge and a compass.

†*48.5 Show that if F is a field of characteristic not dividing n, then

$$x^n - 1 = \prod_{d|n} \Phi_d(x)$$

in $F[x]$, where the product is over all divisors d of n.

*48.6 Find the cyclotomic polynomial $\Phi_n(x)$ over Q for $n = 1, 2, 3, 4, 5$, and 6. [*Hint:* Use Exercise 48.5.]

*48.7 Mark each of the following true or false.

— a) $\Phi_n(x)$ is irreducible over every field of characteristic 0.
— b) Every zero in C of $\Phi_n(x)$ is a primitive nth root of unity.
— c) The group of $\Phi_n(x) \in Q[x]$ over Q has order n.
— d) The group of $\Phi_n(x) \in Q[x]$ over Q is abelian.
— e) The Galois group of the splitting field of $\Phi_n(x)$ over Q has order $\varphi(n)$.
— f) The regular 25-gon is constructible with a straightedge and a compass.
— g) The regular 17-gon is constructible with a straightedge and a compass.

— h) For a prime p, the regular p-gon is constructible if and only if p is a Fermat prime.

— i) All integers of the form $2^{(2^k)} + 1$ for nonnegative integers k are Fermat primes.

— j) All Fermat primes are numbers of the form $2^{(2^k)} + 1$ for nonnegative integers k.

---

*48.8 Find the smallest angle of integral degree, that is, $1°$, $2°$, $3°$, etc., constructible with a straightedge and a compass. [*Hint:* Constructing a $1°$ angle amounts to constructing the regular 360-gon, etc.]

*48.9 Let K be the splitting field of $x^{12} - 1$ over $\mathbf{Q}$.

a) Find $[K : \mathbf{Q}]$.

b) Show that for $\sigma \in G(K/\mathbf{Q})$, σ^2 is the identity automorphism. Classify $G(K/\mathbf{Q})$ according to the Fundamental Theorem of Finitely Generated Abelian Groups.

*48.10 Find $\Phi_3(x)$ over $\mathbf{Z}_2$. Find $\Phi_8(x)$ over $\mathbf{Z}_3$.

*48.11 How many elements are there in the splitting field of $x^6 - 1$ over $\mathbf{Z}_3$?

*48.12 Find $\Phi_{12}(x)$ in $\mathbf{Q}[x]$. [*Hint:* Use Exercises 48.5 and 48.6.]

*48.13 Show that in $\mathbf{Q}[x]$, $\Phi_{2n}(x) = \Phi_n(-x)$ for odd integers $n > 1$. [*Hint:* Use Exercise 48.5 and the factorization $x^{2n} - 1 = -(x^n - 1)((-x)^n - 1)$. Proceed by induction.]

*48.14 Let $n, m \in \mathbf{Z}^+$ be relatively prime. Show that the splitting field in $\mathbf{C}$ of $x^{nm} - 1$ over $\mathbf{Q}$ is the same as the splitting field in $\mathbf{C}$ of $(x^n - 1)(x^m - 1)$ over $\mathbf{Q}$.

*48.15 Let $n, m \in \mathbf{Z}^+$ be relatively prime. Show that the group of $(x^{nm} - 1) \in \mathbf{Q}[x]$ over $\mathbf{Q}$ is isomorphic to the direct product of the groups of $(x^n - 1) \in \mathbf{Q}[x]$ and of $(x^m - 1) \in \mathbf{Q}[x]$ over $\mathbf{Q}$. [*Hint:* Using Galois theory, show that the groups of $x^m - 1$ and $x^n - 1$ can both be regarded as subgroups of the group of $x^{nm} - 1$. Then use Theorem 8.6.]

49 | Insolvability of the Quintic

49.1 THE PROBLEM

The student is familiar with the fact that a quadratic polynomial $f(x) = ax^2 + bx + c$, $a \neq 0$, with real coefficients has $(-b \pm \sqrt{b^2 - 4ac})/2a$ as zeros in $\mathbf{C}$. Actually, this is true for $f(x) \in F[x]$, where F is any field of characteristic $\neq 2$ and the zeros are in $\bar{F}$. The student is asked to show this in Exercise 49.1. Thus, for example, $(x^2 + 2x + 3) \in \mathbf{Q}[x]$ has its zeros in $\mathbf{Q}(\sqrt{-2})$. One wonders whether the zeros of a cubic polynomial over $\mathbf{Q}$ can also always be expressed in terms of radicals. The answer is yes, and, indeed, even the zeros of a polynomial of degree 4 over $\mathbf{Q}$ can be expressed in terms of radicals. After mathematicians had tried for years to find the "radical formula" for zeros of a fifth-degree polynomial, it was a triumph when Abel proved that a quintic need not be solvable by radicals. Our first job will be to describe precisely what this means. The student should be delighted to see what a large amount of the algebra we have developed is used in the forthcoming discussion.

49.2 EXTENSIONS BY RADICALS

Definition. An extension K of a field F is an *extension of F by radicals* if there are elements $\alpha_1, \ldots, \alpha_r \in K$ and positive integers $n_1, \ldots, n_r$ such that $K = F(\alpha_1, \ldots, \alpha_r)$, $\alpha_1^{n_1} \in F$ and $\alpha_i^{n_i} \in F(\alpha_1, \ldots, \alpha_{i-1})$ for $1 < i \leq r$. A polynomial $f(x) \in F[x]$ is *solvable by radicals over F* if the splitting field K of $f(x)$ over F is an extension of F by radicals.

A polynomial $f(x) \in F[x]$ is thus solvable by radicals over F if we can obtain every zero of $f(x)$ by using a finite sequence of the operations of addition, subtraction, multiplication, division, and taking n_ith roots, starting with elements of F. Now to say that the quintic is not solvable in the classical case, i.e., characteristic 0, is not to say that no quintic is solvable, as the following example shows.

Example 49.1 The polynomial $x^5 - 1$ is solvable by radicals over $\mathbf{Q}$. The splitting field K of $x^5 - 1$ is generated over $\mathbf{Q}$ by a primitive 5th root ζ of unity. Then $\zeta^5 = 1$, and $K = \mathbf{Q}(\zeta)$. Similarly, $x^5 - 2$ is solvable by radicals over $\mathbf{Q}$, for its splitting field over $\mathbf{Q}$ is generated by $\sqrt[5]{2}$ and ζ, where $\sqrt[5]{2}$ is the real zero of $x^5 - 2$. ‖

To say that the quintic is insolvable in the classical case means that there exists *some* polynomial of degree 5 with real coefficients which is not solvable by radicals. We shall show this. *We assume throughout this section that all fields mentioned have characteristic 0.*

The outline of the argument is very easy to give, and it is worth while to try to remember it.

1) *We shall show that a polynomial $f(x) \in F[x]$ is solvable by radicals over F (if and) only if its splitting field K over F has a solvable Galois group.* Recall that a solvable group is one having a composition series with *abelian* quotients. While this theorem goes both ways, we shall not prove the "if" part.

2) *We shall show that there is a subfield F of the real numbers and a polynomial $f(x) \in F[x]$ of degree 5 with a splitting field K over F such that $G(K/F) \simeq S_5$, the symmetric group on 5 letters.* Recall that a composition series for S_5 is $\{\iota\} < A_5 < S_5$. Since A_5 is not abelian, we will be done.

The following lemma does most of our work for step (1).

Lemma 49.1 *Let F be a field of characteristic 0, and let $a \in F$. If K is the splitting field of $x^n - a$ over F, then $G(K/F)$ is a solvable group.*

Proof. Suppose first that F contains all the nth roots of unity. By Theorem 45.3 and the comments preceding it, the nth roots of unity form a cyclic subgroup of $\langle F^*, \cdot \rangle$. Let ζ be a generator of the subgroup. (Actually, the generators are exactly the *primitive* nth roots of unity.) Then the nth roots of unity are

$$1, \zeta, \zeta^2, \ldots, \zeta^{n-1}.$$

If $\beta \in \overline{F}$ is a zero of $(x^n - a) \in F[x]$, then all zeros of $x^n - a$ are

$$\beta, \zeta\beta, \zeta^2\beta, \ldots, \zeta^{n-1}\beta.$$

Since $K = F(\beta)$, an automorphism σ in $G(K/F)$ is determined by the value $\beta\sigma$ of the automorphism σ on β. If $\beta\sigma = \zeta^i\beta$ and $\beta\tau = \zeta^j\beta$, where $\tau \in G(K/F)$, then

$$\beta(\sigma\tau) = (\beta\sigma)\tau = (\zeta^i\beta)\tau = \zeta^i(\beta\tau) = \zeta^i\zeta^j\beta,$$

since $\zeta^i \in F$. Similarly,

$$\beta(\tau\sigma) = \zeta^j\zeta^i\beta.$$

Thus $\sigma\tau = \tau\sigma$, and $G(K/F)$ is abelian and therefore solvable.

Now suppose that F does not contain a primitive nth root of unity. Let ζ be a generator of the cyclic group of nth roots of unity under multiplication in $\overline{F}$. Let β again be a zero of $x^n - a$. Since β and $\zeta\beta$ are both in the splitting field K of $x^n - a$, $\zeta = (\zeta\beta)/\beta$ is in K. Let $F' = F(\zeta)$, so we have $F < F' \leq K$. Now F' is a normal extension of F, since F' is the splitting field of $x^n - 1$. Since $F' = F(\zeta)$, an automorphism η in $G(F'/F)$ is determined by $\zeta\eta$, and we must have $\zeta\eta = \zeta^i$ for some i, since all zeros of $x^n - 1$ are powers of ζ. If $\zeta\mu = \zeta^j$ for $\mu \in G(F'/F)$, then

$$\zeta(\eta\mu) = (\zeta\eta)\mu = \zeta^i\mu = (\zeta\mu)^i = (\zeta^j)^i = \zeta^{ij},$$

and, similarly,

$$\zeta(\mu\eta) = \zeta^{ij}.$$

Thus $G(F'/F)$ is abelian. By the Main Theorem of Galois Theory,

$$\{\iota\} \leq G(K/F') \leq G(K/F)$$

is a normal series and hence a subnormal series of groups. The first part of the proof shows that $G(K/F')$ is abelian, and Galois theory tells us that $G(K/F)/G(K/F')$ is isomorphic to $G(F'/F)$, which is abelian. It is easy to see that if a group has a subnormal series of subgroups with abelian quotient groups, then any refinement of this series also has abelian quotient groups. (See Exercise 49.6.) Thus a composition series of $G(K/F)$ must have abelian quotient groups, so $G(K/F)$ is solvable. ∎

The following theorem shows that if the splitting field K of a polynomial $f(x)$ over F is an extension of F by radicals, then $G(K/F)$ is solvable. This theorem will complete part (1) of our program.

Theorem 49.1 *If K is a normal extension by radicals of a field F of characteristic 0, then $G(K/F)$ is solvable.*

Proof. We know that there exist $\alpha_1, \ldots, \alpha_r \in K$ and positive integers $n_1, \ldots, n_r$ such that $K = F(\alpha_1, \ldots, \alpha_r)$, $\alpha_1^{n_1} \in F$ and $\alpha_i^{n_i} \in F(\alpha_1, \ldots, \alpha_{i-1})$ for $1 < i \leq r$. Let $K_0 = F$, and let K_i be the splitting field of $x^{n_i} - \alpha_i^{n_i}$ over K_{i-1}. Then $K \leq K_r$, and Lemma 49.1 shows that $G(K_i/K_{i-1})$ is solvable. Since $G(K_i/K_{i-1}) \simeq G(K_r/K_{i-1})/G(K_r/K_i)$, the normal series

$$\{\iota\} \leq G(K_r/K_{r-1}) \leq G(K_r/K_{r-2}) \leq \cdots \leq G(K_r/K_0) = G(K_r/F)$$

has solvable quotient groups. Exercise 49.7 shows that this series then has a refinement to a composition series with abelian quotients, so $G(K_r/F)$ is solvable. Since $G(K/F) \simeq G(K_r/F)/G(K_r/K)$, Exercise 14.17 shows that $G(K/F)$ is solvable. ∎

49.3 THE INSOLVABILITY OF THE QUINTIC

It remains for us to show that there is a subfield F of the real numbers and a polynomial $f(x) \in F[x]$ of degree 5 such that the splitting field K of $f(x)$ over F has a Galois group isomorphic to S_5.

Let $y_1 \in \mathbf{R}$ be transcendental over $\mathbf{Q}$, $y_2 \in \mathbf{R}$ be transcendental over $\mathbf{Q}(y_1)$, etc., until you get $y_5 \in \mathbf{R}$ transcendental over $\mathbf{Q}(y_1, \ldots, y_4)$. It can easily be shown by a counting argument that such transcendental real numbers exist. Transcendentals found in this fashion are **independent**

transcendental elements over Q. Let $K = Q(y_1, \ldots, y_5)$, and let

$$f(x) = \prod_{i=1}^{5} (x - y_i).$$

Thus $f(x) \in K[x]$. Now the coefficients of $f(x)$ are, except possibly for sign, among the so-called *elementary symmetric functions* in the y_i, namely

$$s_1 = y_1 + y_2 + \cdots + y_5,$$
$$s_2 = y_1 y_2 + y_1 y_3 + y_1 y_4 + y_1 y_5 + y_2 y_3$$
$$+ y_2 y_4 + y_2 y_5 + y_3 y_4 + y_3 y_5 + y_4 y_5,$$
$$\vdots$$
$$s_5 = y_1 y_2 y_3 y_4 y_5.$$

The coefficient of x^i in $f(x)$ is $\pm s_{5-i}$. Let $F = Q(s_1, s_2, \ldots, s_5)$; then $f(x) \in F[x]$ (see Fig. 49.1). Clearly, K is the splitting field over F of $f(x)$. Since the y_i behave as indeterminates over **Q**, for each $\sigma \in S_5$, the symmetric group on 5 letters, σ induces an automorphism $\bar{\sigma}$ of K defined by $a\bar{\sigma} = a$ for $a \in Q$ and $y_i \bar{\sigma} = y_{i\sigma}$. Since $\prod_{i=1}^{5} (x - y_i)$ is the same polynomial as $\prod_{i=1}^{5} (x - y_{i\sigma})$, we have

$K = Q(y_1, \ldots, y_5)$

$$s_i \bar{\sigma} = s_i$$

for each i, so $\bar{\sigma}$ leaves F fixed, and hence $\bar{\sigma} \in G(K/F)$. Now S_5 has order 5!, so

$F = Q(s_1, \ldots, s_5)$

$$|G(K/F)| \geq 5!.$$

Since the splitting field of a polynomial of degree 5 over F has degree at most 5! over F, we see that

Q

$$|G(K/F)| \leq 5!.$$

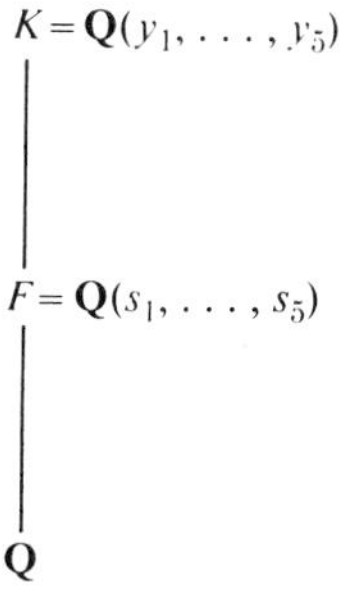

Fig. 49.1

Thus $|G(K/F)| = 5!$, and the automorphisms $\bar{\sigma}$ comprise the full Galois group $G(K/F)$. Therefore, $G(K/F) \simeq S_5$, so $G(K/F)$ is not solvable. By Theorem 49.1, $f(x)$ is not solvable by radicals over F. We summarize this in a theorem.

Theorem 49.2 *Let* $y_1, \ldots, y_5$ *be independent transcendental real numbers over* **Q**. *The polynomial*

$$f(x) = \prod_{i=1}^{5} (x - y_i)$$

is not solvable by radicals over $F = Q(s_1, \ldots, s_5)$, *where* s_i *is the* ith *elementary symmetric function in* $y_1, \ldots, y_5$.

It is evident that a generalization of these arguments shows that (*Final Goal*) a polynomial of degree n need not be solvable by radicals for $n \geq 5$.

In conclusion, we comment that there exist polynomials of degree 5 in $Q[x]$ which are not solvable by radicals over Q. A demonstration of this is left to the exercises (see Exercise 49.9).

EXERCISES

49.1 Let F be a field, and let $f(x) = ax^2 + bx + c$ be in $F[x]$, where $a \neq 0$. Show that if the characteristic of F is not 2, the splitting field of $f(x)$ over F is $F(\sqrt{b^2 - 4ac})$. [*Hint:* Complete the square, just as in your high school work, to derive the "quadratic formula."]

49.2 Can the splitting field K of $x^2 + x + 1$ over Z_2 be obtained by adjoining a square root to Z_2 of an element in Z_2? Is K an extension of Z_2 by radicals?

49.3 Show that if F is a field of characteristic different from 2 and

$$f(x) = ax^4 + bx^2 + c,$$

where $a \neq 0$, then $f(x)$ is solvable by radicals over F.

49.4 Is every polynomial in $F[x]$ of the form $ax^8 + bx^6 + cx^4 + dx^2 + e$, where $a \neq 0$, solvable by radicals over F, if F is of characteristic 0? Why?

49.5 Mark each of the following true or false.

— a) Let F be a field of characteristic 0. A polynomial in $F[x]$ is solvable by radicals if and only if its splitting field in $\bar{F}$ is an extension of F by radicals.

— b) Let F be a field of characteristic 0. A polynomial in $F[x]$ is solvable by radicals if and only if its splitting field in $\bar{F}$ has a solvable Galois group over F.

— c) The splitting field of $x^{17} - 5$ over Q has a solvable Galois group.

— d) The numbers π and $\sqrt{\pi}$ are independent transcendental numbers over Q.

— e) The Galois group of a finite extension of a finite field is solvable.

— f) No quintic polynomial is solvable by radicals over any field.

— g) Every fourth-degree polynomial over a field of characteristic 0 is solvable by radicals.

— h) The zeros of a cubic polynomial over a field F of characteristic 0 can always be attained by means of a finite sequence of operations of addition, subtraction, multiplication, division, and taking square roots starting with elements in F.

— i) The zeros of a cubic polynomial over a field F of characteristic 0 can never be attained by means of a finite sequence of operations of addition, subtraction, multiplication, division, and taking square roots, starting with elements in F.

— j) The theory of normal series of groups plays an important role in applications of Galois theory.

---

*49.6 Show that for a finite group, every refinement of a subnormal series with abelian quotients also has abelian quotients, thus completing the proof of Lemma 49.1. [*Hint:* Use Theorem 13.6.]

*49.7 Show that for a finite group, a subnormal series with solvable quotient groups can be refined to a composition series with abelian quotients, thus completing the proof of Theorem 49.1. [*Hint:* Use Theorem 13.6.]

*49.8 Let K be a normal extension by radicals of a field F of characteristic 0, and let E be a normal extension of F, $F \le E \le K$. Show that $G(E/F)$ is solvable. [*Hint:* Use Galois theory and Exercise 14.17.]

*49.9 This exercise exhibits a polynomial of degree 5 in $\mathbf{Q}[x]$ which is not solvable by radicals over $\mathbf{Q}$.

a) Show that if a subgroup H of S_5 contains a cycle of length 5 and a transposition τ, then $H = S_5$. [*Hint:* Show that H contains every transposition of S_5, and apply the corollary of Theorem 5.1. See Exercise 9.14.]

b) Show that if $f(x)$ is an irreducible polynomial in $\mathbf{Q}[x]$ of degree 5 having exactly two complex and three real zeros in $\mathbf{C}$, then the group of $f(x)$ over $\mathbf{Q}$ is S_5. [*Hint:* Use Sylow theory to show that the group has an element of order 5. Use the fact that $f(x)$ has exactly two complex zeros to show that the group has an element of order 2. Then apply (a).]

c) The polynomial $f(x) = 2x^5 - 5x^4 + 5$ is irreducible in $\mathbf{Q}[x]$, by the Eisenstein criterion, with $p = 5$. Use the techniques of calculus to find relative maxima and minima and to "graph the polynomial function f" well enough to see that $f(x)$ must have exactly three real zeros in $\mathbf{C}$. Conclude from (b) and Theorem 49.1 that $f(x)$ is not solvable by radicals over $\mathbf{Q}$.

Bibliography

Classic Works

1. N. BOURBAKI, *Éléments de Mathématique*, Book II of Part I, *Algèbre*. Paris: Hermann, 1942-58.

2. N. JACOBSON, *Lectures in Abstract Algebra*. Princeton, N.J.: Van Nostrand, vols. I, 1951, II, 1953, and III, 1964.

3. O. SCHREIER and E. SPERNER, *Introduction to Modern Algebra and Matrix Theory* (English translation), second edition. New York: Chelsea, 1959.

4. B. L. VAN DER WAERDEN, *Modern Algebra* (English translation). New York: Ungar, vols. I, 1949, and II, 1950.

General Algebra Texts

5. A. A. ALBERT, *Fundamental Concepts of Higher Algebra*. Chicago: University of Chicago Press, 1956.

6. G. BIRKHOFF and S. MAC LANE, *A Survey of Modern Algebra*, third edition. New York: Macmillan, 1965.

7. R. A. DEAN, *Elements of Abstract Algebra*. New York: Wiley, 1966.

8. I. N. HERSTEIN, *Topics in Algebra*. New York: Blaisdell, 1964.

9. T. W. HUNGERFORD, *Algebra*. New York: Holt, Rinehart and Winston, 1974.

10. R. E. JOHNSON, *University Algebra*. Englewood Cliffs, N.J.: Prentice Hall, 1966.

11. S. LANG, *Algebra*. Reading, Mass.: Addison-Wesley, 1965.

12. N. H. McCOY, *Introduction to Modern Algebra*. Boston: Allyn and Bacon, 1960.

13. G. D. MOSTOW, J. H. SAMPSON, and J. MEYER, *Fundamental Structures of Algebra*. New York: McGraw-Hill, 1963.

14. W. W. SAWYER, *A Concrete Approach to Abstract Algebra*. San Francisco: Freeman, 1959.

15. S. WARNER, *Modern Algebra*. Englewood Cliffs, N.J.: Prentice Hall, vols. I and II, 1965.

Group Theory

16. W. BURNSIDE, *Theory of Groups of Finite Order*, second edition. New York: Dover, 1955.

17. H. S. M. COXETER and W. O. MOSER, *Generators and Relations for Discrete Groups*, second edition. Berlin: Springer, 1965.

18. M. HALL, JR., *The Theory of Groups*. New York: Macmillan, 1959.

19. A. G. KUROSH, *The Theory of Groups* (English translation). New York: Chelsea, vols. I, 1955, and II, 1956.

20. W. LEDERMANN, *Introduction to the Theory of Finite Groups*, fourth revised edition. New York: Interscience, 1961.

21. J. G. THOMPSON and W. FEIT, "Solvability of Groups of Odd Order." *Pac. J. Math.*, **13** (1963), 775–1029.

22. M. A. RABIN, "Recursive Unsolvability of Group Theoretic Problems." *Ann. Math.*, **67** (1958), 172–194.

Ring Theory

23. E. ARTIN, C. J. NESBITT, and R. M. THRALL, *Rings With Minimum Condition*. Ann Arbor: University of Michigan Press, 1944.

24. N. H. McCOY, *Rings and Ideals* (Carus Monograph No. 8). Buffalo: The Mathematical Association of America; LaSalle, Illinois: Open Court, 1948.

25. N. H. McCOY, *The Theory of Rings*. New York: Macmillan, 1964.

Field Theory

26. E. ARTIN, *Galois Theory* (Notre Dame Mathematical Lecture No. 2), second edition. Notre Dame, Indiana: University of Notre Dame Press, 1944.

27. O. ZARISKI and P. SAMUEL, *Commutative Algebra*. Princeton, N.J.: Van Nostrand, vol. I, 1958.

Number Theory

28. G. H. HARDY and E. M. WRIGHT, *An Introduction to the Theory of Numbers*, fourth edition. Oxford: Clarendon Press, 1960.

29. S. LANG, *Algebraic Numbers*. Reading, Mass.: Addison-Wesley, 1964.

30. W. J. LeVEQUE, *Elementary Theory of Numbers*. Reading, Mass.: Addison-Wesley, 1962.

31. W. J. LeVEQUE, *Topics in Number Theory*. Reading, Mass.: Addison-Wesley, 2 vols., 1956.

32. T. NAGELL, *Introduction to Number Theory*. New York: Wiley, 1951.

33. I. NIVIN and H. S. ZUCKERMAN, *An Introduction to the Theory of Numbers*. New York: Wiley, 1960.

34. H. POLLARD, *The Theory of Algebraic Numbers* (Carus Monograph No. 9). Buffalo: The Mathematical Association of America; New York: Wiley, 1950.

35. D. SHANKS, *Solved and Unsolved Problems in Number Theory*. Washington: Spartan Books, vol. I, 1962.

36. B. M. STEWART, *Theory of Numbers*, second edition. New York: Macmillan, 1964.

37. J. V. USPENSKY and M. H. HEASLET, *Elementary Number Theory*. New York: McGraw-Hill, 1939.

38. E. WEISS, *Algebraic Number Theory*. New York: McGraw-Hill, 1963.

Homological Algebra

39. J. P. JANS, *Rings and Homology*. New York: Holt, 1964.

40. S. MAC LANE, *Homology*. Berlin: Springer, 1963.

Other References

41. A. A. ALBERT (editor), *Studies in Modern Algebra* (MAA Studies in Mathematics, vol. 2). Buffalo: The Mathematical Association of America; Englewood Cliffs, N.J.: Prentice Hall, 1963.

42. E. ARTIN, *Geometric Algebra*. New York: Interscience, 1957.

43. R. COURANT and R. ROBBINS, *What is Mathematics?*. Oxford: Oxford University Press, 1941.

44. H. S. M. COXETER, *Introduction to Geometry*. New York: Wiley, 1961.

45. R. H. CROWELL and R. H. FOX, *Introduction to Knot Theory*. New York: Ginn, 1963.

Answers and Comments

Section 0

0.1 $\{-\sqrt{3}, \sqrt{3}\}$ **0.2** $\varnothing$

0.3 $\{1, -1, 2, -2, 3, -3, 4, -4, 5, -5, 6, -6, 10, -10, 12, -12, 15, -15, 20, -20, 30, -30, 60, -60\}$

0.4 $\{0, 1, -1, 2, -2, 3, -3, 4, -4, 5, -5, 6, -6, 7, -7, 8, -8, 9, -9, 10, -10, 11\}$

0.5 Not an equivalence relation

0.6 Not an equivalence relation

0.7 An equivalence relation; $\bar{0} = \{0\}$, $\bar{a} = \{a, -a\}$ for each nonzero $a \in \mathbf{R}$

0.8 Not an equivalence relation

0.9 An equivalence relation; $\bar{1} = \{1, 2, \ldots, 9\}$, $\overline{10} = \{10, 11, \ldots, 99\}$, $\overline{100} = \{100, 101, \ldots, 999\}$, and in general $\overline{10^n} = \{10^n, 10^n + 1, \ldots, 10^{n+1} - 1\}$

0.10 An equivalence relation;
$$\bar{1} = \{1, 11, 21, 31, \ldots\} = \{10(n - 1) + 1 \mid n \in \mathbf{Z}^+\}$$
$$\bar{2} = \{2, 12, 22, 32, \ldots\} = \{10(n - 1) + 2 \mid n \in \mathbf{Z}^+\}$$
$$\vdots$$
$$\bar{9} = \{9, 19, 29, 39, \ldots\} = \{10(n - 1) + 9 \mid n \in \mathbf{Z}^+\}$$
$$\overline{10} = \{10, 20, 30, 40, \ldots\} = \{10n \mid n \in \mathbf{Z}^+\}$$

0.11 An equivalence relation;
$$\bar{1} = \{1, 3, 5, 7, \ldots\} = \{2(n - 1) + 1 \mid n \in \mathbf{Z}^+\}$$
$$\bar{2} = \{2, 4, 6, 8, \ldots\} = \{2n \mid n \in \mathbf{Z}^+\}$$

0.12 For $n = 1$, $\bar{1} = \mathbf{Z}^+$
For $n = 2$, $\bar{1} = \{2(n - 1) + 1 \mid n \in \mathbf{Z}^+\}$
$\bar{2} = \{2n \mid n \in \mathbf{Z}^+\}$
For $n = 3$, $\bar{1} = \{3(n - 1) + 1 \mid n \in \mathbf{Z}^+\}$
$\bar{2} = \{3(n - 1) + 2 \mid n \in \mathbf{Z}^+\}$
$\bar{3} = \{3n \mid n \in \mathbf{Z}^+\}$

0.13 (We don't want to spoil your fun with this one.)

Section 1

1.1 a) e, b, a b) a, a. You can't say. c) a, c. $*$ is not associative.
d) No, $b * e \neq e * b$

1.2 Top row: d. Second row: a. Fourth row: c, b.

1.3 d, c, c, d

1.4   a) No, (2) is violated.   b) Yes   c) Yes   d) Yes   e) No, (1) is violated.
f) No, (2) is violated; $1 * 1$ is not in $\mathbf{Z}^+$.

1.6   a) Not commutative, not associative
  b) Commutative, not associative
  c) Commutative, associative
  d) Commutative, not associative
  e) Not commutative, not associative

1.7   F T F F F T T T T F

1.8   Let $S = \{?, \Delta\}$. Define $*$ and $*'$ on S by $a * b = ?$ and $a *' b = \Delta$ for all
$a, b \in S$. (Other answers are possible.)

1.9   1. 16. 19,683. $n^{(n^2)}$.

1.10   8. 729. $n^{[n(n+1)/2]}$.

1.11   a) Two binary operations $*$ and $*'$ on the same set S give **algebraic structures
of the same type**, if each $x \in S$ has a counterpart $x' \in S$ such that the correspondence
$x \leftrightarrow x'$ is one to one and such that $(a * b)' = a' *' b'$ for all $a, b \in S$. (This may
be expressed in other ways.)   b) 10

Section 2

2.1   a) No. $\mathcal{G}_3$ fails.   b) No. $\mathcal{G}_1$ fails.   c) Yes   d) No. $\mathcal{G}_3$ fails at $a = 0$.
e) Yes   f) Yes

2.2 $\mathcal{G}_1\mathcal{G}_3\mathcal{G}_2$, $\mathcal{G}_3\mathcal{G}_1\mathcal{G}_2$, and $\mathcal{G}_3\mathcal{G}_2\mathcal{G}_1$ are not acceptable. The identity e occurs in
the statement of $\mathcal{G}_3$, which must not come before e is defined in $\mathcal{G}_2$.

2.3   *Partial answer:* $(a * b' * c)' = c' * b * a'$

2.4   The three tables are

I.

	e	a	b	c
e	e	a	b	c
a	a	e	c	b
b	b	c	e	a
c	c	b	a	e

II.

	e	a	b	c
e	e	a	b	c
a	a	e	c	b
b	b	c	a	e
c	c	b	e	a

III.

	e	a	b	c
e	e	a	b	c
a	a	b	c	e
b	b	c	e	a
c	c	e	a	b

The Tables II and III give the same type of group structure on $\{e, a, b, c\}$. In II,
rename a by "$\hat{b}$", b by "$\hat{a}$", and rewrite the resulting Table II in the order $e, \hat{a}, \hat{b}, c$
to obtain the Table III (up to $\hat{\ }$). All groups of 4 elements are commutative.

2.6   F T T F F T T T F T
i) By $\mathcal{G}_2$, every group contains at least one element, the identity element.

2.7

	e	a	b
e	e	a	b
a	a	e	e
b	b	e	e

(Other answers are possible.)

2.8 2. 3. It gets harder for 4 elements, where the answer is *not* 4.

2.9 c) $-\frac{1}{3}$

2.10 c) No   d) It shows that the "formally weaker" axioms for a group must either be all "left axioms" or all "right axioms," and not half and half.

Section 3

3.1 a) Yes b) No c) Yes d) Yes e) Yes f) No

3.2 $G_1 \le G_1, G_1 < G_4$

$G_2 < G_1, G_2 \le G_2, G_2 < G_4, G_2 < G_7, G_2 < G_8$

$G_3 \le G_3, G_3 < G_5$

$G_4 \le G_4$

$G_5 \le G_5$

$G_6 < G_5, G_6 \le G_6$

$G_7 < G_1, G_7 < G_4, G_7 \le G_7$

$G_8 < G_1, G_8 < G_4, G_8 < G_7, G_8 \le G_8$

$G_9 < G_3, G_9 < G_5, G_9 \le G_9$

3.3 a) $0, 25, 50, -25, -50$ b) $1, \frac{1}{2}, 2, 4, \frac{1}{4}$ c) $1, \pi, \pi^2, 1/\pi, 1/\pi^2$ (Other answers are possible.)

3.4 G_1 is cyclic with generators 1 and -1. G_2 is not cyclic. G_3 is not cyclic. G_4 is cyclic with generators 6 and -6. G_5 is cyclic with generators 6 and $\frac{1}{6}$. G_6 is not cyclic.

3.5 a)

	0	1	2	3	4	5
0	0	1	2	3	4	5
1	1	2	3	4	5	0
2	2	3	4	5	0	1
3	3	4	5	0	1	2
4	4	5	0	1	2	3
5	5	0	1	2	3	4

b) $\langle 2 \rangle = \langle 4 \rangle = \{0, 2, 4\}$
$\langle 3 \rangle = \{0, 3\}$
$\langle 1 \rangle = \langle 5 \rangle = \mathbf{Z}_6$

c) 1 and 5

3.7 T F T F F T F F T F

d) A group $\{e\}$ of one element has only one (improper) subgroup.

3.8 If $H = \varnothing$, no $a \in H$ exists.

3.17 The Klein 4-group V is an example.

Section 4

4.1 a) $\begin{pmatrix} 1 & 2 & 3 & 4 & 5 & 6 \\ 1 & 2 & 3 & 6 & 5 & 4 \end{pmatrix}$ b) $\begin{pmatrix} 1 & 2 & 3 & 4 & 5 & 6 \\ 2 & 4 & 1 & 5 & 6 & 3 \end{pmatrix}$

c) $\begin{pmatrix} 1 & 2 & 3 & 4 & 5 & 6 \\ 3 & 4 & 1 & 6 & 2 & 5 \end{pmatrix}$ d) $\begin{pmatrix} 1 & 2 & 3 & 4 & 5 & 6 \\ 5 & 1 & 6 & 2 & 4 & 3 \end{pmatrix}$ e) $\begin{pmatrix} 1 & 2 & 3 & 4 & 5 & 6 \\ 2 & 6 & 1 & 5 & 4 & 3 \end{pmatrix}$

4.2 f_1 and f_3 are permutations.

4.3 a) $\langle \rho_1 \rangle = \langle \rho_2 \rangle = \{\rho_0, \rho_1, \rho_2\}$
$\langle \mu_1 \rangle = \{\rho_0, \mu_1\}$

b)

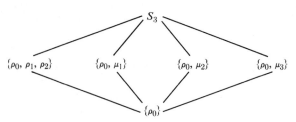

4.4

	ρ^0	ρ^1	ρ^2	ρ^3	ρ^4	ρ^5
ρ^0	ρ^0	ρ^1	ρ^2	ρ^3	ρ^4	ρ^5
ρ^1	ρ^1	ρ^2	ρ^3	ρ^4	ρ^5	ρ^0
ρ^2	ρ^2	ρ^3	ρ^4	ρ^5	ρ^0	ρ^1
ρ^3	ρ^3	ρ^4	ρ^5	ρ^0	ρ^1	ρ^2
ρ^4	ρ^4	ρ^5	ρ^0	ρ^1	ρ^2	ρ^3
ρ^5	ρ^5	ρ^0	ρ^1	ρ^2	ρ^3	ρ^4

No, it is not isomorphic to S_3, for it is abelian. It is isomorphic to the group of Exercise 3.5.

4.6 T F T F T T F F F T

4.7 S_3 is an example. See Exercise 4.3(b).

4.8 $|D_n| = 2n$

4.9 24

4.10 $\mathcal{O}_{1,\sigma} = \{1, 2, 3, 4, 5, 6\}$
$\mathcal{O}_{1,\tau} = \{1, 2, 3, 4\}$
$\mathcal{O}_{1,\mu} = \{1, 5\}$

Section 5

5.1 a) $\begin{pmatrix} 1 & 2 & 3 & 4 & 5 & 6 & 7 & 8 \\ 4 & 5 & 3 & 7 & 1 & 6 & 8 & 2 \end{pmatrix}$ b) $\begin{pmatrix} 1 & 2 & 3 & 4 & 5 & 6 & 7 & 8 \\ 3 & 7 & 2 & 8 & 5 & 4 & 1 & 6 \end{pmatrix}$

c) $\begin{pmatrix} 1 & 2 & 3 & 4 & 5 & 6 & 7 & 8 \\ 5 & 8 & 3 & 2 & 7 & 6 & 1 & 4 \end{pmatrix}$

5.2 a) $(1, 8)(3, 6, 4)(5, 7) = (1, 8)(3, 6)(3, 4)(5, 7)$
b) $(1, 3, 4)(2, 6)(5, 8, 7) = (1, 3)(1, 4)(2, 6)(5, 8)(5, 7)$
c) $(1, 3, 4, 7, 8, 6, 5, 2) = (1, 3)(1, 4)(1, 7)(1, 8)(1, 6)(1, 5)(1, 2)$

5.4 ρ_0, ρ_1, and ρ_2 are even.

	ρ_0	ρ_1	ρ_2
ρ_0	ρ_0	ρ_1	ρ_2
ρ_1	ρ_1	ρ_2	ρ_0
ρ_2	ρ_2	ρ_0	ρ_1

5.5   a) 4   b) A cycle of length n has order n.
c) σ has order 6; τ has order 4.   d) (a) 6   (b) 6   (c) 8
e) The order of a permutation expressed as a product of disjoint cycles is the least common multiple of the lengths of the cycles.

5.6   F T F F F F T T T F
c) It is desirable to show that no permutation can be expressed as a product of both an even and an odd number of transpositions, before defining even and odd permutations.
f) S_1 and S_2 are cyclic. S_n is not cyclic for $n > 2$.

5.7 n

5.9   No. Permutation multiplication is not closed on K.

5.13   If σ is a cycle of length n, then σ^r is also a cycle if and only if n and r are relatively prime, that is, n and r have no common factors greater than 1.

Section 6

6.1   2, 4, 4, 16

6.3   a) 6   b) 7   c) 4   d) 8   e) An infinite number

6.4   a)   b)   c)

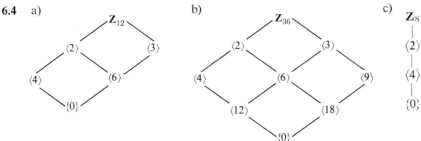

6.5 $\mathbf{Z}_6$: 1, 2, 3, 6
   $\mathbf{Z}_8$: 1, 2, 4, 8
   $\mathbf{Z}_{12}$: 1, 2, 3, 4, 6, 12
   $\mathbf{Z}_{60}$: 1, 2, 3, 4, 5, 6, 10, 12, 15, 20, 30, 60
   $\mathbf{Z}_{17}$: 1, 17

6.6   T F F F T F ? F T T
f) The Klein 4-group V is not cyclic.
g) As the student does mathematics, the answer is probably T. As the mathematical logician does mathematics, the answer may be F. There is probably no uniformity of opinion.

6.7 S_3 gives a counterexample.

6.8 $(p - 1)(q - 1)$

6.9 $p^{r-1}(p - 1)$

6.11 There are exactly d solutions, where d is the greatest common divisor of m and n.

Section 7

7.1 1) Z_4 is cyclic while V is not.

2) Z_4 has only two elements which are solutions of $x + x = 0$, while V has four solutions of the corresponding equation $x^2 = e$.

7.2 $\{Z, 17Z, 3Z, \langle \pi \rangle\}$ is a subcollection consisting of isomorphic groups, as are $\{Z_6, G\}$, $\{Z_2, S_2\}$, $\{S_6\}$, $\{Q\}$, $\{R, R^+\}$, $\{R^*\}$, $\{Q^*\}$ and $\{C^*\}$.

7.4 1, 2, 4, 2, 16

7.5 T T F T F T ? T F T

g) The answer will probably vary as mathematical maturity varies.

7.12 *Partial answer:* $a\psi = a - 1$

7.13 a) $a *_1 b = ab + 4a + 4b + 12$

b) S_2 is the set of all real numbers except $-t$; $a *_2 b = ab + ta + tb + (t^2 - t)$.

c) S_3 is the set of all real numbers except 1; $a *_3 b = ab - a - b + 2$.

7.14 Z_4: $\lambda_0 = \begin{pmatrix} 0 & 1 & 2 & 3 \\ 0 & 1 & 2 & 3 \end{pmatrix}$, $\lambda_1 = \begin{pmatrix} 0 & 1 & 2 & 3 \\ 1 & 2 & 3 & 0 \end{pmatrix}$,

$$\lambda_2 = \begin{pmatrix} 0 & 1 & 2 & 3 \\ 2 & 3 & 0 & 1 \end{pmatrix}, \quad \lambda_3 = \begin{pmatrix} 0 & 1 & 2 & 3 \\ 3 & 0 & 1 & 2 \end{pmatrix}$$

S_3: $\rho_{\rho_0} = \begin{pmatrix} \rho_0 & \rho_1 & \rho_2 & \mu_1 & \mu_2 & \mu_3 \\ \rho_0 & \rho_1 & \rho_2 & \mu_1 & \mu_2 & \mu_3 \end{pmatrix}$,

$\rho_{\rho_1} = \begin{pmatrix} \rho_0 & \rho_1 & \rho_2 & \mu_1 & \mu_2 & \mu_3 \\ \rho_1 & \rho_2 & \rho_0 & \mu_3 & \mu_1 & \mu_2 \end{pmatrix}$

$= (\rho_0, \rho_1, \rho_2)(\mu_1, \mu_3, \mu_2)$,

$\rho_{\rho_2} = (\rho_0, \rho_2, \rho_1)(\mu_1, \mu_2, \mu_3)$,

$\rho_{\mu_1} = (\rho_0, \mu_1)(\rho_1, \mu_2)(\rho_2, \mu_3)$,

$\rho_{\mu_2} = (\rho_0, \mu_2)(\rho_1, \mu_3)(\rho_2, \mu_1)$,

$\rho_{\rho_3} = (\rho_0, \mu_3)(\rho_1, \mu_1)(\rho_2, \mu_2)$

Section 8

8.1

Element	Order	Element	Order
(0, 0)	1	(0, 2)	2
(1, 0)	2	(1, 2)	2
(0, 1)	4	(0, 3)	4
(1, 1)	4	(1, 3)	4

The group is not cyclic.

8.2 24, 60

8.4 $\{(0, 0), (0, 1)\}$, $\{(0, 0), (1, 0)\}$, $\{(0, 0), (1, 1)\}$

8.5 $Z_{20} \times Z_3$, $Z_{15} \times Z_4$, $Z_{12} \times Z_5$, $Z_5 \times Z_3 \times Z_4$

8.6   a) 4   b) 12   c) 12   d) 2, 2   e) 8

8.7   T T F T F F F F F T

8.8   *Partial answer:* There are 14 in all, 7 of order 2 and 7 of order 4.

8.9   *Partial answer:* There are 7 of them.

8.10 Z_2 is an example.

8.12   *Yes* for order 3. *No* for order 4. For n a prime integer.

8.13 S_3

8.14   The groups might have empty intersection. Also, the group operations on the elements in common might not coincide.

8.15   If a group G is the internal direct product of abelian subgroups, then G is abelian.

8.18 $HK = H \vee K = \langle 2 \rangle$

8.19 $HK = \{\rho_0, \rho_2, \mu_1, \mu_2\}$; $H \vee K = S_3$

8.20 $HK = \{\rho_0, \delta_1, \delta_2, \rho_2\} = H \vee K$

Section 9

9.1   For 720: $Z_{16} \times Z_9 \times Z_5 \simeq Z_{720}$,

$Z_2 \times Z_8 \times Z_9 \times Z_5 \simeq Z_2 \times Z_{360}$,

$Z_4 \times Z_4 \times Z_9 \times Z_5 \simeq Z_4 \times Z_{180}$,

$Z_2 \times Z_2 \times Z_4 \times Z_9 \times Z_5 \simeq Z_2 \times Z_2 \times Z_{180}$,

$Z_2 \times Z_2 \times Z_2 \times Z_2 \times Z_9 \times Z_5 \simeq Z_2 \times Z_2 \times Z_2 \times Z_{90}$,

$Z_{16} \times Z_3 \times Z_3 \times Z_5 \simeq Z_3 \times Z_{240}$,

$Z_2 \times Z_8 \times Z_3 \times Z_3 \times Z_5 \simeq Z_6 \times Z_{120}$,

$Z_4 \times Z_4 \times Z_3 \times Z_3 \times Z_5 \simeq Z_{12} \times Z_{60}$,

$Z_2 \times Z_2 \times Z_4 \times Z_3 \times Z_3 \times Z_5 \simeq Z_2 \times Z_6 \times Z_{60}$,

$Z_2 \times Z_2 \times Z_2 \times Z_2 \times Z_3 \times Z_3 \times Z_5$
$$\simeq Z_2 \times Z_2 \times Z_6 \times Z_{30},$$

For 1089: $Z_9 \times Z_{121} \simeq Z_{1089}$,

$Z_3 \times Z_3 \times Z_{121} \simeq Z_3 \times Z_{363}$,

$Z_9 \times Z_{11} \times Z_{11} \simeq Z_{11} \times Z_{99}$,

$Z_3 \times Z_3 \times Z_{11} \times Z_{11} \simeq Z_{33} \times Z_{33}$

9.2   Torsion coefficients: 2, 6, 60. Betti number: 3.

9.3   There are 3 of order 24. There are 2 of order 25. There are 6 of order (24)(25).

9.5   49

9.6   T F F T T F T F T T
h) The Klein 4-group V is a counterexample.

9.7 $\{2, 3\}$ generates Z_{12}. $\{4, 6\}$ generates $\langle 2 \rangle$. $\{8, 6, 10\}$ generates $\langle 2 \rangle$.

9.8   12 for $Z_4 \times Z \times Z_3$. 144 for $Z_{12} \times Z \times Z_{12}$

9.9   a) Yes; G has exactly one subgroup of order 8. It can be characterized as the subgroup of all elements of G of order dividing 8.
b) No. For example $Z_8 \times Z_9$ has only one subgroup of order 4, while $Z_2 \times Z_2 \times Z_2 \times Z_9$ has seven subgroups of order 4.

9.11 For square free integers n, that is, for n not divisible by p^2 for any prime p

9.12 The numbers are the same.

9.13 $T = \{\cos q\pi + i \sin q\pi \mid q \in \mathbf{Q}\}$

9.15 2 for S_3 and D_4. 3 for $\mathbf{Z}_2 \times \mathbf{Z}_2 \times \mathbf{Z}_2$

9.16 A *subgroup* of S_n may not be able to be generated by two elements.

Section 10

10.5 *Partial answer:* The group has order 48.

10.6 c) S_4

10.7 c) 432   d) $|H| = 6; H \simeq S_3$

Section 11

11.1   a) 18   b) 8   c) 6

11.2 *Partial answer:* For both H_1 and H_3 the induced operation is well defined and the left cosets form a group. The induced operation is not well defined for H_2.

11.3 *Partial answer:* The left cosets are

$$
\begin{aligned}
e &= (0, 0) + \langle(1, 2)\rangle = \{(0, 0), (1, 2)\}, \\
a &= (1, 0) + \langle(1, 2)\rangle = \{(1, 0), (0, 2)\}, \\
b &= (0, 1) + \langle(1, 2)\rangle = \{(0, 1), (1, 3)\}, \\
c &= (1, 1) + \langle(1, 2)\rangle = \{(1, 1), (0, 3)\}.
\end{aligned}
$$

The induced operation on the left cosets is well defined, and they do form a group. This group of cosets is isomorphic to $\mathbf{Z}_4$.

11.4   1

11.6   T T T F T F T T F T

i) If A_4 had an element a of order 6, then $\langle a \rangle$ would be a subgroup of A_4 of order 6, contrary to the assertion of the last paragraph in this section.

11.11 For the first subgroup given, the induced operation on left cosets is well defined, and they form a group isomorphic to $\mathbf{Z}_2 \times \mathbf{Z}_3$. For the second subgroup given, the induced operation on left cosets is not well defined.

Section 12

12.1   a) 3   b) 4   c) 8   d) 4   e) 6

12.2 *Partial answer:* $S_n/A_n \simeq \mathbf{Z}_2$

12.3   a) $\mathbf{Z}$ b) $\mathbf{Z}$ c) $\mathbf{Z} \times \mathbf{Z}$

12.4   a) $\mathbf{Z}_4$ b) $\mathbf{Z}_2 \times \mathbf{Z}_2$ c) $\mathbf{Z}_4$

12.5 $\{\rho_0, \mu_1\}$, $\{\rho_0, \mu_2\}$ and $\{\rho_0, \mu_3\}$

12.7   T T T T T F T F T F

f) $\mathbf{Z}$ is torsion free but $\mathbf{Z}/n\mathbf{Z} \simeq \mathbf{Z}_n$, a torsion group.

h) For $n > 2$, S_n is nonabelian, but $S_n/A_n \simeq \mathbf{Z}_2$, and $\mathbf{Z}_2$ is abelian.

j) $n\mathbf{R} = \mathbf{R}$, so $\mathbf{R}/n\mathbf{R}$ is of order 1.

12.8

Subgroup	Factor Group (up to isomorphism)	Subgroup	Factor Group (up to isomorphism)
$\langle(1, 0)\rangle$	$\mathbf{Z}_4$	$\langle 2\rangle \times \langle 2\rangle$	$\mathbf{Z}_2 \times \mathbf{Z}_2$
$\langle(0, 1)\rangle$	$\mathbf{Z}_4$	$\langle(2, 0)\rangle$	$\mathbf{Z}_2 \times \mathbf{Z}_4$
$\langle(1, 1)\rangle$	$\mathbf{Z}_4$	$\langle(0, 2)\rangle$	$\mathbf{Z}_4 \times \mathbf{Z}_2$
$\langle(1, 2)\rangle$	$\mathbf{Z}_4$	$\langle(2, 2)\rangle$	$\mathbf{Z}_2 \times \mathbf{Z}_4$
$\langle(2, 1)\rangle$	$\mathbf{Z}_4$	$\langle(0, 0)\rangle$	$\mathbf{Z}_4 \times \mathbf{Z}_4$
$\langle(1, 3)\rangle$	$\mathbf{Z}_4$		

12.14 *Partial answer:* Let $G = N = S_3$, and let $H = \{\rho_0, \mu_1\}$ (see Example 4.1). Then N is normal in G, but $H \cap N = H$ is not normal in G.

12.20 $G' = \{\rho_0, \rho_2\}$

Section 13

13.1 a) Yes. (Image ϕ) $= \mathbf{Z}$. (Kernel ϕ) $= \{0\}$.
  b) No. $2 = (\frac{3}{2})\phi + (\frac{3}{2})\phi \neq 3\phi = 3$.
  c) Yes. (Image ϕ) $= \mathbf{R}^+$. (Kernel ϕ) $= \{1, -1\}$.
  d) Yes. (Image ϕ) $= \mathbf{Z}_2$. (Kernel ϕ) $= \{0, 2, 4\}$.
  e) No. In $\mathbf{Z}_2$, $0 = 1 + 1 = 5\phi + 5\phi \neq (5 + 5)\phi = 1\phi = 1$.

13.3 2 of $\mathbf{Z}$ onto $\mathbf{Z}$; 2 of $\mathbf{Z}$ into $\mathbf{Z}_2$; 1 of $\mathbf{Z}$ onto $\mathbf{Z}_2$

13.4 8 of $\mathbf{Z}$ into $\mathbf{Z}_8$; 4 of $\mathbf{Z}$ onto $\mathbf{Z}_8$

13.5 0 of $\mathbf{Z}_{12}$ onto $\mathbf{Z}_5$; 6 of $\mathbf{Z}_{12}$ into $\mathbf{Z}_6$; 2 of $\mathbf{Z}_{12}$ onto $\mathbf{Z}_6$; 2 of $\mathbf{Z}_{12}$ into $\mathbf{Z}_{14}$; 4 of $\mathbf{Z}_{12}$ into $\mathbf{Z}_{16}$

13.6 A homomorphism of a simple group either is an isomorphism or maps every element onto the identity of the image group.

13.7 T T F T F F T T T F
e) If ϕ is a homomorphism of G, then $|G\phi| = |G|/|(\text{Kernel } \phi)|$.   f) See (e).
g) A mapping of every element of a group G onto the identity of any group is always a homomorphism.   h) See (g).   i) See Exercise 13.6.   j) See Example 13.1.

13.8 4 of $\mathbf{Z}_2 \times \mathbf{Z}_2$ into $\mathbf{Z}_2$; 3 of $\mathbf{Z}_2 \times \mathbf{Z}_2$ onto $\mathbf{Z}_2$; 4 of $\mathbf{Z}_2 \times \mathbf{Z}_2$ into $\mathbf{Z}_6$; 64 of $\mathbf{Z}_2 \times \mathbf{Z}_2$ into $\mathbf{Z}_2 \times \mathbf{Z}_2 \times \mathbf{Z}_2$; 64 of $\mathbf{Z}_2 \times \mathbf{Z}_2$ into $\mathbf{Z}_2 \times \mathbf{Z}_2 \times \mathbf{Z}_4$

13.9 A_n

13.10 *Partial answer:* (Kernel π_1) $= \{(e_1, y) \mid y \in G_2\} \simeq G_2$

13.11 *Partial answer:* (Image ϕ) $= \langle a\rangle$. (Kernel ϕ) $= n\mathbf{Z}$ for some nonnegative $n \in \mathbf{Z}$ (including $n = 0, 1$)

13.12 b) *Partial answer:* If r is not relatively prime to $|G|$, (kernel ϕ_r) $\neq \{e\}$ and ϕ_r is not a map onto G. Thus for some $a \in G$, $x^r = a$ has no solution. In the extreme case that $|G|$ divides r, $x^r = a$ has no solution for any $a \in G$, $a \neq e$.

13.14 *Partial answer:* ϕ is an isomorphism if (kernel ϕ) $= \{e\}$, that is, if the center of G is the trivial group $\{e\}$.

Section 14

14.1 The refinements $\{0\} < 2940\mathbf{Z} < 60\mathbf{Z} < 20\mathbf{Z} < 4\mathbf{Z} < \mathbf{Z}$ of $\{0\} < 60\mathbf{Z} < 20\mathbf{Z} < \mathbf{Z}$ and $\{0\} < 2940\mathbf{Z} < 980\mathbf{Z} < 245\mathbf{Z} < 49\mathbf{Z} < \mathbf{Z}$ of $\{0\} < 245\mathbf{Z} < 49\mathbf{Z} < \mathbf{Z}$ are isomorphic.

14.2 $\{0\} < \langle 30 \rangle < \langle 15 \rangle < \langle 5 \rangle < \mathbf{Z}_{60}$,
$\{0\} < \langle 30 \rangle < \langle 15 \rangle < \langle 3 \rangle < \mathbf{Z}_{60}$,
$\{0\} < \langle 30 \rangle < \langle 10 \rangle < \langle 5 \rangle < \mathbf{Z}_{60}$,
$\{0\} < \langle 30 \rangle < \langle 10 \rangle < \langle 2 \rangle < \mathbf{Z}_{60}$,
$\{0\} < \langle 30 \rangle < \langle 6 \rangle < \langle 3 \rangle < \mathbf{Z}_{60}$,
$\{0\} < \langle 30 \rangle < \langle 6 \rangle < \langle 2 \rangle < \mathbf{Z}_{60}$,
$\{0\} < \langle 20 \rangle < \langle 10 \rangle < \langle 5 \rangle < \mathbf{Z}_{60}$,
$\{0\} < \langle 20 \rangle < \langle 10 \rangle < \langle 2 \rangle < \mathbf{Z}_{60}$,
$\{0\} < \langle 20 \rangle < \langle 4 \rangle < \langle 2 \rangle < \mathbf{Z}_{60}$,
$\{0\} < \langle 12 \rangle < \langle 6 \rangle < \langle 3 \rangle < \mathbf{Z}_{60}$,
$\{0\} < \langle 12 \rangle < \langle 6 \rangle < \langle 2 \rangle < \mathbf{Z}_{60}$,
$\{0\} < \langle 12 \rangle < \langle 4 \rangle < \langle 2 \rangle < \mathbf{Z}_{60}$

For each series the factor groups are isomorphic to $\mathbf{Z}_2$, $\mathbf{Z}_2$, $\mathbf{Z}_3$, and $\mathbf{Z}_5$ in some order.

14.3 They are all of the form $\{(0, 0)\} < H < \mathbf{Z}_5 \times \mathbf{Z}_5$, where H may be any of the subgroups $\langle (0, 1) \rangle$, $\langle (1, 0) \rangle$, $\langle (1, 1) \rangle$, $\langle (1, 2) \rangle$, $\langle (1, 3) \rangle$, and $\langle (1, 4) \rangle$ of $\mathbf{Z}_5 \times \mathbf{Z}_5$. Thus, there are six in all.

14.4 $\{(\rho_0, 0)\} < A_3 \times \{0\} < S_3 \times \{0\} < S_3 \times \mathbf{Z}_2$,
$\{(\rho_0, 0)\} < A_3 \times \{0\} < A_3 \times \mathbf{Z}_2 < S_3 \times \mathbf{Z}_2$,
$\{(\rho_0, 0)\} < \{\rho_0\} \times \mathbf{Z}_2 < A_3 \times \mathbf{Z}_2 < S_3 \times \mathbf{Z}_2$

14.6 T F T F F T F F T T
i) The Jordan-Hölder theorem applied to the groups $\mathbf{Z}_n$ implies the Fundamental Theorem of Arithmetic.

14.8 $\{\rho_0\} \times A_3 < A_3 \times A_3 < S_3 \times A_3 < S_3 \times S_3$
Yes, $S_3 \times S_3$ is solvable.

14.10 Yes. $\{\rho_0\} < \{\rho_0, \rho_2\} < \{\rho_0, \rho_1, \rho_2, \rho_3\} < D_4$ is a composition (actually a principal) series and all factor groups are isomorphic to $\mathbf{Z}_2$ and are thus abelian.

14.11 $\{\rho_0\} \times \mathbf{Z}_4$

14.13 $\{\rho_0\} \times \mathbf{Z}_4 \leq \{\rho_0\} \times \mathbf{Z}_4 \leq \{\rho_0\} \times \mathbf{Z}_4 \leq \cdots$

Section 15

15.1 a) 3 b) 27 c) 1, 3 d) 1, 85, 1, 51 e) 215 (It might be abelian!)

15.2 a) The conjugate classes are $\{\rho_0\}$, $\{\rho_2\}$, $\{\rho_1, \rho_3\}$, $\{\mu_1, \mu_2\}$, $\{\delta_1, \delta_2\}$.
b) $8 = 2 + 2 + 2 + 2$

15.3 The Sylow 3-subgroups are $\langle (1, 2, 3) \rangle$, $\langle (1, 2, 4) \rangle$, $\langle (1, 3, 4) \rangle$, and $\langle (2, 3, 4) \rangle$. Also $(3, 4)\langle (1, 2, 3) \rangle (3, 4) = \langle (1, 2, 4) \rangle$, etc.

15.6 T T T F T F T T F F
a) Of course, the p is to be the same for the two subgroups.

15.7
$H = \{\iota, (1, 2, 3, 4), (1, 3)(2, 4), (1, 4, 3, 2), (1, 3), (2, 4), (1, 2)(3, 4), (1, 4)(2, 3)\}$,
$K = \{\iota, (1, 3, 2, 4), (1, 2)(3, 4), (1, 4, 2, 3), (1, 2), (3, 4), (1, 3)(2, 4), (1, 4)(2, 3)\}$,
$K = (2, 3)H(2, 3)$

15.11 e) $p(1) = 1$. $p(2) = 2$. $p(3) = 3$. $p(4) = 5$. $p(5) = 7$. $p(6) = 11$. $p(7) = 15$

15.12 $\{\iota\}$,

$\{(1, 2), (1, 3), (1, 4), (2, 3), (2, 4), (3, 4)\}$,

$\{(1, 2)(3, 4), (1, 3)(2, 4), (1, 4)(2, 3)\}$,

$\{(1, 2, 3), (1, 2, 4), (1, 3, 4), (2, 3, 4), (1, 3, 2), (1, 4, 2), (1, 4, 3), (2, 4, 3)\}$,

$\{(1, 2, 3, 4), (1, 2, 4, 3), (1, 3, 2, 4), (1, 3, 4, 2), (1, 4, 2, 3), (1, 4, 3, 2)\}$,

$24 = 1 + 6 + 3 + 8 + 6$

15.13 $120 = 1 + 10 + 15 + 20 + 20 + 30 + 24$,

$720 = 1 + 15 + 45 + 40 + 15 + 120 + 90 + 40 + 90 + 144 + 120$

Section 16

16.3 T T F T T T T T F F

e) This is somewhat a matter of opinion.

i) A group of order 42 can have no subgroup of order 8, for 8 does not divide 42.

Section 17

17.1 a) $a^2b^2a^3c^3b^{-2}$, $b^2c^{-3}a^{-3}b^{-2}a^{-2}$ b) $a^{-1}b^3a^4c^6a^{-1}$, $ac^{-6}a^{-4}b^{-3}a$

17.2 a) a^5c^3, $a^{-5}c^{-3}$ b) $a^2b^3c^6$, $a^{-2}b^{-3}c^{-6}$

17.3 a) 16 b) 36 c) 36

17.4 a) 12 [Your reasoning is probably wrong if you found 12 by computing (4)(3).] b) 24 c) 18

17.5 a) 16 b) 36 c) 18

17.6 a) 12 b) 24 c) 0

17.8 T F F T F F F ? F T

e) An abelian group of finite order is finitely generated but not free abelian.

f) **Z** is a free group which is abelian. It is the only such group (up to isomorphism).

g) **Z** is the only (up to isomorphism) free abelian group which is free.

h) This is a matter of opinion.

17.9 a) *Partial answer:* $\{1\}$ is a basis for $\mathbf{Z}_4$. c) Yes

17.10 c) A blip group of G is isomorphic to the *abelianized version of G*, that is, to G modulo its commutator subgroup.

17.11 c) A blop group on S is isomorphic to the *free group F[S] on S*.

17.12 Change the statement of Exercise 17.11 by requiring that both G and G' be *abelian* groups.

Section 18

18.1 $(a : a^4 = 1)$. $(a, b : a^4 = 1, b = a^2)$. $(a, b, c : a = 1, b^4 = 1, c = 1)$. (Other answers are possible.)

18.2 $(a, b, c : a^3 = 1, b^2 = 1, ba = cb, c = a^2)$. (Other answers are possible.)

18.3　Octic group:

	1	a	a^2	a^3	b	ab	a^2b	a^3b
1	1	a	a^2	a^3	b	ab	a^2b	a^3b
a	a	a^2	a^3	1	ab	a^2b	a^3b	b
a^2	a^2	a^3	1	a	a^2b	a^3b	b	ab
a^3	a^3	1	a	a^2	a^3b	b	ab	a^2b
b	b	a^3b	a^2b	ab	1	a^3	a^2	a
ab	ab	b	a^3b	a^2b	a	1	a^3	a^2
a^2b	a^2b	ab	b	a^3b	a^2	a	1	a^3
a^3b	a^3b	a^2b	ab	b	a^3	a^2	a	1

Quaternion group: The same as the table for the octic group except that the 16 entries in the lower right corner are

a^2	a	1	a^3
a^3	a^2	a	1
1	a^3	a^2	a
a	1	a^3	a^2

18.4　T T F F F T T F T F

j) $(a, b, c : c = 1)$ is a presentation of a free group on two generators.

18.8　Z_{14}. $(a, b : a^7 = 1, b^2 = 1, ba = a^6b)$

18.9　Z_{21}. $(a, b : a^7 = 1, b^3 = 1, ba = a^2b)$

18.11　Z_{12}. $Z_2 \times Z_6$

$A_4 \simeq (a, b, c : a^2 = b^2 = c^3 = 1, ab = ba, ca = bc, cb = abc, cab = ac)$
$\simeq (s, t : s^3 = 1, t^2 = 1, (st)^3 = 1)$
$D_6 \simeq (a, b : a^6 = 1, b^2 = 1, ba = a^5b)$
$(a, b : a^3 = 1, b^4 = 1, ba = a^2b)$
(See Dean [7, p. 246] for a complete solution.)

18.12　Z_{30}. $D_{15} \simeq (a, b : a^{15} = 1, b^2 = 1, ba = a^{14}b)$
$Z_3 \times D_5 \simeq (a, b : a^6 = 1, b^5 = 1, ba = a^4b^4)$
$Z_5 \times S_3 \simeq (a, b : a^{10} = 1, b^3 = 1, ba = ab^2)$
(See Coxeter-Moser [16, p. 134] for a table giving all nonabelian groups of order < 32.)

Section 19

19.1　a) $2P_1P_3 - 3P_1P_4 + P_1P_6 - 3P_2P_3 + 3P_2P_4 - 5P_3P_4 + 4P_3P_6 - 5P_4P_6$
b) No　c) Yes

19.3　$C_i(P) = Z_i(P) = B_i(P) = H_i(P) = 0$ for $i > 0$. $B_0(P) = 0$. $Z_0(P) \simeq Z$ and is generated by the 0-cycle P. $H_0(P) \simeq Z$.

19.4 $C_i(X) = Z_i(X) = B_i(X) = H_i(X) = 0$ for $i > 0$. $B_0(X) = 0$. $Z_0(X) \simeq \mathbf{Z} \times \mathbf{Z}$ and is generated by the two 0-cycles P and P'. $H_0(X) \simeq \mathbf{Z} \times \mathbf{Z}$.

19.5 $C_i(X) = Z_i(X) = B_i(X) = H_i(X) = 0$ for $i > 0$. $B_0(X) \simeq \mathbf{Z}$ and is generated by the 0-chain $P_2 - P_1$. $Z_0(X) \simeq \mathbf{Z} \times \mathbf{Z}$ and is generated by the two 0-cycles P_1 and P_2. Since $Z_0(X)/B_0(X)$ "identifies P_1 with P_2," $H_0(X) \simeq \mathbf{Z}$ and is generated by the coset $P_1 + B_0(X)$.

19.6 T F T T T T F T T T

19.7 a) An **oriented n-simplex** is an ordered sequence $P_1 P_2 \cdots P_{n+1}$.
b) The **boundary of** $P_1 P_2 \cdots P_{n+1}$ is given by

$$\partial_n (P_1 P_2 \cdots P_{n+1}) = \sum_{i=1}^{n+1} (-1)^{i+1} P_1 P_2 \cdots P_{i-1} P_{i+1} \cdots P_{n+1}.$$

c) Each individual summand of the boundary of an oriented n-simplex is a **face of the simplex**.

19.8 Take the definitions in the text exactly as they stand.

19.10 a) $\delta^{(0)}(P_1) = P_2 P_1 + P_3 P_1 + P_4 P_1$, $\delta^{(0)}(P_4) = P_1 P_4 + P_2 P_4 + P_3 P_4$
b) $\delta^{(1)}(P_3 P_2) = P_1 P_3 P_2 + P_4 P_3 P_2$ c) $\delta^{(2)}(P_3 P_2 P_4) = P_1 P_3 P_2 P_4$

19.11 a) $\delta^{(n)} \left(\sum_i m_i \sigma_i \right) = \sum_i m_i \delta^{(n)}(\sigma_i)$

19.12 *Partial answer:* $Z^{(n)}(X)$ is the kernel of $\delta^{(n)}$, and $B^{(n)}(X)$ is the image of $\delta^{(n-1)}$.

19.13 $H^{(n)}(X) = Z^{(n)}(X)/B^{(n)}(X)$
$H^{(0)}(S) \simeq \mathbf{Z}$ and is generated by $(P_1 + P_2 + P_3 + P_4) + \{0\}$
$H^{(1)}(S) = 0$
$H^{(2)}(S) \simeq \mathbf{Z}$ and is generated by $P_1 P_2 P_3 + B^{(2)}(S)$

Section 20

20.1 $H_0(X) \simeq \mathbf{Z}$. $H_1(X) \simeq \mathbf{Z} \times \mathbf{Z}$. $H_n(X) = 0$ for $n > 1$.

20.2 $H_0(X) \simeq \mathbf{Z}$. $H_1(X) = 0$. $H_2(X) \simeq \mathbf{Z} \times \mathbf{Z}$. $H_n(X) = 0$ for $n > 2$.

20.3 $H_0(X) \simeq \mathbf{Z} \times \mathbf{Z}$. $H_1(X) \simeq \mathbf{Z}$. $H_2(X) \simeq \mathbf{Z}$. $H_n(X) = 0$ for $n > 2$.

20.4 $H_0(X) \simeq \mathbf{Z}$. $H_1(X) = 0$. $H_2(X) \simeq \mathbf{Z}$. $H_n(X) = 0$ for $n > 2$.

20.5 $H_0(X) \simeq \mathbf{Z}$. $H_1(X) \simeq \mathbf{Z}$. $H_2(X) \simeq \mathbf{Z}$. $H_n(X) = 0$ for $n > 2$.

20.6 $H_0(X) \simeq \mathbf{Z}$. $H_1(X) \simeq \mathbf{Z} \times \mathbf{Z}$. $H_2(X) \simeq \mathbf{Z}$. $H_n(X) = 0$ for $n > 2$. This space is homeomorphic to a torus.

20.7 T F F F T F T F F T

20.8 $H_0(X) \simeq \mathbf{Z} \times \mathbf{Z}$. $H_1(X) \simeq \mathbf{Z} \times \mathbf{Z} \times \mathbf{Z} \times \mathbf{Z}$. $H_2(X) \simeq \mathbf{Z} \times \mathbf{Z}$. $H_n(X) = 0$ for $n > 2$.

20.9 $H_0(X) \simeq \mathbf{Z}$. $H_1(X) \simeq \mathbf{Z} \times \mathbf{Z} \times \mathbf{Z}$. $H_2(X) \simeq \mathbf{Z} \times \mathbf{Z}$. $H_n(X) = 0$ for $n > 2$.

20.10 $H_0(X) \simeq \mathbf{Z}$. $H_1(X) \simeq \mathbf{Z}$. $H_2(X) \simeq \mathbf{Z} \times \mathbf{Z}$. $H_n(X) = 0$ for $n > 2$.

20.11 $H_0(X) \simeq \mathbf{Z}$. $H_1(X) \simeq \mathbf{Z} \times \mathbf{Z} \times \mathbf{Z} \times \mathbf{Z}$. $H_2(X) \simeq \mathbf{Z}$. $H_n(X) = 0$ for $n > 2$.

20.12 $H_0(X) \simeq \mathbf{Z}$. $H_1(X) \simeq \underbrace{\mathbf{Z} \times \cdots \times \mathbf{Z}}_{2n \text{ factors}}$. $H_2(X) \simeq \mathbf{Z}$. $H_n(X) = 0$ for $n > 2$.

Section 21

21.1 Both counts show that $\chi(X) = 1$.

21.2 a) $n_0 = 10$, $n_1 = 20$, and $n_2 = 10$, so $\chi(X) = 10 - 20 + 10 = 0$,
$\beta_0 = 1$, $\beta_1 = 1$, and $\beta_2 = 0$, so $\chi(X) = 1 - 1 + 0 = 0$
b) $n_0 = 9$, $n_1 = 27$, and $n_2 = 18$, so $\chi(X) = 9 - 27 + 18 = 0$,
$\beta_0 = 1$, $\beta_1 = 2$, and $\beta_2 = 1$, so $\chi(X) = 1 - 2 + 1 = 0$
c) $n_0 = 9$, $n_1 = 27$, and $n_2 = 18$, so $\chi(X) = 9 - 27 + 18 = 0$,
$\beta_0 = 1$, $\beta_1 = 1$, and $\beta_2 = 0$, so $\chi(X) = 1 - 1 + 0 = 0$

21.3 It will hold for a square region, for such a region is homeomorphic to E^2. It obviously does not hold for two disjoint 2-cells, for each can be mapped continuously onto the other, and such a map has no fixed points.

21.4 $H_0(X) \simeq \mathbf{Z}$. $H_1(X) \simeq \mathbf{Z} \times \mathbf{Z}_2$. $H_2(X) \simeq \mathbf{Z}$. $H_n(X) = 0$ for $n > 2$.

21.5 $H_0(X) \simeq \mathbf{Z} \times \mathbf{Z}$. $H_1(X) \simeq \mathbf{Z} \times \mathbf{Z} \times \mathbf{Z}_2 \times \mathbf{Z}_2$. $H_n(X) = 0$ for $n > 1$.

21.6 F T T F T T T T T F
a) $H_0(X)$ is never the group of one element.

21.7 $2 - 2n$

21.8 $H_0(X) \simeq \mathbf{Z}$. $H_1(X) \simeq \mathbf{Z}_2$. $H_n(X) = 0$ for $n > 1$.

21.9 $H_0(X) \simeq \mathbf{Z}$. $H_1(X) \simeq \underbrace{\mathbf{Z} \times \mathbf{Z} \times \cdots \times \mathbf{Z} \times \mathbf{Z}_2}_{(q-1) \text{ factors}}$. $H_n(X) = 0$ for $n > 1$.

21.10 Let Q be a vertex of X, and let c be the 2-chain consisting of all 2-simplexes of X, all oriented the same way, so that $c \in Z_2(X)$.
a) $f_{*0}: H_0(X) \to H_0(X)$ is given by

$$f_{*0}(Q + B_0(X)) = Q + B_0(X),$$

that is, f_{*0} is the identity map.
$f_{*1}: H_1(X) \to H_1(X)$ is defined by

$$f_{*1}((ma + nb) + B_1(X)) = (ma + 2nb) + B_1(X),$$

reflecting the fact that f maps X twice around itself in the θ direction.
$f_{*2}: H_2(X) \to H_2(X)$ is given by

$$f_{*2}(c + B_2(X)) = 2c + B_2(X),$$

reflecting the fact that each point of X is the image of two points of X under f.
b) f_{*0} is as in (a).
$f_{*1}: H_1(X) \to H_1(X)$ is given by

$$f_{*1}((ma + nb) + B_1(X)) = (2ma + nb) + B_1(X),$$

reflecting the fact that f maps X twice around itself in the ϕ direction.
$f_{*2}: H_2(X) \to H_2(X)$ is given by

$$f_{*2}(c + B_2(X)) = 2c + B_2(X),$$

reflecting the fact that each point of X is the image of two points of X under f.
c) f_{*0} is as in (a).

$f_{*1}: H_1(X) \to H_1(X)$ is given by

$$f_{*1}((ma + nb) + B_1(X)) = (2ma + 2nb) + B_1(X),$$

reflecting the fact that f maps X twice around itself in both the θ and ϕ directions.

$f_{*2}: H_2(X) \to H_2(X)$ is given by

$$f_{*2}(c + B_2(X)) = 4c + B_2(X),$$

reflecting the fact that each point of X is the image of four points of X under f.

21.11 Let Q be a vertex of b, and let c be as in the answer to Exercise 21.10.

a) f_{*0} is given by $f_{*0}(Q + B_0(X)) = Q + B_0(b)$.

f_{*1} is given by $f_{*1}((ma + nb) + B_1(X)) = nb + B_1(b)$.

f_{*2} is given by $f_{*2}(c + B_1(X)) = 0$.

b) f_{*0} is as in (a).

f_{*1} is given by $f_{*1}((ma + nb) + B_1(X)) = 2nb + B_1(b)$.

f_{*2} is as in (a).

21.12 The answers are as for Exercise 21.11, replacing $B_i(b)$ by $B_i(X)$.

21.13 Let Q be a vertex on b.

f_{*0} is given by $f_{*0}(Q + B_0(X)) = Q + B_0(b)$.

f_{*1} is given by $f_{*1}((ma + nb) + B_1(X)) = nb + B_0(b)$, where $m = 0, 1$.

f_{*2} is trivial, since both $H_2(X)$ and $H_2(b)$ are 0.

21.14 Let Q be a vertex on a.

f_{*0} is given by $f_{*0}(Q + B_0(X)) = Q + B_0(a)$.

f_{*1} is given by $f_{*1}((ma + nb) + B_1(X)) = ma + B_0(a)$, where $m = 0, 1$.

f_{*2} is trivial, since both $H_2(X)$ and $H_2(a)$ are 0.

Section 22

22.5 For Theorem 22.2, the condition $f_{k-1}\,\partial_k = \partial'_k f_k$ implies that

$$f_{k-1}(B_{k-1}(A)) \subseteq B_{k-1}(A').$$

Then Exercise 13.16 shows that f_{k-1} induces a natural homomorphism of $Z_{k-1}(A)/B_{k-1}(A)$ into $Z_{k-1}(A')/B_{k-1}(A')$. This is the correct way to view Theorem 22.2.

For Theorem 22.3, if we use Exercise 13.16, the fact that $\partial_k(A'_k) \subseteq A'_{k-1}$ shows that ∂_k induces a natural homomorphism $\bar{\partial}_k: (A_k/A'_k) \to (A_{k-1}/A'_{k-1})$.

22.6 $H_0(X, a) = 0$. $H_1(X, a) \simeq \mathbf{Z}$. $H_2(X, a) \simeq \mathbf{Z}$. $H_n(X, a) = 0$ for $n > 2$.

22.7 The exact homology sequence is

$$[H_2(a) = 0] \xrightarrow{i_{*2}} [H_2(X) \simeq \mathbf{Z}] \xrightarrow{j_{*2}} [H_2(X, a) \simeq \mathbf{Z}] \xrightarrow{\partial_{*2}} [H_1(a) \simeq \mathbf{Z}]$$
$$\xrightarrow{i_{*1}} [H_1(X) \simeq \mathbf{Z} \times \mathbf{Z}] \xrightarrow{j_{*1}} [H_1(X, a) \simeq \mathbf{Z}] \xrightarrow{\partial_{*1}} [H_0(a) \simeq \mathbf{Z}]$$
$$\xrightarrow{i_{*0}} [H_0(X) \simeq \mathbf{Z}] \xrightarrow{j_{*0}} [H_0(X, a) = 0].$$

j_{*2} maps a generator $c + B_2(X)$ of $H_2(X)$ onto the generator

$$(c + C_2(a)) + B_2(X, a)$$

of $H_2(X, a)$ and is an isomorphism. Thus (kernel j_{*2}) = (image i_{*2}) = 0.

∂_{*2} maps everything onto 0, so (kernel ∂_{*2}) = (image j_{*2}) $\simeq$ **Z**.

i_{*1} maps the generator $a + B_1(a)$ onto $(a + 0b) + B_1(X)$, so i_{*1} is an isomorphism *into*, and (kernel i_{*1}) = (image ∂_{*2}) = 0.

j_{*1} maps $(ma + nb) + B_1(X)$ onto $(nb + C_1(a)) + B_1(X, a)$, so (kernel j_{*1}) = (image i_{*1}) $\simeq$ **Z**.

∂_{*1} maps $(nb + C_1(a)) + B_1(X, a)$ onto 0, so (kernel ∂_{*1}) = (image j_{*1}) $\simeq$ **Z**. For a vertex Q of a, i_{*0} maps $Q + B_0(a)$ onto $Q + B_0(X)$, so i_{*0} is an isomorphism, and (kernel i_{*0}) = (image ∂_{*1}) = 0.

j_{*0} maps $Q + B_0(X)$ onto $B_0(X, a)$ in $H_0(X, a)$, so (kernel j_{*0}) = (image i_{*0}) $\simeq$ **Z**.

22.8 $H_0(X, a) = 0$. $H_1(X, a) \simeq$ **Z**. $H_2(X, a) \simeq$ **Z**. $H_n(X, a) = 0$ for $n > 2$.

22.9   The answer is formally identical with that in Exercise 22.7.

22.10 $H_0(X, Y) = 0$. $H_1(X, Y) \simeq$ **Z**. $H_2(X, Y) \simeq$ **Z**. $H_n(X, Y) = 0$ for $n > 2$.

22.11   *Partial answer:* The exact homology sequence is

$$[H_2(Y) = 0] \xrightarrow{i_{*2}} [H_2(X) = 0] \xrightarrow{j_{*2}} [H_2(X, Y) \simeq \mathbf{Z}] \xrightarrow{\partial_{*2}} [H_1(Y) \simeq \mathbf{Z} \times \mathbf{Z}]$$
$$\xrightarrow{i_{*1}} [H_1(X) \simeq \mathbf{Z}] \xrightarrow{j_{*1}} [H_1(X, Y) \simeq \mathbf{Z}] \xrightarrow{\partial_{*1}} [H_0(Y) \simeq \mathbf{Z} \times \mathbf{Z}]$$
$$\xrightarrow{i_{*0}} [H_0(X) \simeq \mathbf{Z}] \xrightarrow{j_{*0}} [H_0(X, Y) = 0].$$

The verification of exactness is left to you. Note that the edge P_1Q_1 of Fig. 20.3 gives rise to a generator of $H_1(X, Y)$. Starting with ∂_{*2}, these maps are very interesting.

Section 23

23.1   a) Yes   b) No. $\mathbf{Z}^+$ has no identity for addition.   c) Yes   d) Yes
e) Yes   f) Yes   g) No.   Multiplication is not closed on $\{ri \mid r \in \mathbf{R}\}$.

23.2   a) Commutative, no unity, not a field
        c) Commutative, unity $(1, 1)$, not a field
        d) Commutative, no unity, not a field
        e) Commutative, unity $1 + 0\sqrt{2}$, not a field
        f) Commutative, unity $1 + 0\sqrt{2}$, it is a field

23.3   a) $1, -1$ b) $(1, 1), (1, -1), (-1, 1), (-1, -1)$ c) $1, 2, 3, 4$
d) All nonzero $q \in \mathbf{Q}$
e) $(1, q, 1), (1, q, -1), (-1, q, 1), (-1, q, -1)$ for all nonzero $q \in \mathbf{Q}$ f) $1, 3$

23.5   T F F F T F T T T T
c) $\mathbf{Z}_2$ has only one unit, namely 1.

23.7 $\mathbf{Z} + \mathbf{Z}$ has unity $(1, 1)$. The subring $\mathbf{Z} + \{0\}$ has unity $(1, 0) \neq (1, 1)$.

23.16

+	$\varnothing$	$\{a\}$	$\{b\}$	S
$\varnothing$	$\varnothing$	$\{a\}$	$\{b\}$	S
$\{a\}$	$\{a\}$	$\varnothing$	S	$\{b\}$
$\{b\}$	$\{b\}$	S	$\varnothing$	$\{a\}$
S	S	$\{b\}$	$\{a\}$	$\varnothing$

$\cdot$	$\varnothing$	S	$\{a\}$	$\{b\}$
$\varnothing$	$\varnothing$	$\varnothing$	$\varnothing$	$\varnothing$
S	$\varnothing$	S	$\{a\}$	$\{b\}$
$\{a\}$	$\varnothing$	$\{a\}$	$\{a\}$	$\varnothing$
$\{b\}$	$\varnothing$	$\{b\}$	$\varnothing$	$\{b\}$

Section 24

24.1   0, 3, 5, 8, 9, 11

24.2   3 in $\mathbf{Z}_7$. $-7 = 16$ in $\mathbf{Z}_{23}$.

24.3   a) 0   b) 0   c) 0   d) 3   e) 12   f) 30

24.4   4

24.6   F T F F T T F T F F
g) $\mathbf{Z} + \mathbf{Z}$ is not an integral domain.
i) $n\mathbf{Z}$ does not have unity for $n > 1$.

24.7   There are no solutions of $x^2 + 2x + 2 = 0$.
2 is a solution of $x^2 + 2x + 4 = 0$.

24.14

	1	5	7	11
1	1	5	7	11
5	5	1	11	7
7	7	11	1	5
11	11	7	5	1

It is isomorphic to $\langle \mathbf{Z}_2 \times \mathbf{Z}_2, + \rangle$.

24.15

$\varphi(1) = 1$	$\varphi(11) = 10$	$\varphi(21) = 12$
$\varphi(2) = 1$	$\varphi(12) = 4$	$\varphi(22) = 10$
$\varphi(3) = 2$	$\varphi(13) = 12$	$\varphi(23) = 22$
$\varphi(4) = 2$	$\varphi(14) = 6$	$\varphi(24) = 8$
$\varphi(5) = 4$	$\varphi(15) = 8$	$\varphi(25) = 20$
$\varphi(6) = 2$	$\varphi(16) = 8$	$\varphi(26) = 12$
$\varphi(7) = 6$	$\varphi(17) = 16$	$\varphi(27) = 18$
$\varphi(8) = 4$	$\varphi(18) = 6$	$\varphi(28) = 12$
$\varphi(9) = 6$	$\varphi(19) = 18$	$\varphi(29) = 28$
$\varphi(10) = 4$	$\varphi(20) = 8$	$\varphi(30) = 8$

24.16 1

Section 25

25.1 $\begin{pmatrix} 7 & 13 \\ 6 & 2 \end{pmatrix}$ and $\begin{pmatrix} -8 & 63 \\ 29 & 2 \end{pmatrix}$

25.2   The unity element (δ_{ij}) of $M_n(F)$ has $\delta_{ii} = 1$ for $i = 1, \ldots, n$ and $\delta_{ij} = 0$ for $i \neq j$.

25.4   a) $1e + 0a + 3b$ b) $2e + 1a + 2b$ c) $2e + 2a + 2b$

25.5   a) $-6 + i + 13j + 2k$ b) j c) $-\frac{1}{2}i - \frac{1}{2}j$ d) $\frac{1}{50}j - \frac{3}{50}k$

25.6   F F F F F F T F T F
c) If $|A| = 1$, then $Hom(A) = \{0\}$.
e) $0 \in Hom(A)$ is not in $Iso(A)$.

25.9 $\begin{pmatrix} 1 & 0 \\ 0 & 1 \end{pmatrix}$, $\begin{pmatrix} 0 & 1 \\ 1 & 0 \end{pmatrix}$, $\begin{pmatrix} 1 & 0 \\ 1 & 1 \end{pmatrix}$, $\begin{pmatrix} 0 & 1 \\ 1 & 1 \end{pmatrix}$, $\begin{pmatrix} 1 & 1 \\ 0 & 1 \end{pmatrix}$,   and $\begin{pmatrix} 1 & 1 \\ 1 & 0 \end{pmatrix}$

are units in $M_2(F)$.

25.12 $1\rho_0 + 0\rho_1 + 1\rho_2 + 0\mu_1 + 1\mu_2 + 1\mu_3$

25.14 $\{a_1 + a_3 j \mid a_1, a_3 \in \mathbf{R}\}$ and $\{a_1 + a_4 k \mid a_1, a_4 \in \mathbf{R}\}$

25.15 $x^2 = -1$ in $\mathbb{Q}$ is an example.

25.17 $\mathbf{R}^*$, that is, $\{a_1 + 0i + 0j + 0k \mid a_1 \in \mathbf{R}, a_1 \neq 0\}$

Section 26

26.1 $\{q_1 + q_2 i \mid q_1, q_2 \in \mathbf{Q}\}$

26.2 $\{q_1 + q_2\sqrt{2} \mid q_1, q_2 \in \mathbf{Q}\}$

26.3 $D = \mathbf{Q}$ and $D' = \mathbf{Z}$ gives an example.

26.5 T F T F T T F T T

26.13 4, for 1 and 3 are already units in $\mathbf{Z}_4$.

26.14 $\left\{\dfrac{m}{2^n} \,\middle|\, m \in \mathbf{Z}, n \in \mathbf{Z}^+\right\}$

26.15 $\left\{\dfrac{m}{6^n} \,\middle|\, m \in \mathbf{Z}, n \in \mathbf{Z}^+\right\}$

26.16 The analog of Lemma 26.1 does not hold, i.e., $\sim$ does not give a partition of S. For $\mathbf{R} = \mathbf{Z}_6$ and $T = \{1, 2, 4\}$, $(3, 1) \sim (0, 2)$, and $(0, 1) \sim (0, 2)$, but $(3, 1)$ is not equivalent to $(0, 1)$. Thus $[(3, 1)] \cap [(0, 1)] \neq \varnothing$, but $[(3, 1)] \neq [(0, 1)]$.

Section 28

28.1 $N_1 = \{0\}, \mathbf{Z}_{12}/N_1 \simeq \mathbf{Z}_{12}$;
$N_2 = \{0, 2, 4, 6, 8, 10\}, \mathbf{Z}_{12}/N_2 \simeq \mathbf{Z}_2$;
$N_3 = \{0, 3, 6, 9\}, \mathbf{Z}_{12}/N_3 \simeq \mathbf{Z}_3$;
$N_4 = \{0, 4, 8\}, \mathbf{Z}_{12}/N_4 \simeq \mathbf{Z}_4$;
$N_5 = \{0, 6\}, \mathbf{Z}_{12}/N_5 \simeq \mathbf{Z}_6$;
$N_6 = \mathbf{Z}_{12}, \mathbf{Z}_{12}/\mathbf{Z}_{12} \simeq \{0\}$

28.2

+	8Z	2 + 8Z	4 + 8Z	6 + 8Z
8Z	8Z	2 + 8Z	4 + 8Z	6 + 8Z
2 + 8Z	2 + 8Z	4 + 8Z	6 + 8Z	8Z
4 + 8Z	4 + 8Z	6 + 8Z	8Z	2 + 8Z
6 + 8Z	6 + 8Z	8Z	2 + 8Z	4 + 8Z

·	8Z	2 + 8Z	4 + 8Z	6 + 8Z
8Z	8Z	8Z	8Z	8Z
2 + 8Z	8Z	4 + 8Z	8Z	4 + 8Z
4 + 8Z	8Z	8Z	8Z	8Z
6 + 8Z	8Z	4 + 8Z	8Z	4 + 8Z

$2\mathbf{Z}/8\mathbf{Z}$ is not isomorphic to $\mathbf{Z}_4$, for $2\mathbf{Z}/8\mathbf{Z}$ has no unity.

28.3 $\{(n, n) \mid n \in \mathbf{Z}\}$.   (Other answers are possible.)

28.5   F T F T T T T F F F

28.10   They are the subrings of the form $m\mathbf{Z} + n\mathbf{Z}$ for $m, n \in \mathbf{Z}$.

28.12   The radical of $\mathbf{Z}_{12}$ is $\{0, 6\}$.   The radical of $\mathbf{Z}$ is $\{0\}$.   The radical of $\mathbf{Z}_{32}$ is $\{2n \mid n \in \mathbf{Z}, 0 \le n \le 15\}$.

28.15   *Partial answer:* By the definition in Exercise 28.15, $\sqrt{R} = R$ for every ring R.   However, according to the definition in Exercise 28.11, the radical of R is not always all of R.   Thus this terminology is inconsistent if $N = R$.

28.16   a) Let $R = \mathbf{Z}$, and let $N = 4\mathbf{Z}$.   Then $\sqrt{N} = 2\mathbf{Z} \ne N$.
b) Let $R = \mathbf{Z}$, and let $N = 2\mathbf{Z}$.   Then $\sqrt{N} = 2\mathbf{Z} = N$.
(Other examples are possible.)

28.17   If $\sqrt{N}/N$ is viewed as a subring of R/N, then it is the radical of R/N, in the sense of the definition in Exercise 28.11.

Section 29

29.1 ϕ_1 such that $1\phi_1 = 1$. ϕ_2 such that $1\phi_2 = 0$

29.2 ϕ_1 such that $(1, 0)\phi_1 = 1$, $(0, 1)\phi_1 = 0$;
ϕ_2 such that $(1, 0)\phi_2 = 0$, $(0, 1)\phi_2 = 1$;
ϕ_3 such that $(1, 0)\phi_3 = 0$, $(0, 1)\phi_3 = 0$

29.3 $\{0, 2, 4, 6, 8, 10\}$ and $\{0, 3, 6, 9\}$ are the prime ideals, and they are also the maximal ideals.

29.4 $2\mathbf{Z} + \mathbf{Z}$ is a maximal ideal of $\mathbf{Z} + \mathbf{Z}$. $\mathbf{Z} + \{0\}$ is a prime ideal which is not maximal. $4\mathbf{Z} + \{0\}$ is an ideal which is not prime.

29.6   T F T T T F T T F T
d) A field has no proper ideals, and a simple group has no proper normal subgroups.   In both cases all quotient structures are either trivial or isomorphic to the original structure.

29.7 ϕ_1 such that $(1, 0)\phi_1 = (1, 0)$, $(0, 1)\phi_1 = (0, 1)$;
ϕ_2 such that $(1, 0)\phi_2 = (0, 1)$, $(0, 1)\phi_2 = (1, 0)$;
ϕ_3 such that $(1, 0)\phi_3 = (0, 0)$, $(0, 1)\phi_3 = (1, 1)$;
ϕ_4 such that $(1, 0)\phi_4 = (0, 0)$, $(0, 1)\phi_4 = (1, 0)$;
ϕ_5 such that $(1, 0)\phi_5 = (0, 0)$, $(0, 1)\phi_5 = (0, 1)$;
ϕ_6 such that $(1, 0)\phi_6 = (1, 1)$, $(0, 1)\phi_6 = (0, 0)$;
ϕ_7 such that $(1, 0)\phi_7 = (1, 0)$, $(0, 1)\phi_7 = (0, 0)$;
ϕ_8 such that $(1, 0)\phi_8 = (0, 1)$, $(0, 1)\phi_8 = (0, 0)$;
ϕ_9 such that $(1, 0)\phi_9 = (0, 0)$, $(0, 1)\phi_9 = (0, 0)$

29.11   Yes.   Consider $\mathbf{Z}_2 + \mathbf{Z}_3$ for both questions.

29.12   No.   Enlarging the domain to a field of quotients, you would have to have a field containing two different prime fields $\mathbf{Z}_p$ and $\mathbf{Z}_q$, which is impossible.

Section 30

30.1 $f(x) + g(x) = 3x^4 + 2x^3 + 4x^2 + 1$,
$f(x)g(x) = x^7 + 2x^6 + 4x^5 + x^3 + 2x^2 + x + 3$

30.2 $(y + 1)x^4 + (3y^3)x^3 + (y^2 - 3)x^2 + (2y^3 - 6y^2 - 2y)x + (y^2 + 2)$

30.3   a) 0   b) 2   c) 2   d) −1

30.4   0, $x - 5$, $2x - 10$, $x^2 - 25$, $x^2 - 5x$, $x^4 - 5x^3$.   (Other answers are possible.)

30.5   0, 4 = −1

30.7   T T T T F F T T F T
f) In $\mathbf{Z}_6[x]$, $(2x^3)(3x^4) = 0$.

30.9   1

30.10   0, 1, 2, 3

30.11   a) They are the units of D.   b) 1, −1   c) 1, 2, 3, 4, 5, 6

30.12   a) Let F be a subfield of a field E, let $\alpha_1, \ldots, \alpha_n$ be any elements of E, and let $x_1, \ldots, x_n$ be indeterminates. The map $\phi_{\alpha_1,\ldots,\alpha_n} : F[x_1, \ldots, x_n] \to E$ defined by

$$\left(\sum a_{m_1,\ldots,m_n} x_1^{m_1} \cdots x_n^{m_n}\right)\phi_{\alpha_1,\ldots,\alpha_n} = \sum a_{m_1,\ldots,m_n} \alpha_1^{m_1} \cdots \alpha_n^{m_n}$$

for

$$\left(\sum a_{m_1,\ldots,m_n} x_1^{m_1} \cdots x_n^{m_n}\right) \in F[x_1, \ldots, x_n]$$

is a homomorphism of $F[x_1, \ldots, x_n]$ into E. Also, $x_i \phi_{\alpha_1,\ldots,\alpha_n} = \alpha_i$, and $\phi_{\alpha_1,\ldots,\alpha_n}$ maps F isomorphically, by the identity map. $\phi_{\alpha_1,\ldots,\alpha_n}$ is an **evaluation homomorphism**.
b) 558
c) Let F be a subfield of a field E. Then $(\alpha_1, \ldots, \alpha_n) \in \underbrace{(E \times E \times \cdots \times E)}_{n \text{ factors}}$ is a **zero of** $f(x_1, \ldots, x_n) \in F[x_1, \ldots, x_n]$, if $f(x_1, \ldots, x_n)\phi_{\alpha_1,\ldots,\alpha_n} = 0$.

30.17   a) 4, 27   b) $\mathbf{Z}_2 \times \mathbf{Z}_2$, $\mathbf{Z}_3 \times \mathbf{Z}_3 \times \mathbf{Z}_3$

Section 31

31.1 $q(x) = x^4 + x^3 + x^2 + x - 2$, $r(x) = 4x + 3$

31.2 $q(x) = 5x^4 + 5x^2 - x$, $r(x) = x + 2$

31.3 $(x - 1)(x + 1)(x - 2)(x + 2)$

31.4   No. It has -1 as a zero and hence $x + 1$ as a factor. $(x + 1)^2(x - 2)$

31.5   *Partial answer:* $f(x)$ is not irreducible over $\mathbf{R}$, and is not irreducible over $\mathbf{C}$.

31.6   *Partial answer:* $g(x)$ is irreducible over $\mathbf{R}$, but is not irreducible over $\mathbf{C}$.

31.9   T T T F T F T T T F

31.10   a) Yes. $p = 3$ b) Yes. $p = 3$ c) No   d) Yes. $p = 5$

31.11   Yes. It is of degree 3 with no zeros in $\mathbf{Z}_5$. $2x^3 + x^2 + 2x + 2$

31.12 $\frac{2}{3}$, $-\frac{5}{2}$

31.14   In $\mathbf{Z}_2[x]$: $x^2 + x + 1$, $x^3 + x + 1$, $x^3 + x^2 + 1$
In $\mathbf{Z}_3[x]$: $\pm(x^2 + 1)$, $\pm(x^2 + x + 2)$, $\pm(x^2 + 2x + 2)$,
$\pm(x^3 + 2x + 1)$, $\pm(x^3 + 2x + 2)$, $\pm(x^3 + x^2 + 2)$,
$\pm(x^3 + x^2 + x + 2)$, $\pm(x^3 + x^2 + 2x + 1)$,
$\pm(x^3 + 2x^2 + 1)$, $\pm(x^3 + 2x^2 + x + 1)$,
$\pm(x^3 + 2x^2 + 2x + 2)$

31.15 No. $\langle x^2 - 5x + 6 \rangle$ is not a maximal ideal of $\mathbf{Q}[x]$, since $x^2 - 5x + 6 = (x - 3)(x - 2)$ is not irreducible over $\mathbf{Q}$. Yes, $\langle x^2 - 6x + 6 \rangle$ is a maximal ideal of $\mathbf{Q}[x]$.

Section 32

32.1 a) Yes b) Yes c) No d) Yes e) No f) Yes g) Yes h) Yes

32.2 In $\mathbf{Z}[x]$: only $2x - 7$, $-2x + 7$
In $\mathbf{Q}[x]$: $4x - 14$, $x - \frac{7}{2}$, $6x - 21$, $-8x + 28$
In $\mathbf{Z}_{11}[x]$: $2x - 7$, $10x - 2$, $6x + 1$, $3x - 5$, $5x - 1$
(Other answers are possible.)

32.3 In $\mathbf{Z}[x]$: $(2)(2)(x^2 - x + 2)$
In $\mathbf{Q}[x]$: $4x^2 - 4x + 8$
In $\mathbf{Z}_{11}[x]$: $(4x + 2)(x + 4)$

32.4 a) $(6)(3x^2 - 2x + 8)$ b) $(1)(18x^2 - 12x + 48)$ or as in (a)
c) $(1)(2x^2 - 3x + 6)$ d) $(1)(2x^2 - 3x + 6)$

32.6 T T T F T F F T F T
i) Either p or one of its associates must appear in every factorization into irreducibles.

32.8 They are the irreducibles of D, together with the irreducibles of $F[x]$ which are in $D[x]$ and are furthermore primitive polynomials in $D[x]$.

32.9 Partial answer: $D^* - U$ is not a group under multiplication for $1 \notin (D^* - U)$.

32.10 $2x + 4$ is irreducible in $\mathbf{Q}[x]$ but not in $\mathbf{Z}[x]$.

32.11 Not every nonunit $\neq 0$ of $\mathbf{Z} + \mathbf{Z}$ has a factorization into irreducibles. For example, $(1, 0)$ is not a unit, and every factorization of $(1, 0)$ has a factor of the form $(\pm 1, 0)$, which is not irreducible, since $(\pm 1, 0) = (\pm 1, 0)(1, 50)$. The only irreducibles of $\mathbf{Z} + \mathbf{Z}$ are $(\pm 1, p)$ and $(q, \pm 1)$, where p and q are irreducibles in $\mathbf{Z}$.

32.14 Partial answer: $x^3 - y^3 = (x - y)(x^2 + xy + y^2)$

32.17 $\mathbf{Z}$

Section 33

33.1 a) Yes b) No. (1) is violated. c) No. (1) is violated.
d) No. (2) is violated. e) Yes

33.2 61

33.3 $x^3 + 2x - 1$

33.4 $23 = 25(22{,}471) + (-172)(3{,}266)$

33.5 a) Yes. $\mathbf{Z}$ is a UFD, and Theorem 32.3 applies. c) No
d) No. By Theorem 33.1, $\mathbf{Z}[x]$ a Euclidean domain would imply $\mathbf{Z}[x]$ a PID, contradicting (c).

33.7 T F T F T T T F T T

33.8 No. The arithmetic structure of D is completely determined by the binary operations of addition and multiplication on D.

33.9 $61 = 29(49{,}349) + (-92)(15{,}555)$

33.17 Partial answer: The equation $ax = b$ has a solution in $\mathbf{Z}_n$ for nonzero $a, b \in \mathbf{Z}_n$ if and only if the positive gcd of a and n in $\mathbf{Z}$ divides b.

33.18 Find the positive gcd d of a and n by the Euclidean algorithm. If d does not divide b, $ax \equiv b \pmod{n}$ has no solution. If d divides b, find λ and μ in **Z** such that $d = \lambda a + \mu n$. Then $\lambda(b/d)$ is a solution of the congruence.

A solution of $12x \equiv 18 \pmod{42}$ is -9. All other solutions are of the form $-9 + 7k$ for $k \in$ **Z**.

Section 34

34.1   a) $5 = (1 + 2i)(1 - 2i)$ b) 7 is irreducible in $\mathbf{Z}[i]$
c) $4 + 3i = (1 + 2i)(2 - i)$ d) $6 - 7i = (1 - 2i)(4 + i)$

34.2 $6 = (2)(3) = (-1 + \sqrt{-5})(-1 - \sqrt{-5})$

34.3 $\sigma = 1 + i$, $\rho = 3i$

34.4 $7 - i$

34.6   T T T F T T T F T T

34.12 $1 + 2i$

34.13   c) (i) Order 9, characteristic 3. (ii) Order 2, characteristic 2. (iii) Order 5, characteristic 5

Section 35

35.1   a) $x^2 - 2x - 1$ b) $x^4 - 10x^2 + 1$ c) $x^2 - 2x + 2$
d) $x^6 - 3x^4 + 3x^2 - 3$ e) $x^{12} + 3x^8 - 4x^6 + 3x^4 + 12x^2 + 5$

35.2   a) $\mathrm{Irr}(\alpha, \mathbf{Q}) = x^4 - 6x^2 + 3$, $\deg(\alpha, \mathbf{Q}) = 4$
b) $\mathrm{Irr}(\alpha, \mathbf{Q}) = x^4 - \frac{2}{3}x^2 - \frac{62}{9}$, $\deg(\alpha, \mathbf{Q}) = 4$
c) $\mathrm{Irr}(\alpha, \mathbf{Q}) = x^4 - 2x^2 + 9$, $\deg(\alpha, \mathbf{Q}) = 4$

35.3   a) Algebraic, $\deg(\alpha, F) = 2$ b) Algebraic, $\deg(\alpha, F) = 2$
c) Transcendental   d) Algebraic, $\deg(\alpha, F) = 1$
e) Algebraic, $\deg(\alpha, F) = 2$ f) Algebraic, $\deg(\alpha, F) = 4$
g) Transcendental   h) Algebraic, $\deg(\alpha, F) = 1$
i) Algebraic, $\deg(\alpha, F) = 3$ j) Algebraic, $\deg(\alpha, F) = 6$

35.4 $x^2 + x + 1 = (x - \alpha)(x + 1 + \alpha)$

35.5   It is the monic polynomial in $F[x]$ of *minimal* degree having α as a zero.

35.7   b)

$+$	0	1	2	α	2α	$1 + \alpha$	$1 + 2\alpha$	$2 + \alpha$	$2 + 2\alpha$
0	0	1	2	α	2α	$1 + \alpha$	$1 + 2\alpha$	$2 + \alpha$	$2 + 2\alpha$
1	1	2	0	$1 + \alpha$	$1 + 2\alpha$	$2 + \alpha$	$2 + 2\alpha$	α	2α
2	2	0	1	$2 + \alpha$	$2 + 2\alpha$	α	2α	$1 + \alpha$	$1 + 2\alpha$
α	α	$1 + \alpha$	$2 + \alpha$	2α	0	$1 + 2\alpha$	1	$2 + 2\alpha$	2
2α	2α	$1 + 2\alpha$	$2 + 2\alpha$	0	α	1	$1 + \alpha$	2	$2 + \alpha$
$1 + \alpha$	$1 + \alpha$	$2 + \alpha$	α	$1 + 2\alpha$	1	$2 + 2\alpha$	2	2α	0
$1 + 2\alpha$	$1 + 2\alpha$	$2 + 2\alpha$	2α	1	$1 + \alpha$	2	$2 + \alpha$	0	α
$2 + \alpha$	$2 + \alpha$	α	$1 + \alpha$	$2 + 2\alpha$	2	2α	0	$1 + 2\alpha$	1
$2 + 2\alpha$	$2 + 2\alpha$	2α	$1 + 2\alpha$	2	$2 + \alpha$	0	α	1	$1 + \alpha$

·	0	1	2	α	2α	1+α	1+2α	2+α	2+2α
0	0	0	0	0	0	0	0	0	0
1	0	1	2	α	2α	1+α	1+2α	2+α	2+2α
2	0	2	1	2α	α	2+2α	2+α	1+2α	1+α
α	0	α	2α	2	1	2+α	1+α	2+2α	1+2α
2α	0	2α	α	1	2	1+2α	2+2α	1+α	2+α
1+α	0	1+α	2+2α	2+α	1+2α	2α	2	1	α
1+2α	0	1+2α	2+α	1+α	2+2α	2	α	2α	1
2+α	0	2+α	1+2α	2+2α	1+α	1	2α	α	2
2+2α	0	2+2α	1+α	1+2α	2+α	α	1	2	2α

35.8 T T T T F T F T F T

35.11 a) $Q(\pi^3)$ b) $Q(\pi, e^5)$. (Other answers are possible.)

35.13 b) $x^3 + x^2 + 1 = (x - \alpha)(x - \alpha^2)(x - (1 + \alpha + \alpha^2))$

35.15 $\langle Z_2(\alpha), + \rangle \simeq Z_2 \times Z_2 \times Z_2$. $\langle Z_2(\alpha)^*, \cdot \rangle \simeq Z_7$

Section 36

36.1 $\{(0, 1), (1, 0)\}$, $\{(1, 1), (-1, 1)\}$, $\{(2, 1), (1, 2)\}$. (Other answers are possible.)

36.2 a) Yes b) No. $2(-1, 1, 2) - 4(2, -3, 1) + (10, -14, 0) = (0, 0, 0)$

36.3 $x^2 + x + 1$

36.4 a) $\{1, \sqrt{2}\}$ b) $\{1\}$ c) $\{1, \sqrt[3]{2}, (\sqrt[3]{2})^2\}$ d) $\{1, i\}$ e) $\{1, i\}$
f) $\{1, \sqrt[4]{2}, \sqrt{2}, (\sqrt[4]{2})^3\}$

36.6 T F T T F F F T T T

36.7 a) A subset W of a vector space V over a field F is a **subspace of V over F** if the induced operations of vector addition and scalar multiplication are closed on W and W is a vector space over F under these operations.

36.8 a) The **subspace of V generated by** S is the intersection of all subspaces of V containing S.

36.9 *Partial answer:* The **direct sum** $V_1 + \cdots + V_n$ **of the vector spaces** V_i is the set $V_1 \times \cdots \times V_n$ of vectors, together with the operations of vector addition and scalar multiplication defined by

$$(\alpha_1, \ldots, \alpha_n) + (\beta_1, \ldots, \beta_n) = (\alpha_1 + \beta_1, \ldots, \alpha_n + \beta_n)$$

and $a(\alpha_1, \ldots, \alpha_n) = (a\alpha_1, \ldots, a\alpha_n)$ for $\alpha_i, \beta_i \in V_i$ and $a \in F$.

36.10 *Partial answer:* A basis for F^n over F is

$$\{(1, 0, \ldots, 0), (0, 1, \ldots, 0), \ldots, (0, 0, \ldots, 1)\},$$

where 1 is the multiplicative identity of F.

36.12 A vector space V over F is **isomorphic to a vector space** V' **over F** if there exists a one-to-one function $\phi: V \xrightarrow{\text{onto}} V'$ such that $(\alpha + \beta)\phi = \alpha\phi + \beta\phi$ and $(a\alpha)\phi = a(\alpha\phi)$ for all $\alpha, \beta \in V$ and $a \in F$.

36.15 a) A homomorphism

b) *Partial answer:* The **kernel** (or **nullspace**) of ϕ is $\{\alpha \in V \mid \alpha\phi = 0\}$.

c) ϕ is an isomorphism of V with V' if (kernel ϕ) $= \{0\}$ and ϕ maps V onto V'.

36.16 *Partial answer:* The **quotient space** V/S is the abelian group $\langle V, + \rangle/\langle S, + \rangle$, together with the operation of scalar multiplication defined by $a(\alpha + S) = a\alpha + S$ for $\alpha \in V$ and $a \in F$.

36.18 a) *Partial answer:* The **join** $S \vee T$ of S and T is the intersection of all subspaces of V containing both S and T.

b) The elements of $S \vee T$ are of the form $\sigma + \tau$ for $\sigma \in S$ and $\tau \in T$.

Section 37

37.1 $\langle G, \mathcal{O}, *, \cdot \rangle$. (Other notations are possible.)

37.2 $\langle M, R, \oplus, +, \cdot, \times \rangle$. (Other notations are possible.)

37.3 $\langle V, F, \oplus, \odot, +, \cdot, \times \rangle$. (Other notations are possible.)

37.5 $\mathbf{Z}_2 \times \{0\}$ is not a characteristic subgroup of $\langle \mathbf{Z}_2 \times \mathbf{Z}_2, + \rangle$, for there exists an automorphism of $\mathbf{Z}_2 \times \mathbf{Z}_2$ mapping $\mathbf{Z}_2 \times \{0\}$ onto $\{0\} \times \mathbf{Z}_2$.

37.7 *Partial answer:* A map $\phi: G \to G'$ is an $\mathcal{O}$-**homomorphism of the** $\mathcal{O}$-**group** G **into the** $\mathcal{O}$-**group** G' if for all $\alpha, \beta \in G$ and $a \in \mathcal{O}$, both $(\alpha + \beta)\phi = \alpha\phi + \beta\phi$ and $(\alpha a)\phi = (\alpha\phi)a$.

37.10 A **submodule** N of a **(left)** R-**module** M is a subset of M that is a (left) R-module under induced operations of addition in M and external multiplication by R.

A **quotient module** M/N of a **(left)** R-**module** M **modulo a submodule** N is $\langle M, + \rangle/\langle N, + \rangle$, together with the operation of external multiplication by elements of R defined by $r(\alpha + N) = r\alpha + N$ for $\alpha \in M$ and $r \in R$.

37.11 A **homomorphism of a (left)** R-**module** M **into a (left)** R-**module** M' is a function $\phi: M \to M'$ such that $(\alpha + \beta)\phi = \alpha\phi + \beta\phi$ and $(r\alpha)\phi = r(\alpha\phi)$ for all $\alpha, \beta \in M$ and $r \in R$.

Section 38

38.1 a) 2, $\{1, \sqrt{2}\}$

b) 4, $\{1, \sqrt{2}, \sqrt{3}, \sqrt{6}\}$

c) 8, $\{1, \sqrt{3}, \sqrt{5}, \sqrt{15}, \sqrt{2}, \sqrt{6}, \sqrt{10}, \sqrt{30}\}$

d) 6, $\{1, \sqrt[3]{2}, (\sqrt[3]{2})^2, \sqrt{3}, \sqrt[3]{2}\sqrt{3}, (\sqrt[3]{2})^2(\sqrt{3})\}$

e) 6, $\{1, \sqrt{2}, \sqrt[3]{2}, \sqrt{2}(\sqrt[3]{2}), (\sqrt[3]{2})^2, \sqrt{2}(\sqrt[3]{2})^2\}$

38.2 a) 4 b) 2 c) 6 d) 9 e) 2 f) 2 g) 1 h) 2

38.3 e) $\{1, \sqrt{2}\}$ f) $\{1, \sqrt{2}\}$ g) $\{1\}$ h) $\{1, \sqrt{2}\}$

38.6 T F T F F T F F F F

38.8 *Partial answer:* Extensions of degree 2^n for $n \in \mathbf{Z}^+$ are obtained.

Section 39

39.5 T T T F T F T T T F

f) It is true that every such real number is constructible, but we have not shown this.

Section 40

40.1 a) $\sqrt{2}, -\sqrt{2}$ b) $\sqrt{2}$ c) $3 + \sqrt{2}, 3 - \sqrt{2}$

d) $\sqrt{2} - \sqrt{3}, \sqrt{2} + \sqrt{3}, -\sqrt{2} - \sqrt{3}, -\sqrt{2} + \sqrt{3}$

e) $\sqrt{2} + i, \sqrt{2} - i, -\sqrt{2} + i, -\sqrt{2} - i$

f) $\sqrt{2} + i, \sqrt{2} - i$

g) $\sqrt{1 + \sqrt{2}}, -\sqrt{1 + \sqrt{2}}, \sqrt{1 - \sqrt{2}}, -\sqrt{1 - \sqrt{2}}$

h) $\sqrt{1 + \sqrt{2}}, -\sqrt{1 + \sqrt{2}}$

40.2 a) $\sqrt{3}$ b) $-\sqrt{2} + \sqrt{5}$ c) $-\sqrt{2} + 3\sqrt{5}$

d) $\dfrac{\sqrt{2} + 3\sqrt{5}}{-2\sqrt{3} - \sqrt{2}}$ e) $-\sqrt{2} + \sqrt{45}$ f) $\sqrt{2} + \sqrt{3} - \sqrt{30}$

40.3 a) $3 - \sqrt{2}$ b) They are the same maps.

40.4 a) Q b) $Q(\sqrt{6})$ c) Q

40.5 a) $Q(\sqrt{2}, \sqrt{5})$ b) $Q(\sqrt{2}, \sqrt{3}, \sqrt{5})$ c) $Q(\sqrt{5})$

d) $Q(\sqrt{3}, \sqrt{10})$ e) $Q(\sqrt{6}, \sqrt{10})$ f) Q

40.7 F F T T F T T T T T

40.8 b) *Partial answer:* $H = \{\iota, \tau_2, \tau_3, \tau_5, \tau_2\tau_3, \tau_2\tau_5, \tau_3\tau_5, \tau_2\tau_3\tau_5\}$. The group is abelian and all elements are of order 2 so that it is isomorphic to $\langle Z_2 \times Z_2 \times Z_2, + \rangle$. It is essentially the group $\{\iota, \tau_2\} \times \{\iota, \tau_3\} \times \{\iota, \tau_5\}$. You should have no trouble making the table.

40.9 $0\sigma_2 = 0, 1\sigma_2 = 1, \alpha\sigma_2 = 1 + \alpha, (1 + \alpha)\sigma_2 = \alpha.$ $Z_2(\alpha)_{\{\sigma_2\}} = Z_2$

40.10 $0\sigma_3 = 0, 1\sigma_3 = 1, 2\sigma_3 = 2, \alpha\sigma_3 = -\alpha, (2\alpha)\sigma_3 = -2\alpha, (1 + \alpha)\sigma_3 = 1 - \alpha,$
$(1 + 2\alpha)\sigma_3 = 1 - 2\alpha, (2 + \alpha)\sigma_3 = 2 - \alpha, (2 + 2\alpha)\sigma_3 = 2 - 2\alpha$
$Z_3(\alpha)_{\{\sigma_3\}} = Z_3$

40.11 Let $F = Z_p(x)$, where x is an indeterminate. Then the image of F under σ_p is $Z_p(x^p)$, a proper subfield of $Z_p(x)$. While σ_p is an isomorphic mapping for all fields E of characteristic p, σ_p *need not be onto* E, so it need not be an automorphism.

40.16 Yes

40.17 Every automorphism of $F(x)$ maps x onto some $y = (ax + b)/(cx + d)$, where $a, b, c, d \in F$ and $ad - bc \neq 0$. Conversely, each such $y \in F(x)$ gives rise to an automorphism of $F(x)$ mapping x onto y and leaving F fixed.

Section 41

41.1 a) The identity map of E into $\overline{Q}$.

τ given by $\sqrt{2}\tau = \sqrt{2}, \sqrt{3}\tau = -\sqrt{3}, \sqrt{5}\tau = -\sqrt{5}$

b) τ_1 given by $\sqrt{2}\tau_1 = \sqrt{2}, \sqrt{3}\tau_1 = \sqrt{3}, \sqrt{5}\tau_1 = -\sqrt{5};$

τ_2 given by $\sqrt{2}\tau_2 = \sqrt{2}, \sqrt{3}\tau_2 = -\sqrt{3}, \sqrt{5}\tau_2 = \sqrt{5}$

c) τ_1 given by $\sqrt{2}\tau_1 = \sqrt{2}, \sqrt{3}\tau_1 = \sqrt{3}, \sqrt{5}\tau_1 = -\sqrt{5};$

τ_2 given by $\sqrt{2}\tau_2 = \sqrt{2}, \sqrt{3}\tau_2 = -\sqrt{3}, \sqrt{5}\tau_2 = \sqrt{5};$

τ_3 given by $\sqrt{2}\tau_3 = -\sqrt{2}, \sqrt{3}\tau_3 = \sqrt{3}, \sqrt{5}\tau_3 = \sqrt{5};$

τ_4 given by $\sqrt{2}\tau_4 = -\sqrt{2}, \sqrt{3}\tau_4 = -\sqrt{3}, \sqrt{5}\tau_4 = -\sqrt{5}$

41.2 a) The identity map of $Q(\sqrt[3]{2})$ into itself.

τ_1 given by $\alpha_1\tau_1 = \alpha_2$, that is, $\tau_1 = \psi_{\alpha_1,\alpha_2}$. $\tau_2 = \psi_{\alpha_1,\alpha_3}$

b) The identity map of $\mathbf{Q}(\sqrt[3]{2}, \sqrt{3})$ into itself.

τ_1 given by $\alpha_1\tau_1 = \alpha_1$, $\sqrt{3}\tau_1 = -\sqrt{3}$;

τ_2 given by $\alpha_1\tau_2 = \alpha_2$, $\sqrt{3}\tau_2 = \sqrt{3}$;

τ_3 given by $\alpha_1\tau_3 = \alpha_2$, $\sqrt{3}\tau_3 = -\sqrt{3}$;

τ_4 given by $\alpha_1\tau_4 = \alpha_3$, $\sqrt{3}\tau_4 = \sqrt{3}$;

τ_5 given by $\alpha_1\tau_5 = \alpha_3$, $\sqrt{3}\tau_5 = -\sqrt{3}$

c) τ_1 given by $i\tau_1 = i$, $\sqrt{3}\tau_1 = -\sqrt{3}$, $\alpha_1\tau_1 = \alpha_1$;

τ_2 given by $i\tau_2 = -i$, $\sqrt{3}\tau_2 = -\sqrt{3}$, $\alpha_1\tau_2 = \alpha_1$;

τ_3 given by $i\tau_3 = i$, $\sqrt{3}\tau_3 = -\sqrt{3}$, $\alpha_1\tau_3 = \alpha_2$;

τ_4 given by $i\tau_4 = -i$, $\sqrt{3}\tau_4 = -\sqrt{3}$, $\alpha_1\tau_4 = \alpha_2$;

τ_5 given by $i\tau_5 = i$, $\sqrt{3}\tau_5 = -\sqrt{3}$, $\alpha_1\tau_5 = \alpha_3$;

τ_6 given by $i\tau_6 = -i$, $\sqrt{3}\tau_6 = -\sqrt{3}$, $\alpha_1\tau_6 = \alpha_3$

41.3 a) $\mathbf{Q}(\pi^2)$ b) τ_1 given by $\sqrt{\pi}\tau_1 = i\sqrt{\pi}$; τ_2 given by $\sqrt{\pi}\tau_2 = -i\sqrt{\pi}$

41.5 F T F T F T T T T F

e) $\mathbf{Q}$ and $\mathbf{Q}(\sqrt{2})$ are not isomorphic, but have isomorphic algebraic closures.

Section 42

42.1 a) 2 b) 2 c) 4 d) 6 e) 2 f) 12

42.2 $1 \le [E : F] \le n!$

42.3 a) 1 b) 6 c) 2

42.4 α^2 and $1 + \alpha + \alpha^2$ are also zeros of $x^3 + x^2 + 1$.

42.6 T F T T T T F F T T

42.7 Let $F = \mathbf{Q}$ and $E = \mathbf{Q}(\sqrt{2})$. Then

$$f(x) = x^4 - 5x^2 + 6 = (x^2 - 2)(x^2 - 3)$$

has a zero in E, but does not split in E.

42.8 a) Not necessarily. For example,

$6 = |G(\mathbf{Q}(\sqrt[3]{2}, i\sqrt{3})/\mathbf{Q})|$

$\ne |G(\mathbf{Q}(\sqrt[3]{2}, i\sqrt{3})/\mathbf{Q}(\sqrt[3]{2}))| \cdot |G(\mathbf{Q}(\sqrt[3]{2})/\mathbf{Q})| = (2)(1) = 2.$

b) Yes. Each field is a splitting field of the one immediately under it. If E is a splitting field over F, $|G(E/F)| = \{E : F\}$, and the index is multiplicative.

42.14 a) 6

Section 43

43.1 $f(x) = x^4 - 4x^2 + 4 = (x^2 - 2)^2$. Here $f(x)$ is not an irreducible polynomial. Every irreducible factor of $f(x)$ has zeros of multiplicity 1 only.

43.3 F T T F F T T T T T

43.8 b) The field F c) $F[x^p]$

43.15 Compute a gcd of $f(x)$ and $f'(x)$ using the Euclidean algorithm. Then $f(x)$ has a zero of multiplicity >1 if and only if this gcd is of degree >0.

43.16 $\alpha = \sqrt{2} + \sqrt{3}$. $\sqrt{2} = (1/2)\alpha^3 - (9/2)\alpha$. $\sqrt{3} = (11/2)\alpha - (1/2)\alpha^3$.
(Other answers are possible.)

Section 44

44.1 $Z_3(y^3, z^9)$

44.2 $Z_3(y^3, z^{27})$

44.4 $Z_3(y^4, z^2)$

44.5 $Z_3(y^4, z^4)$

44.6   F T F F F F T F T T

Section 45

45.1   a) 3, 5 b) 3, 10, 5, 11, 14, 7, 12, 6
c) 5, 10, 20, 17, 11, 21, 19, 15, 7, 14

45.2   a) 4 b) 6 c) 8 d) 0

45.5   T F T F T F T T F T

45.9   b) It is irreducible.

Section 46

46.1   a) 8 b) 8 c) 8 d) 2 e) 4 f) 4 g) 2 h) 1

46.2   The group has two elements, the identity automorphism ι of $Q(i)$ and σ
such that $i\sigma = -i$.

46.3   The order is 3.   A generator of the group is σ_9, where $\alpha\sigma_9 = \alpha^9$ for
$\alpha \in GF(729)$.

46.4 $F = Q$, $K_1 = Q(\sqrt{2})$, and $K_2 = Q(\sqrt{3})$

46.5   a) Let $\alpha_1 = \sqrt[3]{2}$, $\alpha_2 = \sqrt[3]{2}\,\dfrac{-1 + i\sqrt{3}}{2}$,   and $\alpha_3 = \sqrt[3]{2}\,\dfrac{-1 - i\sqrt{3}}{2}$.

The maps are:

ρ_0, where ρ_0 is the identity map;
ρ_1, where $\alpha_1\rho_1 = \alpha_2$ and $i\sqrt{3}\rho_1 = i\sqrt{3}$;
ρ_2, where $\alpha_1\rho_2 = \alpha_3$ and $i\sqrt{3}\rho_2 = i\sqrt{3}$;
μ_1, where $\alpha_1\mu_1 = \alpha_1$ and $i\sqrt{3}\mu_1 = -i\sqrt{3}$;
μ_2, where $\alpha_1\mu_2 = \alpha_3$ and $i\sqrt{3}\mu_2 = -i\sqrt{3}$;
μ_3, where $\alpha_1\mu_3 = \alpha_2$ and $i\sqrt{3}\mu_3 = -i\sqrt{3}$.

b) S_3.   The notation in (a) was chosen to coincide with the notation for S_3 in
Example 4.1.

c)

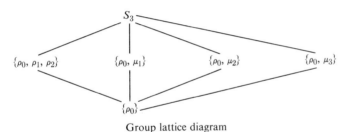

Group lattice diagram

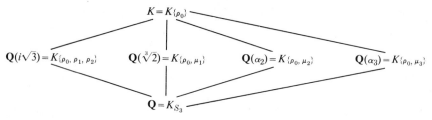

Field lattice diagram

46.7   F F T T T F F T F T

46.9   a) 4 b) 1 c) 36 d) 16 e) 0 f) 0 g) 0 h) 8

46.11   The splitting field of $(x^4 - 5x^2 + 6) \in \mathbf{Q}[x]$ is $\mathbf{Q}(\sqrt{2}, \sqrt{3})$, and the group is that of Example 40.4.

46.12   The splitting field of $(x^3 - 1) \in \mathbf{Q}[x]$ is $\mathbf{Q}(i\sqrt{3})$, and the group is cyclic of order 2 with elements: ι, where ι is the identity map of $\mathbf{Q}(i\sqrt{3})$, and σ, where $(i\sqrt{3})\sigma = -i\sqrt{3}$.

46.18 $G(K/(E \vee L)) = G(K/E) \cap G(K/L)$

46.19 $G(K/(E \cap L)) = G(K/E) \vee G(K/L)$

Section 47

47.3 $\mathbf{Q}(\sqrt[4]{2}, i)$: $\sqrt[4]{2} + i$, $x^8 + 4x^6 + 2x^4 + 28x^2 + 1$;
  $\mathbf{Q}(\sqrt[4]{2})$: $\sqrt[4]{2}$, $x^4 - 2$;
  $\mathbf{Q}(i(\sqrt[4]{2}))$: $i(\sqrt[4]{2})$, $x^4 - 2$;
  $\mathbf{Q}(\sqrt{2}, i)$: $\sqrt{2} + i$, $x^4 - 2x^2 + 9$;
  $\mathbf{Q}(\sqrt[4]{2} + i(\sqrt[4]{2}))$: $\sqrt[4]{2} + i(\sqrt[4]{2})$, $x^4 + 8$;
  $\mathbf{Q}(\sqrt[4]{2} - i(\sqrt[4]{2}))$: $\sqrt[4]{2} - i(\sqrt[4]{2})$, $x^4 + 8$;
  $\mathbf{Q}(\sqrt{2})$: $\sqrt{2}$, $x^2 - 2$;
  $\mathbf{Q}(i)$: i, $x^2 + 1$;
  $\mathbf{Q}(i\sqrt{2})$: $i\sqrt{2}$, $x^2 + 2$;
  $\mathbf{Q}$: 1, $x - 1$

47.4   c) $G(K/\mathbf{Q})$ is isomorphic to $\{1, 2, 3, 4\}$ under multiplication modulo 5, that is, to $\langle \mathbf{Z}_5^*, \cdot \rangle$. This group is abelian and is cyclic of order 4. A table describing the elements is given in (d).

d) $G(K/\mathbf{Q}) = \{\iota, \sigma_2, \sigma_3, \sigma_4\}$, where

	ι	σ_2	σ_3	σ_4
$\zeta \rightarrow$	ζ	ζ^2	ζ^3	ζ^4

$G(K/\mathbf{Q}) = \{\iota, \sigma_2, \sigma_3, \sigma_4\}$
|
$\{\iota, \sigma_4\}$
|
$\{\iota\}$

Group lattice diagram

$K = K_{\{\iota\}}$
|
$\mathbf{Q}(\sqrt{5}) = \mathbf{Q}(\cos 72°) = K_{\{\iota, \sigma_4\}}$
|
$\mathbf{Q} = K_{\{\iota, \sigma_2, \sigma_3 \sigma_4\}}$

Field lattice diagram

47.5   The group is cyclic of order 5, and its elements are

	ι	σ_1	σ_2	σ_3	σ_4
$\sqrt[5]{2}\rightarrow$	$\sqrt[5]{2}$	$\zeta(\sqrt[5]{2})$	$\zeta^2(\sqrt[5]{2})$	$\zeta^3(\sqrt[5]{2})$	$\zeta^4(\sqrt[5]{2})$

where $\sqrt[5]{2}$ is the real 5th root of 2.

47.6   c) $G(K/\mathbf{Q})$ is isomorphic to $\{1, 2, 3, 4, 5, 6\}$ under multiplication modulo 7, that is, to $\langle \mathbf{Z}_7{}^*, \cdot \rangle$. This group is abelian, and is cyclic of order 6. A table describing the elements is given in (d).

d) $G(K/\mathbf{Q}) = \{\iota, \sigma_2, \sigma_3, \sigma_4, \sigma_5, \sigma_6\}$, where

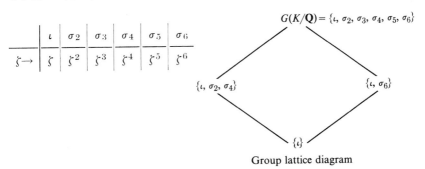

	ι	σ_2	σ_3	σ_4	σ_5	σ_6
$\zeta\rightarrow$	ζ	ζ^2	ζ^3	ζ^4	ζ^5	ζ^6

$G(K/\mathbf{Q}) = \{\iota, \sigma_2, \sigma_3, \sigma_4, \sigma_5, \sigma_6\}$

$\{\iota, \sigma_2, \sigma_4\}$ $\{\iota, \sigma_6\}$

$\{\iota\}$

Group lattice diagram

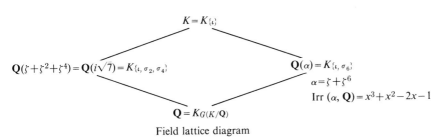

$K = K_{\{\iota\}}$

$\mathbf{Q}(\zeta+\zeta^2+\zeta^4) = \mathbf{Q}(i\sqrt{7}) = K_{\{\iota,\, \sigma_2,\, \sigma_4\}}$ $\mathbf{Q}(\alpha) = K_{\{\iota,\, \sigma_6\}}$

$\alpha = \zeta + \zeta^6$

$\text{Irr}\,(\alpha, \mathbf{Q}) = x^3 + x^2 - 2x - 1$

$\mathbf{Q} = K_{G(K/\mathbf{Q})}$

Field lattice diagram

47.7   The splitting field of $x^8 - 1$ over $\mathbf{Q}$ is the same as the splitting field of $x^4 + 1$ over $\mathbf{Q}$, so a complete description is contained in Example 47.2. (This is the easiest way to answer the problem.)

47.8 $K = \mathbf{Q}(\alpha_1, \alpha_2)$,   where $\alpha_1 = \sqrt{\sqrt{5} + 2}$ and $\alpha_2 = i\sqrt{\sqrt{5} - 2}$. $[K : \mathbf{Q}] = 8$. The eight elements of $G(K/\mathbf{Q})$ are given by the following table.

	ρ_0	ρ_1	ρ_2	ρ_3	μ_1	μ_2	δ_1	δ_2
$\alpha_1\rightarrow$	α_1	α_2	$-\alpha_1$	$-\alpha_2$	α_2	$-\alpha_2$	$-\alpha_1$	α_1
$\alpha_2\rightarrow$	α_2	$-\alpha_1$	$-\alpha_2$	α_1	α_1	$-\alpha_1$	α_2	$-\alpha_2$

The group is isomorphic to D_4, and the notation here is taken to coincide with that in Example 4.2. The group lattice diagram is identical with that in Fig. 47.3. Let H_i be as given in Fig. 47.3.

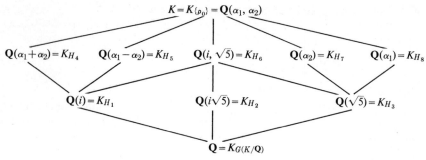

Field lattice diagram

47.9 a) $s_1{}^2 - 2s_2$

b) $s_1{}^2 s_2{}^2 - 4s_1{}^3 s_3 - 4s_2{}^3 + 18 s_1 s_2 s_3 - 27 s_3{}^2$

c) $\dfrac{s_1 s_2 - 3s_3}{s_3}$

47.10 a) $x - 4$ b) $x^3 - 4x^2 + 20x - 4$ c) $x^3 + 4x^2 + 4x + 152$

Section 48

48.2 $\langle \mathbf{Z}_4 \times \mathbf{Z}_2, + \rangle$

48.3 a) 16 b) 400 c) 2160

48.4

3	12	30	60	102
4	15	32	64	120
5	16	34	68	128
6	17	40	80	136
8	20	48	85	160
10	24	51	96	170

48.6 $\Phi_1(x) = x - 1$

$\Phi_2(x) = x + 1$

$\Phi_3(x) = x^2 + x + 1$

$\Phi_4(x) = x^2 + 1$

$\Phi_5(x) = x^4 + x^3 + x^2 + x + 1$

$\Phi_6(x) = x^2 - x + 1$

48.7 T T F T T F T T F T

48.8 $3°$

48.9 a) 4   b) *Partial answer:* $G(K/\mathbf{Q}) \simeq \langle \mathbf{Z}_2 \times \mathbf{Z}_2, + \rangle$

48.10 $\Phi_3(x)$ over $\mathbf{Z}_2$ is $x^2 + x + 1$.

$\Phi_8(x)$ over $\mathbf{Z}_3$ is $x^4 + 1 = (x^2 + x + 2)(x^2 + 2x + 2)$.

48.11 3. $\mathbf{Z}_3$ is the splitting field.

48.12 $x^4 - x^2 + 1$

Section 49

49.2 No. Yes, K is an extension of $\mathbf{Z}_2$ by radicals.

49.4 Yes. If α is a zero of $f(x) = ax^8 + bx^6 + cx^4 + dx^2 + e$, then α^2 is a zero of $g(x) = ax^4 + bx^3 + cx^2 + dx + e$. Since $g(x)$ is a quartic, $F(\alpha^2)$ is an extension of F by radicals, and thus $F(\alpha)$ is an extension of F by radicals.

49.5 T T T F T F T F F T

i) $x^3 - 2x$ over $\mathbf{Q}$ gives a counterexample.

Notations

S_A	group of permutations of A, 39
ι	identity map, 39
S_n	symmetric group on n letters, 39
$n!$	n factorial, 39
D_n	nth dihedral group, 40
A_n	alternating group on n letters, 49
$\mathbf{Z}_n$	cyclic group $\{0, 1, \ldots, n-1\}$ under addition modulo n, 55
	group of residue classes modulo n, 104
	ring $\{0, 1, \ldots, n-1\}$ under addition and multiplication modulo n, 195
	ring of residue classes modulo n, 233
gcd	greatest common divisor, 56, 280
$\simeq, G \simeq G'$	isomorphic groups, 59
S^*	nonzero elements of S, 63
$\underset{i=1}{\overset{n}{\times}} S_i,$ $S_1 \times S_2 \times \cdots \times S_n$	Cartesian product of sets, 69
$\underset{i=1}{\overset{n}{\times}} G_i$	direct product of groups, 70
$\sum_{i=1}^{n} G_i$	direct sum of groups, 70
$\overline{G}_i$	natural subgroup of $\times_{i=1}^{n} G_i$, 72
$\bigcap_{i \in I} S_i,$ $S_1 \cap S_2 \cap \cdots \cap S_n$	intersection of sets, 72
HK	product set, 73
$H \vee K$	subgroup join, 74
$aH, a + H$	left coset, 93
$Ha, H + a$	right coset, 93
$(G : H)$	index of H in G, 98
φ	Euler phi-function, 99, 207
i_g	conjugation by g, 102
$G/N; R/N$	factor group, 104; factor ring, 232
G'	commutator subgroup, 108
γ	canonical residue class map, 113, 236
$\mathcal{I}_G$	set of inner automorphisms of G, 119
$Z(G)$	center of G, 125
$C[a]$	conjugate class of a, 128
$N[a]$	normalizer of a, 131
$N[S]$	normalizer of S, 132
$F[A]$	free group on A, 141
$(x_j : r_i), (x_j : r_i = 1)$	group presentation, 150
$C_n(X)$	group of n-chains of X, 160
∂_n	boundary homomorphism, 160, 183
$Z_n(X), Z_n(A)$	group of n-cycles, 161, 184
$B_n(X), B_n(A)$	group of n-boundaries, 161, 184
$H_n(X), H_n(A)$	n-dimensional homology group, 163, 184

S^n	n-sphere, 167
E^n	n-cell, 167
$\chi(X)$	Euler characteristic of X, 177
f_{*n}, ∂_{*k}	induced map of homology groups, 178, 189
$\langle A, \partial \rangle$	chain complex, 183
$\bar{\partial}_k$	induced map on quotient groups, 186
$H_k(A/A')$	kth relative homology group, 186, 187
$\langle R, +, \cdot \rangle$	ring, 195
$\mathfrak{R}_1, \mathfrak{R}_2, \mathfrak{R}_3$	ring axioms, 195
$n \cdot a$	n summands of a, 196
$R_1 + R_2 + \cdots + R_n$	direct sum of rings, 198
b/a	quotient, 204
(a_{ij})	matrix, 209
$M_n(F)$	ring of $n \times n$ matrices over F, 209
$Hom(A)$	ring of endomorphisms of A, 211
$R(G)$	group ring of G over R, 213
$F(G)$	group algebra of G over F, 214
$\mathbb{Q}$	quaternions, 215
$f(x)$	polynomial in x, 245
$R[x]$	ring of polynomials in x over R, 246
$R[x_1, \ldots, x_n]$	ring of polynomials in $x_1, \ldots, x_n$ over R, 247
$F(x)$	field of rational functions in x over F, 247
$F(x_1, \ldots, x_n)$	field of rational functions in $x_1, \ldots, x_n$ over F, 247
ϕ_α	homomorphism, 247
$f(\alpha)$	image of $f(x)$ under ϕ_α, 250
$a \mid b$	a divides b, 266
UFD	unique factorization domain, 266
PID	principal ideal domain, 267
$\bigcup_{i \in I} A_i$	set union, 268
ACC	ascending chain condition, 268
ν	valuation, 278
$\mathbf{Z}[i]$	ring of Gaussian integers, 286
$N(\alpha)$	norm of α, 286, 289
$\mathrm{irr}(\alpha, F)$	monic irreducible polynomial for α over F, 297
$\deg(\alpha, F)$	degree of α over F, 297
$F(\alpha)$	simple extension of F by α, 297
$\mathcal{V}_1, \mathcal{V}_2, \mathcal{V}_3, \mathcal{V}_4, \mathcal{V}_5$	vector space axioms, 302
$[E : F]$	degree of E over F, 317
$F(\alpha_1, \ldots, \alpha_n)$	extension of F by $\alpha_1, \ldots, \alpha_n$, 319
$\bar{F}_E$	algebraic closure of F in E, 321
$\bar{F}$	algebraic closure of F, 322
$\psi_{\alpha, \beta}$	basic isomorphism, 335, 336
$E_{\{\sigma_i\}}$	fixed field of $\{\sigma_i\}$, 339

$G(E/F)$ group of E over F, 340
σ_p Frobenius automorphism, 341
$\{E : F\}$ index of E over F, 348
$\mathrm{GF}(p^n)$ Galois field of order p^n, 370
$\Phi_n(x)$ nth cyclotomic polynomial, 393

Index